Jacques E Schouten
4302 Ve
Carpinter
August 6, 1994

Perspectives of nonlinear dynamics

To my lieveling, Cindi, and
our wonderful sons, Eric and Mark.
This study is dedicated to their
growing appreciation of the wonders
and beauties in life

VOLUME 1

Perspectives of nonlinear
dynamics

E. ATLEE JACKSON

CAMBRIDGE
UNIVERSITY PRESS

Published by the Press Syndicate of the University of Cambridge
The Pitt Building, Trumpington Street, Cambridge CB2 1RP
40 West 20th Street, New York, NY 10011-4211, USA
10 Stamford Road, Oakleigh, Melborne 3166, Australia

First published 1991
First paperback edition (with corrections) 1991
Reprinted 1992, 1994

Printed in the United States of America

Library of Congress Cataloging in Publication Data is available

A catalogue record for this book is available from the British Library

ISBN 0-521-34504-9 hardback
ISBN 0-521-42632-4 paperback

Contents of Volume 1

Preview of coming attractions   251

Preface

Ah, but a man's reach should exceed his grasp, or what's a heaven for?
Robert Browning

This book represents an attempt to give an introductory presentation of a variety of complementary methods and viewpoints that can be used in the study of a fairly broad spectrum of nonlinear dynamic systems. The skeleton of this organization consists of the three perspectives afforded by classic and some modern analytic methods, together with topological and other global viewpoints introduced by the genius of Poincaré around the turn of the century and, finally, the computational and heuristic opportunities arising from modern computers, as partially foreseen by von Neumann in 1946. On a more profound level, the interplay between computational concepts and physical theories, and what they may teach each other, has become a subject of growing interest since von Neumann's and Ulam's introduction of cellular automata.

Filling out this skeleton are different viewpoints which stimulate other perspectives, such as: bifurcation concepts; the beautiful, if often ethereal, visions of catastrophe theory as conceived by Thom and practiced, with varying degrees of abandon, by his disciples; a variety of mapping concepts, dating again back to Poincaré, ultimately giving rise to such abstract perspectives as symbolic dynamics; the mind-stretching world of chaotic dynamics, uncovered first by Poincaré's imagination, and abstractly studied by Birkhoff; 'curious' attractors, first discovered in the solutions of 'physical' differential equations by Cartwright and Littlewood and by Levinson in the 1940s; the subsequent discovery of numerous physical strange attractors; the complementary situation of persisting oscillatory, non-chaotic dynamics (Fermi–Pasta–Ulam and Kolmogorov, in the 1950s) and its relationship to chaos in conservative system (the Kolmogorov–Arnold–Moser theorem); the coherent dynamics of integrable systems and solitary waves (Russell, 1844; Korteweg and deVries, 1895); the remarkable joint discovery of the inverse scattering transforms, and its yet mystical connection with other methods of solution, such as the Painlevé property; the development (morphogenesis) and dynamics of spatial structures in reactive chemical and biological systems, initiated by Turing (1952); the rudimentary dynamic modeling of 'living' and 'cognitive' systems. A more complete outline of the development of dynamic concepts can be found in the Historical outline.

It is clear that any book of the present length can only present such a variety of concepts at a rather superficial level. Moreover, as the opening quotation suggests, my grasp frequently falls short of my reach. Nonetheless I think there is a need to introduce students and researchers to this broad spectrum of concepts, if only to make them aware that such ideas exist and can be useful, hopefully to stimulate their imaginations to more profound studies and applications. Indeed the *raison d'être* for this book is to afford an introductory access to concepts which will stimulate imaginations in the future.

Since one of the significant impediments to the introduction of these concepts to nonmathematicians (the writer, and the intended readers of this book) is the technical jargon, which tends to obscure many presentations in the mathematical literature, an attempt has been made to lower this barrier without losing too much precision. The necessary technical terms needed for clarity, and to make it possible to read the general literature with some degree of ease, have been collected in a simplified glossary (Appendix A), or can be found in the index. However, it should be made abundantly clear that this book is in no sense a pure mathematical text, despite the mathematical terms which are retained for clarity.

The topics which are discussed have been selected with the hope that they will prove useful (perhaps in the distant future!) in analyzing some dynamic effect arising in models which, hopefully, bears some relationship to observed phenomena. Indeed the interrelation between models and observed phenomena, and the degree to which models engender an 'understanding' of the phenomena, is becoming a more serious question with our growing understanding of the richness and variety of the dynamics described by even 'simple' models. This is illustrated by the surprising fact that all complicated (prescribed) dynamics, $x(t)$, can be 'modeled' by solutions of one 'universal' low order differential equation, to any prescribed accuracy. Clearly such a 'model' says nothing about the underlying physical causes in the real world, and is an extreme example of the precaution which we need to take in ascribing physical or observational significance to mathematical concepts, regardless of how 'beautiful' they appear to be.

This ever-increasing encroachment of nonempirical mathematical concepts into empirical sciences is a phenomenon which should be recognized and at least viewed with caution. Therefore the bias of this book is strongly on the 'pragmatic' side of exploring mathematical concepts which are likely to be useful in describing observable phenomena, rather than 'pure' mathematics. As examples, such beautiful concepts as Cantor sets, fractals, and asymptotic features need to be examined in physical (empirical) contexts, which is often a challenging process.

In keeping with the introductory level of this book, the exercises are generally intended to be simple, frequently requiring only a few minutes thought. They are discussed in the comments at the end of each chapter.

Those who wish to read more about the mathematical refinements and proofs will find discussions in the cited literature. While this list is not encyclopedic, it is at least representative of the rapidly growing literature in this area.

Acknowledgements

I am indeed indebted to many people for their contribution to my knowledge and awareness in this area, not the least being my students over the past fifteen years who have endured my groping presentations of new ideas. Many colleagues have suggested physical and mathematical models which I have found fruitful, or have attempted to explain concepts and to correct numerous misconceptions on my part. Unfortunately they were not always successful, but their patience and generosity is warmly appreciated. I have been particularly fortunate to have had the opportunity to interact with a number of such knowledgeable and generous people. In particular I would like to acknowledge my indebtedness to: F. Albrecht, A. Bondeson, S.J. Chang, J. Dawson, J. Ford, G. Francis, J. Greene, Y. Ichikawa, M. Kruskal, D. Noid, Y. Oono, N. Packard, J. Palmore, J. Pasta, R.E. Peirels, M. Raether, L. Rubel, R. Shaw, R. Schult, M. Toda, N. Wax, M.P.H. Weenink, H. Wilhelmson, J. Wright, and N. Zabusky as well as a number of industrious students, among them K. Miura, R. Martin, A. Mistriotis, P. Nakroshis, F. Nori, S. Puri, and M. Zimmer. Finally, the typing of this manuscript benefited significantly from the dedication, fast fingers, and keen eyes of Mary Ostendorf, to whom I am sincerely indebted.

Special thanks are due to R. Schult for detecting a number of typographical errors in the first printing.

Concepts related to nonlinear dynamics

A BRIEF HISTORICAL OUTLINE

The analytic period (before 1880) – characterized by the search for analytic solutions and perturbation methods; searches for *integrals of the motion*, particularly *time independent, algebraic* integrals.

Main areas: classical mechanics, celestial mechanics – Newton, Euler, Lagrange, Laplace, Jacobi *et al.* Hydrodynamics – e.g., Rayleigh (who briefly considered limit cycles and bifurcation concept) Kinetic theory of gases – Boltzmann equation: stosszahlansatz (implied concept: complexity arises from the interaction of many particles); the *H*-function, statistical concept of entropy.

Relevant abstract mathematical concepts: non-Euclidean geometry; set theory; Cantor sets; transfinite numbers; the continuous, space-filling curves of G. Peano; Painlevé transcendentals.

S. Lie (1879–1900): A general principle for obtaining integrals of nonlinear *partial differential equations*, by determing the invariance properties under a *continuous* group (Lie groups); a frequent application is the invariance under some *scaling* ('similarity transformations').

Three historical theorems:

Bruns (1887): the only independent *algebraic integrals* of the motion of the three-body problem (which has 18 integrals) are the *ten 'classic integrals'* (energy, total linear momentum, total angular momentum, and the time-dependent equations for the motion of the center of mass).

Poincaré (1890): if the Hamiltonian of a system, when expressed in terms of action-angle variables (J, θ), is of the form $H(J, \theta, \lambda) = H_0(J) + \lambda H_1(J, \theta)$, where $H_1(J, \theta)$ is periodic in every θ_i $(i = 1, \ldots, N)$, and if the Hessian does not vanish identically, $|\partial^2 H_0 / \partial J_i \partial J_k| \not\equiv 0$, then there exists *no analytic, single-valued integral of the motion*, $I(J, \theta, \lambda) = \sum_n \lambda^n I_n(J, \theta)$, which are periodic in θ, other than the Hamiltonian, $H(J, \theta, \lambda)$.

Painlevé (1898): the only independent integrals of the motion of the *N-body problem*, which involve the *velocities algebraically* (regardless how the spatial coordinates enter), are the classic integrals.

Stability of motion – results of A.M. Lyapunov (1892); Lyapunov exponents.

Korteweg and deVries demonstrated the existence of *finite amplitude solitary water waves.*

Poincaré (1880–1910): emphasized the study of the *qualitative, global aspects* of dynamics in *phase space*; developed *topological analysis*; generalized *bifurcation concept*; introduced *mappings in phase space* (difference equations); *surface of section*; introduced *rotation numbers* of maps; *index of a closed curve* in a vector field; initiated the *recursive method of defining dimensions*.

Whittaker: obtained the *adelphic integrals* for coupled harmonic oscillators, where the integrals are *nowhere analytic functions of the frequencies* (1906).

1920–1930

Mathematics: the *theory of dimensions* (Poincaré, Brouwer, Menger, *Hausdorff, et al.*); *fixed point theorems* (Brouwer, Poincaré–Birkhoff); the development of *topology, differential geometry* (*Bäcklund transformations*); Birkhoff studied the *abstract dynamics of analytic one-to-one transformations*, emphasized the various categories of asymptotic sets (*alpha and omega limit sets*, various periodic sets, hyperbolic and elliptic fixed point neighborhoods, *recurrent motions of a discontinuous type*, etc.).

Numerical computations by Størmer, and students (!), of the dynamics of solar particles in the dipole magnetic field of the Earth (a nonintegrable system), during 1907–30.

The Madelstam–Andronov school of applied nonlinear analysis; replacement of nonlinear system by a set of linear segments.

E. Fermi (1923) *attempted* to generalize Poincaré's theorem in order to prove *ergodicity* in some systems.

van der Pol: the extensive *study of limit cycles, relaxation oscillations*, leading to *singular perturbation theory*. Studied the forced van der Pol oscillator with van der Mark (1927); observed *subharmonic generation, hysteresis, 'noisy' regions* in *parameter space*. A *variety of bifurcation phenomena*.

The Andronov–Poincaré bifurcation (1930)
The averaging method of perturbation theory is further refined (Bogoliubov–Krylov–Mitropolsky).

Mathematics: the introduction of the concept of structural stability of equation of motion by Andronov and Pontriagan (1937); gradient dynamics; symbolic dynamics;

embedding concepts; logical foundations (K. Gödel, 1931); computational foundations (A.M. Turing, 1936).

The birth of *mathematical biophysics*; Lotka, Volterra, Fisher, Rashevsky.

E. Schrödinger's book, *What is Life? The Physical Aspect of the Living Cell.*

Kolmogorov's spectrum for the case of *homogeneous turbulence* in fluids.

The digital computer: the ENIAC, built at the Moore School of Electrical Engineering,
the University of Pennsylvania (1943–46).

1945–55

The studies of Cartwright and Littlewood, and of Levinson (around 1950): gave a
mathematical proof that the forced van der Pol oscillator has a *family of solutions
which is as 'chaotic' as the family of all sequences of coin tosses*; a physical dynamic
example of Birkhoff's abstract discontinuous dynamics; the first physical demonstration of the existence of a *curious* 'attractor'.

von Neumann investigated the problem of *self-reproducing automata*; with Ulam,
introduced *cellular automata*, whose dynamics is exact (no roundoff errors). He
emphasized the *heuristic use of computers*, to discover general dynamic
characteristics.

S. Ulam emphasized the interaction between man and computer ('*synergesis*'); looked
for the asymptotic properties of certain nonlinear maps; studied the *growth of
patterns* in cellular automata.

The quantitative description of membrane currents; Hodgkin and Huxley.

The *Hopf bifurcation*: a local bifurcation from a fixed point to a limit cycle in R^n.

The *Fermi–Pasta–Ulam* computer study of lattice dynamics: the search for relaxation
to equilibrium; found *no simple relaxation* (non-Boltzmann, non-Fermi), but nearly-
periodic behavior (*simple motion in a 'complex' system*). This is known as the *FPU
phenomena*. (Fermi: 'A minor discovery').

The Kolmogorov–Arnold–Moser theorem: proves that, for a special class of solutions of
systems whose Hamiltonian satisfies *Poincaré's theorem*, a canonical transformation
exists to new action-angle variables, when the Hamiltonian is *weakly* perturbed; this
class contains *most solutions, as the perturbation tends to zero*; briefly, most tori which
are ergodically covered by solutions only become distorted, but not 'destroyed', by
sufficiently small perturbations; these *tori are preserved* in phase space. The
preserved tori are known as *KAM surfaces*.

von Neumann's proof of the existence of universal, self-reproducing automata;
manuscript completed after his death (1957) by A.W. Burks.

The Turing instability: the instability of a homogeneous system of dynamic cells coupled by diffusion; *morphogenesis of spatial structures.*

Mathematics: Kolmogorov–Sinai concept of *dynamic entropy*; concept of *mixing*; Arnold's '*cat map*'; Smale's '*horseshoe map*' (inspired by strange attractor dynamics of the *forced van der Pol oscillator*).

1960–1970

The computer studies of the *continuum lattice* (*Korteweg–deVries equation*), by Kruskal and Zabusky, inspired by the *FPU phenomena*; rediscovery of solitary waves in nonlinear dispersive media; discovery of '*soliton*' (stability) property in multiple-soliton configurations; nonlinear 'basis' set.

Coherent, periodic oscillations in chemical systems – the Belousov–Zhabotinskii oscillations; *low dimensional attractor in a high dimensional phase space* (around 30 chemical compounds).

Computer study of the Bénard problem by Saltzman; the discovery of sometimes 'erratic' dynamics in solutions of the Navier–Stokes equations.

The Lorenz equations; an ordinary differential equation approximation of the Navier–Stokes equations for the Bénard problem. Solutions bifurcate to '*chaotic dynamics*' – a '*strange attractor*' in an *autonomous system*. Also has *homoclinic orbits*, '*preturbulence*', and stable limit cycles.

The *bifurcation sequence* of *general one-dimensional, single-maxima maps* of the interval into itself (Sharkovsky, 1964). The *logistic map*, developed in biology; *period-two bifurcations*, chaotic regions, *windows of periodicity*.

Inverse cascading (to shorter wavelengths) in two-dimensional hydrodynamics.

The breakup of KAM surfaces: the area preserving map of Hénon–Heiles, motivated by astronomical problem. The estimates of breakup, based on overlap of resonances, by Chirikov.

The further development of the concept of *fractal structures* – sets with *fractional dimensions*, by Mandelbrot.

The introduction of the concept of *topological entropy*.

The *heuristic use of the computer*, by Codd, to simplify von Neumann's self-reproducing automata.

Smale's result; structurally stable systems are not dense (1966).

Catastrophe theory, both elementary and general, as visualized by R. Thom; In part, a study of the structurally stable sets in parameter space where a system is structurally unstable (!); many ethereal and imaginative generalizations are visualized by Thom and others. Roundly criticized by many!

The inverse scattering transformation, due to Gardner, Greene, Kruskal, and Miura: a method for obtaining the *general solution of a particular* ('*integrable*') *class of partial differential equations*; this discovery proves that not all analytic methods have been discovered!

The proof of the existence of *Lyapunov exponents* for systems of ordinary differential equations (Oseledec, 1967).

1970–1980

The concept of '*Synergetics*' becomes more diversified, expanded: Zabusky; Haken, *et al.*

Self-organization of matter; Biological evolution; Eigen (1971), Smale (1975).

The Newhouse, Ruelle, Takens theorem – roughly, 'most' systems which are nearly the same as a system whose dynamics consists of three or more periodic components, will have a strange attractor. This suggests that the bifurcation sequence to chaos is from a fixed point, to periodic, then doubly periodic, and then 'turbulence' (a strange attractor). This theorem was preceeded by the *Ruelle–Takens theorem* (1971).

Solitons found in the discrete *Toda lattice*.

Computer and Poincaré map used to test for integrability: Ford predicts the Toda lattice is integrable.

Toda lattice is proved to be integrable – but no use is made of all those integrals of the motion, even when they are known explicitly! Why not? Better yet, *how*? Are they 'macroscopically controllable'?

'Direct method' of obtaining soliton solutions – another analytic method, by Hirota.

Bennett's introduction of *logical reversibility* in computations (1973).

Strange attractor in the two-dimensional map of Hénon's; explicit example of Birkhoff's dynamics.

Possible mechanics for organization of *memory and learning*; Cooper (1973).

Ruelle–Takens introduce the '*strange attractor*' characterization and definition (variously modified later).

The cellular automata game of '*Life*' is invented by J.H. Conway.

Qualitative 'universal' features of the bifurcation patterns of *many* one-dimensional maps is discovered by Metropolis, Stein, and Stein.

Solitons found in many partial differential equations; *generalizations of the inverse scattering transformation* (Zakharov–Shabat, Ablowitz–Kaup–Newell–Segur).

The *logistic map is 'discovered'* by many people, thanks to the article by R. May (1976).

Quantitative 'universal' features are discovered in the bifurcation sequence of the logistic and similar maps, by Feigenbaum; importance of *renormalization concepts*.

The dynamo problem: advances are made in the self-consistent theory of geomagnetic dynamics; simplified models immitate the chaotic flip-flop of the Earth's magnetic field (Lorenz equations).

Experimental determination of bifurcation sequences in hydrodynamic systems (Gollub, Swinney, Ahlers *et al.*) spatial patterns, intermittent spatial patterns; bifurcation sequences differ from theoretical 'generic' predictions.

Protein molecules: possible soliton energy transmission (Davydov); experimentally determined 'fractal dimension' (Stapleton *et al.*).

The *semi-periodic dynamics* of the logistic map – similarity with weather 'periodicity'.

The *homoclinic bifurcation in the Lorenz system* – 'perturbulence'.

Conjecture on the relationship between the *capacity* of an attractor and the spectrum of the *Lyapunov exponents* (Kaplan and Yorke, 1979).

The possible relationship between the *Painlevé property* and integrability.

Many mathematical models of biological systems; Eigen and Schuster's *hypercycle*; Generalized Lotka–Volterra systems.

1980–

Theorems concerning attractors in *infinite dimensional systems*.

Conjectured criteria concerning the *breakup of KAM surfaces* in the standard map. The possible use of *embedding concepts* in chaotic dynamics – Takens.

Experiments on the bifurcations in homogeneous chemical oscillations – *embedding dimension* of attractor.

The study of the *KAM breakup* in the standard map using *renormalization methods* – Kadanoff–Greene–MacKay.

Studies of *'soliton' interactions* in higher dimensions; 'Resonances'.

Nonlinear (3D) instability ('hard' loss of stability) of Poiseuille flow in the *Navier–Stokes equations*.

Cellular atuomata studies – spatial patterns and growth; self-reproduction which is simpler than 'universal' type, but not trivial; statistical characteristics of dynamics by Wolfram.

'non-universal' behavior of bifurcations in solid state devices, etc.; experimental dimensions of chaos.

Generalization of Hirota's *direct method* – analytical extensions of soliton solutions.

Reversible cellular automata: conservative logic; *Digital Information Mechanics* (Fredkin, 1982) has the maximum number of constants of the motion. Is it *a basic description of nature?*; *Can quantum phenomena be described* in a cellular automata scheme?

Many types of chemical oscillations; biological oscillations.

Nondiffusive behavior in the chaotic region of standard map – *'sticky island'* effect.

Topological character of the homoclinic bifurcation in the Lorenz equations; *fractal basin boundaries.*

Spatial order vs. temporal chaos; spatial pattern competition leading to chaos; space–time 'entropies'.

Neural network dynamics –
Where and what is quantum 'chaos'?

1

In the beginning...

1.1 ... there was Poincaré

If modern nonlinear dynamics has a father, it is Henri Poincaré (1854–1912). Dynamic studies, prior to his studies in the 1880s, concentrated on obtaining analytic solutions of dynamic equations, as characterized by many astronomical investigations of planetary motions and by Lord Rayleigh's ubiquitous studies of nearly every moving mechanical system. Many great names in analysis are associated with these studies – Newton, Leibniz, Euler, Gauss, Lagrange, Laplace, Jacobi, Lie and, of course Poincaré, among others. One of the 'grand problems' of classical dynamics, whose solution had withstood the efforts of many people, was the gravitational three-body problem. While ten integrals of the motion for this system had long been known, all attempts to find any of the remaining eight integrals had ended in failure. In 1887 Bruns proved that the ten classic integrals are, in fact, the only algebraic integrals of this system. This was followed by Poincaré's more famous, and frequently misunderstood, theorem (1890) concerning the nonexistence of integrals which are analytic in a perturbation parameter (see Historical outline). Finally, Painlevé (1898) extended Bruns' theorem to N-bodies, and generalized the spatial possibilities. These theorems, which represent a turning point in the analysis of dynamic systems, are discussed in some detail by Whittaker (1944). One can say that the theorems represented a certain loss of innocence. No longer was there any hope of solving all dynamic systems in terms of indefinite integrals involving elementary functions and uniformly valid power series. Other approaches had to be developed to understand the secrets hidden within the dynamic equations.

It was the genius of Poincaré which supplied us with many of our present methods for exploring the unexpected wonders of even 'simple' (low-order) dynamic systems. In particular, he emphasized the importance of obtaining a global, qualitative understanding of the character of a system's dynamics. Many of his suggestions were subsequently refined and extended by others, but our debt to his imagination and insight can hardly be overstated. Unfortunately, his contributions in this area were largely ignored by scientists for about 50 years, but fortunately not by many mathematicians, who extended his concepts to such areas as topology, dimension theory, asymptotic series, various maps and their fixed points, bifurcations, and also proved a number of his conjectures. This wealth of ideas was subsequently enriched by the studies of Birkhoff, who further characterized possible dynamic complexities, and by such concepts as the structural stability of equations, introduced by Andronov and Pontriagan. While the strict application of these concepts proved to have a limited practical impact, they produced an appreciation for the concept of 'robustness' and stimulated other avenues of investigation whose import are still being determined. Some of Birkhoff's abstract concepts were found to occur in surprisingly simple physical systems, as first experimentally detected by van der Pol and van der Mark (1927), and analyzed by Cartwright and Littlewood, and by Levinson in the 1940s. This in turn stimulated talented mathematicians, such as Smale, and thus produced a variety of new concepts.

In the 1940s an entirely new tool of analysis also came on the scene – the digital computer. In addition to its obvious brute-force capabilities of grinding out numerical solutions of differential equations, both von Neumann and Ulam foresaw some of its more subtle applications, as a flexible, interactive tool for the purpose of discoveries. This interplay between computations and analysis, which Ulam called 'synergetics', has indeed proved to be of great importance, and represents one of the major methods of uncovering dynamic properties. In more recent years, the term 'synergetics' has also been applied to a field of nonlinear dynamics, in which the system consists of many subsystems, of possibly very different characters (see 'coexistence' in Section 1.5). Indeed, the area of computer science has rapidly progressed to a state in which it can now make fundamental contributions to our knowledge, rather than acting only as the servant of other methods of analysis.

The period following 1950 has been particularly rich in new ideas and in the growing application of these ideas to a wide variety of disciplines, such as physics, chemistry, biology, neurology, astronomy, geophysics, meteorology, aeronomy, economics. In addition, many new perspectives have been introduced from mathematics, along with the innovative and potentially basic contributions of computer science. The purpose of this book is to stimulate your imagination, by surveying a wide variety of these concepts, and to illustrate them as explicit as possible by simple dynamic models (some of which, hopefully, are relevant to the real word!). Examples will be drawn from any of the above areas which will help to illustrate the relevance of some concept to real dynamic problems.

Unfortunately, the present limited survey omits concepts which may prove to be of great importance but it is hoped that the flavor and thrust of modern developments will be made clear in what follows. The objective here is not to present an established and completed format for analyzing nonlinear dynamics, for that does not exist but, rather, to illustrate the rich variety of concepts which are presently known, in the hope that it will stimulate new ideas and experiments which will further delineate those dynamic aspects that are of basic importance. This book is therefore to be viewed as a 'living' text in the process of growth; it is full of questions whose answers are unknown to the author and which, he trusts, will be both stimulating and fun for the reader.

1.2 What are 'nonlinear phenomena'?

Before considering any details of equations which describe 'nonlinear dynamics', the meaning of a 'nonlinear process', or 'nonlinear phenomena', should be examined. The latter expressions are introduced to make it clear that what is of greatest interest in the present study are 'phenomena'; that is to say observable physical variables and the dynamics of these variables. This is to draw the important distinction between the physical question: 'What are nonlinear phenomena?'; and the relatively trivial mathematical question: 'What are nonlinear equations'? In particular, a response to the former question which might first come to mind:

> nonlinear phenomena are those phenomena which are described
> by nonlinear equations (1.2.1)

is not necessarily correct and, even when true, fails to clarify a basic aspect of the orginal question.

In order that there is no misunderstanding concerning the mathematical side, recall that a linear operator $L(\phi)$ is defined to be one such that linear superposition holds:

$$L(a\phi + b\psi) = aL(\phi) + bL(\psi) \qquad (1.2.2)$$

for any two constants a, b and all (vector) functions ϕ, ψ with suitable regularity properties (e.g., if $L(\phi)$ is a differential operator, then functions need to be in C^m, the space of m-continuously differentiable functions).[†] We shall then define a nonlinear operator $N(\phi)$ to be any operator which does not satisfy (1.2.2) for some a, b or some ϕ, ψ in C^m.

While this might appear to be an obvious definition for a nonlinear operator, the literature does not conform uniformly to this definition. For example, linear parametric equations are frequently included in studies of nonlinear oscillations, whereas nonlinear equations which are known to be linearizable by transformations, such as 'quasilinear' partial differential equations (PDEs), are sometimes considered to be linear (see below). We shall, however, use the above definition for nonlinear operators.

Now, returning to the original physical question, we can see that (1.2.1) is not generally a satisfactory answer because of the simple observation that the fundamental many-body Schrödinger equation is a linear equation. Moreover, in the classical

---

[†] Mathematical notation (e.g. C^m will be kept to a minimum but some is necessary for a degree of precision. A glossary defining the notation is provided is Appendix A

For nonlinear operators: $N(a\phi + b\psi) \neq aN(\phi) + bN(\psi)$

If $L(\phi + \psi) - \{L(\phi) + L(\psi)\} = \Delta(\phi, \psi)$ is not equal to zero, i.e., $\Delta(\phi, \psi) \neq 0$, then the operator is non-linear. Linear operators result in $\Delta(\phi, \psi) = 0$. See H.T. Davis, Introduction to Nonlinear etc...., p. 2

context, the Liouville equation, which encompasses all of Hamiltonian dynamics, is also a linear equation. Since any(!) mechanical phenomena can presumably be described by such basic linear equations, it follows that nonlinear equations are in principle not necessary to describe nonlinear phenomena! What then are 'nonlinear phenomena'? Moreover, this raises the interesting question:

> Why use nonlinear equations if the nonlinear phenomena can be described by linear equations?

This confronts a common misconception which we will discuss in Section 1.3.

As noted before, the key word is 'phenomena', or even 'phenomena of interest', with the implication that the variables are observable, and 'of interest'. It has of course been appreciated for a long time that the Schrödinger equation or the Liouville equation for many interacting particles contains dynamical information which probably will never (or, possibly, can never) be of any observational interest, and certainly is not of present interest. The phenomena of interest usually involve the behavior of certain macroscopic properties, which can be viewed as 'projections' or 'reductions' from the detailed microscopic states of the system. For example, in equilibrium systems we typically describe the system in terms of its density, pressure, volume, and temperature (Fig. 1.1).

Fig. 1.1

'N'
'P', 'V', 'T'

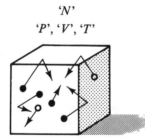

These concepts are all 'projections' of the microscopic state of those systems involving various space–time average properties. However, we will be interested in time dependent systems, for which such projections are more difficult. In any case, it is the spatial–temporal behavior of projections which forms the nonlinear dynamics of our present interest.

A fundamental theoretical problem arising from such a reduced description (projection) of the underlying microscopic dynamics is the determination of the collection of projected variables whose dynamics can be 'approximately' described by a (deterministic) system of nonlinear equations. Simple examples of such systems of equations occur in the fields of mechanics, or electrical circuits.

Exercise 1.1. The following systems may contain more than 10^{25} particles, which we typically project down to two or three variables, plus assorted constants.

(a) What is a set of dynamic variables, and a set of related physical parameters, which are presumably deterministic in simple electric circuits?

(b) A mechanical system consists of a solid wheel, which can rotate about a central axle (Fig. 1.2). An off-center mass and another mass hanging from a rope wrapped around the wheel are attached. Name a set of presumably deterministic dynamic variables, and associated parameters. State two major assumptions about these 10^{25} particles?

Fig. 1.2

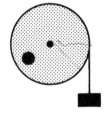

A system can have both nonlinear and linear properties:

(i) A nonlinear property in the present system is to determine its equilibrium configuration. Does it exist? If so, what is it?

(ii) A linear property is to determine the frequency of small oscillations about its equilibrium configuration. Determine this angular frequency.

More sophisticated projections occur in various fluid equations of motion, which result from different methods of truncating (closing) an infinite system of moment equations (see Exercise 1.4). The limitations implicit in such truncations are not usually known with any precision, but are more or less loosely formulated. However, the physical nonlinear equations of interest are obtained by some 'projection and truncation' procedure, which yields a dynamic model of the system. The most common of types of models are:

> *ordinary differential equations* (ODE); e.g. $dx/dt = ax + bx^2$;
> *partial differential equations* (PDE); e.g. $(\partial u/\partial t) + u(\partial u/\partial x) + (\partial^3 u/\partial x^3) = 0$; KdV
> *difference equations* (DE), *or Maps*; e.g. $x(t+1) = ax(t) + bx^2(t)(t = 0, 1 \ldots)$;
> *cellular Automata* (CA): discrete time, space, and functional values.

Attempts to obtain more fundamental projections, particularly in the area of nonequilibrium solids, liquids and gases can be found in the literature (e.g., see Grabert, 1982). Such a fundamental systematic approach is impossible in more complicated systems (chemically reactive, biological, sociological, economic, etc.), and we must rely on fairly simple models (Fig. 1.3).

In any case these ideas yield one (somewhat abstract) definition of what we will mean by 'nonlinear phenomena', namely

Fig. 1.3

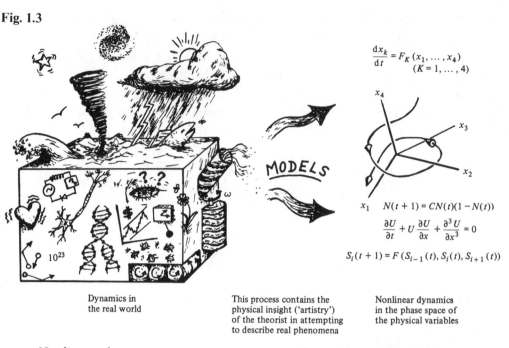

$$\frac{\mathrm{d}x_k}{\mathrm{d}t} = F_K(x_1, \ldots, x_4)$$
$$(K = 1, \ldots, 4)$$

$$N(t+1) = CN(t)(1 - N(t))$$

$$\frac{\partial U}{\partial t} + U\frac{\partial U}{\partial x} + \frac{\partial^3 U}{\partial x^3} = 0$$

$$S_i(t+1) = F(S_{i-1}(t), S_i(t), S_{i+1}(t))$$

| Dynamics in the real world | This process contains the physical insight ('artistry') of the theorist in attempting to describe real phenomena | Nonlinear dynamics in the phase space of the physical variables |

> *Nonlinear phenomena concern processes involving 'physical' variables, which are governed by nonlinear equations. These models have been obtained, by some approximate 'projection' rationale from presumably more fundamental microscopic dynamics of the system.* (1.2.3)

It is possible, of course, that a reasonable projection may yield simple linear equations in some approximation, and that indeed is usually the first approximation that is attempted. Classic examples of this are sound waves in gases, water waves, the 'phonons' in solids, or 'plasmons' in ionized systems, which define 'collective' variables. Note that these approximations yield linear equations in the physical variables. This is to be distinguished from linear equations in nonphysical variables which may be obtained from a mathematical transformation of the original physical nonlinear equations. The phenomenon is linear only if the equations are linear in the physical variables.

It is useful to augment the rather abstract definition (1.2.3) with a more physical, operational definition. One possibility is:

> *Physical phenomena concern the interrelationship of a set of physical variables which are deterministic (within some accuracy). Nonlinear phenomena involve those sets of variables such that an initial change of one variable does not produce a proportional change in the behavior of that variable, or some other variable. In other words, the ratio (action/reaction) is not constant.* (1.2.4)

To further emphasize (and clarify) some of the mathematical interrelationships between linear and nonlinear equations, consider the following facts:

(A) The general solution of the linear PDE

$$\frac{\partial \phi}{\partial t} + \sum_{k=1}^{n} F_k(x, t)\frac{\partial \phi}{\partial x_k} = 0 \quad (x \in R^n) \tag{1.2.5}$$

for the function $\phi(x, t)$ of $(n+1)$ independent variables is (locally in space/time) equivalent to solving the nonlinear system of ODE

$$\dot{x}_k = F_k(x, t) \quad (k = 1, \ldots, n) \tag{1.2.6}$$

(e.g., see R. Courant and D. Hilbert, *Methods of Mathematical Physics*, Vol. 2, Chapter II, §2). That is, any function $\phi(K(x, t))$ of the constants of the motion K_i, of (1.2.6) (so that $d\phi/dt \equiv 0$; see (2.4.4)) is a solution of (1.2.5), and any solution of (1.2.5) yields a constant of the motion of (1.2.6).

(B) A system of nonlinear PDE of the special form

$$\sum_{i=1}^{n} A_i(x, \phi) - \partial\phi_j/\partial x_i = B_j(x, \phi) \quad (j = 1, \ldots, m) \tag{1.2.7}$$

is called a quasi-linear system ($x \in R^n$, $\phi \in R^m$). If one defines $x_{n+s} \equiv \phi_s$ and $A_{n+s} \equiv -B_s(s = 1, \ldots, m)$, then (1.2.7) is equivalent to the single homogeneous linear PDE (see Courant and Hilbert)

$$\sum_{i=1}^{n+m} A_i(x)\partial\phi/\partial x_i = 0 \quad (x \in R^{n+m}), \tag{1.2.8}$$

which, of course, is the same as (1.2.5). Hence (1.2.7) is equivalent to (1.2.5).

(C) This yields the theory of characteristics for general first order PDE, $F(x, u, u_{x_1}, \ldots, u_{x_n}) = 0$ (where $u_x \equiv \partial u/\partial x$). This can be replaced by a system of $(n+1)$ quasi-linear PDE, of the form (1.2.7); n equations are obtained by considering $dF/dx_i = 0$ $(i = 1, \ldots, n)$ and setting $\phi_i \equiv u_{x_i}$. Then $dF/dx_i = \partial F/\partial x_i + \partial F/\partial u\phi_i + \sum \partial F/\partial\phi_k \partial u_{x_k}/\partial x_i = 0$. But $\partial u_{x_k}/\partial x_i = \partial\phi_i/\partial x_k$. So there are n equations of the form (1.2.7), namely

$$\sum_k \left(\frac{\partial F}{\partial \phi_k}\right)\frac{\partial \phi_i}{\partial x_k} + \left(\frac{\partial F}{\partial u}\right)\phi_i + \frac{\partial F}{\partial x_i} = 0 \quad (i = 1, \ldots, n).$$

Since $F = F(x, u, \phi)$ we need one more equation for u. This is obtained trivially, because

$$\sum \left\{\left(\frac{\partial F}{\partial \phi_i}\right)\frac{\partial u}{\partial x_i} - \left(\frac{\partial F}{\partial \phi_i}\right)\phi_i\right\} = 0.$$

Thus the general (and hence possibly very nonlinear) first order PDE

$$F\left(x, u, \frac{\partial u}{\partial x_1}, \ldots, \frac{\partial u}{\partial x_n}\right) = 0 \tag{1.2.9}$$

can be related to the linear PDE (1.2.5).

More interconnections between linear and nonlinear equations are known (e.g., see the Riccatti equation), and are continually being uncovered (e.g., the Hopf–Cole transformation, the inverse scattering transform, etc.).

Exercise 1.2. (This exercise requires some background in statistical mechanics). Write the linear PDE (1.2.5) for a system of N noninteracting particles with charges $q_i(i = 1, \ldots, N)$ in an electric field E (constant). What does this equation refer to in the field of statistical mechanics (i.e., what does $\phi(\mathbf{r}_1, \ldots, \mathbf{r}_N, \mathbf{v}_1, \ldots, \mathbf{v}_N, t)$ represent)? By solving equations (1.2.6) obtain the general solution of (1.2.5), using the $2N$ constants of the motion (initial conditions $(\mathbf{r}_i^0, \mathbf{v}_i^0)$ – see Chapter 2 if you have difficulty). Note that in this simple case even (1.2.6) is a linear system. Does ϕ (in its statistical mechanical context) represent an observable ('physical') variable? What types of quantities, related to ϕ, are common physical variables? What are the 'projections' in these cases? Obtain some PDE equations of motion for these physical variables from (1.2.5). Are they deterministic? If not, what must one do to get a deterministic system? Is that physically reasonable for the above system?

1.3 Two myths

It was noted in the last section that the clarification of the meaning of nonlinear phenomena raised the point that these phenomena presumably can be described by more basic linear equations (however, not in the 'physical' variables). If nonlinear phenomena can be described by linear equations, why would we want to use a nonlinear equation? After all, it is widely 'known' that

MYTH 1: linear equations are easier to solve than nonlinear equations.

This widely held misconception may have its origin in the fact that 'linear equations', to many scientists, connote equations for harmonic oscillators, free particles, the (very special!) Schrödinger equations for the hydrogen atom or a particle in a box, or possibly even a system of ordinary linear equations with constant coefficients. These linear equations are, of course, elementary to solve, but they are obviously not characteristic of most linear equations.

The property of linear superposition (1.2.2) is generally useful only when one can obtain a complete set of solutions $\{\phi_n\}$ of $L(\phi_n) = 0$, from which one can then construct general solutions. At least one can construct a series representation of the solution, but its general properties may remain unknown. In any case, the determination of this basis set $\{\phi_n\}$ is generally not aided in the slightest by the linearity of the operator. The point to be emphasized is that the fundamental linear equations of physics must in fact generally be more difficult to solve than their 'projected' nonlinear equations which describe the nonlinear phenomena of interest. (Those who may still doubt this should peruse the book, *Perturbation Theory of Linear Operators*, by T. Kato.)

Finally, the last paragraph raises the important fact that it is not generally true that

> *MYTH 2: an analytic solution of an equation gives, if it exists, the most useful information.*

One may have an analytic solution and yet have a very incomplete understanding of the behavior of the system. To make this point clear, it is only necessary to consider the simple equation

$$\frac{d^2x}{dt^2} + x = 0. \tag{1.3.1}$$

This has the solution

$$x = A\cos(t) + B\sin(t), \tag{1.3.2}$$

where A and B are arbitrary constants, provided that the functions in (1.3.2) are defined by

$$\cos(t) = \sum_{n=0}^{\infty} (-1)^n t^{2n}/(2n)!$$

$$\sin(t) = \sum_{n=0}^{\infty} (-1)^n t^{2n+1}/(2n+1)! \tag{1.3.3}$$

It is, of course, 'known' that these functions are periodic, satisfying

$$\cos(t + 2\pi) = \cos(t); \quad \sin(t + 2\pi) = \sin(t). \tag{1.3.4}$$

This periodicity is obviously a very basic and important feature of the solutions of (1.3.1), and establishing it is therefore of crucial importance.

Exercise 1.3. It is left as a challenge for you to prove that (1.3.4) follows from the solutions (1.3.3) (i.e., to show that a 'π' exists such that (1.3.4) holds).

This 'simple' feature (1.3.4) does not come most readily from the analytic expressions (1.3.3) but, rather, from other methods. These methods may, moreover, be applied as easily to the nonlinear oscillator (e.g., if $F(-x) = -F(x)$ and $F(x) > 0$ if $x > 0$)

$$\frac{d^2x}{dt^2} + F(x) = 0 \tag{1.3.5}$$

as they can be applied to (1.3.1), and hence a basic property of $x(t)$ satisfying (1.3.5) is not made any more transparent if $F(x) = x$ (i.e., if the equation is linear).

It might be tempting to argue that, given (1.3.3) and a computer, it should be relatively easy to establish (1.3.4) (and determine π?!). It is obvious, however, that a computer (alone) can never prove such an equality. It can only show that

$$\cos(t + 2'\pi') \simeq \cos(t) \tag{1.3.6}$$

for some 'π', and to within some accuracy. The difference between (1.3.6) and (1.3.4) becomes of more than theoretical interest if we want to know if

$$x(t + n2'\pi') \simeq x(t)$$

for a large number, say $n = 10^5$. Moreover, since the answer can be established without using a 'sledgehammer on a peanut' approach, it is clear that the computer (alone!; see below) is quite inappropriate for such questions, and indeed for many 'qualitative' features of solutions.

This point illustrates the importance of using a variety of approaches to analyzing systems, as outlined in the last section. In particular, the periodicity property of (1.3.5) and (1.3.1) is established by a simple phase plane ('topological') analysis. However, particularly in more complicated cases, the use of the computer together with continuity and topological considerations can be a very useful blending of the approaches in analyzing systems. The computer can be used to determine effectively whether some condition has changed between two regions of space, from which conclusions can be drawn about the situation at some intermediate point, based on continuity arguments. Numerous examples of this useful method will be discussed later.

Exercise 1.4. Higher order equations require more ingenuity. For example, three independent solutions of $d^3x/dt^3 + x = 0$ are, in analogy with (1.3.3), $\text{ain}(t) \equiv t - t^4/4! + t^7/7! - \cdots$, $\text{bin}(t) \equiv t^2/2! - t^5/5! + t^8/8! - \cdots$, and $\text{cin}(t) \equiv 1 - t^3/3! + t^6/6! - t^9/9! - \cdots$. Are any or all solutions of this equation periodic?

1.4 Remarks on modeling

The art of modeling physical systems by suitable systems of equations is, of course, an old and yet surprisingly new practice. It is not only that new systems on which to practice this art are always being considered, but also that the possible objectives of such modeling have taken on a new dimension in recent years. As any good book on modeling will tell you (see the references for a start), we begin by trying to identify the physical variables which we believe are responsible for the phenomena in question, and their interrelations, in order to construct a deterministic system of equations. This step is clearly of major importance. We need only to consider the complexity of gases and fluids, economic predictions, weather forecasting, complex chemical reactions, neurological networks (the brain!), to appreciate this step. The idea is then to check predictions of the model with known results ('validating' the model) and, finally, to use the model to predict new results.

What is relatively new about the potential use of models is that they may also be used to show that one cannot predict some observable results, even when these results are deterministically related to the variables of the model. The idea that there can be a lack of determinism in a deterministic system of equations appears at first to be a contradiction in terms. What enters here is the important distinction between mathematical precision and physical observations. The determinism we speak about in models is the property of the mathematical uniqueness of the solution, namely, given the initial conditions, $x(0) = x_0 (x \in R^n)$, there is precisely one solution $x(t) = F(t; x_0)$ satisfying both the equations of the model and the initial conditions. The model is then mathematically deterministic.

This mathematical determinism does not always transcribe into observational (experimental) determinism, and there is indeed a well-defined class of models which lacks such physical predictability. The physical unpredictability results from the impossibility of making physical measurements, such as initial conditions, with infinite precision. For some deterministic models, a sequence of measurements with finite precision cannot yield predictions with equal precision in the future. That this can happen for simple classical models, rather than the complicated systems envisaged in statistical mechanics, or in quantum mechanical systems with their familiar uncertainty effects, is a surprising and potentially very important aspect of future models. We can envisage future models in which some observations are predictable while other are not. Note that the lack of predictability is not a fault of the model, but an expression of a real physical fact of life. Clearly, the knowledge that some system has unpredictable aspects can be of great importance, if for no other reason than it saves futile efforts to make such predictions. On the positive side, this knowledge may also suggest other new observations which are both important and predictable. Also, as will be seen, simple models may have unpredictable aspects under certain conditions, only to become predictable under other conditions (they are predictably unpredictable!), and may simultaneously have other features which always behave predictably.

This distinction between mathematical precision and finite experimental accuracy should be continually kept in mind. Many beautiful mathematical concepts which can be applied to dynamic systems involve sets of points or dynamic properties which involve continually decreasing scales (in some space) as time increases, and these aspects therefore necessarily lose relevance to real observations at some level.

The converse can also occur, where 'nongeneric' equations or situations (i.e., 'untypical' or 'zero measure' mathematical situations) are of great physical importance. Examples of physically important 'nongeneric' concepts are periodic solutions, Hamiltonian equations, and harmonic oscillators. Thus beautiful mathematical concepts do not always yield useful physical insights and, indeed, may be misleading if the connotation of a mathematical concept (e.g., nongeneric) is ascribed to physical observations. This mismatch will be noted as examples arise.

1.5 The ordering and organization of ideas

The logical ordering of any 'holistic' study of nonlinear dynamics (i.e., a broad survey, intended for diverse applications) is presently impossible. A beginning approach is to order the topics mathematically, according to whether the system is described by ordinary differential equations (ODE), partial differential equations (PDE), maps and difference equations (DE), integro-differential equations (IDE), differential delay equations (DDE), cellular automata (CA), or whatever. The present study is restricted to models described by ODE and DE (maps), with a very limited discussion of PDE and CA models. This certainly does not imply that systems described by IDE (e.g., the Boltzmann equation, various non-Markoffian processes, etc.), or DDE (e.g., control problems, delayed response systems, etc.) are not important, but it is to say that they are generally much more difficult to analyze (See Hale (1977) and Pippard (1985) for useful reviews and applications of such equations). In any case, the wealth of phenomena described by ODE and DE is sufficiently challenging for our present purposes.

For systems described by ODE or DE, the ordering of the chapters is based on the number of functions of the independent 'variable' (t or n). Specifically, we will consider the cases of one, two, three, 'four and more', and N (large) functions, leading into the limited study of PDE (solitons), and CA. The number of functions will be referred to as the number of 'dynamic dimensions' (the order of a system of ODE). This is a generalization of the concept of 'degrees of freedom', used in classical mechanics. Note that degrees of freedom have historically always referred to systems with an even number of dynamic dimensions (even ordered ODE, for $x(t)$ and $v(t)$).

The ordering of our study in terms of increasing dynamic dimensions is only partially successful in bringing some coherence to a field which is notoriously fragmented. The difficulty is that *DE* (maps) are dynamically much richer than ODEs, with the same number of dynamic dimensions. The reason for this is that DE can arise from the 'condensation' of the dynamical information contained in systems of ODE with more dynamic dimensions (e.g., see Poincaré maps), and hence contain many of the dynamic complexities of these more sophisticated differential systems. The range of dynamic 'complexity' for increasing dynamic dimensions, of both continuous ODE (C) and discrete maps (D), (e.g., $2D$ = two-dimensional maps), as well as our limited PDE models, and CA possibilities are schematically illustrated in Fig. 1.4. The great irregularity of the complexities is the essential point to note. We have, for lack of a better idea, chosen the dimensional ordering, alternating between DE and ODE, moving to the right along the axis of the above diagram, in our attempt to 'sneak up' on the nonlinearities.

Fig. 1.4

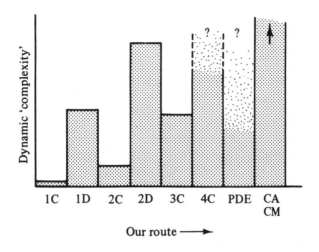

In addition to the 'vertical' ordering just described, it is very important to emphasize the 'lateral' development which should be kept in mind at each level. This lateral study concerns the general problem of understanding the conditions which are required for systems to have contrasting dynamic 'states', and to understand the possible transitions between, or *coexistence* of, these states (Fig. 1.5).

Fig. 1.5

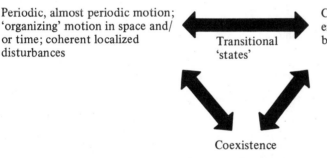

Periodic, almost periodic motion; 'organizing' motion in space and/ or time; coherent localized disturbances

Transitional 'states'

Chaotic, stochastic, turbulent, ergodic motion; irreversible behavior; 'statistical' mechanics

Coexistence

Inhomogeneous 'turbulence'; living cells; cogitation; evolution; ecological, social, economic systems

To describe this variety of 'states', and, indeed, many refinements only vaguely appreciated at present, it is obviously essential to use as broad an approach to their analysis as possible. One of the objectives of this book will be to develop an assortment of concepts and techniques which can be used to describe these states, and which will hopefully prove useful in formulating future analyses and experiments. A list of some of these concepts, which can be applied at most 'vertical' (dimension) levels of complexity, is (don't worry about their meaning at this point – just note that there are many concepts which can be applied):

Theoretical		
Mathematical abstractions	Analytic methods	Topological concepts
abstract dynamical systems, ergodic, mixing, and K systems; symbolic dynamics; entropies; KAM theorem; Arnold dynamic diffusion; measure-dependent features; strange attractors; fractals; mapping theorems; dimensions of sets; complexity; logical depth	Explicit solutions, standard transcendental functions; averaging methods; perturbation methods, bifurcation conditions; inverse scattering methods; direct methods; Lie groups; Painlevé property; Bäcklund transformations.	characteristics of flows in phase space; surfaces of sections; catastrophe concepts; homeomorphism; topological equivalences; structural stability; winding numbers; Poincaré index; topological entropy

Experimental	
Computational studies	Laboratory studies
'Synergetics': the intelligent interplay between ideas and computations; special detailed solutions of simple arbitrary systems; exploration for irreversibility, integrable systems, strange attractors, bound states; limited information effects; study of bifurcation sequences; scaling and self-similarity properties	Frequency spectrum; correlation functions in space/time; transitions to turbulence; aspect ratio and intermittancy effects; response and transport measurements; dimension of chaotic states; entrainment; hysteresis, and jump phenomena; bifurcation sequences; coherent pulse transmission/interactions

Even the above incomplete listing gives some indication of the recent approaches which should help in the future to dispel some of our present ignorance about these systems. It also helps to explain the titillation which is causing the rapid growth of research in the general area of nonlinear dynamics.

In connection with those methods which describe only the qualitative features of dynamic systems (e.g., topological considerations), it is interesting to recall Rutherford's purported remark, 'Qualitative is nothing but poor quantitative', which confuses 'qualitative' with 'inexact'. Thom (1975) has pointed out the example in Fig. 1.6, where two theories (A and B) are developed to explain the indicated experimental data. While theory A is much closer to the experimental values, at least in the mean, it is reasonably certain that the underlying ideas which went into theory B are closer to the truth, because of the qualitative agreement with the experimental trend. It is much better to have *any* global qualitative understanding of the dynamics than one more quantitative page of computer output! In a broader context, see Feynman's remarks in the following section.

Fig. 1.6

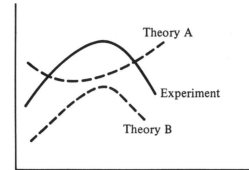

1.6 Some thoughts

'Imagination is more important than knowledge.'

(A. Einstein: *On Science.*)

'Man is not a circle with a single centre; he is an ellipse with two foci. Facts are one, ideas are the other.'

(Victor Hugo: *Les Miserables.*)

'Of course every physicist wishes it were possible to construct ab initio the real world from the fundamental laws, but few have the imagination to make even a tentative start, and without imagination nothing is possible, since systematic procedures are unavailable.'

(A.B. Pippard: *Response and Stability.*)

The great richness of dynamical models, which remains to be discovered, has been expressed rather nicely, and with his usual flare, by R. Feynman (in his *Lectures in Physics*):

'The next great era of awakening of human intellect may well produce a method of understanding the qualitative content of equations. Today we cannot. Today we cannot see that the water flow equations contain such things as the barber pole structure of turbulence that one sees between rotating cylinders. Today we cannot see whether Schrödinger's equation contains frogs, musical composers, or morality – or whether it does not. We cannot say whether something beyond it like God is needed or not. And so we can all hold strong opinions either way.'

Back to the master...

'The scientist does not study nature because it is useful; he studies it because he delights in it, and he delights in it because it is beautiful. If nature were not beautiful, it would not be worth knowing, and if nature were not worth knowing, life would not be worth living.'

(Henri Poincaré.)

Comments on exercises

(1.1) (a) Voltages, currents, charges (on capacitors); parameters: resistance, inductance, capacitance.

(b) Dynamic variables might be θ and $\dot{\theta}$ (they are independent); parameters m_1, m_2, r, R, and I (moment of inertia of the wheel about axle). Major assumptions: the masses are rigid, and the rope does not change its length (Fig. 1.7).

Fig. 1.7

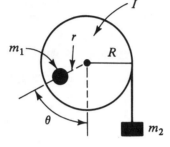

(i) $m_1 r \sin\theta_0 = m_2 R$ provided that $m_1 r \geqslant m_2 R$; independent of I.

(ii) $\omega^2 = m_1 gr \cos\theta_0 / [I + m_1 r^2 + m_2 R^2]$.

(1.2) In this case $k = 1, \ldots, 6N$, namely the $(\mathbf{r}_j, \mathbf{v}_j)\, j = 1, \ldots, N$. The equations (1.3.8) are then $\dot{\mathbf{r}}_j = \mathbf{v}_j$, $\dot{\mathbf{v}}_j = (q_j/m_j)\,\mathbf{E}$, where (q_j/m_j) is the (charge, mass) of particle j. Thus (1.3.7) becomes $\partial\phi/\partial t + \sum_j \mathbf{v}_j \cdot \partial\phi/\partial\mathbf{r}_j + (q_j/m_j)\,\mathbf{E}\cdot\partial\phi/\partial\mathbf{v}_j = 0$. This is an example of Liouville's equation in statistical mechanics for the (N particle) distribution function (see any advanced text in statistical mechanics). One grand 'projection' scheme is the 'BBGKY hierarchy' (e.g. see G.E. Uhlenbeck and G.W. Ford, 'Lectures in statistical mechanics' (*Amer. Math. Soc.*, 1963)). Since there is no interaction in this example, $\phi = \prod_{j=1}^{N} f_j(\mathbf{r}_j, \mathbf{v}_j, t)$, and the simplest 'projections' are obtained by velocity integrals of the f_j (e.g., giving the number density, drift velocities, etc.).

The PDE equations for these variables are the continuity equations, equations of motion, energy equations, etc. (These are not deterministic, because there are more unknown functions than equations.) Some *ad hoc* truncation scheme must be introduced, which then yield Euler's perfect fluid, or perfect magneto-hydrodynamic equations, or the Navier–Stokes equations, etc. This truncation is necessary because, without interactions no relationship need exist between these variables, so an *ad hoc* scheme must be used. For the Boltzmann equation (containing interactions) there is an Enskog–Chapman iteration scheme, which systematically gives these models (see e.g., S. Chapman and T.G. Cowling, *The Mathematical Theory of Non-Uniform Gases*, 3rd ed., Cambridge University Press, 1970.)

(1.3) To quote G.H. Hardy (*Pure Mathematics*, Cambridge University Press, p. 433): 'The property of periodicity is a little more troublesome.' The proof of periodicity, using (1.3.3), can be found in E.T. Whittaker and G.N. Watson, *A Course of Modern Analysis* (Cambridge University Press, 1962), Appendix A.5. p. 585

(1.4) The functions ain(t), bin(t), and cin(t) are not very helpful. Consider the characteristic solutions in the phase space $x_1 \equiv x$, $x_2 \equiv \dot{x}$, $x_3 \equiv \ddot{x}$. You should be able to prove that *no* nontrivial solution is periodic (see Chapter 2).

If $e^{z_1} = e^{z_2}$ multiplying by e^{-z_2} then $e^{z_1 - z_2} = e^{z_2} e^{-z_2} = e^0 = 1$, also letting

$z_1 - z_2 = \gamma$

$\exp(z + n\gamma) = \exp z \cdot \exp(n\gamma) = e^z \exp(\gamma)^n = \exp z$ since $\gamma = 1$

2

A potpourri of basic concepts

The study of nonlinear dynamics is dependent on the knowledge of at least the elementary aspects of a variety of mathematical concepts. This chapter is intended to be an elementary introduction to some of these concepts. The references give more detailed discussions of these topics. While the latter sections of this chapter may be initially omitted, the first two or three sections should be studied before proceeding to other chapters.

2.1 Dynamic equations: topological orbital equivalence

Many systems consist of a finite number (n) of discrete elements (e.g., particles, voltages, currents, chemical concentrations, species populations, etc.), which are described by real-valued functions $x_i(t)(i = 1, \ldots, n)$ of one real independent variable which we will denote by t. We will say that such a system has n dynamic dimensions. It is also understood throughout that no $x_i(t) \equiv t$ (see 'nonautonomous' below). We will use the common, and rather abbreviated, notation

$$x(t) \equiv x = (x_1(t), \ldots, x_n(t)),$$

which is also described by the notation $x \in R^n (R = \text{real})$. Note that the frequently used bold-letter vector notation is dropped, in keeping with most modern literature. The equations of motion governing these functions will be assumed to be of the form

$$\frac{\mathrm{d}x_i}{\mathrm{d}t} = F_i(x, t; c), \quad (i = 1, \ldots, n); \quad c = (c_1, \ldots, c_k),$$

which will be written in the corresponding abbreviated form

$$\dot{x} = F(x, t; c) \quad (x \in R^n, c \in R^k). \tag{2.1.1}$$

Note that any ODE which is first *degree* (power) in its highest order derivative, namely

$$\frac{\mathrm{d}^n y}{\mathrm{d}t^n} = G\left(t, y, \dot{y}, \ldots, \frac{\mathrm{d}^{n-1}y}{\mathrm{d}t^{n-1}}\right) \quad (y \in R^1), \tag{2.1.1a}$$

can be put in the form (2.1.1), simply by setting

$$x_k \equiv d^{k-1}y/dt^{k-1} \qquad (k = 1, \ldots, n),$$
$$F_n = G, \quad F_l = x_{l+1} \quad (l = 1, \ldots, n-1).$$

While equations of the form (2.1.1) clearly do not cover all dynamic systems, they do describe a large group of models.

If the functions F_i do not depend explicitly on time, the system is said to be *autonomous*, otherwise it is *nonautonomous*. Of course, any nonautonomous system in R^n can be made autonomous in R^{n+1}, trivially, by setting $x_{n+1} \equiv t$. That mathematical legerdemain ignores the physical distinction of the time variables (expressed by $\dot{x}_{n+1} = 1$, a positive constant), and will not be used (see, however, 'extended phases space'). The real constants in (2.1.1) $(c_1, \ldots, c_k)$ are physical parameters which appear in the functions (e.g., particle masses, resistances, spring constants, the strength and frequencies of applied force fields, reactivity, toxicity, metabolic parameters, etc.) and may be considered to be variable. In some fields of study, the variables $x_k(t)$ are referred to as *endogenous variables*, while the c_k are called *exogenous variables*. We will refer to the c_k as *control parameters*, and they will be understood to be independent of t (changeable only for distinct dynamic 'runs'). The occurrence of such variable control parameters allows us to account for 'outside' influences on the dynamics in a model. In a physical sense, then, the system is not autonomous, even though it may be referred to as such (see above definition). Considerable interest has been centered on this *control space C* in recent years, particularly as it relates to the *structural stability* of models defined by (2.1.1). These considerations will arise in such areas as *bifurcations*, and '*catastrophes*', which may be picturesequely described as '*dynamic phase transitions*', to be considered in later sections.

As Poincaré taught us, it is very useful to view (2.1.1) as describing the motion of a point in an *n-dimensional space*, which will be called the *phase space* of the system. The space, whose points are the ordered sets of real numbers $(x_1 x_2, \ldots, x_n)$ is denoted as R^n (Fig. 2.1). We will take this to be a Euclidean space (sometimes denoted as E^n), in which

Fig. 2.1

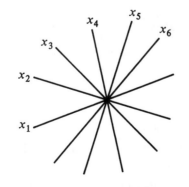

Schematic (!) R^6

the metric (distance) is defined between any two points,

$$d(x, y) = \left(\sum_{i=1}^{n} (x_i - y_i)^2 \right)^{1/2}.$$

Note that the phase space, which can have even or odd values of n, is more general than the phase space (p, q) associated with *Hamiltonian systems*, which always has even dimensions $n = 2N$,

$$\dot{p} = - \partial H(p, q, t)/\partial q, \quad \dot{q} = \partial H(p, q, t)/\partial p \quad (p, q \in R^N).$$

Moreover, Hamiltonian systems represent only one extreme group of equations of the form (2.1.1), and do not include any of the dissipative (nor many conservative) models which are so important in a variety of fields, where there are strong influences due to 'outside' factors. At the opposite extreme from the Hamiltonian systems are the so-called gradient systems, for which $F = -\nabla V$, which include over-damped (inertialess) dynamics. Gradient systems have been conspicuous in recent years due to their prominence in *catastrophe theory*, which will be discussed later.

This general phase space (P) is often viewed in conjunction with the '*control space*' (C), giving rise to the 'Cartesian product' space $(P \times C)$ in R^{n+k} (i.e., the Euclidean space $(x_1, \ldots, x_n, c_1, \ldots, c_k)$, or *phase-control space*). However, the dynamics only occur in the phase space – at constant values of C, as illustrated in Fig. 2.2.

Fig. 2.2

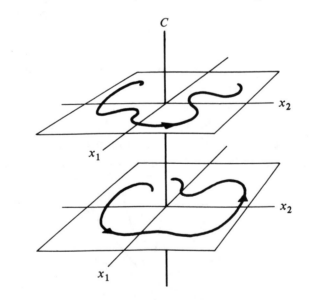

The importance of the phase space, or the phase-control space, as a visual/conceptual tool for analyzing the dynamics of systems (2.1.1) can hardly be overstated. It frees us from the detailed aspects of analytic or numerical solutions and focuses our attention on the *global characteristics* of the motion (e.g., is the motion recurrent, confined to one region, attracted toward some region, tends to 'fill out' some region; do nearby solutions rapidly diverge, etc.?).

If $F(x, t)$ in (2.1.1) is a function of time (nonautonomous system), more than one solution, say $x^{(1)}(t)$, $x^{(2)}(t)$, can 'pass through' the same point of phase space (i.e., $x^{(1)}(t_0) = x^{(2)}(t_0)$ for some t_0). To ensure that only one solution passes through each point, it is necessary to consider autonomous systems (see Section 2.2). We therefore make this restriction for the present

$$\dot{x} = F(x; c) \quad x \in R^n \text{ (autonomous)} \tag{2.1.2}$$

The solutions of (2.1.2) involving different initial conditions generate the *family of oriented phase curves* in the phase space which is called the *phase portrait* of the system. The vector field $F(x; c)$ is everywhere tangent to these curves, and their orientation is defined by the direction of the associated tangent vector, $F(x; c)$. Note that the phase portrait (Fig. 2.3) consists of oriented curves through *all* points of the phase space, where the functions $F(x; c)$ in (2.1.2) are defined.

Fig. 2.3

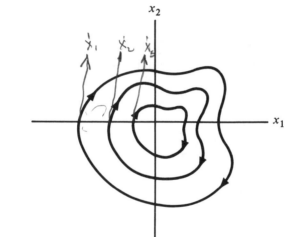

Different systems (2.1.2), involving very different functions $F(x)$, may have solutions which, although analytically quite different, have a family of trajectories in phase space that possess features which are 'similar' to each other. One type of 'similarity' involves the question of whether one family of trajectories can be *continuously deformed* into another family, retaining the sense of motion (their orientation) in the phase space. Systems with this type of similarity will be referred to (Arnold, 1983) as being *topologically orbital equivalent* (TOE). Thus (Fig. 2.4) illustrates two families, both in R^2, which can clearly be continuously deformed into each other ('stretching' or 'contracting', but not 'cutting' or 'connecting' orbits). While this type of similarity is geometrically very appealing, its physical significance should be considered in all of the

Fig. 2.4

Fig. 2.4

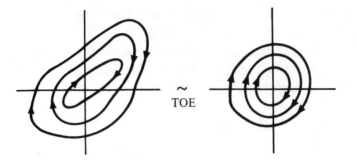

$\underset{\text{TOE}}{\sim}$

following examples (e.g., what is the dynamical similarity of the two systems in the last figure!).

TOE can be expressed in another fashion, which is equivalent, although it is not obvious that this is so. A *homeomorphism* is an association of the points $x \in R^n$ and $y \in R^n$ which is a one-to-one, continuous map with a continuous inverse. That is,

$$f : X \to Y \text{ means } y = f(x) \quad (x \in R^n, y \in R^n) \tag{2.1.3}$$

where $f(x)$ is single-valued and continuous. Moreover, $x = f^{-1}(y)$ is likewise single-valued and continuous. A set in X is said to be *topologically equivalent* to a set in Y if the two sets can be mapped into each other by a homeomorphism. This equivalence is too general for dynamic purposes – in particular, for TOE.

Fig. 2.5

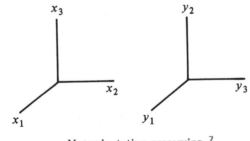

Not orientation preserving

A homeomorphism is called a *diffeomorphism* if $f(x)$ and $f^{-1}(x)$ are differentiable at all points. A homeomorphism is *orientation-preserving* if a right-handed coordinate system in X is mapped into a right-handed system in Y (Fig. 2.5). If the map is a diffeomorphism, then it is orientation-preserving if the Jacobian is everywhere positive,

$$|\partial f_i(x)/\partial x_j| > 0. \tag{2.1.4}$$

If the phase portrait (i.e., all orbits) of $\dot{x} = F(x)(x \in R^n)$ and $\dot{y} = G(y)(y \in R^n)$ can be related by an orientation preserving homeomorphism, then the two systems are TOE. Given the previous definition of TOE, this statement is by no means obvious, but it is true (see Appendix B for further details).

Fig. 2.6

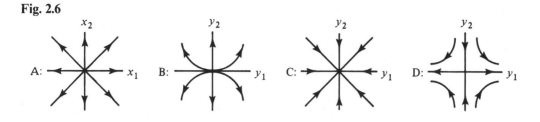

To illustrate this equivalence, consider the phase portraits in the (Fig. 2.6): It is clear that portrait A can be continuously deformed into portrait B. Moreover, A and B are readily transformed into each other by (for example) the following homeomorphism

$$h: y_1 = x_1; \ y_2 = x_2^2 (x_2 \geqslant 0), \quad y_2 = -x_2^2 (x_2 \leqslant 0),$$

so that

$$h^{-1}: x_1 = y_1; \ x_2 = +y_2^{1/2}(y_2 \geqslant 0), \quad x_2 = -(-y_2)^{1/2}(y_2 \leqslant 0).$$

Neither A nor B is equivalent to C, because the orientation cannot be mapped continuously between them. For example, the map

$$h: C \to A, \quad x_1 = 1/y_1; \ x_2 = 1/y_2 (y_1, y_2 \neq 0); \ x_k = 0 (y_k = 0)$$

is not continuous at the origin. Moreover, it is clear that the portrait D is 'altogether different' from the others.

One difference in the nonequivalent phase portraits is that the set of (inward, outward) solutions to the origin have dimensions $(0, 2), (0, 2), (2, 0)$ and $(1, 1)$ in cases A, B, C, and D respectively. These dimensions are invariant under an orientation preserving homeomorphisms, as just illustrated. These dimensions are therefore an example of a *topological invariant* (i.e., a property which is common to all topologically equivalent sets). Other invariants are discussed in Section 2.6.

Fig. 2.7

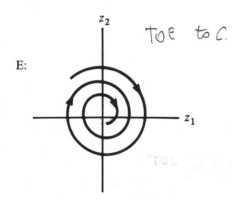

E:

$\mathsf{TO\,E \ to\, C}$

A less obvious case involves the portrait E shown in Fig. 2.7. It appears to be possibly equivalent to C in the previous diagram, but not to A because of the orientation difference. Picturesequely, E looks like it came from C by attaching an electric drill or mixer to the origin and letting it run forever. Forever is a long time, but it is required in order to eliminate any limiting direction as the phase curves approach the origin. To accomplish this effect in a map, we must arrange it so that as $y_1^2 + y_2^2 \to 0$, 'time' (or the rotation angle) goes to infinity. That, however, is easy to accomplish, for

$$\left. \begin{aligned} z_1 &= h_1(y) \equiv y_1 \sin\left((y_1^2 + y_2^2)^{-1}\right) - y_2 \cos\left((y_1^2 + y_2^2)^{-1}\right) \\ z_2 &= h_2(y) \equiv y_1 \cos\left((y_1^2 + y_2^2)^{-1}\right) + y_2 \sin\left((y_1^2 + y_2^2)^{-1}\right) \end{aligned} \right\} \qquad (2.1.5)$$

defines $h\colon Y \to Z$, and the inverse is simple because $z_1^2 + z_2^2 = y_1^2 + y_2^2$. Since this map is also continuous everywhere, including the origin, it is a homeomorphism. Moreover, (2.1.5) is orientation preserving. Thus C and E are indeed topological orbital equivalent systems. Again, it is useful to consider the dynamical difference between these systems.

It is important to note that this type of similarity does not imply that the speed is the same along equivalent curves after the mapping. Hence the existence of such an equivalence does not imply that there is a continuous transformation which takes the solutions of $\dot{x} = F(x)$ into the solutions of $\dot{y} = G(y)$.

Exercise 2.1. Consider the two systems

$$(A)\,\dot{x}_1 = -x_2, \quad \dot{x}_2 = x_1 \quad \text{and} \quad (B)\,\dot{y}_1 = y_2, \quad \dot{y}_2 = -2y_1.$$

Draw the phase portrait for both systems in their phase spaces, (x_1, x_2) or (y_1, y_2). Show that there is a homeomorphism $h\colon X \to Y$ connecting these portraits. Determine whether (A) and (B) are TOE. Determine if this mapping takes the solutions of $\dot{x} = F(x)$ into those of $\dot{y} = G(y)$.

A useful theorem here is:

> *Two linear systems, $\dot{x}_i = \sum_j A_{ij}x_j$ and $\dot{y}_i = \sum_j B_{ij}y_j$ (or $\dot{x} = Ax$, $\dot{y} = By$) all of whose eigenvalues (λ, μ), where $Ax = \lambda x$, $By = \mu y$, have nonzero real parts, are TOE if and only if the number of positive (and therefore negative) real parts are the same for both systems. The same is true for nonlinear systems in the neighborhood of a fixed point (for more details see Arnold, 1978).*

The dynamics in the phase space may be restricted to *manifolds* which are embeded in this space. By 'manifolds' we will simply mean k-dimensional, continuous regions embedded in the n-dimensional Euclidean phase space, defined by some equations $M_j(x_1,\ldots,x_n) = 0$ $(j = 1,\ldots,n-k)$. If the functions $M_j(x)$ are differentiable, then the manifold is called a differentiable manifold. More general definitions exist, but we will not need them. A few more topological details are given in Appendix B.

The harmonic oscillator provides a simple example of such a manifold. The equations of motion are

$$\dot{x} = v, \quad \dot{v} = -(k/m)x.$$

The phase space is two-dimensional R^2, but each solution lies on a one-dimensional (differentiable) manifold

$$H(x, v) \equiv \tfrac{1}{2}mv^2 + \tfrac{1}{2}kx^2 = \text{constant}. \tag{2.1.6}$$

Because this manifold is simply an integral of the equations of motion (i.e., $dH(x(t), v(t))/dt \equiv 0$ for all solution), it is referred to as an *integral manifold*. All these manifolds (for different constants H) are topologically equivalent to the *one-sphere* S^1 (or circle), defined in R^2 by

$$S_1 = \{x \in R^n; \; x_1^2 + x_2^2 = 1\}. \tag{2.1.7}$$

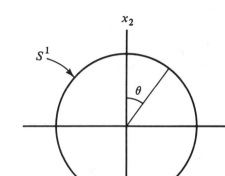

Fig. 2.8

This is clearly so, because the manifold (2.1.6) can be mapped onto this circle (Fig. 2.8) by

$$x_1 = \left(\frac{k}{2H}\right)^{1/2} x; \quad x_2 = \left(\frac{m}{2H}\right)^{1/2} v.$$

Moreover, the angle variable θ can be introduced by

$$x_1 \equiv \sin\theta, \quad x_2 \equiv \cos\theta. \tag{2.1.8}$$

The variable θ can be limited to the interval $0 \leqslant \theta < 2\pi$.

Alternatively θ can be viewed as taking on all values in R^1 with the understanding that $\theta + 2\pi$ and θ are the same point of the manifold, sometimes written $\theta \in R^1 \bmod 2\pi$. This manifold can also be described as a *one-torus*, $T^1 \equiv S^1$.

A slightly more involved case is a simple rigid pendulum (Fig. 2.9). The mass at the end moves in a plane, (x, y), so that its phase space is generally R^4, $(x, y, \dot{x}, \dot{y})$. However, it is subjected to a 'constraint'

$$x^2 + y^2 = l^2,$$

(margin note, handwritten) If the Hamiltonian is independent of time (autonomous) then $\dot{H} = 0$ and $H = \text{const}$.

Fig. 2.9

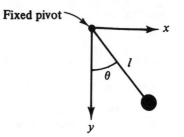

or $x = l\sin\theta$, $y = l\cos\theta$. The configuration of a system in our usual R^3-space, is described by its *configuration space*. In the present system the configuration space (x, y), is restricted to the manifold S^1, whereas its velocity (e.g., $\dot\theta$) can take all values in R^1. Thus its dynamics takes place on a two-dimensional manifold, called a cylinder $(\theta, \dot\theta)$, which is the *cartesian product space*, $S^1 \times R^1$. There is no particular benefit in considering this manifold to be in the original phase space, R^4, since we cannot picture it there. Fortunately, we can represent ('embed') such a cylinder in our configuration space, which is R^3, a portion of which is illustrated in Fig. 2.10 (it extends to $\pm\infty$ along

Fig. 2.10

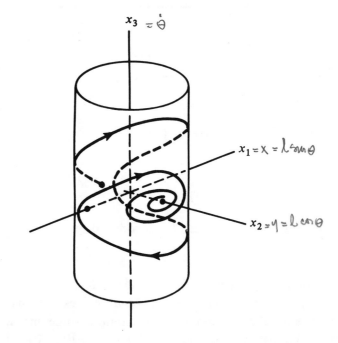

$x_3 = \dot\theta$). The trajectory which is illustrated is for a damped oscillator. Thus all of the topological properties of the constrained motion in R^4 can be faithfully represented on this cylinder in the space (x_1, x_2, x_3). A mapping from R^4 to the cylinder is

$$x_1 = x, \quad x_2 = y, \quad x_3 = -\dot{y}/x = \dot{x}/y.$$

The manifold, $S^1 \times R^1$, may also be viewed as a (non-Euclidean) phase space of this system. Because we cannot view the backside of a cylinder, it is sometimes 'split' along the x_3 direction, and spread out, as illustrated in Fig. 2.11. Finally, it may be simply represented periodically, with $S \equiv l\theta \in R^1$.

Fig. 2.11

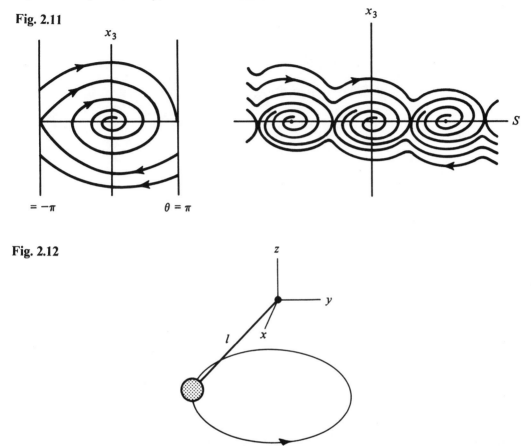

Fig. 2.12

The more general 'spherical' pendulum is not restricted to a plane (Fig. 2.12), but is required to satisfy

$$x^2 + y^2 + z^2 = l^2,$$

Thus the configuration manifold is topologically a *two-sphere* $S^2 = \{x \in R^2 : x_1^2 + x_2^2 + x_3^2 = 1\}$, that is to say, the surface of an ordinary sphere. The velocities are in R^2, so the dynamic manifold in R^6 is $R^2 \times S^2$, which cannot be represented (embedded) in our configuration space, R^3.

Another interesting case is a planar double pendulum, whose configuration space (x_1, y_1, x_2, y_2) is constrained by two conditions (Fig. 2.13):

$$x_1^2 + y_1^2 = l_1^2; \quad (x_1 - x_2)^2 + (y_1 - y_2)^2 = l_2^2.$$

Fig. 2.13

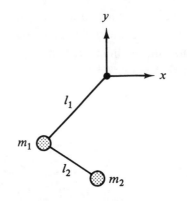

Fig. 2.14

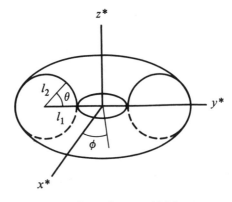

A configuration manifold

This manifold in R^4 is $S^1 \times S^1$, which is called a *two-torus*, T^2. It can be represented (embedded) in R^3. It is the surface of a doughnut (Fig. 2.14), namely

$$x^* = (l_1 + l_2 \cos \theta) \cos \phi;$$

$$y^* = (l_1 + l_2 \cos \theta) \sin \phi, \quad z^* = l_2 \sin \theta;$$

where

$$l_1 > l_2 \quad \text{and} \quad 0 \leqslant \theta < 2\pi, \quad 0 \leqslant \phi < 2\pi.$$

Exercise 2.2. Consider now some phase portraits on these manifolds (Fig. 2.15). Determine if any of the following phase portraits are topologically orbital equivalent. At this level of description it suffices to simply see (in your mind's eye) whether or not the orbits can be continuously deformed into one another, preserving the orientation. Beware, however! There is one pair which is ambiguous, try to identify it. For TOE cases obtain a possible homeomorphism (in R^n).

Fig. 2.15

<u>*n* = 1</u>

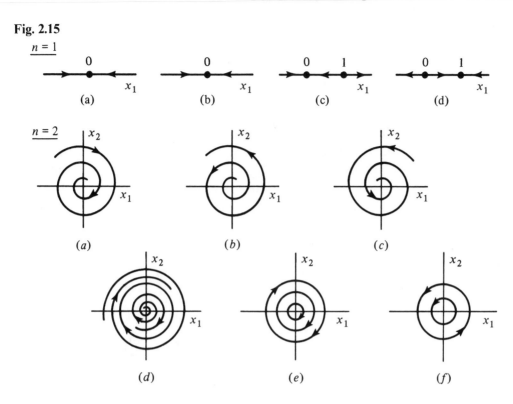

(a)   (b)   (c)   (d)

<u>*n* = 2</u>

(*a*)   (*b*)   (*c*)

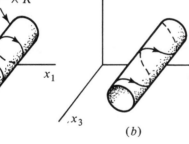 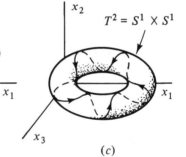

(*d*)   (*e*)   (*f*)

<u>*n* = 3</u>

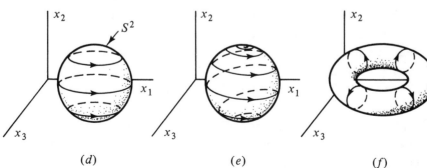

(*a*)   (*b*)   (*c*)

(*d*)   (*e*)   (*f*)

Generalizations of the above manifolds are:

$$n\text{-sphere, } S^n = \{x_1^2 + x_2^2 + \cdots + x_{n+1}^2 = 1\}$$

$$n\text{-torus, } T^n = S^1 \times S^1 \times \cdots \times S^1 (n \text{ products}).$$

Only S^1, S^2, T^1 and T^2 can be embedded in R^3, so that we can represent them in our configuration space. Other manifolds which occur in dynamics are the Möbius strip, which is a non-orientable surface with one edge, and various knotted torii (a two-dimensional example is illustrated). Finally, the famous Klein bottle, obtained by identifying the opposite ends of a finite cylinder, but with the orientations of the circles reversed, is an example of a two-dimensional manifold which can only be embedded in R^4.

Fig. 2.16

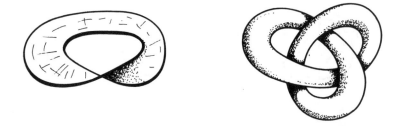

Fig. 2.16 illustrate special cases of a general *embedding theorem*, discussed in more detail in Appendix B. Briefly, the theorem (Whiteney, 1936, 1944) states that any C^r manifold of dimension m can be mapped by a C^r homeomorphism into R^n ('embedded' in R^n) if $n = 2m$. Some of the above 'exotic' manifolds have been found to be of physical importance, and the embedding theorem has been applied to other experimental investigations (see the index), indicating that these concepts may have useful physical applications.

2.2 Existence, uniqueness and constants of the motion

We will now take up (very briefly) the two questions: (a) does (2.1.1) have a unique solution if $x(0)$ or a periodic condition is given, and (b) does the solution exist for all t, as $t \to +\infty$? Two useful sufficient (but not necessary) conditions serve to give some idea of what is required in these areas, without getting involved in numerous refinements which exist (see references).

If the solutions of (2.1.2) are unique, then no two trajectories of this autonomous system can intersect in the phase space. For a nonautonomous system (2.1.1) if the solutions are unique, two trajectories can only intersect for two values of t which yield different values of $F(x, t)$. Such an intersection does not imply nonuniqueness, since uniqueness deals with comparing solutions at the same value of F. While more refined conditions are known (e.g., see Ince, 1927, Chapter 3), a widely used uniqueness-criterion is based on the *Lipschitz condition*.

Theorem 1. *Let y, and z be any two vectors (in R^n), satisfying*

$$|y - a| < a_1, |z - a| < a_1$$

which defines a region $\Omega(a, a_1)$ in the phase space (Fig. 2.17). Also let $|t - t_0| \leqslant T$

Fig. 2.17

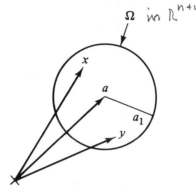

which, together with Ω, defines a region D in R^{n+1} around (a, t_0). If the function $F(x, t)$ satisfies the **Lipschitz condition**

$$|F(y, t) - F(z, t)| < a_2|y - z| \quad (y, t) \in D; (z, t) \in D,$$

where a_2 depends only on $\Omega(a, a_1)$ (not on t), then there exists a unique solution of

$$\dot{x} = F(x, t), \quad x(t_0) = a$$

for $\min(T, a_1/a_3) > t \geqslant t_0$, where $a_3 = \max\limits_{D}|F(x, t)|$.

Here the *norm* of x can be taken to be $|x| \equiv \sum_{i=1}^{n}|x_i|$, or the Euclidean norm $|x^2| = \sum x_i^2$. In particular, if all derivatives $(\partial F_i/\partial x_j)$ are finite everywhere, then F satisfies the Lipschitz condition. However, piecewise linear functions, as well as discontinuous functions, can also satisfy the Lipschitz condition (at least locally), allowing for the

Fig. 2.18

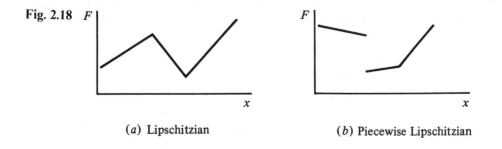

(*a*) Lipschitzian   (*b*) Piecewise Lipschitzian

analysis of near-discontinuous dynamics (e.g., bowed violin strings, clocks, etc.). Note that this uniqueness in the phase space holds for a nonautonomous system. Also the existence and uniqueness is for a limited period of time (Fig. 2.18).

Exercise 2.3. Do the following equations have unique solutions for all (x_1, x_2)?

$$\dot{x}_1 = 3x_1^2 + (x_1 - x_2)^{1/2}, \quad \dot{x}_2 = (x_1 + x_2)^4.$$

Examples of nonuniqueness in physical examples appear to occur most commonly (excluding poor models!) when we use constants of the motion as new equations of

Fig. 2.19

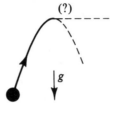

motion. For example, we throw a ball up in the air. Its vertical velocity is governed by $m\ddot{z} = -mg$, or $m/2\,(\dot{z})^2 = E - mgz$, where E is its 'vertical energy'. Thus its upward velocity is given by $\dot{z} = +[(2E/m) - gz]^{1/2}$. If this is used as an 'equation of motion', it must be used with care (e.g., in numerical integrations), because it does not satisfy the Lipschitz condition at $z = 2E/mg$. At this point the 'equation of motion' will put us on the spurious branch $z(t) \equiv 2E/mg$ ($\dot{z} = 0$), rather than the physical branch $\dot{z} = -[(2E/m) - gz]^{1/2}$ required by the original equation of motion. More subtle examples of this type occur when action-angle variables, (J, θ), are used to study the dynamics of coupled nonlinear oscillators – again because $J_k^{1/2}$ terms can appear in the equations of motion. Then, if some $J_k(t) \simeq 0$, and numerical integration method will become inaccurate, reflecting the nonuniqueness at $J_k = 0$.

Theorem 1 concerns only the question of uniqueness for the *initial value problem* (sometimes called 'Cauchy data', $x(0) = x^0$). It is possible, however, to have problems in which other conditions are required of the solutions of (2.1.1). One such possibility is the requirement that the solution be periodic

$$x(t + T) = x(t), \tag{2.2.1}$$

where T may either be unspecified, or may be required to take a given value. Such solutions may either be nonunique or may not exist. Indeed, in higher order systems, periodic orbits are generally very uncommon, relative to all of the orbits in the phase space. If one measures the 'volume' occupied by all of these orbits in R^n, it is generally zero (they form a set of 'measure zero'). Even so, they may be found to pass through any neighborhood of any point in the phase space (they may be 'everywhere dense'). Perhaps the simplest illustration of this feature can be found in the standard map model (see the index). Due to their many important applications, periodic solutions have historically been the most commonly studied dynamic case. This illustrates the important fact that a set of mathematical measure zero may be of great physical importance (it has a 'physical application measure' of about 0.9 to date!).

Periodic solutions are frequently nonunique in important physical situations (even when T is specified in (2.2.1)). This non-uniqueness is one case which arises in *bifurcation theories*, to be considered later. To illustrate this, and at the same time emphasize that 't' in (2.1.1) need not be the time variable, consider the simple wave equation (in one dimension)

$$\frac{\partial^2 y}{\partial t^2} - c^2\frac{\partial^2 y}{\partial x^2} = 0$$

and look for a 'normal mode' (standing wave) solution

$$y(x, t) = A(x)\cos(\omega t + \phi),$$

so that the amplitude satisfies the ODE

$$\lambda^2\frac{\mathrm{d}^2 A}{\mathrm{d}x^2} + A = 0 \quad (\lambda = c/\omega), \tag{2.2.2}$$

and the independent variable is now x.

Exercise 2.4. Determine for what values of λ equation (2.2.2) has periodic solutions satisfying $A(x + L) = A(x)$ for specified (given) L, and determine the number of solutions.

The question of uniqueness of solutions is more involved if the equations are not of the form (2.1.1). Thus $(\dot{x})^2 = x$, which is first order but second degree, can split into any of three solutions at $x = 0$ (see Exercise 2.5). More startling yet is the following theorem due to Rubel, which shows that there are '*universal differential equations*' (of low order) which have solutions arbitrarily close to any prescribed function (i.e., one equation for all functions!). The precise theorem (Rubel, 1981) is:

There exists a nontrivial fourth-order differential equation

$$P(y', y'', y''', y'''') = 0 \quad (y' = \mathrm{d}y/\mathrm{d}t, \text{ etc.}), \tag{A}$$

where P is a polynomial in four variables, with integer coefficients, such that for any continuous function $\phi(t)$ on $(-\infty, \infty)$ and for any positive continuous function $\varepsilon(t)$ on $(-\infty, \infty)$, there exists a C^∞ solution $y(t)$ such that $|y(t) - \phi(t)| < \varepsilon(t)$ for all t on $(-\infty, \infty)$. In other words, you specify any continuous $\phi(t)$, and any small $\varepsilon(t) > 0$, then there is a C^∞ solution of A which stays within $\varepsilon(t)$ of $\phi(t)$!

A specific example of such an equation is

$$3y'^4y'''^2 - 4y'^4y''^2y''' + 6y'^3y''^2y'''y'''' + 24y'^2y''^4y''''$$
$$- 12y'^3y''y''' - 29y'^2y''^3y'''^2 + 12y''^7 = 0$$

(note that it is seventh degree). Since the universal equation is only fourth order, one can relax the condition on the solution from C^∞ to C^4, and a simpler equation can then be obtained (Duffin, 1981)

$$2y'''y'^2 - 5y'''y''y' + 3y''^3 = 0.$$

Obviously no model of a real process should be based on these equations! Similarly dense solutions can occur for smooth PDE (Buck, 1981). The purpose in pointing out these universal equations is to illustrate the amazing complexity that the solutions of low order differential equations can have – a feature which is not uncommon, even for 'realistic' models.

To answer the question of the *existence* of a solution for all t, as $t \to +\infty$, a useful theorem is (e.g., see Cesari, p. 3):

Theorem 2

Let $\dot{x} = F(x, t)$, where $F(x, t)$ is continuous for all x and $t \geqslant t_0$. Assume that

$$M(|x|) \geqslant |F(x, t)| \quad \text{for all} \quad t \geqslant t_0.$$

If, for any r_0 such that $M(r_0) \neq 0$,

$$\int_{r_0}^\infty \frac{dr}{M(r)} = \infty \quad \text{(Wintner's condition)},$$

then the solution $x(t)$ exists for all $t \geqslant t_0$ and all $x_0 \equiv x(t_0)$.

Some examples involving these theorems are given as exercises. Fig. 2.20 illustrates an $F(x)$ which does not satisfy each theorem.

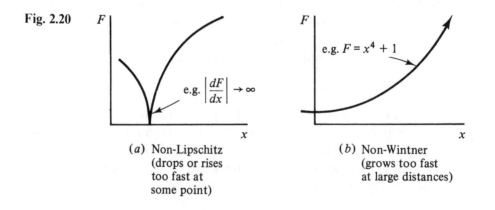

Fig. 2.20

(a) Non-Lipschitz (drops or rises too fast at some point)   e.g. $\left|\dfrac{dF}{dx}\right| \to \infty$

(b) Non-Wintner (grows too fast at large distances)   e.g. $F = x^4 + 1$

Whereas existence and uniqueness are taken for granted in all 'well-formulated' physical models, some cases which warrant care will be noted in Chapter 3. Indeed, there are many useful models which have solutions that do not exist after a finite time. The Riemann shock singularity in a perfect fluid is one classic example in PDEs (see Chapter 9 and Exercise 2.5 for others). This limitation may occur because some physical effect has been ignored (e.g., viscosity, in the Riemann shock). In other cases it may be a realistic representation of what happens to the 'system' in Fig. 2.21 (e.g.,

Fig. 2.21   'System'

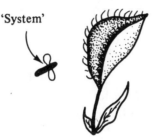

what happens to a fly, a corporation stock, a star, or a civilization when (respectively) a venus flytrap closes, there is a stockmarket crash, it becomes a supernova, or a nuclear war occurs!?). In the mathematical literature one of the conditions which is frequently required in order for a solution to be described as a 'flow' is that the solution exists for all $t \to +\infty$ (or possibly, all $t \in R$). In keeping with this, we will say that, if the solutions of the autonomous system (2.1.2) are unique and exist for all $t \in R$, these solutions define a *flow* in the phase space.

Exercise 2.5. Determine whether the following equations satisfy the Lipschitz and/or Wintner conditions for all initial conditions $x(0) = x_0$. If the Lipschitz condition is not satisfied for some x_0 determine if there are two solutions and if the Wintner condition is not satisfied determine if the solution does not exist for all $t > 0$.

(a) $\dot{x} = Ax^{1/2}$ (there are action-angle equations similar to this);
(b) $\dot{x} = x^2$ (there are 'explosive instabilities' similar to this);
(c) $\ddot{r} = -r^{-2}$ $(r > 0)$ (e.g., radial motion with gravitational attraction)

Exercise 2.6. To illustrate the fact that in the Lipschitz condition the constant a_2 cannot depend on t, consider the equation

$$\dot{x} = \frac{4t^3 x}{x^2 + t^4}, \quad \text{if} \quad (xt) \neq 0; \quad \dot{x} = 0; \quad \text{if} \quad (xt) = 0.$$

Evaluate $|F(x, t) - F(y, t)|$ along the lines $x = at^2$, $y = bt^2$ and show that the Lipschitz condition is not satisfied near $t = 0$. The equation admits an infinite number (continuum) of solutions satisfying $x(0) = 0$ (namely, $x = c - (c^2 + t^4)^{1/2}$, arbitrary c).

Exercise 2.7.

  (a) Does the harmonic oscillator satisfy Wintner's condition?
  (b) Does the nonlinear oscillator $\dot{x} = -y^3$, $\dot{y} = x^3$ satisfy Wintner's condition? Do you believe every solution exists for all t in this case? Why?

For the remainder of this section we will consider only *autonomous systems,* $\dot{x} = F(x; c)$. *The general solution* of this equation is of the form

$$x(t) = X(t; x^0) \tag{2.2.3}$$

where

$$x(0) = X(0; x^0) = x^0$$

are the prescribed initial conditions ('Cauchy conditions'). If $F(x^0; c) = 0$ at some point x^0 in the phase space, then $\dot{x} = F(x; c)$ has a constant solution $X(t; x^0) \equiv x^0$. Such points, x^0, are variously called *singular* (*critical*), or *fixed points*, and physically represent equilibrium (i.e., stationary) states of the system.

Equation (2.2.3) can be viewed as a *continuous mapping* in the phase space $X^t: x^0 \to x$ (another notation is $X^t: R^n \to R^n$), which carries the initial point x^0 to the point $x(t)$. If the dynamical solution exists and is unique, then to any given x and t there must correspond a unique initial point x^0 – that is, one can write

$$x^0 = K(x, t). \tag{2.2.4}$$

This is the inverse mapping of (2.2.3), and since the mapping is one-to-one and continuous it is referred to as a *homeomorphism* (and if it is differentiable, a *diffeomorphism*) in the phase space, as discussed in the last section. Note that, while $X(t; x^0)$ must be differentiable in t, it is not necessarily differentiable in x^0, and hence not necessarily a diffeomorphism.

The functions $K(x, t)$ in (2.2.4) are n constants of the motion, since the initial conditions x^0 are obviously constants. Thus

$$K(x(t), t) = K(x(t'), t'), \quad \text{or} \quad dK(x(t), t)/dt \equiv 0. \tag{2.2.5}$$

Any function $K(x(t), t)$ which satisfies (2.2.5) (identically), when $x(t)$ satisfies (2.1.1), will be called *a constant (or an integral) of the motion* of (2.1.1).

It should be emphasized that this definition of the integral of the motion, which has a long usage (e.g. Whittaker, 1944), is the most generous and most inclusive possible definition. It does not agree with the more restrictive definition frequently found in the mathematical literature (e.g., Arnold, 1978), which requires that the integral be a differentiable function of its variables throughout phase space. The distinction is made clear by the simple example

$$\dot{x}_1 = x_1, \quad \dot{x}_2 = x_2. \tag{A}$$

A strange constant of the motion for this system, which, however, satisfies (2.2.5), is

$$K \equiv 0 \quad \text{if} \quad x_1 = x_2 = 0, \, K = x_1/x_2 (x_1 \leqslant x_2, x_1 \neq 0),$$
$$K = x_2/x_1, (x_2 \leq x_1, x_2 \neq 0). \tag{B}$$

This function is not continuous at the origin, nor differentiable anywhere on the diagonals $(x_1 = \pm x_2)$. However, if we substitute any solution of (A) into (B) it is differentiable in t and equals zero. The condition of (general) differentiability is more restrictive than is warranted by physical interests, but it does warn of the need for more refined concepts. Some of these refinements have been distinguished by adding adjectives (e.g., 'isolating') to the term integral of the motion. This will be discussed later in this section.

The constants of the motion (2.2.5), however, are nearly always explicit functions of the time (time-dependent constants of the motion). There are clearly always n such constants, but they are, generally speaking, not the most useful (i.e., 'informative') constants.

Example. For $\dot{x} = 1, x(0) = x^0$, we have $x = x^0 + t$ so that $x^0(x, t) = x - t$, considered as a function of x and t, is a time dependent constant of the motion. It is the only (independent) constant of the motion for this system.

Of particular interest are the *time-independent constants of the motion*, $K^*(x; x^0) = 0$, since each such constant defines a manifold in the phase space of dimension $(n - 1)$, in which $x(t)$ will remain for all t. These manifolds are referred to as *integral manifolds*. Such constants are obtained by some process which is equivalent to eliminating the explicit time dependence between the constants in (2.2.5). Sometimes, however, they can be obtained more directly, as the following exercises illustrate.

Exercise 2.8 For $\dot{x}_1 = x_2, \dot{x}_2 = x_1^2$ obtain a time independent constant of the motion without obtaining $x_1(t), x_2(t)$ explicitly (hint: take the ratio of these equations, and eliminate 'dt').

$$\frac{dx_1}{dx_2} = \frac{F_1}{F_2}, \quad \text{or} \quad F_2 dx_1 = F_1 dx_2.$$

Exercise 2.9. Show that if *any* two variables in R^n, say (x_1, x_2), enjoy the 'conjugate relationship' $F_2 = \partial \Phi(x_1, x_2)/\partial x_1, F = -\partial \Phi(x_1, x_2)/\partial x_2$, then $\Phi(x_1, x_2)$ is a constant of the motion.

The possibility of eliminating time over some interval to obtain the local time independent integrals depends on the implicit function theorem. This gives the conditions required for the local inversion, $t = T(x^0, x)$, of (2.2.4). This inversion

Fig. 2.22

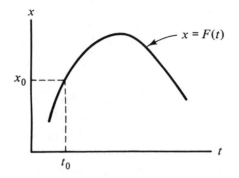

problem is illustrated by the simple example shown in Fig. 2.22. Given this $x = f(t)$, clearly we can obtain a unique $t = g(x; x_0, t_0)$, where $g(x_0; x_0, t_0) = t_0$, in the neighborhood of x_0, provided $(\partial f / \partial t)_{t_0} \neq 0$. Since this problem also has important applications in other contexts, it will be stated here in a general form.

Implicit function theorem

Let $F(x, y) = 0$ be a system of n equations, and let $A \in R^n$ and $B \in R^m$ be regions such that, for $x \in A$ and $y \in B$, this system has a solution, and the Jacobian determinant does not vanish,

$$\mathrm{Det} \left| \frac{\partial F_i}{\partial x_j} \right| \equiv \left| \frac{\partial(F_1, \ldots, F_n)}{\partial(x_1, \ldots, x_n)} \right| \neq 0.$$

Then there is a region $C \subset B$ such that for $y \in C$ there is a unique solution

$$x = g(y) \quad (x \in A),$$

which is continuous in y, such that

$$F(g(y), y) \equiv 0 \quad (y \in C).$$

Note that the size of the region C is not specified, only that it exists. Therefore the unique inversion may only be local.

Exercise 2.10. In the following two sets of equations

$$\text{(a)} \quad y_1 = x_1^2 - x_2^2, \quad y_2 = x_1 x_2$$
$$\text{(b)} \quad y_1 = \cos x_2 \ \exp(x_1), \quad y_2 = \sin x_2 \ \exp(x_1),$$

the values of (y_1, y_2) are determined by the values of (x_1, x_2). These equations therefore define two maps $f_i: X \to Y$. Apply the implicit function theorem to determine if they are everywhere locally invertible (i.e., there is a locally unique map $g: Y \to X$). Are the maps, f_i, globally one-to-one?

In order to eliminate t from the constants of the motion (2.2.5), we need only one equation ($n = 1$ in the theorem) such that $\partial K_i / \partial t \neq 0$ for some regions A and B, $x \in A$, $t \in B$. Then, for some subregion $t \in C \subset B$, one can obtain a unique relationship

$$t = T(x, x_1^0) \tag{2.2.6}$$

where T is continuous in its variables. We can then use this to eliminate t from the remaining $(n - 1)$ expressions (2.2.5) to obtain

$$x^0 = K(x, T(x, x_i^0)) \tag{2.2.7}$$

(the component i gives the trivial equation $x_i^0 = x_i^0$). Thus it is always possible (in principle) to locally obtain $(n - 1)$ continuous time independent constants of the motion (2.2.5) which, however, need not be smooth (differentiable) functions. Unfortunately, this method cannot generally yield *global constants* (defined uniquely by some function). This failure clearly occurs if $\partial K_i / \partial t$ vanishes for some t, which frequently happens, making it possible to obtain (2.2.3) only for a limited range of t. However, even if this does not occur, the local uniqueness does not imply global uniqueness (Exercise 2.10).

Exercise 2.11. Consider the case of two (decoupled) harmonic oscillators

$$\dot{x}_i = \omega_i y_i; \quad \dot{y}_i = -\omega_i x_i \quad (i = 1, 2).$$

This system has four constants of the motion, and one can therefore obtain three time independent constants, at least locally. Prove that this is true (using the implicit function theorem) and show that there are two which are global, but that one is only local. If you are ambitious, show that if $3\omega_1 = 2\omega_2$, then there is a third global time independent constant of the motion.

Thus, in general, the trajectories in phase space for an autonomous system of ODE do not lie on a manifold which is determined by the level of any smooth (differentiable) function, $K(x) = $ constant. Obviously, Hamiltonian systems are exceptions to this general rule. The important question of the relationships between the 'type' of constants of the motion (e.g., algebraic, additive, etc.) and the qualitative characteristics of the motions of the system, remains one of the baffling and intriguing areas of research, which has been heightened by the dynamic insights that have been obtained since 1960. These basic considerations will be discussed further in the section on general dynamics (see Poincaré's theorem on analytic integrals, Chapter 8, and Whittaker's adelphic integrals (Whittaker, 1944)).

For the present we simply mention the interesting discussion by Contopoulos (1963) who introduces the distinction between constants of the motion and 'constraints on the motion' or 'isolating integrals'. These are time independent constants which are also

Fig. 2.23

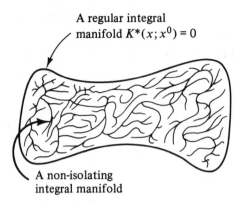

A regular integral manifold $K^*(x; x^0) = 0$

A non-isolating integral manifold

sufficiently smooth to isolate 'macroscopic' regions of the phase space (Fig. 2.23). In a similar vein, Dragt and Finn (1976) discuss the concept of 'regional constants'. There are also, however, regular constants of the motion which fail to 'isolate', or demarcate regions of different dynamical behavior (a simple example can be found in Chapter 5, Section 5.2). The occurrence of isolating or regional constants means that the possibility of ergodic behavior is either eliminated or restricted to a lower dimensional manifold in the phase space. Even when such isolating regions do not occur, the migration of a solution through phase space may be surprisingly slow (see Arnold diffusion; the Fermi–Pasta–Ulam result; Fermi's conjecture; 'stochastic metastability'). The topic of constants of the motion is one which has misled many people, but we must move on to other topics for the present.

Exercise 2.12. Consider a classical system of N particles (masses, m_i) interacting with potentials $V_{ij}(|r_i - r_j|)$, but free of any 'external' force. Determine one (or more) time dependent constants of the motion K, which is linear in t (not one of those in (2.2.4), (2.2.5)) and hence trivially gives a relationship (2.2.6) (hint: it is one of the ten 'classic' constants). Note that the motion of this system is not bounded in phase space (its energy surface is infinite).

In addition to the general solutions (2.2.3), nonlinear equations may also possess *singular solutions* which are distinct from the general solutions. A good treatment of this subject can be found in the book by Ince (1927). While these solutions do not represent any of the general solutions, they can be related to observable phenomena (e.g., caustics in optics or electron optics). We will limit our discussion of singular solutions to some comments in the chapter on first order equations, largely as a warning that some extraneous solutions can be introduced by simple, and common, analytic manipulations.

2.3 Types of stabilities

An important question concerning any particular solution of (2.1.1), say $x = u(t)$, is whether it is stable. Before this can be answered we must obviously define stability. There are in fact over fifty generally accepted definitions of stability in the literature (e.g., see Szebehely, 1984). From these many possibilities we will select only a few, which we will find to be the most useful:

Stability in the sense of Lyapunov

(a) The solution $u(t)$ is said to be *uniformly stable* if there exists a $\delta(\varepsilon) > 0$ for every $\varepsilon > 0$, such that any other solution $v(t)$ for which $|u - v| < \delta(\varepsilon)$ at $t = t_0$, satisfies $|u(t) - v(t)| < \varepsilon$ for all $t \geqslant t_0$. If no such $\delta(\varepsilon)$ exists then $u(t)$ is said to be unstable.

(b) If $u(t)$ is uniformly stable and, in addition,

$$\lim_{t \to \infty} |u(t) - v(t)| \to 0,$$

then $u(t)$ is said to be *asymptotically stable*.

These stability criteria are quite restrictive for they require that $u(t)$ and $v(t)$ remain close to each other for the same values of t in both solutions (time 'ticks' are illustrated in Fig. 2.24. This is sometimes referred to as an *isochronous correspondence* of the two solutions. For example, it will be shown below that even an anharmonic oscillator is unstable in this sense. To relax this restriction and thereby include an important type of stability, Poincaré introduced the concept of orbital stability.

Fig. 2.24

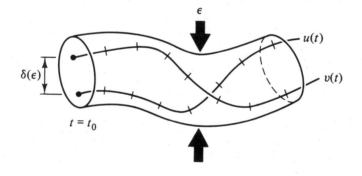

Stability in the sense of Poincaré

(c) Let Γ be the orbit defined by $u(t)$ for all t, and Γ' be the orbit defined by $v(t)$ for all t. We say that Γ is *orbitally stable* if, for any $\varepsilon > 0$, there exists a $\delta(\varepsilon) > 0$ such that, if $|u(0) - v(\tau)| < \delta(\varepsilon)$ for some τ, then there exists a $t'(t)$ such that $|u(t) - v(t')| < \varepsilon$ for all $t > 0$.

(d) The *orbit* Γ is said to be asymptotically stable if Γ' tends toward Γ as $t \to \infty$.

What orbital stability implies is that the two solutions will 'experience the same history, but possibly on different time scales', related by $t(t')$. Note the different density of time 'tick' on the orbits in Fig. 2.25 (exaggerated). In many cases this is physically the most significant test of stability, but it is rarely the one which is investigated mathematically because it can be very difficult to establish.

Fig. 2.25

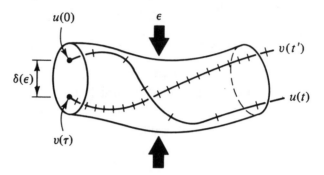

Perhaps the 'coarsest' kind of stability is:

Stability in the sense of Lagrange (bounded stability)

(e) The solutions of $\dot{x} = F(x, t)$ are said to be boundedly stable if $M < \infty$, $|x| \leqslant M$ for all t.

When we speak of 'stability' we will usually mean stability in the sense of Lyapunov. The others shall be designated explicitly as orbital stability or bounded stability.

Exercise 2.13. Determine which of the five stability conditions (a–e) are satisfied by a free particle, $\dot{x} = v$, $\dot{v} = 0$. Do the same for a harmonic oscillator, $\dot{x} = v$, $\dot{v} = -x$. This gives some appreciation of the physical implication of each condition.

Exercise 2.14. Obtain the general solution of the nonlinear oscillator $\dot{x} = -R(x, y)y$, $\dot{y} = R(x, y)x$, where $R = (x^2 + y^2)^{1/2}$ and determine if it is (Lyapunov) stable, or orbitally stable. Suppose $\dot{x} = ax + y - xf(x, y)$; $\dot{y} = ay - x - yf(x, y)$ where $f(x, y) \to +\infty$ as $x^2 + y^2 \to \infty$. Describe the character of the solutions for all $-\infty \leqslant a \leqslant +\infty$.

A nice method for establishing the stability of a fixed point, which does not involve solving equations of motion, is to make use of the following simple theorem:

Theorem 3. (Lyapunov's first theorem for stability).

Let $\dot{x} = F(x)$, and x_0 be a fixed point $(F(x_0) = 0)$. Assume there is a function $L(x)$ such that (a) $L(x_0) = 0$, and $L(x) > 0$ if, $x \neq x_0$ in some region about x_0, and (b) $F \cdot \nabla L \leqslant 0$ in that region. Then x_0 is stable.

Fig. 2.26

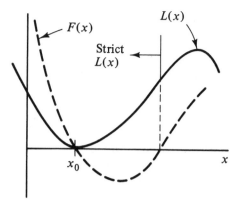

Fig. 2.26

The function $L(x)$ which satisfies these conditions is called a *Lyapunov function* (Fig. 2.26). If $F \cdot \nabla L < 0$ it is called a strict Lyapunov function. The conclusion of this theorem follows simply from the fact that $\dot{L} = F \cdot \nabla L$, so the point $x(t)$ moves along a path where L does not increase in value (because of (b)). This insures, for small enough perturbations from x_0, that $x(t)$ will remain close (because of property (a)), or return to x_0 if $F \cdot \nabla L$ is strictly negative. In practice the trick is to find the function $L(x)$, which may not be easy to do. An exception is conservative systems where L is frequently just the energy. To illustrate:

Exercise 2.15. For $\dot{x}_1 = 2x_2(x_3 - 1)$; $\dot{x}_2 = -x_1(x_3 - 1)$; $\dot{x}_3 = x_1 x_2$, try $L = \sum a_i x_i^2$, and determine whether $x = 0$ is asymptotically stable (by a judicious choice of a_1, a_2, a_3).

Exercise 2.16. Obtain a strict Lyapunov function for the system $\dot{x}_1 = -2x_1 - x_2^2$; $\dot{x}_2 = -x_2 - x_1^2$.

Exercise 2.17. Consider a Hamiltonian system

$$\dot{p} = -\partial H/\partial q; \quad \dot{q} = \partial H/\partial p \qquad (p, q \in R^n)$$

which has a fixed point which can be taken to be at $(p, q) = (0, 0)$ with $H(0, 0) = 0$. Using a Lyapunov function, show that if the fixed point is a local minimum of H, then the fixed point is stable.

Exercise 2.18. Consider the Hamiltonian

$$H = \tfrac{1}{2}(p_1^2 + p_2^2) + \alpha(p_1 q_2 - p_2 q_1) + \tfrac{1}{2}(q_1^2 + q_2^2),$$

which can be associated with a symmetric top, with (q_1, q_2) being the direction cosines of the symmetry axis with respect to horizontal axes. Show that the fixed point $(p, q) = 0$ is not a minimum of H, but that one can easily obtain a Lyapunov function $L(p, q)$

which establishes that the fixed point is stable. What combination of L and H has particular significance?

Exercise 2.19. Euler's equations of motion for a free rigidly rotating body can be written

$$I_i\dot{\omega}_1 = \omega_2\omega_3(I_2 - I_3); \quad I_2\dot{\omega}_2 = \omega_1\omega_3(I_3 - I_1); \quad I_3\dot{\omega}_3 = \omega_1\omega_2(I_1 - I_2);$$

where $I_1 > I_2 > I_3 > 0$ are the principal moments of inertia. Now all points on each ω_k axis are fixed points (why?). Obtain a Lyapunov function for this system which establishes that the fixed points $(\omega_1^0, 0, 0)$ and $(0, 0, \omega_3^0)$ are stable. Explain what this means if we spin a book around each of its principal axes. Now add a frictional torque to the ω_1 motion only, by adding a term $-\mu\omega_1 I_1$ to the right side of the $\dot{\omega}_1$ equation, with $\mu > 0$. Show that $\omega_1 = 0$ is now asymptotically stable.

It is not difficult to see how this approach could be generalized to establish bounded stability, which is a point of considerable interest for so-called *explosive instabilities*. These point will be elaborated upon in later sections.

2.4 Integral invariants

An important concept which should be discussed, at least briefly, is that of an *integral invariant*, due to Poincaré, and discussed in some detail by Whittaker. The most important of these (apparently!), for it applies to all equations of the form (2.1.1) (rather than just to Hamiltonian systems), is a *generalized Liouville's theorem*.

Let $\Omega(0)$ represent an arbitrary domain ('volume') in phase space at $t = 0$, and let $\Omega(t)$ represent this domain at the time t; so $\Omega(t)$ is the set of all point $x(t, x^0)$ which propagate according to $\dot{x} = F(x, c, t)$ and such that $x(0) \equiv x^0$ is contained in $\Omega(0)$. We then consider the integral of some function $\rho(x, t): R^n \times R \to R$ over this moving domain $\Omega(t)$,

$$I(t) \equiv \int \cdots \int_{\Omega(t)} \rho(x, t)\, dx_1, \ldots, dx_n. \tag{2.4.1}$$

If $\rho(x, t) > 0$, we might view $\rho\, dx_1 \cdots dx_n$ as an ensemble probability in this phase space. $I(t)$ is called an *integral invariant* of (2.1.1) if its value is constant in time,

$$dI/dt = 0. \tag{2.4.2}$$

The trick in evaluating dI/dt is to change the variable limits of integration in (2.4.1) to the fixed limits of the initial region $\Omega(0)$. This means one uses the variable x^0 by introducing the Jacobian $|\partial(x_1, \ldots, x_n)/\partial(x_1^0, \ldots, x_n^0)|$. Then

$$\frac{dI}{dt} = \int \cdots \int_{\Omega(0)} \frac{d}{dt}\left\{\rho(x, t)\left|\frac{\partial(x_1, \ldots, x_n)}{\partial(x_1^0, \ldots, x_n^0)}\right|\right\}\, dx_1^0 \cdots dx_n^0, \tag{2.4.3}$$

where it is now understood that all x_i in the integrand are the functions $x_i(t, x^0)$

satisfying $\dot{x}_i = F_i(x, c, t)$. Since $\Omega(0)$ is arbitrary, (2.4.2) can only be satisfied if the integrand of (2.4.3) vanishes identically. It is then rather straightforward to establish: *Liouville's theorem.*

> *The integral $I(t)$ is an integral invariant of (2.1.1) if and only if $\partial\rho/\partial t + \nabla\cdot(\rho(x, t)F(x, t)) \equiv 0$.*

Exercise 2.20. Prove this theorem. Note that F can depend explicitly on t, but what condition must be assumed about the Jacobian? What does the condition mean in light of the implicit function theorem?

If $\rho(x, t)$ is a probability density, (2.4.2) must hold because of probability conservation. Liouville's theorem then yields an equation for $\rho(x, t)$, for any $F(x, t)$, including dissipative systems (see Appendix C).

Exercise 2.21. Show that $\nabla\cdot F = 0$ for all Hamiltonian systems, so that $\rho(x, t) = 1$ can be used in (2.4.1).

Another beautiful (but apparently unapplied) integral invariant of Poincaré is $I_2 = \sum_k \iint_{\Omega_k} dp_k dq_k$ (for Hamiltonian systems only). Picturesquely: the sum of the 'shadow areas' $\Omega_k(t)$ (on all planes (p_k, q_k)) of a two-dimensional region in R^{2n}, $\Omega(t)$, is constant, see Fig. 2.27 (see Whittaker (1944)).

Fig. 2.27

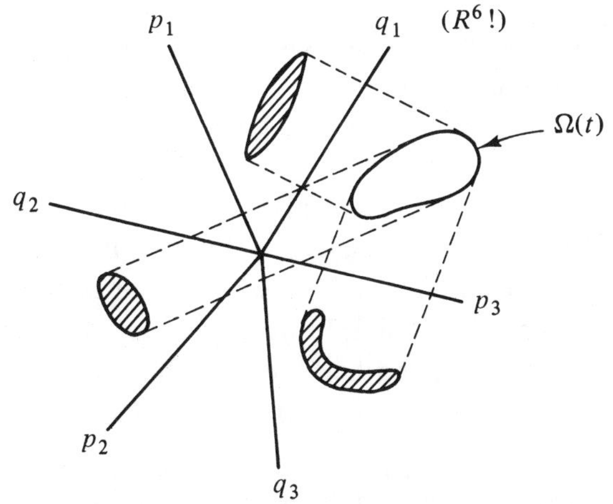

Exercise 2.22. Consider a particle moving in one dimension, $\dot{x} = v$, $\dot{v} = F(x, t)$ with the phase space (x, v). The region $\Omega(t)$ is initially $\Omega(0) = \{0 \leqslant x \leqslant 1, 1 \leqslant v \leqslant 2\}$. Prove by explicit calculations that $I(t) = I(0)$ (where $\rho = 1$), by determining $\Omega(t)$. To do this, first compute where the corners of $\Omega(t)$ are located, then obtain $\Omega(t)$. In parts (a) and (b) assume that the particle experiences an elastic reflection ($v \rightarrow -v$) at $x = \pm 2$ (it is in a

box). (a) First take $F = 0$ if $|x| < 2$, and $t = 1$; (b) $F = 0$ if $|x| < 2$, and $t = 5$; (c) $F = +1$, and $t = 1$ (no box); (d) $F = -x$, and $t = 2$; (e) $F \equiv -t$, and $t = 2$. Draw pictures of these moving regions, $\Omega(t)$ (Fig. 2.28). You can learn much from this simple exercise.

Fig. 2.28

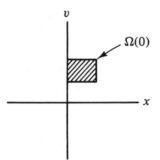

If one views $\rho(x)$ as the 'weight' ascribed to different parts of phase space in a measure $\mathrm{d}\mu = \rho(x)\,\mathrm{d}x_1 \cdots \mathrm{d}x_n$, then Liouville's theorem says that the total measure of $\Omega(t)$ $(\mu(\Omega(t)) = \int_\rho \prod_i \mathrm{d}x_i)$ is constant in time (i.e., $I(t) \equiv \mu(\Omega(t))$). Such *'measure-preserving'* flows, $\mu(\Omega(0)) = \mu(\Omega(t))$, are of interest because they imply a basic and fairly common constraint on the trajectories in phase space.

Exercise 2.23. The weight $\rho(x)$ need not be unity. Show that, if $F_1 = 5x_1 - 6x_1^2 x_2$, $F_2 = -3x_2 + 4x_1 x_2^2$, then $\rho \equiv x_1^2 x_4^2$ is the required weight for the Liouville theorem. What does this imply about the variation in the volume of a region $\Omega(t)$?

The above elementary comments on systems of ordinary differential equations are simply meant to introduce some of the basic ideas which will be required shortly. This subject, particularly as it applies to Hamiltonian systems, has been refined to a fine art over the last one hundred years.[†]
It might be pointed out that (2.4.1) may also be used to determine if most solutions are unbounded, in which case, of course, a Lyapunov function doesn't exist. The idea is to pick a $\rho(x)$ which is such that I is finite for any finite Ω. Then, if $\mathrm{d}I/\mathrm{d}t \geq \varepsilon > 0$, most solutions must be unbounded. This follows simply from the fact that $I(t)$ can only increase indefinitely if Ω becomes infinite, and hence most solutions are unbounded. Hence if $\nabla \cdot (\rho F) \geq \delta > 0$ for all x, then most solutions are unbounded. Even if we cannot find a ρ_1

[†] In this context there exists a rather extraordinary series of books ($\simeq 6000$ pages!), *Celestial Mechanics* (Vols. I–V) by Y. Hagihara (1970), which appears to be an essentially complete review of all major methods and results obtained for *Hamiltonian systems*. This series, which deals with many non-celestial problems, is far and away the most extensive single reference that I know of for classical mechanics. In particular, for general properties of Hamiltonian equations see Vol. III (part 2!), which contains (among other things) an extensive discussion of types of stabilities, and methods for constructing Lyapunov functions.

such that $\nabla \cdot (\rho_1 F) > 0$ for all x, it may be possible to find a ρ_2 such that $\nabla \cdot [F\nabla \cdot (\rho_2 F)] > 0$, where $\nabla \cdot (\rho_2 F)$ is not necessarily > 0. In this case most solutions are still unbounded. This follows from the fact that

$$\frac{\mathrm{d}^2 I}{\mathrm{d}t^2} = \frac{\mathrm{d}}{\mathrm{d}t} \int_\Omega \nabla \cdot (\rho F) \, \mathrm{d}x_1 \cdots \mathrm{d}x_n = \int \nabla \cdot [F\nabla \cdot (\rho F)] \, \mathrm{d}x_1 \cdots \mathrm{d}x_n.$$

Obviously, if we can find a ρ_2 for all x, we do not need ρ_1 to prove unboundedness. This method of testing may be extended by considering $\nabla \cdot \{ F\nabla \cdot [F\nabla \cdot (\rho F)] \}$, related to $\mathrm{d}^3 I / \mathrm{d}t^3$, etc.

Exercise 2.24. Consider a special 'mode-coupling' or 'Lotka–Volterra' system

$$\dot{x}_i = \sum_j a_{ij} x_j + \sum_{j,k} b_{ijk} x_j x_k \quad (a_{ij}, \, b_{ijk}, \, x_k \in R^1),$$

where the $b_{iij} = 0 = b_{iji}$. Determine a simple sufficient condition for most solutions to be unbounded. Are all solutions unbounded?

Another integral invariant which is very useful when combined with Poincaré return map (see Section 2.5), can only be accurately portrayed by the intimidating description: An *integral invariant* on an *integral submanifold*. The Liouville theorem gives the condition for the integral $I(t)$, equation (2.4.1), to be constant in time. In this case the domain of integration, $\Omega(t)$, is n-dimensional. We now want to consider, instead of this domain, a domain, $\Omega_k(t)$, which is $(n-1)$-dimensional, and which moves on a surface (manifold) that is defined by some time-independent (and differentiable) constant of the motion of the system (2.1.2)

$$K(x) = K_0 \quad (x \in R^n). \tag{2.4.4}$$

This defines an integral manifold in the phase space. The question is, if we have a system which satisfies the Liouville equation, namely

$$\dot{x} = F(x), \quad \nabla \cdot (\rho F) = 0,$$

so that

$$I(t) = \int_{\Omega(t)} \rho(x) \, \mathrm{d}x_1 \cdots \mathrm{d}x_n$$

is an integral invariant, can we obtain an integral invariant for a domain $\Omega_{K_0}(t)$ which moves on the manifold (2.4.4) according to the equations of motion $\dot{x} = F(x)$? That is, can we obtain an integral

$$I_{K_0}(t) = \int_{\Omega_{K_0}} \Delta(x_1, \ldots, x_{n-1}) \, \mathrm{d}x_1, \ldots, \mathrm{d}x_{n-1}, \tag{2.4.5}$$

such that

$$\mathrm{d}I_{K_0}(t)/\mathrm{d}t = 0? \tag{2.4.6}$$

Fig. 2.29

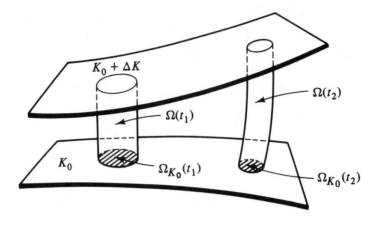

To answer this, and show that it is reasonable, it is useful to draw a diagram, such as that shown in Fig. 2.29, illustrating what is involved.

First of all the variable x_n which is missing from the integral in (2.4.5) is assumed to be such that $(\partial K(x)/\partial x_n) \neq 0$; in other words this direction is not 'tangent' to the manifold defined by (2.4.4). This coordinate is then said to be transverse to the manifold (2.4.4). This may not be possible for all times without a readjustment of the coordinates (see below). We next consider adjacent manifolds defined by $K = K_0$ and $K = K_0 + (\partial K/\partial x_n)\Delta x_n \equiv K_0 + \Delta K$, and let the domain $\Omega(t)$ of the Liouville theorem lie between these adjacent manifolds (see the diagram). Because the 'ends' of the region $\Omega(t)$ are at a constant difference in the value of $K(x)$, namely $\Delta K = (\partial K/\partial x_n)\Delta x_n$, the 'height' of $\Omega(t)$ is

$$\Delta x_n = \Delta K/(\partial K/\partial x_n). \tag{2.4.7}$$

The value of x_n on the manifold $K(x) = K_0$ can be obtained in terms of $K_0, x_1, \ldots, x_{n-1}$ (using the implicit function theorem), so that it equals some function

$$x_n = x_n^0(K_0, x_1, \ldots, x_{n-1}). \tag{2.4.8}$$

Therefore, in Liouville's theorem, the integral involving x_n ranges from this x_n^0 to $x_n^0 + \Delta x_n$, and this height depends on the value of ΔK, by (2.4.7). Different values of ΔK, of course, define different regions, but each region has its own integral invariance, $I(t, \Delta K)$. We now consider the limit of small ΔK or, more precisely, differentiate the Liouville integral $I(t, \Delta K)$ with respect to ΔK, and by what was just said, this must also be an invariant in time. This differentiation readily yields (2.4.5), where

$$\Delta(K_0, x_1, \ldots, x_{n-1}) = \rho(x_1, \ldots, x_n^0(K_0, x_1, \ldots, x_{n-1}))/(\partial K/\partial x_n)_{x_n = x_n^0}. \tag{2.4.9}$$

Therefore, if $(\partial K/\partial x_n)_{x_n^0} \neq 0$, the integral (2.4.5) exists, where Δ is given by (2.4.9), and the integral satisfies (2.4.6). This amounts to a Liouville theorem on the manifold (2.4.5).

In other words, for

$$\dot{x} = \hat{F}(x) \equiv F(x_1, \ldots, x_{n-1}, x_n^0(K_0, x_1, \ldots, x_{n-1})) \quad (x \in R^{n-1}),$$

we now have $\Delta \cdot (\Delta(K_0, x_1, \ldots, x_{n-1}) \hat{F}(x)) = 0$.

To illustrate this, consider a particle (of unit mass) acted on by a constant force, g, so that

$$\ddot{x} = g.$$

Set $x_1 = \dot{x}$, $x_2 = x$, then

$$\dot{x}_1 = g, \quad \dot{x}_2 = x_1.$$

A time independent constant of the motion (the energy) is

$$K(x_1, x_2) = \tfrac{1}{2}x_1^2 - g\,x_2.$$

We note that

$$\partial K/\partial x_2 = -g \neq 0 \quad \text{(for any } x_1, x_2\text{)}.$$

Moreover, corresponding to (2.4.8),

$$x_2 = \frac{1}{g}\left(\frac{1}{2}x_1^2 - K_0\right) \equiv x_2^0(K_0, x_1).$$

Finally, for the present Hamiltonian system,

$$\nabla \cdot (\rho F) = \frac{\partial}{\partial x_1}(g) + \frac{\partial}{\partial x_2}(x_1) = 0 \quad \text{(for } \rho = 1\text{)}.$$

Therefore the Δ of (2.4.9) is simply $-1/g$ and

$$I_{K_0} = \int_{\Omega_{K_0}} \mathrm{d}x_1/(-g) \tag{2.4.10}$$

is the integral invariant. This is essentially the projected length of the region on K_0 along the x_1 axis. Two such projections $\bar{\Omega}_{K_0}$ of Ω_{K_0} along x_1 are shown in Fig. 2.30: $\bar{\Omega}_0(t)$ for $K_0 = 0$, and $\bar{\Omega}_{-1}(t)$ for $K_0 = -1$. These are illustrated for $t = 0$ and $t = t_1 = 3.5$. We see that $\bar{\Omega}_{K_0}(t)$ is constant.

On the other hand, if we try to eliminate x_1, we note that

$$\partial K/\partial x_1 = x_1 \neq 0 \quad \text{(if } x_1 \neq 0\text{)},$$

then

$$I_{K_0} = \int_{\Omega_{K_0}} \frac{\mathrm{d}x_2}{x_1^0(K_0, x_2)}, \tag{2.4.11}$$

where now

$$x_1^0 = \pm[2(K_0 + g\,x_2)]^{1/2} \tag{2.4.12}$$

The double-valued nature of $x_1^0(x_2)$ is the source of the difficulty at $x_1 = 0$. If the region

Fig. 2.30 $\boxed{g = 1}$

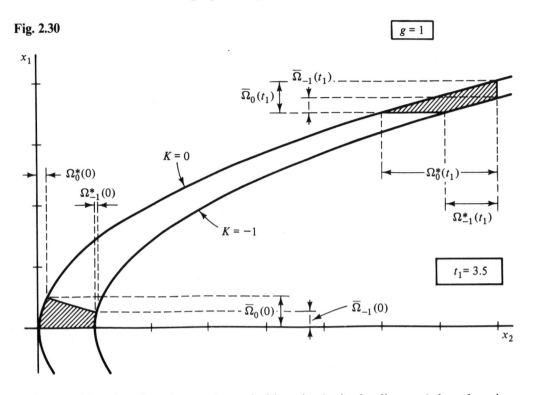

under consideration doesn't pass through this point (as in the diagram) then there is no problem with changing signs and the limits of integration. Note that now, with (2.4.12) substituted into (2.4.11) the projected lengths of Ω_{K_0} along x_2, denoted as $\Omega_{K_0}^*$ in the diagram, vary dramatically, in contrast with (2.4.10).

Exercise 2.25. Obtain an integral invariant on the manifold $p^2 + \omega^2 q^2 = K$ for a harmonic oscillator ($\dot{q} = p$, $\dot{p} = -\omega^2 q$), and indicate its meaning in a sketch. Note the 'awkward' points, if p or q are used as the 'transverse' variable. Can you obtain a bettter variable?

2.5 More abstract dynamic systems

With the exception of some initial considerations of topological equivalence, the point of view presented in the previous sections is largely the classic, or historical, viewpoint, where we solve a system of differential equations and then represent these solutions in a phase space. From this perspective, the phase space is secondary and the process of solving the differential equations is of primary concern. In the case of 'nice' (integrable) problems, this was historically the ultimate objective, but in order to obtain a more flexible and imaginative perspective of the dynamics which might occur, without being burdened by the task (impossible?) of actually solving some system of equations, it is useful, and possibly crucial, to view dynamics in a more abstract form. We have in fact already introduced some of this viewpoint by considering dynamics as a diffeomorphism (a one-to-one map, $f:R^n \to R^n$, such that both f and f^{-1} are at least once differentiable, C^1). Once we become accustomed to that perspective of dynamics, it is not too difficult to take it as the starting point and to view the specific differential equations as being of secondary concern. Even more abstract 'dynamic systems' are often considered (e.g., symbolic dynamics), but we will save those for later.

As an example of this point of view, here is a definition of a *classical dynamical system* (Arnold and Avez (1968)) which is hardly the 'classical system' of old:

Let M be a smooth manifold, μ a measure on M defined by a continuous positive density, $f^t:M \to M$ a one-parameter group of measure-preserving diffeomorphisms. The collection (M, μ, f^t) is called a *classical dynamical system*. [Newton may have just turned over!]

The *measure preserving* requirement of the mapping f^t, is of course a continuation of the content of the Liouville theorem, which restricts us to models such that $\nabla \cdot (\rho F) = 0$ for some $\rho(x) > 0$. More general dynamic systems can be considered, and then some features are lost while new features are gained. Understanding these interrelations is one of the advantages of abstract dynamics, since we can focus attention outside the analytic and classic 'problem solving' arena.

In this section we will limit our considerations to a few basic, and quite useful, concepts for future use. As discussed in Section 2.1, the starting point which is most familiar is a *phase space R^n* (which we can take to be Euclidean), or some *manifold, M*, which is some subset of R^n. In some cases, such as the surface of a cylinder, $S^1 \times R^1$, or of a torus, $S^1 \times S^1$, we can embed M in R^3 (section 1), but that is an unnecessary luxury for what follows.

We next consider smooth maps, f^t, (diffeomorphisms) which map the manifold, M, onto itself, which depend on one continuous parameter (t) and satisfy group properties ($f^0 =$ identity, f^{-t} exists, and $f^{t'} \cdot f^t = f^t \cdot f^{t'} = f^{t+t'}$). For a continuous parameter t, this mapping defines a *flow* on the manifold, and the set of points generated by this flow, containing some point x, is then called the *orbit* $\Gamma(x) = \{ f^t x | t \in R \}$. It will be noted that, at this level of abstraction, there is no longer a reference to the autonomous differential equations (2.1.2), nor to their associated phase space. The discussion can therefore be abstracted from such considerations, although our ultimate objective is to return to physical examples.

Some manifolds, such as n-spheres, S^n, or an n-torus, T^n, are bounded (compact) sets, whereas other manifolds, such as R^k or cylinders, $S^1 \times R^1$, are unbounded. We will consider a bounded manifold, M and assume that a measure, $\mu(x)$, exists on M which is an invariant of the flow, or schematically $\mu(\Omega) = \mu(f^t\Omega)$ for any set $\Omega \in M$. Moreover, we can normalize $\mu(M) = 1$.

An interesting classic result for such bounded manifolds ($\mu(M) = 1$), *Poincaré recurrence theorem*, can be established quite easily within this very general (abstract) context. There is a variety of theorems related to the concept of recurrence (e.g., see Furstenberg (1983) for a nice summary). The simplest theorem is:

'Recurrence I'. For any region $A \in M$, such that $\mu(A) > 0$, there exists some point $x \in A$ such that $f^n x \in A$ for some integer n (Fig. 2.31).

Fig. 2.31

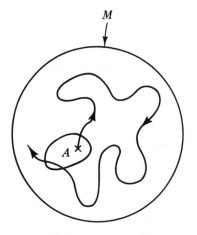

The proof is too simple to omit! Assume that no point $x \in A$ returns to A. Then all points which map into A are not contained in A, so $f^{-n} A \cap A = \emptyset$ (null set), for any n. (Note: $f^{-n} A$ maps onto A in time n.) Moreover, consider $f^{-m} A$, where $m \neq n$. We must have $f^{-m} A \cap f^{-n} A = \emptyset$, for otherwise this intersection would map into A at two distinct times. For example, if $m > n$, this would mean that there is a set in A which returns to A in a time $(m - n)$, contradicting the assumption. Thus the above intersection is indeed null. But $\mu(f^{-n} A) = \mu(A)$ for all n, so all the sets $f^{-n} A$ cannot be

Fig. 2.32

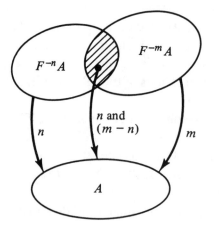

disjoint (as just concluded) since the manifold has a finite measure, which requires

$$\mu\left(\bigcup_{n=1}^{\infty} f^{-n}A\right) \leqslant \mu(M) = 1.$$

This completes the proof (Fig. 2.32).

From 'Recurrence I' we easily conclude:

'Recurrence II'. This recurrence occurs for all points in A, except possibly a set of measure zero.

For, if $A' \subset A$, $\mu(A') > 0$, is a set which does not recur in A, then no point in A' can recur in A'. But, by 'Recurrence I', we conclude that $\mu(A') = 0$, contradicting our assumption.

It should be clear from these proofs that recurrence does not occur only once, but an infinite number of times (determine how large N can be, and still satisfy for $f^{-n}A \cap A \neq \varnothing$ for all $n \leqslant N$).

Now returning to general considerations of flows on manifolds, we observe that associated with a flow is a *velocity vector field*

$$\dot{x} \equiv v(x) = (\mathrm{d}\,f^t x/\mathrm{d}t)_{t=0} \equiv F(x) \tag{2.5.1}$$

which assigns to each point in the phase space a vector, $v(x) \in R^n$, so these vectors are associated with locations in the phase space. Note that this 'velocity' is not generally a physical velocity (e.g., components can have a variety of dimensional properties – such as cm/sec^2). the 'tangent space' is the set of all possible velocity vectors at x for orbits in M – and designated by TM_x (tangent to M at x). By *trajectories* in the phase space (and here the notation does not seem to be uniform in the literature), we will simply mean the *directed orbits*, indicating the sense of the local velocity. The family of all such trajectories in the phase space is the *phase portrait*, discussed in Section 2.1, and the comparison of such portraits led us to the concept of *topological orbital equivalence* (TOE), which was also discussed.

A very useful concept is that of *Poincaré maps*. Consider first a vector field near some nontrivial orbit (not a fixed point). This vector field is 'transverse' to some manifold

Fig. 2.33

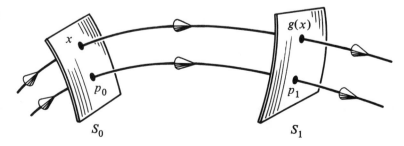

which varies along the orbit. The 'transverse manifold' is simply a 'surface' which is not tangent to the local vector field, as shown in Fig. 2.33. The dimension of the transverse manifold is one less than that of the original manifold (it is said to have *codimension one*). Now let p_0 and p_1 be two points on the orbit corresponding to the time t_0 and t_1, and let S_0 and S_1 be the two transverse manifolds containing these points (respectively), as illustrated in Fig. 2.33. If $x \in S_0$, there will be some minimum time $\tau(x)$ for this point to intersect S_1 (provided that x is close enough to p_0), so that

$$f^\tau x \in S_1.$$

The map which prescribes where the points of S_0 (near p_0) end up on S_1, that is, the map

$$g(x) = f^\tau x : S_0 \to S_1 \tag{2.5.2}$$

is called a *Poincaré map*.

The most common use of this map is when S_0 and S_1 are the same submanifold, which clearly can be arranged if p_0 is a periodic point. That is, if there is a periodic orbit, it is always possible to obtain a map of this type ($S_0 \to S_1 \equiv S_0$) in the neighbourhood of the periodic point, p_0. In this case $g(x)$, in (2.5.2), is called Poincaré's *first-return map*. However, it is important to note that not all flows have first return maps. A simple example of such a flow is shown in Fig. 2.34.

Fig. 2.34

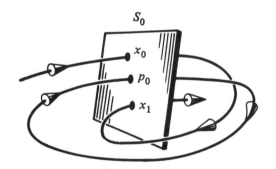

Fig. 2.35

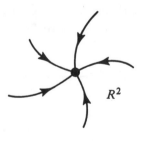

More generally, there may be a closed manifold of R^n (Fig. 2.35), denoted by S^*, such that all points intersect S^* in the same directional sense within some fixed, sufficiently large, interval of time, even if there is no periodic motion. This implies that 'all' points intersect S^* an infinite number of times. The map $P\colon S^* \to S^*$ is a Poincaré map, and this manifold, S^*, is called *Poincaré surface of section*.

An example which can be easily visualized in R^3, has all solutions confined to a manifold T^2. Moreover, assume that all solutions on T^2 are nonperiodic,

$$\dot{\theta} = \alpha \neq 0; \quad \dot{\phi} = \beta \neq 0; \quad \alpha/\beta = \text{irrational number.}$$

We can take the surface of section to be a closed curve which cuts the torus into a finite cylinder, but there are, of course, an infinite number of other possibilities. In this case the surface of section is a circle, S^1, and the Poincaré map is $P\colon S^1 \to S^1$. Note that the Poincaré map does not imply that different points require the same time interval to produce the dynamic 'map' (Fig. 2.36).

Fig. 2.36

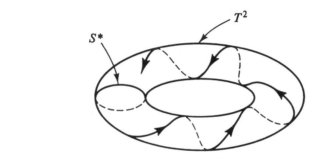

Exercise 2.26. A simple (because it is a solvable linear system!), but useful, example of Poincaré first return map can be obtained by considering a two-dimensional harmonic oscillator. This Hamiltonian system is defined by

$$\dot{p}_i = -\partial H/\partial q_i, \quad \dot{q}_i = \partial H/\partial p_i \quad (i = 1, 2),$$

$$H = \tfrac{1}{2}(p_1^2 + p_2^2) + \tfrac{1}{2}(\omega_1^2 q_1^2 + \omega_2^2 q_2^2)$$

First of all sketch the closed surface, $H(p_1, p_2 = 0, q_1, q_2) = K$ (a constant) in the space $(p_1, \omega_1 q_1, \omega_2 q_2)$. All motions with $p_2 \neq 0$ lie inside this surface, and the trajectories *must* periodically cross the plane $q_2 = 0$. Therefore this plane is transverse to the orbits, and can be used for a Poincaré map (the real Poincaré surface is in R^4, but that is not important). Illustrate the map defined by $q_2 = 0$, $p_2 > 0$, which produces $(q_1, p_1) \to (q_1^*, p_1^*)$ for successive intersections with the plane (note: $p_2 > 0$). Obtain the analytic expression for this map, $p_1^* = f(p_1, q_1)$, $q_1^* = g(p_1, q_1)$. Indicate what special property this map has if ω_1/ω_2 is rational.

Exercise 2.27. Show that the map of the last exercise is area preserving, by showing that $dp_1 dq_1 = dp_1^* dq_1^*$.

Exercise 2.28. In exercise 2.26 we can interchange 1 and 2 and obtain a similar map $(q_2, p_2) \to (q_2^*, p_2^*)$. Consider instead the Hamiltonian $H = \frac{1}{2}(p_1^2 + p_2^2) + \frac{1}{2}(\omega_1^2 q_1^2 - \omega_2^2 q_2^2)$. Show that there is only one map in this case, obtain it, and illustrate it on an appropriate plane.

Fig. 2.37

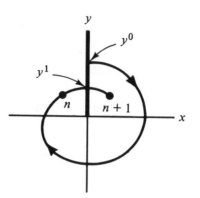

In contrast with these exercises, Poincaré's first return map frequently can only be obtained by numerically integrating the equations of motion. Starting from x_0, on S^*, the problem is to determine x^1 on S^*. The trick is to end up on S^*, not just near it, using an integration scheme with a finite time step. A method to accomplish this has been given by Hénon (1982). A simple example of this method is illustrated for the system (Fig. 2.37)

$$\dot{x} = F(x, y), \quad \dot{y} = G(x, y),$$

which maps $y^0 \to y^1$ on the surface $S^* = \{y > 0, \quad x = 0\}$. Using some time step, Δt, one obtains a series of values $x_n = x(n\Delta t)$ from some integration scheme (e.g., a fourth order Runge–Kutta method). For some n we find $x_n < 0$ and $x_{n+1} > 0$, implying that $x(t)$ has crossed S^* in the time step $(n+1)$. To end exactly on S^*, starting from (x_n, y_n), we consider the equations

$$dt/dx = 1/F(x, y); \quad dy/dx = F(x, y)/G(x, y).$$

We integrate these equations one step, $\Delta x = |x_n|$, starting at (x_n, y_n). This will yield $(0, y^1 = y_n + \Delta y)$, and hence the map $y^0 \to y^1$. The accuracy depends, of course, on the original time step Δt. The generalization of this method to arbitrary systems, $\dot{x} = F(x)$ $(x \in R^n)$, are given by Hénon (1982).

It will be noticed that, in the above processes, a flow (associated with a continuous time) has been related to a *discrete map*, and from a point x_0 one obtains a point x_1, which in turn maps into x_2, and so on. In other words,

$$x_{n+1} = g(x_n) = g(g(x_{n-1})), \quad \text{etc.,} \tag{2.5.3}$$

Fig. 2.38

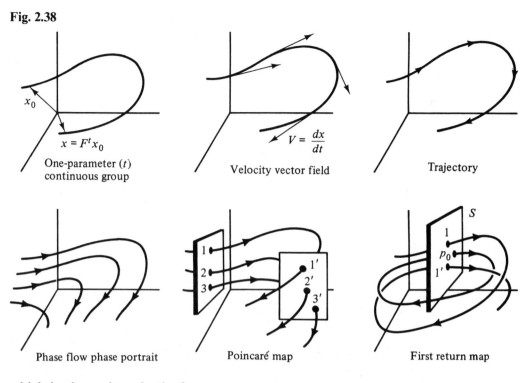

One-parameter (t) continuous group

Velocity vector field

Trajectory

Phase flow phase portrait

Poincaré map

First return map

which is also written in the form

$$x_{n+1} = g(x_n) = g^2(x_{n-1}) = \cdots. \tag{2.5.4}$$

To summarize, we have seen how the following connections shown in Fig. 2.38 can be made. These connections shows that, associated with a system of ODE there is a Poincaré map, or, in other words, a system of difference equations:

$$x_{k+1} = F(x_k; c) \quad (k = 1, 2, \ldots; x \in R^n; c \in R^m). \tag{2.5.5}$$

Of greatest interest, however, are first return maps whose existence can generally only be assured in the neighborhood of a periodic point p_0, or the map on a surface of section, S^*.

What is also very important is whether the inverse connection from a map to the flow system of ODE can also be made. This operation is referred to as a 'suspension' of the map. In the special case of a diffeomorphism $g: S \to S$, for a compact region $S \in R^{n-1}$, this map can always be associated with the first return map of some (nonunique) vector field in a manifold of one higher dimension than S (e.g., Nitecki (1971)). Moreover, this vector field defines a system of differential equations, namely

$$\dot{x} = v(x)$$

whose solution $\phi(t) = f^t x$ yields this vector field, by (2.5.1) (e.g., Sverdlove (1977),

Mayer-Kress and Haken (1987). However, if the map is not a diffeomorphism, then the process of obtaining an associated flow of a system of ODE is much more difficult, if indeed it is possible (see, 'suspension' of a tent map, Section 4.10).

To summarize this sequence of ideas:

Difference equation (diffeomorphism) → Poincaré map on S (dim(n)) →

Vector field in dim($n + 1$) → System of ODE (suspension)

It should be emphasized that, even when such a suspension exists, the flow (and hence the system of ODE) is not unique – and, of course, may have no clear connection with any physical system of interest. The nonuniqueness is made clear by the simple example in Fig. 2.39. Both of these systems have the same Poincaré map (an identity map), but have different equations of motion.

Fig. 2.39

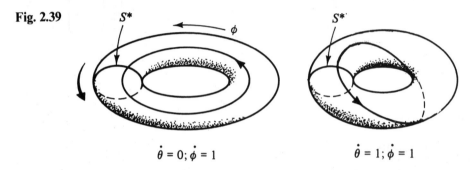

$\dot{\theta} = 0; \dot{\phi} = 1$ $\dot{\theta} = 1; \dot{\phi} = 1$

Therefore the features of solutions of ODE of the form (2.1.2) can also be faithfully represented by diffeomorphisms in a space of one lower dimension. This reduction by one dimension, while it may not seem very impressive, means (for example) that we can 'faithfully' represent the features of motion in a four-dimensional phase space using a diffeomorphism in R^3, which we can picture ('picture' is the important word). 'Faithfully' means that systems of ODE that have topologically orbitally equivalent (TOE) phase portraits will produce maps which are also *topologically conjugate*. Two maps $F: R^n \to R^n$ and $G: R^n \to R^n$ are said to be conjugate if $G = hFh^{-1}$ for some homeomorphism $h: R^n \to R^n$. It is obvious, however, by the above simple examples that a discrete map cannot tell us everything about the dynamics between the intersections with the manifold S. In particular, conjugate (e.g., identical) maps do not imply that their associated flows are TOE. One gives up some detailed knowledge from maps, that gains a dimensional reduction.

2.6 Dimensions and measures of sets

It is useful to have various ways of characterizing sets of points in phase space, for purposes of comparison, and one such characterization is the dimension of the set.

Consider a set of points, S, in a Euclidean space, R^n, in which a distance is defined (see Appendix A). The problem is to determine the 'dimension' of this set of points. We are most familiar with the concept of dimension when it is applied to some continuum of points, such as a line, a plane, or some volume.

The historic (analytic) definition of dimension for such continua was that they have dimension n if their elements (points) can be 'specified', in some vague sense, by n independent quantities (coordinates), which can assume all real values over some range. This viewpoint, however, ran into difficulties with the introduction of sets by Cantor (around 1880) and in particular with his one-to-one association of the points on the interval $[0, 1]$, and those in a square. Moreover, Peano demonstrated his famous space-filling curve, which maps the unit interval continuously onto the unit square. Thus the above vague concept of 'specification' clearly required refinement. (How many coordinates specify the points in a square?) Poincaré was the first (1912) to introduce a topological and recursive definition of dimension, briefly: a continuum is of dimension n if it can be cut into disjoint sets by a continuum of dimension $(n - 1)$, where the null set has dimension -1 (by definition). Brouwer (1913) pointed out that this definition is imperfect, and gave the first modern topological and recursive definition. His idea was rediscovered independently by Menger and by Uryshon (1922), who gave the generally accepted modern definition for metric spaces:

> *The dimension of a space is the least integer n for which every point has arbitrarily small neighborhoods whose boundaries have dimension less than n, and the null set has dimension −1.*

It should be noted that all of these dimensions have integer values. What we want to do is to extend the concept of dimensionality from continua to any set of points, which may be neither finite in number nor a continuum (simple examples are the sets of rational or of irrational points on the interval $(0, 1)$, but mappings frequently produce more interesting sets). For an excellent general introduction to dimension theory see Hurewicz and Wallman (1948). For a few more details on the above remarks, see Appendix B.

There are a number of possible definitions of dimension, but we will focus on two types which appear to be potentially most significant physically (for a review, see Farmer, Ott & Yorke, 1983; also see Mandelbrot's historical sketch (1983), p. 409 ff.). The first dimension, d_c, is called the *capacity*, and is due to Kolmogorov (1958). Let the set of points be in a space R^n, and consider n-dimensional cubes with sides of length ε (volume ε^n). Let

> $N(\varepsilon) = $ *The minimum number of cubes of size ε which is required to enclose all points of a bounded set S.*

Fig. 2.40

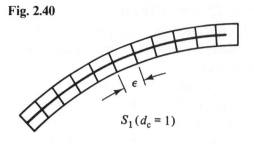

$S_1(d_c = 1)$

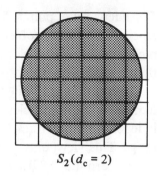

$S_2(d_c = 2)$

The dimension, d_c, of the set S will now be related to the behavior of $N(\varepsilon)$ as $\varepsilon \to 0$. The fact that the set is bounded insures that $N(\varepsilon)$ is finite.

To motivate this definition, consider two familiar sets in a plane R^2, namely a curve (call it S_1) and the points inside a disk $S_2 = \{x, y | x^2 + y^2 \leqslant M^2\}$ (Fig. 2.40). As discussed above, these are normally assigned the dimensions 1 and 2 respectively. If the curve has a length L, it usually takes $N(\varepsilon) = 1 + \text{int}(L/\varepsilon)$ two-dimensional cubes of size ε to enclose these points, where $\text{int}(z)$ is the integer part of z. Hence, for small ε, $N(\varepsilon) \simeq \text{int}(L/\varepsilon)$. On the other hand, to enclose the points in the disk requires $N(\varepsilon) \simeq \text{int}(\pi M^2/\varepsilon^2)$ if ε is small. Note that L and M are not the important aspects here, since all regular curves have the same dimension, as do all disks. What is important is the power of ε which occurs in $N(\varepsilon)$, since that is clearly the dimension of these sets (i.e., 1 for the curve, 2 for the disk). This suggests that if

$$N(\varepsilon) \simeq A/\varepsilon^k \qquad (\text{as } \varepsilon \to 0),$$

then we should identify k with the dimension (capacity). Note that

$$\log N(\varepsilon) = \log A - k \log \varepsilon$$

or

$$k = \frac{\log N(\varepsilon)}{-\log \varepsilon} + \frac{\log A}{\log \varepsilon}$$

and as $\varepsilon \to 0$ we want to identify k with the dimension d_c. This leads to the definition

$$d_c = \lim_{\varepsilon \to 0} \log N(\varepsilon) / \log(\varepsilon^{-1}). \qquad (2.6.1)$$

Note that the value of A is irrelevant, as desired. This dimensionality is a simplified (and much more useable) version of the Hausdorff dimension (Appendix A).

A second measure of sets is called the *information dimension* d_I, which generalizes the concept of capacity by taking into account the probability p_i that a point of the set S is in the cell i. In other words, it takes into account how the points are distributed among the cells. We first introduce the concept of *information*, following Shannon (1948) (also see Brillouin (1963, 1964))

$$I(\varepsilon) = - \sum_{i=1}^{N(\varepsilon)} p_i \log p_i. \qquad (2.6.2)$$

$I(\varepsilon)$ is the information obtained by a measurement of the actual distribution over the $N(\varepsilon)$ cells, when the *a priori* probability of occurrence in these cells is p_i (see below). Thus, assuming uniform *a priori* probabilities in the above sets, the probabilities in S_1 are, $p_i = \varepsilon/L$, and $I(\varepsilon) = N(\varepsilon)(\varepsilon/L)\log(L/\varepsilon) \approx \log(L/\varepsilon)$, whereas for the set S_2, $I(\varepsilon) \approx \log(\pi M^2/\varepsilon^2)$. We see that the usual dimension is obtained by dividing $I(\varepsilon)$ by $\log(1/\varepsilon)$, and taking the limit $\varepsilon \to 0$. This suggests that an information dimension can be defined according to

$$d_{\mathrm{I}} = \lim_{\varepsilon \to 0} I(\varepsilon)/\log(\varepsilon^{-1}). \qquad (2.6.3)$$

The largest value of $I(\varepsilon)$ occurs for equal values of $p_i (= 1/N(\varepsilon))$ (i.e., one can learn the most from an experiment if the *a priori* probabilities are equal – the uncertainty is the greatest prior to the experiment). Therefore $\log N(\varepsilon) \geqslant I(\varepsilon)$, so

$$d_{\mathrm{c}} \geqslant d_{\mathrm{I}}. \qquad (2.6.4)$$

From the operational point of view, an argument can be made that this dimension should be referred to as an uncertainty dimension, because what is operationally determined in the present case are the values $\{p_i\}$ (i.e., they are not known *a priori* in experiments, numerical or otherwise). Therefore it is not a question of making observations from some *a priori* base of knowledge, and thus obtaining information about the actual vs. the possible state of a system ('gaining information'). Instead, the experiment establishes how 'uncertain' is the set of points S (its 'entropy') by determining the $\{p_i\}$. From this point of view the dimension (uncertainty) of a set is less than or equal to its capacity

$$d_{\mathrm{c}} \geqslant d_{\mathrm{u}},$$

which is perhaps more intuitive than calling it an information dimension.

If the set S has '*self-similar*' subsets (subsets which have the same structure, but are scaled down in size by some factor Δ), and these subsets in turn have similar scaled subsets, and so on, then d_{c} can be easily computed (numerous *self-similar* '*fractal*' examples can be found in Mandelbrot (1983), together with their *fractal dimensions*, $d_{\mathrm{f}} = d_{\mathrm{c}}$). Assume that each 'mother' set has K subsets, each identical to the 'mother' set when the distance between their points is increased by a factor $(1/\Delta) > 1$. Then, if we normalize the initial scale to unity and take $\varepsilon = \Delta^n$, the number of cubes needed to cover the complete set S is $N(\varepsilon) = K^n$. Note that the selection $\varepsilon = \Delta^n$ is chosen because it makes it simple to determine $N(\varepsilon)$. Therefore the capacity is obtained simply (without even taking the limit $n \to \infty$, since for each n the ratio is the same due to the initial normalization),

$$d_c = \log K/\log(1/\Delta); \quad K = \text{number of subsets each scaled}$$
$$\text{down by } \Delta \text{ from the previous 'mother' set.} \quad (2.6.5)$$

Such self-similarity is particularly useful in physical cases where, because of the discrete nature of systems (e.g., atoms, molecules, finite vortices in fluids, stars in galaxies, etc.), the mathematical limits ($\varepsilon \to 0$, $n \to \infty$) are not realizable. Nonetheless, a 'capacity' can be assigned to the system which has a self-similar scaling over at least a 'reasonable' number of levels (since, as just noted, no limit is then required). However, it may be more physically revealing to simply note the existence of scaling and its magnitude, rather than an approximate value of d.

To illustrate these, and other concepts, we will consider some *Cantor sets*, which we will designate as $C(\Delta)$. We begin with the points on the closed unit interval, $[0, 1]$, and then discard the points in the (open) middle segment of length $(1 - 2\Delta)$, where $\frac{1}{2} > \Delta$. This leaves two closed sets of points, each of length Δ, as shown in Fig. 2.41. From each of these intervals we next discard their open middle segments of *fractional* length $(1 - 2\Delta)$, as was done in the first step. This leaves four closed intervals, each of length Δ^2, as shown. This process is then continued indefinitely and the Cantor set $C(\Delta)$ is the set of points which remain. Since each interval is divided into two parts when segments are discarded, we have $K = 2$ in (2.6.5) and, clearly, the scaling of the segments in this process is Δ. Hence the capacity of this Cantor set is $d_c(\Delta) = \log(2)/\log(1/\Delta)$.

Fig. 2.41

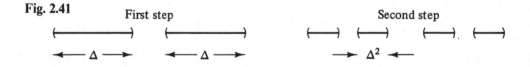

Exercise 2.29. Consider a generalization of the above Cantor set to one in which each interval is divided into two segments of fractional lengths $\Delta_1 = K\Delta_2$ (K: integer), where $1 > \Delta_1 + \Delta_2$ (and discarding the fraction $(1 - \Delta_1 - \Delta_2)$). What is d_c in this case? This, of course, is not a simple similarity case.

Cantor sets have some very interesting and delightfully surprising properties, which requires us to think more carefully about concepts such as dimensions, dense sets, and measure. On the one hand, the set has an 'abundant' number of points, since it is nonnumerable and is closed (because its complement is open). That is, every point of the set is a limit (accumulation) point for the set. On the other hand, the set is nowhere dense in the interval $[0, 1]$. This means that for any open interval within $[0, 1]$, there is a subinterval which does not contain a point of this set (also see Appendix A). This follows from the construction of the set, namely that the discarded segments are 'essentially everywhere' in the interval $[0, 1]$. Such a 'thinly spread' set would seem to be of little physical importance, but we will soon see that this may not be the case (see the even more surprising 'fat' Cantor sets below).

Before discussing this, we should consider the most famous Cantor set is the so-called *middle third set*, which is the special case $\Delta = \frac{1}{3}$ (above). This set is particularly nice since its points can be readily written in base three representation. The Cantor middle third set are the points

$$x = \sum_{k=1}^{\infty} (a_k/3^k), \qquad (2.6.6)$$

where the a_k either equal zero or two (but not one). In base three notation this becomes $x = \cdot(a_1, a_2, \ldots)$. Note, for example, that the point $x = \frac{1}{3}$, which is in the Cantor set, is given by $x = \cdot(0, 2, 2, \ldots)$, because the above sum then becomes $2[(1/(1 - \frac{1}{3})) - 1 - \frac{1}{3}] = \frac{1}{3}$. This set, and its representation in base three, is particularly nice for describing some strange functions, such as the Cantor–Lebsegue ('devil's staircase') function discussed later.

Because of the discrete nature of physical systems, it is often more useful to define a dimension which involves an expanding scaling rather than a reduction in scale lengths ($\varepsilon \to 0$). The idea is quite simple. Consider a finite set of points $N(L)$ enclosed by an n-cube (in some R^n space) with sides of length L (or one can consider an n-ball of radius L). If the points are on average homogeneous, then $N(L) = \rho L^d$ where ρ is an appropriate average density and d is the dimension of the set of points. This holds provided that L is large compared with the separation length (or scale of inhomogeneties) of the points in the set. To formalize this condition one can write

$$d_c = \lim_{L \to \infty} \frac{\log N(L)}{\log L} = \lim_{N \to \infty} \frac{\log N}{\log L(N)}. \qquad (2.6.7)$$

Of course this limit is again only formal, but the point of view of an expanding scale ($L \to \infty$) is simple but important and useful.

Exercise 2.30. What is d_c for the rational points on the interval $[0, 1]$? What is d_c for the irrational points on this inverval? What generalization can be readily obtained?

Exercise 2.31. Sometimes the connection between ε and $N(\varepsilon)$ (the minimum number) is not immediately obvious. Consider the set of points $S = \{x \mid x = 1/n, n = 1, 2, \ldots\}$. Show that $d_c = \frac{1}{2}$.

Exercise 2.32. The points of the set S are obtained by the following operations. A unit square is divided into l^2 squares, and m^2 squares are discarded ($l \geqslant m + 1$; l, m, integers). The process is repeated for each remaining square (scale: $1/l$). Show that $d_c = \log(l^2 - m^2)/\log l$.

Exercise 2.33. The following set (each segment is length $\frac{1}{3}$) is used as a 'generator' on a unit interval (Mandelbrot) to first produce (Fig. 2.42). The generator is then used on each of these straight segments (Fig. 2.43) and the process is continued indefinitely to form a 'Koch triadic curve'. Show that $d_c = \log 4/\log 3$ for the final set.

Fig. 2.42

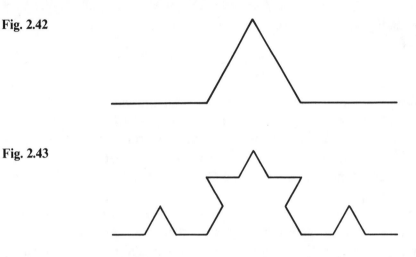

Fig. 2.43

While the above fractal structure is rather contrived, in that it has a simple similarity basis, the occurrence of fractal-like structures in nature is quite common (see Mandelbrot, 1982); Pietronero and Tosatti, 1986; and Thom's quotation, p. 135). An interesting example is the experimental determination of the fractal dimension (capacity) of proteins, both by measuring electron spin relaxations, and from X-ray data of myoglobin (Stapleton *et al.* (1980), Colvin and Stapleton (1985), and Frauenfelder *et al.* (1979)). In this case the expanding form (2.6.7) is used to determine d_c. The results conform nicely with the model of a *self-avoiding (random) walk*, where it is believed that the mean square length, after n steps, is given by

$$\langle R_n^2 \rangle = A n^\Delta$$

where $\Delta = \frac{6}{5}$ (in R^3) and $\Delta = \frac{3}{2}$ (in R^2). In R^3 this gives a capacity of $\frac{5}{3}$, compared with the experimental value of 1.65 ± 0.04. For a review of the theory see, e.g., Barber and Ninham (1970).

It should be emphasized that the *dimension* of a set is independent of, but not unrelated to, its *measure*. This point is elaborated upon in Appendix B. The independence can also be illustrated with the help of various Cantor sets.

Consider a set of disjoint open intervals, I_K, on R^1, $S = \bigcup_k I_k$. The usual Lebesgue measure of this set is the sum of the lengths of the intervals $\mu(S) = \sum_k \mu(I_k)$. If X is some closed set in the interval $[0, 1]$, and $\hat{X}$ is its complement $\hat{X} = [0, 1] - X = [0, 1]/X$, then the measure of X is simply $\mu(X) = 1 - \mu(\hat{X})$. The Cantor set discussed above is a closed set which can be defined in terms of the open sets which are discarded from the interval $[0, 1]$. In the case of the middle-third Cantor set

$$C(\tfrac{1}{3}) = [0, 1] - (\tfrac{1}{3}, \tfrac{2}{3}) - (\tfrac{1}{9}, \tfrac{2}{9}) - (\tfrac{7}{9}, \tfrac{8}{9}) - \cdots,$$

so that its measure is

$$\mu(C) = 1 - \sum_{n=1}^{\infty} 2^{n-1}(\tfrac{1}{3})^n = 1 - \frac{1}{2}\left(\frac{1}{1-(2/3)} - 1\right) = 0.$$

Exercise 2.34. Determine the measure of the above Cantor sets, $C(\Delta)$, for arbitrary $\Delta < \tfrac{1}{2}$.

The Cantor set $C(\tfrac{1}{3})$ has measure zero, and is therefore sometimes referred to as a *'thin' Cantor set*. Since a Cantor set is nowhere dense, it is perhaps not surprising to find that its measure is zero. What usually is found to be quite surprising is that a Cantor set, which is nowhere dense, can have a nonzero measure. These sets are sometimes called *'fat' Cantor sets*.

To illustrate a fat Cantor set, we begin by discarding the open $\tfrac{1}{3}$ interval from the center of $[0, 1]$. Next, however, we discard only segments of length $(\tfrac{1}{4})(\tfrac{1}{3})$ from the two remaining intervals. We proceed then to discard segments of length $(\tfrac{1}{4})^2 (\tfrac{1}{3})$ from the four remaining intervals. Thus, in the nth step of this process, we discard 2^n segments of length $(\tfrac{1}{4})^n (\tfrac{1}{3})$. The measure of this Cantor set is

$$\mu(C) = 1 - \sum_{n=0}^{\infty} (\tfrac{1}{3}) 2^n (\tfrac{1}{4})^n = 1 - \frac{1}{3}\left(\frac{1}{1-(1/2)}\right) = \tfrac{1}{3}.$$

Moreover, not only is this measure nonzero, but it can be made as near to unity as desired; for, if we had begun the above process with a segment of length L rather than $\tfrac{1}{3}$, we would have obtained $\mu(C) = 1 - 2L, (L < \tfrac{1}{2})$. Note also that all these Cantor sets are homeomorphic (topologically equivalent), illustrating that the measure is not invariant under a homeomorphism.

Exercise 2.35. The rationals are everywhere dense in the interval $[0, 1]$, but that set has measure zero, whereas fat Cantor sets are nowhere dense but have finite measure! Use the enumerable property of rational numbers, and sufficiently small intervals (say with lengths $\varepsilon 2^{-n}$) around successive points, to show that the measure of the rational set is zero.

Now we consider the dimension (capacity) of these fat Cantor sets. If we remove L in the first step, it takes two cells of length $\varepsilon = (1 - L)/2$ to cover the remaining set of points. After the next step, it takes 2^2 cells of length $\varepsilon = \tfrac{1}{2}((1 - L)/2 - (L/4)) = (2 - 3L)/8$. In general, after $(n + 1)$ steps

$$N(\varepsilon) = 2^{n+1}; \quad \varepsilon = [2^n - (2^{n+1} - 1)L]/(2 \times 4^n).$$

For large n, we find

$$d_c = \lim_{n \to \infty} \log(2^n)/[\log(2 \times 4^n) - \log(2^n(1 - 2L))]$$
$$= \lim_{n \to \infty} \log(2^n)/[\log 4^n - \log 2^n] = 1.$$

Therefore the dimension of all of these fat Cantor sets is one, regardless of the value of L, whereas their measure varies as $\mu(C) = 1 - 2L$. Moreover, it should be noted that d_c is different for different Cantor sets, and hence, just like $\mu(C)$, is *not* a topological invariant. That is, d_c can be changed by a homeomorphism. The same is true of the Hausdorff dimension (appendix A), and it distinguishes these dimensions from the topological dimensions of Poincaré and Brouwer. The origin of this lack of invariance is the introduction of a distance (metric) which is not a topological invariant.

The question can be reasonably raised as to what, if anything, these mathematical results mean in any particular physical context. It certainly means, at the very least, that we need to be careful in applying these concepts (dimensions, measures, and dense sets), and perhaps we need to be even more careful about drawing conclusions or inferences which may be prejudiced by our unfounded 'understanding' of these concepts. Many of these mathematical concepts, which are based on limiting concepts, need to be applied with care to empirical sciences. Whereas Leibniz noted that mathematics is the science of the infinite, we should note that empirical sciences are the sciences of the finite, be it time or space intervals (see Section 4.11).

Comments on exercises

(2.1) The portraits are simple families of circles or ellipses, but note their orientation is opposite. A homeomorphism from (A) to (B) is $y_1 = -x_1, y_2 = 2^{1/2}x_2$, but this does not take the solutions of $\dot{x} = F(x)$ into those of $\dot{y} = G(y)$. Moreover, the Jacobian of this map is $-2^{1/2}$, so (A) and (B) are not TOE.

(2.2) For $n = 1$, none of the portraits are TOE. For $n = 2$, (e) and (f) may be connected by the homeomorphism $h_1(x) = -x_1, h_2(x) = +x_2$, but this is not orientation

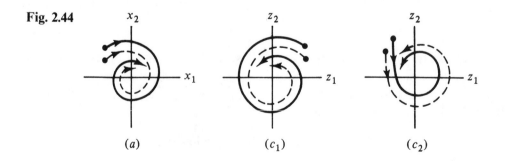

Fig. 2.44

(a) (c_1) (c_2)

preserving. Hence (e) and (f) are not TOE. The portraits (a) and (c) are ambiguous. We need more information to decide whether or not they are TOE. Let the portrait (a) now have two orbits, as illustrated. The corresponding portrait (c) might be either (c_1) or (c_2). The portraits (a) and (c_1) can be connected by the homeomorphism $z_1 = -x_1, z_2 = x_2$, so they are not TOE. The portraits (a) and (c_2) can be connected by two homeomorphisms of the form (2.1.5); namely its inverse, followed by one in which the coefficients of the sine functions have minus signs. Thus (a) and (c_2) are TOE. The lesson here is that orientation refers to the relationship between neighboring orbits, not to a single orbit. For $n = 3$, no flows are TOE.

(2.3) No. $F_2(x_1, x_2)$ is differentiable, and hence satisfies the Lipschitz condition, but $F_1(x_1, x_2)$ does not satisfy the Lipschitz condition on the lines $x_1 = x_2$.

(2.4) L is the control parameter, and the solutions exist only if $L/2\pi\lambda \equiv N$ is in integer. If the allowed amplitude vs L is plotted in this linear case, one simply has vertical lines (arbitrary amplitudes) at equally spaced values of L(n-fold degenerate). Now you might consider the modification of this figure if A is replaced by $A + \varepsilon A^2$ in the second term of (2.2.2).

(2.5) (a) At $x = 0$ the function $F = Ax^{1/2}$ does not satisfy the Lipschitz condition because $|x^{1/2} - y^{1/2}| > a_2|x - y|$ for all $|x|, |y|$ sufficiently small. However, $\int x^{-1/2} dx = \infty$ so solutions exist for all $t \geq t_0$. If $x(t_0) = 0$ the two solutions are $x(t) \equiv 0$ and $x(t) = \frac{1}{4}A^2(t - t_0)^2$.

(b) $dF/dx = 2x$ is finite, so there is a unique solution. However, $\int^\infty dx/x^2$ is finite, so the solution may not exist for all times. Indeed that is so in this case for $x(t) = x_0/(1 + x_0(t_0 - t))$, so we must have $t < (1/x_0) + t_0$.

(c) $\ddot{r} = -r^{-2}$ $(r > 0)$ can be written $\dot{r} = v(= F_1)$, $\dot{v} = -r^{-2}(= F_2)$, so F is differentiable for $r > 0$, and the solution is unique. Wintner's condition however cannot be applied, since we cannot obtain a function of $z \equiv |v| + |r|$ (or $(v^2 + r^2)^{1/2}$) such that $(M(z) \geq |v| + |r^{-2}|$ for all v and $r > 0$. In fact, the solution does not exist for all t, whereas it does if $\ddot{r} = +r^{-2}$.

(2.6) $|F(at^2, t) - F(bt^2, t)| = |(4at/a^2 + 1) - (4bt/b^2 + 1)| > a_2|a - b|t^2$ as $t \to 0$.

(2.7) (a) $\ddot{x} = -x \to \dot{x}_1 = -x_2$, $\dot{x}_2 = x_1$. If $M(r) = r(\equiv (x_1^2 + x_2^2)^{1/2})$, then $M(r) = |F|$ and $\int^\infty dr/M(r) = \infty$, so Wintner's condition is satisfied.

(b) Now $M(r) \geq r^3$, so $\int^\infty dr/M(r)$ is finite. However, $x^3\dot{x} + y^3\dot{y} = 0$ so $x^4 + y^4 = $ constant, the motion is bounded and periodic, hence it exists for all t.

(2.8) $dx_1/dx_2 = x_2/x_1^2$ or $x_1^2 dx_1 - x_2 dx_2 = 0$, or $d(\frac{1}{3}x_1^3 - \frac{1}{2}x_2^2) = 0$; $K(x_1 x_2) = \frac{1}{3}x_1^3 - \frac{1}{2}x_2^2$.

(2.9) $d\Phi/dt = \partial\Phi/\partial x_1 \dot{x}_1 + \partial\Phi/\partial x_2 \dot{x}_2 = \partial\Phi/\partial x_1(-\partial\Phi/\partial x_2) + \partial\Phi/\partial x_2(\partial\Phi/\partial x_1) = 0$.

(2.10) (a) The Jacobian is $2(x_1^2 + x_2^2) \neq 0$ if $(x_1, x_2) \neq 0$. $x_1 = y_2/x_2$, $y_1 = (y_2/x_2)^2 - x_2^2$, so $x_2^4 + x_2^2 y_1 - y_2^2 = 0$, $x_2 = \pm(-y_1/2 + \frac{1}{2}(y_1^2 + 4y_2^2)^{1/2})^{1/2}$. While this is double-valued, it is locally invertible in any region not containing the origin (read the implicit function theorem again if you have any trouble). The double-valuedness does mean, however, that it is not globally invertible – but $(x_1^2 + x_2^2)$ does vanish somewhere.

(2.11) The two global time independent constants are $K_i = x_i^2 + y_i^2$ $(i = 1, 2)$. From the solution $x_i = x_{i0} \cos \omega_i t + y_{i0} \sin \omega_i t$ we obtain $x_i^2 - 2x_i x_{i0} \cos \omega_i t + x_{i0}^2 \cos^2 \omega_i t = y_{i0}^2 (1 - \cos^2 \omega_i t)$. From this we can obtain $\cos \omega_i t$ as two possible functions of the other factors, and thereby t as an arc cosine of these functions (multivalued), which can then be eliminated from another function. This yields only a local time independent constant.

But what if $n\omega_1 = m\omega_2$, where n and m are integers (say $n > m$)? Then we can obtain another global time independent integral of the motion. First note, from above, that (A) $R_i^2 \cos \omega_i t = (x_{i0} x_i + y_{i0} y_i)$; $R_i^2 \sin \omega_i t = (y_{i0} x_i - x_{i0} y_i)$ where $R_i^2 \equiv x_{i0}^2 + y_{i0}^2$. Consider the case $3\omega_1 = 2\omega_2$. In general, for example, $\sin 3\phi = 3 \sin \phi - 4 \sin^3 \phi$. If $\phi = \omega_1 t$, then $3\phi = 2\omega_2 t$, and $\sin 2\omega_2 t = 2 \sin \omega_2 t \cos \omega_2 t$. Therefore $2 \sin \omega_2 t \cos \omega_2 t = 3 \sin \omega_1 t - 4 \sin^3 \omega_1 t$. Now (A) can be used to eliminate the trigonometric functions, yielding a polynomial, time independent (global) integral of the motion $K_3(x_1, y_1, x_2, y_2, x_{10}, y_{10}, x_{20}, y_{20}) = 0$. If we had used $\cos 3\phi = 4 \cos^3 \phi - 3 \cos \phi$, the difference would have been some polynomial in the above K_1, K_2 and constants (x_{i0}, y_{i0}). Clearly, with labor, this can be extended to any case $n\omega_1 = m\omega_2$. Also, as a special case, see Chebyshev polynomials.

(2.12) The three constants are $K = Pt - MX$ where $P = \sum m_i \dot{x}_i$, $M = \sum m_i$, $X = \sum m_i x_i / M$, corresponding to the uniform motion of the center of mass.

(2.13) Free particles: (a) No; (c) Yes; (d) No; (e) No. Harmonic oscillator: (a) yes (only for harmonic case); (b) No; (c) Yes; (d) No; (e) Yes.

(2.14) $x^2 + y^2 \equiv A^2$ is a constant of the motion, hence $x = A \cos(At + \alpha)$, $y = A \sin(At + \alpha)$ (note the amplitude and frequency are now related). This is orbitally stable but not Lyapunov stable.

(2.15) $F \cdot \nabla L = 4a_1 x_1 x_2 (x_3 - 1) - 2a_2 x_1 x_2 (x_3 - 1) + 2a_3 x_1 x_2 x_3 = 2x_1 x_2 x_3 (2a_1 - a_2 + a_3) + 2x_1 x_2 (-2a_1 + a_2)$. For $F \cdot \nabla L \leqslant 0$ requires $a_2 = 2a_1$ and $a_3 = 0$, so $\dot{L} = 0$. Therefore $x = 0$ may be stable, but it is not asymptotically stable (because $x_1^2 + 2x_2^2$ cannot be constant and, at the same time, have $x \to 0$). Because $a_3 = 0$ this function is not a Lyapunov function, since $L(x_1 = 0, x_2 = 0, x_3 \neq 0) = 0$, and no conclusion can be drawn from the above theorem.

(2.16) Try $L = a_1 x_1^2 + a_2 x_2^2$, so $F \cdot \nabla L = - 2x_1^2 (2a_1 + a_2 x_2) - 2x_2^2 (a_2 + a_1 x_1)$. So, if a_1 and a_2 are positive, $L > 0$ for $x \neq 0$ and $\dot{L} < 0$ in the neighbourhood $|x_2| < 2a_1/a_2$ and $|x_1| < a_2/a_1$ (which is all that is required).

(2.17) If $H(p, q) > 0$ in a neighborhood of $(0, 0)$, then take $L = H$ for $F \cdot \nabla L = (-\partial H/\partial q)(\partial H/\partial p) + (\partial H/\partial p)(\partial H/\partial q) = 0$, so the fixed point is stable.

(2.18) Along the axes $p_1 = 0$, $q_2 = 0$, $H = -\alpha p_2 q_1 + \frac{1}{2} q_1^2$, which is negative for $\alpha P_2 > \frac{1}{2} q_1$ (for example), so the origin is not a minimum. A Liapunov function

Fig. 2.45

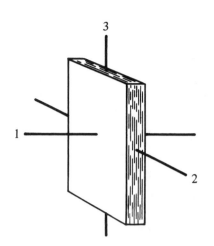

Fig. 2.46

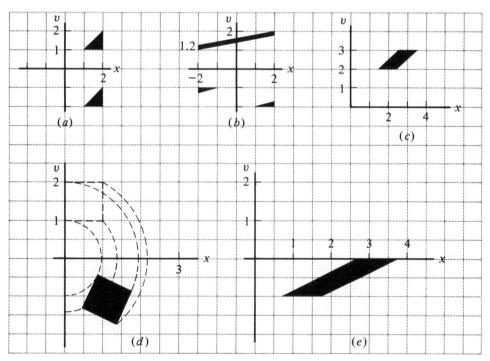

is $L = \frac{1}{2}(p_1^2 + p_2^2) + \frac{1}{2}(q_1^2 + q_2^2)(\dot{L} = 0)$, so H is the sum of this and the constant angular momentum $H - L$.

(2.19) If we take $L = a_1[I_1\omega_1^2/(I_2 - I_3)] + a_2[I_2\omega_2^2/(I_1 - I_3)] + a_3[I_3\omega_3^2/(I_1 - I_2)]$, where all $a_k \geqslant 0$, then $L \geqslant 0$. Moreover, if $a_1 - a_2 + a_3 = 0$, we have $F \cdot \nabla L = 0$. For the fixed points $(\omega_1^0, 0, 0)$ we can take $a_1 = 0$, so that $L(x^0) = 0$. Similarly, take $a_3 = 0$ for the fixed points $(0, 0, \omega_3^0)$. A book is stable when spun about 1 or 3, but not 2 (Fig. 2.45). If $I_1\dot{\omega}_1 = \omega_2\omega_3(I_2 - I_3) - \mu\omega_1 I_1$, we obtain $F \cdot \nabla L = -2\mu a_1 [I_1\omega_1^2/(I_2 - I_3)] < 0$ if $\omega_1 \neq 0$, so $\omega_1 = 0$ is asymptotically stable.

(2.20) See Appendix C.

(2.21) For Hamiltonian systems $\nabla \cdot F = \sum[\partial(-\partial H/\partial q_k)/\partial p_k + \partial(\partial H/\partial p_k)/\partial q_k] = 0$.

(2.23) Because ρ is large where $x_1^2 x_2^4$ is large, $\Omega(t)$ must become correspondingly small in this part of phase space, and become large near either axis.

(2.24) Take $\rho = 1$, $\nabla \cdot F = \sum a_{ii}$, so if $\sum a_{ii} > 0$ most solutions are unbounded. We can, however, take $a_{11} < 0$ (say) and $x_2^0 = x_3^0 = 0$. This motion of x is bounded.

(2.25) The action-angle variables are useful: $p = (2\omega J)^{1/2}\cos\theta$, $q = (2J/\omega)^{1/2}\sin\theta$ so that $\frac{1}{2}(p^2 + \omega^2 q^2) = \omega J$ and $J = (\frac{1}{2\pi})\int_0^{2\pi} p\,dq$. Since $\dot{J} = 0$ and $\dot{\theta} = \omega$, the angular arclength is an integral invariant, $I = \int_{\theta_1(t)}^{\theta_2(t)} d\theta$, $dI/dt = 0$. If the variable q is used $d\theta = dq/\pm[(2J/\omega) - q^2]^{1/2}$ and the sign must be changed at the turning points (where $dq \to 0$).

(2.26) In the space $(p_1, \omega_1 q_1, \omega_2 q_2)$ the energy 'surface' is the interior region of this sphere of radius $(2K)^{1/2}$ (for $p_2 \neq 0$), plus the surface of the sphere (if $p_2 = 0$). This is not a 'proper' representation (embedding) of the energy surface in R^3, because each point corresponds to p_2 and $-p_2$. The map is $q_1^* = q_1\cos(2\pi\omega_1/\omega_2) + \omega_1^{-1}p_1\sin(2\pi\omega_1/\omega_2)$, $p_1^* = p_1\cos(2\pi\omega_1/\omega_2) - \omega_1 q_1\sin(2\pi\omega_1/\omega_2)$. If $\omega_1/\omega_2 = m/n$ (integers), then after n 'steps' the mapped point returns to the start (it is periodic). Note that $(p_1, \omega_1 q_1) \to (p_1^*, \omega_1 q_1^*)$ is a simple rotation by $2\pi m/n = \Delta\theta$.

(2.27) Show that $\partial(p_1^*, q_1^*)/\partial(p_1, q_1) = 1$.

(2.28) There is only periodic motion in the (p_1, q_1) variables, so we consider $q_1 = 0$, $p_1 > 0$ intersections. Now $p_2 = A_2\cosh(\omega_2 t + \phi_2)$, $q_2 = \omega_2^{-1}A_2\sinh(\omega_2 t + \phi_2)$. and $p_2^* = p_2\cosh(2\pi\omega_2/\omega_1)$ $q_2^* = q_2\cosh(2\pi\omega_2/\omega_1) + \omega_2^{-1}p_2\sinh(2\pi\omega_2/\omega_1)$.

(2.29) This exercise is more subtle than it may appear at first. You will probably only deduce the correct answer if you recognize the fact that the value of d_c is independent of the *scale* of a set. For example, if you uniformly stretch a rubber band, the dimension of a set of points on the band does not change (Fig. 2.47). The hidden difficulty in a simple approach is in determining the minimum number of cells $N(\varepsilon_n)$, of size ε_n, which is required to cover the set. This is generally not easy to determine. What is much easier is to use the invariance property of d_c to the scale of the system, together with the self-similarity of the Cantor set.

Fig. 2.47

Scaling × 2

While it is fairly easy to give a simple formulation of the answer (below), let us first see why a cell-counting approach is very difficult (in general). The set is contained in the intervals Δ_1 and $\Delta_2 = K\Delta_1$, so if $\varepsilon_1 = \Delta_1$ then $N(\Delta_1) = (1 + K)$

cells are required to cover the set. From the next stage in the construction of the set, we know that it is contained in the intervals Δ_1^2, Δ_2^2, and in two $\Delta_1 \Delta_2$ intervals, for a total length of $(\Delta_1 + \Delta_2)^2$. Hence, if we take $\varepsilon_2 = \Delta_1^2$, we can certainly cover the set with $(1 + K)^2$ cells, but this is not generally the minimum number. If we continue, and take $\varepsilon_n = (\Delta_1)^n$, then we can conclude that $(1 + K)^n$ cells will certainly cover the set and, therefore, this is in upper bound on $N(\varepsilon_n)$. We can therefore obtain an upper bound on d_c.

$$d_c \leqslant \lim_{n \to \infty} \log(1 + K)^n / \log(1/\Delta_1^n) = \log(1 + K)/\log(1/\Delta_1).$$

We next want to see why this cannot be an equality if $K \neq 1$. Assume that the above is an equality. Then we would obtain

$$(1 + K)\Delta_1^{d_c} = 1.$$

A little thought shows that this expression cannot be correct. This Cantor set was constructed from the unit interval, using the intervals Δ_1 and $K\Delta_1$ as the self-similar generators of the set. Since only these three scales are involved, namely 1, Δ_1, and $K\Delta_1$, the dimension should be a function of these quantities only. If that were not so, then d_c would not be invariant under a scale change (i.e., all intervals multiplied by the same constant). However, the above equality is not a function of these three quantities. If it were, then the second term would be replaced by $(K\Delta_1)^{d_c}$. In order for that to be true, it can be seen from above that the correct asymptotic value for the number of cells is $N(\varepsilon_n) = (1 + K^{d_c})^n$. Clearly, this would be a very difficult conclusion to come to by actually covering the set with cells of length $\varepsilon_n = (\Delta_1)^n$.

The more sensible approach is to make use of the invariance of d_c to scaling, and the self-similarity of the Cantor set. Consider a interval of length L_0 which contains a set of points. As $\varepsilon \to 0$, the number of cells required to cover that set has the asymptotic form

$$N_0(\varepsilon) \sim A(L_0/\varepsilon)^{d_c},$$

The length L_0 is exhibited explicitly, to make it clear that if all lengths are scaled (L_0, the distance between points, and ε), then $N(\varepsilon)$ and d_c are not changed. Because L_0 is explicitly exhibited, the above constant A depends only on the structure of the set, and not its scale.

Next we consider a self-similar Cantor set, in the interval L_0. Assume that it is self-similar in the two subintervals L_1 and L_2, as illustrated in Fig. 2.48. Because the set in L_1 is in every way the same as the set in L_0, except that it has been scaled down to L_1, the number of cells required to cover it must have the asymptotic form

$$N_1(\varepsilon) \sim A(L_1/\varepsilon)^{d_c}.$$

Note that ε has, of course, not been scaled (it is an independent variable). It is important to note that the constant A is the same because the set is indistinguishable from the set in L_0, except for its scale (it is self-similar). similar expression holds for $N_2(\varepsilon)$ in the interval L_2. If we use the same ε for all the

Fig. 2.48

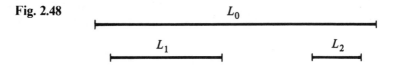

intervals L_0, L_1, and L_2, then, clearly, we must have

$$N_0(\varepsilon) \sim N_1(\varepsilon) + N_2(\varepsilon).$$

That is, the minimum number of cells required to cover the set in L_0 equals $N_1 + N_2$, because there are no points of the set outside of L_1 and L_2. Using the above expressions, and the fact that A is the same in all cases, we obtain the desired result for d_c

$$L_0^{d_c} = L_1^{d_c} + L_2^{d_c}.$$

In the exercise, $L_0 = 1$, $L_1 = \Delta$, and $L_2 = K\Delta_1$, confirming the previous analysis. Note that the case $K = 1$ is an exceptionally simple case, because $N(\Delta_1^n) = (1 + K^{d_c})^n = 2^n$, which can be readily obtained by counting cells. This is the reaon that the Cantor middle-third set is so easy to evaluate.

Clearly, if we consider another self-similar set in L_0 which is self-similar in a series of subintervals $L_1, L_2, \ldots, L_n$, then the dimension of this set must satisfy

$$L_0^{d_c} = \sum_{k=1}^{n} L_k^{d_c}.$$

A nice application of this expression is discussed in Section 4.7 (Chang and McCown, 1984).

(2.30) Both of these sets are dense, so $N(\varepsilon) = 1/\varepsilon$, giving $d_c = 1$.

(2.31) Consider the first m points ($n \leqslant m$) and note that the distance between successive points is $1/(n-1) - 1/n = 1/n(n-1) \geqslant 1/m(m-1) > 1/m^2$. Therefore, if we take $\varepsilon = 1/m^2$ the minimum number of cells required to cover these points is m. The remaining points lie between $[0, 1/m]$ and therefore are covered by m additional cells. Hence $N(\varepsilon) = 2m$ and $d_c = \lim_{m \to \infty} \log(2m)/\log(m^2) = \frac{1}{2}$.

(2.32) $\varepsilon = (1/l)^n$, $N(\varepsilon) = (l^2 - m^2)^n$.

(2.33) $K = 4$, $l = 3$, so $d_c = \log 4/\log 3$.

(2.34) $\mu(C(\Delta)) = 1 - (1 - 2\Delta) - 2 \times \Delta(1 - 2\Delta) - 4 \times \Delta^2(1 - 2\Delta) - \cdots = 1 - (1 - 2\Delta)$
$\sum_{n=0}^{\infty} 2^n \Delta^n = 1 - (1 - 2\Delta)(1/1 - 2\Delta) = 0$; so all $C(\Delta)$ are 'thin'.

(2.35) Let n be the nth rational number (they may be enumerated in any manner, even in a redundant fashion). Consider the open intervals $\varepsilon 2^{-n}$ about these points. The total measure of these intervals is $\sum_{n=0}^{\infty} 2^{-n} = \varepsilon/[1 - (1/2)] = 2\varepsilon$. The measure of the rationals is less than this for any ε, and hence zero.

3

First order differential systems ($n = 1$)

3.1 Selected dynamic aspects

The simplest models of systems are those consisting of a first order differential equation

$$\dot{x} = F(x, t, c) \quad (x \in R^1, c \in R^k). \tag{3.1.1}$$

In the autonomous case (only), (3.1.1) is dynamically very restrictive, since its phase space is just the real axis, through each point of which $x(t)$ can pass only once. In fairly simple non-autonomous cases, on the other hand, the dynamics of (3.1.1) can be very rich, and nearly intractable (of the character of a random walk on the x axis). In this section we will consider some of the dynamic properties of a few classic systems, which can represent interesting physical cases.

Equation (3.1.1) can also represent the simplest type of *bifurcation* phenomena and *catastrophe systems*, in which the combined *control-phase space* representation is both useful and possible (if the dimension, $k + 1$, is not greater than 3). Thus (3.1.1) can be used to illustrate in a relatively simple manner a number of features unique to nonlinear systems. Following the dynamics of this section, we will examine some simple bifurcation features in the following sections, and contrast these later (Chapter 4) with the relatively rich dynamics of autonomous first order *difference equations* (*mappings*).

Any equation (3.1.1) is trivially a *gradient system*, since it can be put in the form

$$\dot{x} = - \, dV/dx.$$

Frequently, V is referred to as a potential, although it is clearly not the potential of classical mechanics, since this equation is first order. V can be related to the usual potential energy only in the inertialess or 'overdamped' limit of the special Newton's equation

$$m\ddot{x} = - \, \dot{x} - dV/dx,$$

which contains a dissipative term, $- \, \dot{x}$. Then (3.1.1) can be viewed as the limit $m \to 0$ of this equation in which V is indeed a potential (force $\equiv - \, dV/dx$). This, however, is a little simplistic because the equation with $m \neq 0$ is a second order differential equation, and the connection between the solution of these two equations involves aspects of *singular perturbation* theory, which we defer for the present.

It should be mentioned first that if (3.1.1) is of the form

$$\dot{x} = f(x)\, g(t),$$

where $f(x)$ and $g(t)$ are elementary functions, then the equation is said to be '*solvable by quadratures*', meaning simply that the problem has been reduced to ordinary indefinite integrals. The problem is then frequently considered to be 'solved', which is particularly reasonable with present day computers.

Among the more common types of equations (3.1.1) are those with polynomial nonlinearities

$$\dot{x} = \sum_{n=0}^{N} a_n(t) x^n, \tag{3.1.2}$$

several of which bear names:

$$\dot{x} = a_0(t) + a_1(t)x + a_2(t)x^2 + a_3(t)x^3 \quad (\textit{Abel equation}), \tag{3.1.2a}$$

$$\dot{x} = a_0(t) + a_1(t)x + a_2(t)x^2 \quad (\textit{Riccati equation}), \tag{3.1.2b}$$

$$\dot{x} = a_1(t)x + a_m(t)x^m \quad (\textit{Bernoulli equation}). \tag{3.1.2c}$$

The Riccati equation is unique in that it is simply related to the general homogeneous linear differential equation of second order (the Sturm–Liouville equation)

$$\ddot{u} + A(t)\,\dot{u} + B(t)u = 0. \tag{3.1.3}$$

The first order term can be removed by the standard transformation $w = u\exp(\int^t A(t')\,dt'/2)$, yielding $\ddot{w} + (B - (A/2) - (A^2/4)w) = 0$. Using the general transformation $x = \dot{w}/w$, which reduces the order of an arbitrary homogeneous linear equation, at the expense of introducing nonlinearities, we find $\ddot{w}/w = \dot{x} + x^2$, so (3.1.3) becomes

$$\dot{x} + x^2 + B - (\dot{A}/2) - A^2/4 = 0, \tag{3.1.4}$$

which is a Riccati equation. Conversely, given (3.1.2b), we can obtain (3.1.3) by the transformation

$$x = -\frac{1}{a_2}(\dot{u}/u), \tag{3.1.5}$$

where $A = -(a_1 + (\dot{a}_2/a_2))$, and $B = a_0 a_2$.

The association of (3.1.3) and (3.1.2b) means that the Riccati equation must obey a simple nonlinear superposition principle. By *nonlinear superposition* is meant that the general solution of an equation can be expressed in some nonlinear fashion in terms of some set of 'fundamental' solutions, $x_1, x_2, \ldots, x_n$, so $x = N(x_1, \ldots, x_N)$. The example most commonly quoted is the '*cross-ratio*' *theorem* for the Riccati equation, namely any solution of (3.1.2b) can be expressed in terms of three other solutions x_1, x_2, x_3 by

$$\frac{(x - x_2)(x_1 - x_3)}{(x - x_1)(x_2 - x_3)} = k \quad \text{(arbitrary constant)}.$$

Exercise 3.1. Using the property of the general solutions of (3.1.3), show that the general solution of (3.1.2b) can be expressed in terms of only two particular solutions of (3.1.2b) (another nonlinear superposition principle).

Exercise 3.2. If a_0, a_1, a_2 are constants, obtain the general solution of the Riccati equation if the two roots of $a_0 + a_1 x + a_2 x^2 = 0$ are distinct and if they are equal. The Riccati equation arises in many unexpected contexts, as you can verify in the index.

The interesting concept of nonlinear superposition, which dates from M.E. Yessiot's research in 1893 (see Ames, 1978), has been receiving increased attention in recent years. A different type of *dynamic nonlinear superposition* occurs in particular solutions of nonlinear PDE (*solitons*), which suggests that this concept in some generalized form may prove to be useful in the future.

There are two special examples of the *Riccati equation* which have been widely used, namely the so-called *logistic equation*

$$\frac{dn}{dt} = cn(1 - n) \quad (n \geqslant 0), \tag{3.1.6}$$

and the *Landau equation*

$$d|A|^2/dt = 2\gamma|A|^2 - l|A|^4, \tag{3.1.7}$$

where $|A|^2 = A(t)A^*(t)$ ($A(t)$ is complex). These equations are obviously the same as (3.1.2b), with suitable scalings of $n(t)$ or $|A|^2$, but they are intended to model entirely different systems. The logistic equation comes from ecological considerations, where $n(t)$ represents the number of some biological species, normalized to its asymptotic value (so $n \to 1$ as $t \to +\infty$). The species reproduces at a rate c, causing an exponential increase, $\exp(ct)$, if n is small. The number $n(t)$ is ultimately limited by self-interaction (n^2) effects, such as competition for limited food supply, or toxicity effects (the 'finite-world' effect).

The equation (3.1.6) can be linearized by dividing by n^2, yielding

$$\frac{dx}{dt} = c(1 - x) \quad (x \equiv 1/n),$$

which can be integrated to give the solution

$$n(t) = \frac{n_0}{n_0(1 - \exp(-ct)) + \exp(-ct)}. \tag{3.1.8}$$

This solution is illustrated in Fig. 3.1 for two initial states. The lower solution is the one which is most commonly considered in the limitation of populations or the onset of turbulence.

Fig. 3.1

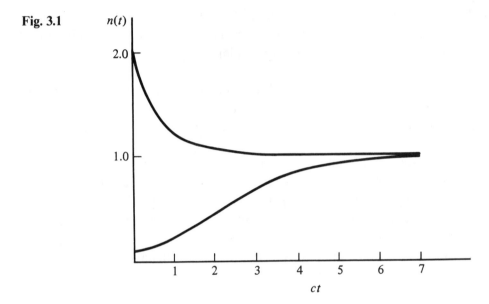

The Landau equation (Landau and Lifshitz, 1959), on the other hand, is a model for the onset of turbulence in a fluid, where the flow velocity is represented as $v(x, t) = A(t)f(x) + A^*(t)f^*(x)$. In the linear approximation this particular disturbance grows as $A(t) = A_0 \exp(\gamma t) + i\omega t$ and $\omega \gg \gamma > 0$ near the onset of instability ($\gamma = 0$). As the amplitude $|A(t)|$ grows it is limited by nonlinear modifications, and the term $-l|A|^4$ represents the time-averaged nonlinear stabilization of the linear instability (cubic terms in A time-average to zero). The physical ideas of Landau are closely related to the mathematical results obtained by Hopf (1948), which will be discussed in Chapter 5. We will return to the logistic equation, for yet another very different system, in the next section. In any case, we have presently examples of two very different physical systems which are modeled by essentially the same equations (3.1.6) and (3.1.7).

The Bernoulli equation (3.1.2c) can be linearized without changing its order simply by setting $x = u^\alpha$ and $\alpha = (1 - m)^{-1}$, where m need not be an integer. Then one obtains

$$\dot{u} = a_1(t)(1 - m)u + a_m(t)(1 - m), \tag{3.1.9}$$

which can be easily integrated. This again illustrates the 'happy' circumstance where a nonlinear transformation to a readily solvable linear system. No such transformation is known, for example, even in the case of Abel's equation, unless $a_0 = 0$, $a_1 \neq 0$ (e.g., Davis, 1962). At present there is very little known about obtaining such linearizing transformations systematically. For a much more complete discussion, see Ince (1927). One classic example for the case of a PDE is the Hopf–Cole transformation of the Burgers equation (see the index).

The equation

$$dx/dt = F(x, t) \tag{3.1.1}$$

can be written in the differential form

$$dx - F(x, t) \, dt = 0. \tag{3.1.10}$$

If the left side is an exact differential of some function $G(x, t)$,

$$dG \equiv \frac{\partial G}{\partial x} \, dx + \frac{\partial G}{\partial t} \, dt, \tag{3.1.11}$$

then the solution is simply $G(x, t) = $ constant. This is, of course, usually not the case, but we can look for a function $\mu(x, t)$ such that

$$\mu \, dx - \mu F \, dt = dG. \tag{3.1.12}$$

Such a function is called an *integrating factor* and, comparing (3.1.11) and (3.1.12), we see that it must satisfy the partial differential relations

$$\frac{\partial G}{\partial x} = \mu, \quad \frac{\partial G}{\partial t} = -\mu F,$$

and hence the PDE

$$\frac{\partial \mu}{\partial t} + \frac{\partial (\mu F)}{\partial x} = 0. \tag{3.1.13}$$

Note that we have gone from the nonlinear ODE (3.1.1), in the 'Lagrangian' variable $x(t)$, to the linear PDE (3.1.13), which involves the 'Eulerian' variables (x, t). You might also refer back to the comments at (1.3.12) and (1.3.13). The general solution of (3.1.13) is more difficult than the original problem (3.1.1), but what is needed to integrate (3.1.1) is any particular solution of (3.1.13). A particular solution of (3.1.13) is most readily obtained when $F(x, t)$ has certain homogeneity or algebraic properties, such as having the form $F = P(x, t)/Q(x, t)$ where P and Q are polynomials. These properties are not particularly common in the dynamic systems (3.1.10), but they are very common when we want to obtain the time independent constant of the motion of a second order autonomous system. Therefore, while this topic occurs mathematically in first order systems, its interesting dynamic application is in integrals of second order systems, so we defer the discussion until then.

Example. If you want to try a rather trivial (linear) example, let $F(x, t) = A(t)x + B(t)$, show that a particular solution of (3.1.13) is $\mu = \mu(t)$, and readily obtain $\mu(t)$ and $G(x, t)$. You can discover for yourself the interesting first order equations by looking back at Exercise (2.8).

While equations which represent physical situations presumably have unique solutions, one may in the course of integrating them, introduce spurious solutions. As noted in the last section, these are called *singular solutions*, which can be demonstrated by first order equations. Consider the simple case (second order, briefly)

$$m\ddot{x} = -\,\mathrm{d}V(x)/\mathrm{d}x, \tag{3.1.14}$$

where $V(x)$ is some potential. Following the standard procedure, if we multiply by $\dot{x}$, we obtain

$$\frac{1}{2}\frac{\mathrm{d}}{\mathrm{d}t}(m\dot{x}^2) = -\frac{\mathrm{d}}{\mathrm{d}t}V(x),$$

which yields the integral or constant of the motion

$$(1/2)m\dot{x}^2 + V(x) = E. \tag{3.1.15}$$

We then obtain the first order nonlinear equation(s)

$$\frac{\mathrm{d}x}{\mathrm{d}t} = \pm\left[\frac{2}{m}(E - V(x))\right]^{1/2}. \tag{3.1.16}$$

Each equation has only one solution except in the case where $(\mathrm{d}x/\mathrm{d}t)_{t_0} = 0$. In that case $E = V(x^0)$, where $x^0 = x(t_0)$, so

$$\frac{\mathrm{d}x}{\mathrm{d}t} = \pm\left[\frac{2}{m}(V(x^0) - V(x))\right]^{1/2}.$$

In addition to the physical (general) solution, there is also the *singular solution*, $x(t) = x^0$, which is not one of the dynamic solutions. The reason this occurs is because the original equation (3.1.14) was multiplied by the 'integrating factor' $\dot{x}$ to obtain the first integral (3.1.15). Thus if the initial condition is $\dot{x} = 0$, then, clearly, $x(t) \equiv x^0$ is now an allowed solution – but it is a spurious solution. While this is reasonably obvious in analytic solutions, one must be careful if numerical integrations begin or approach a point where $F(x)$ (in $\dot{x} = F(x)$) does not satisfy the *Lipschitz condition*.

To illustrate that the singular solution is a tangent curve to the family of general solutions (and is not one of the general solutions), consider the simple gravitational case $V(x) = mgx$. If $\dot{x}(t_0) = 0$, the general solution is

$$x(t) = (E/mg) - 1/2g(t - t_0)^2$$

Fig. 3.2

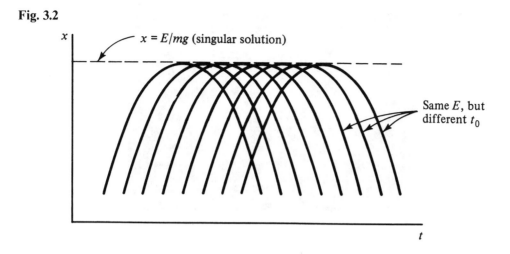

The singular solution of (3.1.16) (Fig. 3.2), $x(t) = E/mg$, can be seen to be tangent to all of the particular solutions of the same energy E. It occurs at the values of x for which (3.1.6) fails to satisfy the Lipschitz condition, yielding a nondynamic solution.

Exercise 3.3. The two-body central force motion yields the energy integral

$$E = m/2\,(\dot{r}^2 + r^2\,\dot{\phi}^2) + V(r) = m/2\,(\dot{r}^2 + (M^2/m^2r^2)) + V(r),$$

where $M = mr^2\,\dot{\phi} = $ constant. Show that if $V = +\,a/r$ ($a > 0$), corresponding to a repulsive force, there is a singular solution of the above differential equation ($E > 0$) which represents periodic motion. Explain the relationship of this periodic orbit to the physically meaningful solutions, and determine the uniqueness condition which is not satisfied. If $a < 0$, how many periodic orbits are there for given ($E < 0$, M)? Are any of the periodic orbits physical solutions?

It should be noted that although singular solutions are not one of the general solutions, they may predict physically observable phenomena. This occurs in the case of the characteristic equations for the 'rays' in wave optics, where such singular solutions are the so-called light *caustics* (Bobbitt and Cumberbatch, 1976; Berry, 1976, 1981, Bruce, Giblin and Gibson 1984). Along these lines, or surfaces, the light intensity increases dramatically. Some far-field caustics are illustrated in Fig. 3.3(a) for the case of an irregular water droplet on a flat plate (after Berry, 1976). Figure 3.3(b) traces the caustics.

In a much simpler fashion, we can use the above gravitational example of a singular solution to obtain a 'density caustic'. To do this, we simply regard the set of parabolas in Fig. 3.2 as being produced by a flux, J_0^+, of particles injected at $x = 0$, all with the same speed v_0, so $J_0^+ = n_0^+ v_0$ and $v_0 > 0$. In the steady state, the total flux of particles at

Fig. 3.3

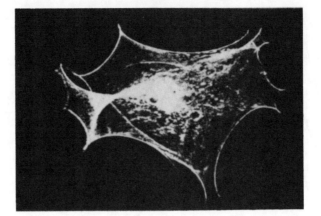

(a)

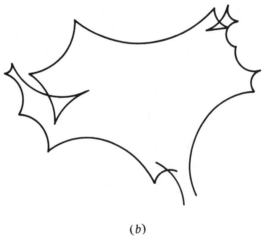

(b)

Fig. 3.4

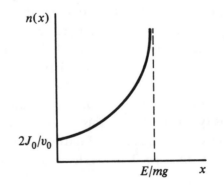

any point is clearly zero, $J = J^+ + J^- = 0$, where $J^\pm = n^\pm(x)\,v_x^\pm(x)$ with $-v_x^- = v_x^+ > 0$, and $n^+(x) = n^-(x)$. Moreover, the total density is $n(x) = n^+(x) + n^-(x) = 2n^+(x)$. To obtain $n(x)$ we use the conservation of particles (continuity equation) so $J_0^+ = J^+(x) = n^+(x)\,v^+(x) = \frac{1}{2}n(x)\,v_x^+(x)$, and from the conservation of energy

$$\tfrac{1}{2}mv_x^2 + mqx = \tfrac{1}{2}mv_0^2.$$

Therefore the density is given by

$$n(x) = 2J_0^+ / [v_0^2 - 2gx]^{1/2}$$

which is singular at the location of the singular solution, $x = E/mg$. The density singularity is similar to the intensity singularity of a light caustic. Of course such a singularity is removed by collisions, but nonetheless the density can become very large.

Fig. 3.5

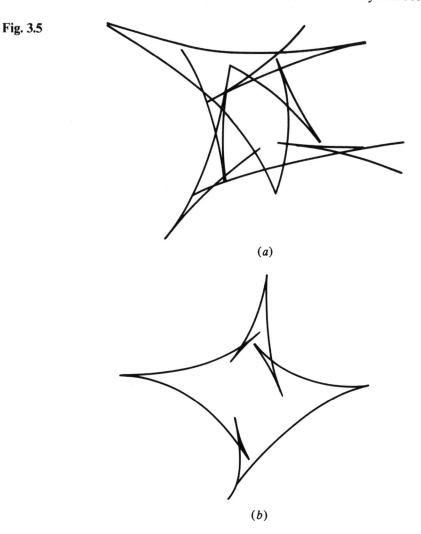

(a)

(b)

An example of this are the caustics which can arise in electron optics, as illustrated in Fig. 3.5 (after Leisegang, 1953).

In this section we have considered some aspects of the nonlinear dynamics in R^1. The phenomena become much more interesting when we consider the changes which can occur in these dynamics as we vary the control parameters of the system. This leads us into the study of elementary bifurcation theory, which we will consider in the remainder of this chapter.

3.2 Control space effects: simple bifurcations

The concept of 'branching off' (abzweigung) of solutions is apparently due to C. Jacobi ('Uber die Figur des Gleichgewichts,' *Pogg. Ann.*, **32**, 229, 1834), and the present terminology 'bifurcation' is due to (who else?!) H. Poincaré ('Sur l'equilbre d'une masses fluide animée d'un mouvement de rotation,' *Acta Math.* **7**, 259–380, 1885). These early studies of equilibrium situations involved a division, branching, or separation of equilibrium states into several parts. The study of bifurcations has since developed into a major area of mathematical and applied research. The term bifurcation is now applied to at least two rather distinct situations involving (a) the study of equilibrium situations (in other words, the change in the number of fixed points as control parameters are changed), and (b) the dynamic studies, which are concerned with the change in the topology (phase portraits) as the control parameters are changed. In these studies, one considers the change in the stability of the fixed points and the introduction of other limit sets of the trajectories in the phase space. This change in the limit properties may occur without any 'bifurcation' (division) of the fixed points, as we will soon see, but it is still considered a bifurcation because of the change in the phase portrait of the solutions. From the present dynamic point of view, in which fixed points are simply a special set in a phase space, it is reasonable to say that

> *a bifurcation occurs any time the phase portrait is changed to a topologically nonequivalent portrait by a change of the control parameters.*

Clearly, if the number of fixed points is changed in a dynamic system, there must be such a change in the portrait, so this definition includes the equilibrium situation. The situations involving periodic boundary conditions, rather than Cauchy (initial) conditions, are somewhat distinct from this dynamic viewpoint unless one is concerned with PDEs. This will be discussed briefly at the end of this section.

A less technical, but physically more suggestive description, of bifurcations would be to call them *dynamic phase transitions*, brought about by changing the control parameters. This, moreover, draws attention to the fact that physical phase transitions (e.g., liquid to gas, changes in crystal or biological structure, etc.) are the end product (involving dissipative effects) of these dynamic phase transitions, which are therefore more fundamental. Some simple types of transitions have nothing to do with non-linearity, *per se*, but the bifurcations become much more varied and interesting when the equations are nonlinear.

Let us begin, like the past masters, with the consideration of equilibrium states (fixed points) of an autonomous system

$$\dot{x} = F(x, c) = 0, \quad \text{for} \quad x = x_s(c) \quad \text{or} \quad c = c_s(x). \tag{3.2.1}$$

If the number of fixed points changes when c is varied, then the phase portrait also changes to a nonequivalent portrait, so the system has by definition gone through a bifurcation. The function $F(x, c)$ may have no real root, or more than one, depending on the value of c. Thus

$$F(x, c) = x^2 - c \tag{3.2.2}$$

has no real root for $c < 0$, but two real roots for $c > 0$. In this case 'nothing' branches into two solutions of (3.2.1) at $c = 0$. We will consider this to be an example of a bifurcation point (defined below), although some authors do not, because the branching is not from an existing root (when $c < 0$).

The bifurcation is most simply pictured in the combined *control-phase space*, (x, c). In this space we can plot the points $x = x_s(c)$ or equivalently $c = c_s(x)$ which satisfy (3.2.1). This is sometimes referred to as the *solution set*, hence the subscripts. This is illustrated on the left of Fig. 3.6 for the case (3.2.2). It is a little more interesting if $F(x, c) = (x - 1)^2 - c - 2$, shown in the right-hand diagram of Fig. 3.6 but, clearly, there is no

Fig. 3.6

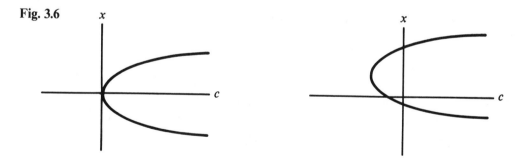

topological difference between these solution sets (the physical difference between the systems, however, may be important). Since the two solution sets can be mapped onto each other by a homeomorphism, they are called topologically equivalent.

More generally, we can imagine that the solution set of (3.2.1) looks something like that shown in Fig. 3.7, which is clearly not equivalent to the previous solution set. You might even question whether there exists a function with such a complicated solution set, but it is rather simple to construct such functions (see Exercise 3.10).

Before we consider further examples, let us define a bifurcation point, $c = c_0$, for

Fig. 3.7

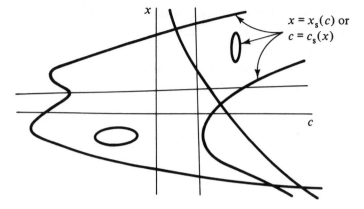

solution sets (as illustrated above). A point c_0 is a *bifurcation point* of the fixed points (3.2.1) if there are two (or more) distinct solutions, $F(x_1, c) = 0$, $F(x_2, c) = 0$ when (x, c) is in any R^2 neighborhood of (x_0, c_0), where $F(x_0, c_0) = 0$. Note that there is one value of c, and two values of x.

The *implicit function theorem* gives a necessary condition for c_0 to be a bifurcation point. Recall that if the Jacobian determinant does not vanish, $|\partial F_i/\partial x_j| \neq 0$, in some region of (x_0, c_0) then there is a subregion where we can uniquely determine $x_s(c)$ which satisfies $F(x_s(c), c) = 0$. Clearly, if the solution $x_s(c)$ is unique, c_0 cannot be a bifurcation point. In the present simple case, where $x \in R^1$, the determinant of the Jacobian is simply the derivative of $F(x, c)$. We therefore have:

> If $\partial F/\partial x \neq 0$ *for any* x *such that* $F(x, c) = 0$, (3.2.3)
> *then* c *is not a bifurcation point.*

On the other hand it does not follow that, if $F(x_0, c_0) = 0$ and $(\partial F(x, c_0)/\partial x)_{x_0} = 0$, c_0 is a bifurcation point. A little graphical analysis makes this simple to see. Consider first case (3.2.2). The graph (Fig. 3.8(a)) of F vs. x for $c < 0$ shows that there is no intersection with the x axis $F = 0$, and that there are two intersections (the roots $x_s(c)$) of $F(x_s(c), c) = 0$) if $c > 0$ (Fig. 3.8(b)). It is, moreover, obvious that $c = 0$ is the bifurcation point. This is perhaps even clearer in the three dimensional diagram (Fig. 3.8(c)) above. In this case $\partial F/\partial x = 2x$ *vanishes only for* $x = 0$ (regardless of the value of c) and, on the other hand, this $(x = 0, c)$ is a solution of $x^2 - c = 0$ only for $c = 0$; hence that is the bifurcation point.

Now consider

$$F(x, c) = c + x - x^2 + \tfrac{1}{3}x^3.$$ (3.2.4)

We have

$$\frac{\partial F}{\partial x} = 1 - 2x + x^2 = 0,$$

only if $x = 1$. But $F(1, c)$ is in the solution set only if $c + 1 - 1 + \tfrac{1}{3} = 0$, or $c = -\tfrac{1}{3}$. One can conclude that this is the only possible bifurcation point, but one cannot conclude

Fig. 3.8

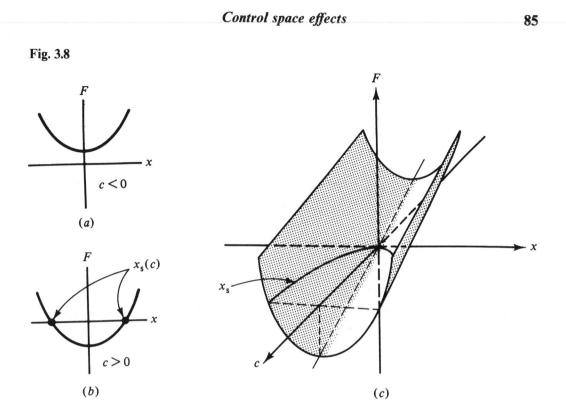

(a)

(b)

(c)

that it is such a point. The graph of F vs. x readily shows what is involved in this decision (Fig. 3.9). One can see that in this case $x = 1$ is simply an inflection point $(\partial^2 F/\partial x^2 = 0$ at $x = 1)$, so that, as c passes through $c = -\frac{1}{3}$, there remains only one root of $F(x, c) = 0$, hence no bifurcation. A vanishing first derivative and a nonvanishing second derivative does, on the other hand, indicate a bifurcation point c_0.

Exercise 3.4. Determine if there are bifurcation points if (3.2.4) is multiplied by x, so that $F = cx + x^2 - x^3 + \frac{1}{3}x^4$.

Fig. 3.9

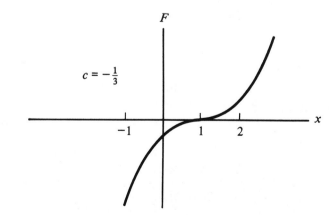

It is rather clear from this discussion that, if the first and second derivatives vanish on the solution set, for some c_0, then it is necessary to consider other derivatives to determine whether c_0 is a bifurcation point. These general considerations would take us too far afield, so we restrict our discussion to more common situations and, particularly, their relationship to dynamics, rather than just fixed points.

Exercise 3.5. You might like to discover for yourself what happens when there are two control parameters (c_1, c_2). Bifurcation points now become bifurcation curves in the (c_1, c_2) plane. Obtain this curve for $F(x, c_1, c_2) = x^2 + c_2 x - c_1$.

Exercise 3.6. Determine what the following modifications of (3.2.2) do to the bifurcation properties locally and globally (that is, for all values of x, rather than just near $x = 0$). (a) $F = x^4 + x^2 - c$ and (b) $F = x^3 + x^2 - c$.

Exercise 3.7. Can you think of a simple mechanical system which could have the form (3.2.2), provided that it is obtained from Newton's equations by the 'overdamped' method discussed at the beginning of the chapter?

We will next consider some aspects of the dynamics of the system (3.1.1). Because the phase space is simply the x axis (R^1), the topological character of the trajectories is completely determined by the fixed points and their stability. Therefore, one of the more interesting dynamical questions which can be asked is: *how does the stability of a fixed point change along a solution set as we pass through a bifurcation point?* The question may seem rather contrived, and it is, but it also has useful physical applications.

The idea is fairly straightforward, but requires some analysis which will only be outlined. To study the dynamics near a point on the solution set, $x = x_s(c)$, we set

$$x = x_s(c) + \delta x \tag{3.2.5}$$

and expand $F(x, c)$ about this point, using (3.2.1) to give

$$\delta \dot{x} = (\partial F / \partial x)_{x_s} \delta x. \tag{3.2.6}$$

Here the higher order terms in δx have been neglected, which is justified provided that $(\partial F / \partial x)_{x_s} \neq 0$; in other words, if c is not a bifurcation point. However, (3.2.6) is valid on either side of a bifurcation point. We can view the coefficient of δx in (3.2.6) to be either a function of c, or of x, by taking $c = c_s(x)$, where $F(x, c_s(x)) \equiv 0$. Therefore, let

$$\mu(c) = (\partial F / \partial x)_{x_s}(c) \quad \text{or} \quad \mu(x) = (\partial F / \partial x)_{c_s}(x). \tag{3.2.7}$$

Then (3.2.6) yields

$$\delta x = \delta x(0) \exp(\mu t),$$

Fig. 3.10

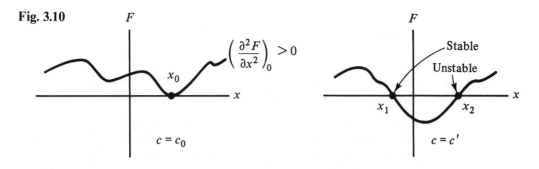

so the stability at $x = x_s(c)$ or, equivalently, $c = c_s(x)$, depends on the sign of μ.

In the case of the simplest bifurcation $((\partial F/\partial x)_{x_0} = 0$ and $(\partial^2 F/\partial x^2)_{x_0} \neq 0$ at $c = c_0)$, it is easy to see that $\mu = (\partial F/\partial x)_{x_s}$ has different signs for the two roots (x_1, c') and (x_2, c'), when c has been changed to c', as (Fig. 3.10) shows. Therefore the root x_1 is stable and x_2 is unstable. This can also be pictured in the control-phase space, where the unstable branch is represented by a dashed curve. As noted above, this solution set can also be represented as the single-valued curve $c = c_s(x)$. Then, as x is varied through the bifurcation value x_0, the stability changes. Physically, it is c which is varied, of course, rather than x, but mathematically it is useful to use the single-valued representation (Fig. 3.11).

Fig. 3.11

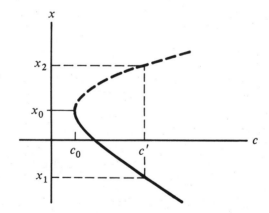

Exercise 3.8. Draw a similar figure for $F = c - x^2$ and for $F = c + x^2$, labeling the stable and unstable branches.

This is an elementary example of what is called an *exchange of stability* (here it is really only a change, a real exchange comes shortly).

If $(\partial F/\partial c) \neq 0$, so that $c_s(x)$ can be obtained, and if $(\partial c_s/\partial x)$ changes sign as x passes through x_0, then the point $(x_0, c_0 \equiv c_s(x_0))$ is called a *regular turning point*. The trajectories in phase space (R^1) are shown in Fig. 3.12. These 'family' (one-member!)

Fig. 3.12

$c = c_0$ $c = c'$

portraits are clearly not equivalent, so this dynamic concept of bifurcation agrees with the fixed point definition (above). This is not always the case, as will be found later. It might also be mentioned that the above bifurcation in which a stable and an unstable fixed point come together and 'annihilate' one another, is a one-dimensional version of what is called a *saddle-node bifurcation* in second order systems.

An example of this type of bifurcation involves the equilibrium points of a cylindrical wheel, of mass M and radius R, which rolls without slipping on an inclined plane, and has a mass, m, attached to a rod which bisects the wheel. The mass m is at a distance r from the center, which is variable ($r = $ control parameter). The center of mass of this system is at a distance $r_{cm} = m/(m + M)r$ from the center. If the gravitational potential energy is taken to be zero at $\theta = 0$ (see Fig. 3.13) then we readily find that the potential energy is

$$V(\theta) = -Mg\,R\theta\sin\phi + mg\,(r\sin\theta - R\theta\sin\phi),$$

Fig. 3.13

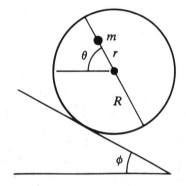

because $R\theta\sin\phi$ is the vertical distance moved by the center of the wheel (no slippage). The equilibrium points of this system are given by $dV(\theta)/d\theta = 0$, or $(1 + (M/m))$ $\cdot\sin\phi = (r/R)\cos\theta$. This condition simply states that the center of mass must be directly above the point of contact between the wheel and the plane ($R\sin\phi = r_{cm}\cos\theta$). Clearly, there are solutions only if $1 \geqslant r/R \geqslant (1 + (M/m))\sin\phi$ and the upper branch is unstable (as in the above discussion) (Fig. 3.14).

A more elaborate bifurcation point is a *singular point*, where

$$\left(\frac{\partial F}{\partial x}\right)_0 = 0, \quad \left(\frac{\partial F}{\partial c}\right)_0 = 0, \tag{3.2.8}$$

and the subscript 0 stands for (x_0, c_0), on the solution set. In the neighborhood of the bifurcation point we can set $x = x_0 + dx$ and $c = c_0 + dc$, and expand $F(x, c)$. Then

Fig. 3.14

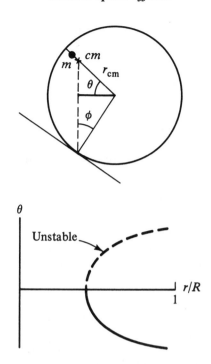

because of (3.2.8), the lowest order terms remaining in the Taylor expansion are

$$F(x_0 + dx, c_0 + dc) = \frac{1}{2}\left(\frac{\partial^2 F}{\partial x^2}\right)_0 (dx)^2 + \left(\frac{\partial^2 F}{\partial x \partial c}\right)_0 dxdc + \frac{1}{2}\left(\frac{\partial^2 F}{\partial c^2}\right)_0 (dc)^2$$

$$\equiv \alpha(dx)^2 + 2\beta\, dxdc + \gamma(dc)^2. \tag{3.2.9}$$

If (dx, dc) are along the solution set, so that $F(x, c) = 0$, then (3.2.9) yields two distinct roots for the tangents (e.g., dc/dx) if and only if

$$D = \beta^2 - \alpha\gamma > 0. \tag{3.2.10}$$

If (3.2.10) is satisfied the (singular) bifurcation point is called a *double point*. What this means is that there are two curves passing through (x_0, c_0) with distinct tangents.

Fig. 3.15

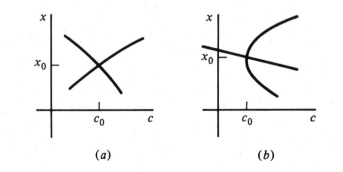

(a)   (b)

There are two possibilities: (a) neither curve has a turning point at (x_0, c_0) or (b) one curve has a turning point at (x_0, c_0) (see Fig. 3.15). Since the tangents to the solution set are obtained when (3.2.9) is set equal to zero, clearly $dc/dx = 0$ is one root (case b) only if $\alpha = 0$. So these two possibilities are characterized by

$$(a)\ (\partial^2 F/\partial x^2)_0 \neq 0, \quad (b)\ (\partial^2 F/\partial x^2)_0 = 0.$$

The linear *stability of the roots* of $F(x, c) = 0$ is determined by the sign of $\mu = (\partial F/\partial x)_{c_s}(x)$. According to (3.2.6) $\mu > 0$ implies instability, $\mu < 0$ implies stability, and when $\mu = 0$ the linear analysis fails. We will find it useful to consider μ as a function of x, near x_0. Since $F(x, c_s(x)) = 0$, so is its derivative $(\partial F/\partial x)_{c_s} + (\partial F/\partial c)_{c_s} dc_s/dx = 0$. Therefore,

$$\mu(x) = -(\partial F/\partial c)_{c_s}(dc_s/dx)$$

and, expanding the first factor about the bifurcation value, so that $x = x_0 + dx$, gives

$$\mu(x_0 + dx) \simeq -[(\partial^2 F/\partial c^2)_0(dc_s/dx)_0 + (\partial^2 F/\partial c\partial x)_0]\,dx\,(dc_s/dx). \quad (3.2.11)$$

But, from (3.2.9) equated to zero we obtain the roots for $(dc_s/dx)_0$, namely

$$\left(\frac{\partial^2 F}{\partial c^2}\right)_0\left(\frac{dc_s}{dx}\right)_0 = -\left(\frac{\partial^2 F}{\partial x\partial c}\right)_0 \pm D^{1/2},$$

and substituting this into (3.2.11) yields

$$\mu(x_0 + dx) = \mp D^{1/2}(dc_s/dx)\,dx. \quad (3.2.12)$$

This shows that if (dc_s/dx) does not change sign going through x_0 (case a above) then μ does change sign, because dx changes sign. Conversely if dc_s/dx changes sign (a turning point) then μ does not change sign. This yields a variety of possible types of *exchange of stabilities*. Several examples are illustrated in Fig. 3.16 (the dashed curves are unstable).

Fig. 3.16

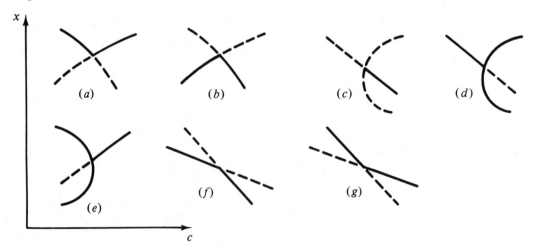

Fig. 3.17

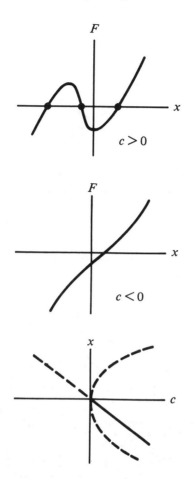

Example. Consider the case $F = (x^2 - c)(x + c)$ near $c = 0$. The solution sets are $x_s(c) = \pm c^{1/2}$ (if $c > 0$, and $x_s(c) = -c$. A sketch of F vs x is shown in Fig. 3.17 for $c > 0$, illustrating the three roots of $F = 0$. The outside roots are unstable, whereas the central root is stable. If $c < 0$, the only root of $F = 0$ is unstable. Therefore, in the control-phase space, this bifurcation looks like the illustration. It has the opposite stability from (d) in Fig. 3.16. Changing the present F to $-F$ therefore produces the bifurcation diagram, Fig. 3.16(d), with the (x, c) axis shifted appropriately, of course.

Notice that the flow on either side of the bifurcation point in cases (a), (b), (f) and (g) are TOE. It is only at the bifurcation point where this equivalence is broken. This can be seen clearly by including the flow vectors in the diagram, as illustrated in case (a) (Fig. 3.18). You should sketch the corresponding diagram for (d) and note that the flow is 'permanently' changed at this bifurcation point. This flow is illustrated later in this chapter.

Exercise 3.9. Why does the case illustrated in Fig. 3.19 not occur?

Fig. 3.18

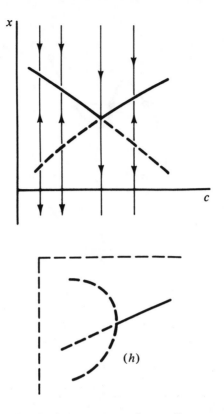

Fig. 3.19

Exercise 3.10. Complicated solution sets can be easily constructed by setting $F = \prod_i g_i(x, c)$, where each function $g_i(x, c)$ is a simple polynomial. For example, associate each of the following cases with the equivalent diagram $(a–g)$, in Fig. 3.16.

$$(A) \quad F = (x^2 - c)(x + c); \quad (B) \quad F = (x^2 + c)(c - x);$$
$$(C) \quad F = (x - c)(2c + x); \quad (D) \quad F = (c + 2x)(x + c).$$

To establish the stability, it is useful to sketch F vs. x. Note any double crossings which occurs.

A simple physical example of a double point bifurcation consists of a light, stiff plastic (or metal) strip with its bottom edge attached to a table, and two magnets on opposite sides of the strip (to maintain symmetry), which can be slid along the strip. This is a simple variation of a classic structural problem, sometimes called an *Euler strut* (Fig. 3.20). The strip is initially vertical with the weights at the bottom ($d = 0$), and the control parameter is the distance, d, of the magnets from the bottom. Assume first that the torque generated by the strip about the bottom axis, when it is displaced by an angle θ, is simply proportional to θ

$$\text{strip's torque} = -\eta(d)\,\theta. \tag{3.2.13}$$

Fig. 3.20

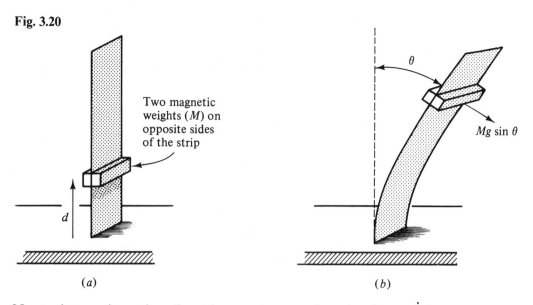

Two magnetic weights (M) on opposite sides of the strip

$Mg \sin \theta$

(*a*)   (*b*)

Newton's equation, when the strip experiences a damping force $\mu\dot\theta$, is

$$Md^2\ddot\theta = -\mu\dot\theta - \eta\theta + Mgd \sin \theta$$

and ignoring the inertial effect (assume $|\mu\dot\theta| \gg |Md^2\ddot\theta|$), this gives the approximate 'overdamped' equation

$$\dot\theta = (-\eta/\mu)\theta + (Mg/\mu)\,d \sin \theta \equiv F(\theta, d). \qquad (3.2.14)$$

The equilibrium states (fixed points) are then given by

$$\theta = (d/D)\sin\theta; \quad D^{-1} = Mg/\eta. \qquad (3.2.15)$$

One readily concludes that $d = D$ is the bifurcation point, and the solution set is shown in Fig. 3.21(*b*) in the control-phase space.

The symmetry is broken if the strip is not oriented vertically when $d = 0$, as illustrated in Fig. 3.21(*a*). In this case (3.2.15) is replaced by

$$\theta = (d/D)\sin(\theta + \psi), \qquad (3.2.16)$$

and this leads to the solution set shown on the right, in Fig. 3.21(*c*). This is an example of what is referred to as an *imperfect bifurcation*, where a 'real-word imperfection' (here any small symmetry breaking, $\psi \neq 0$) causes a change to a nonequivalent solution set (a double point bifurcation is changed to a simple bifurcation).

It is sometimes useful to also think of these bifurcations in terms of the potential, V, of the gradient system

$$\dot x = F(x, c) \equiv -\frac{\partial V}{\partial x}. \qquad (3.2.17)$$

Fig. 3.21

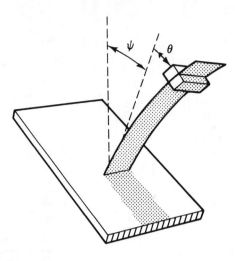

(a)

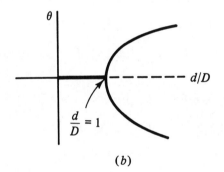

(b)

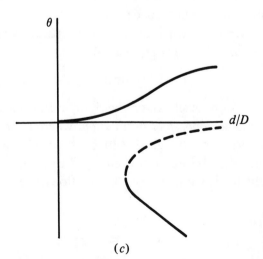

(c)

Fig. 3.22

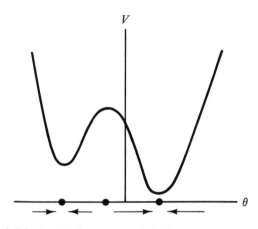

For (3.2.16) the potential is (aside from a multiplicative constant)

$$V(\theta, d) = \tfrac{1}{2}\theta^2 + (d/D)\cos(\theta + \psi)$$

which is illustrated on in Fig. 3.22. In a gradient system the motion is always toward the local minimum, as is clear from (3.2.17). These minima therefore represent *attractors* of the motion, a concept which arises in many physical contexts which involve dissipation. Recall again the approximate relationship between these gradient systems and overdamped Newtonian systems, discussed in Section 3.1.

In (3.2.13), the torque acting on the strip is a linear function of θ. More interesting effects can arise if the strip's torque is not simply linear in θ. To illustrate this, assume that the restoring torque of the strip has a decreasing coefficient as θ increases, but that it only decreases by a finite fraction. As an example, assume that

$$\eta(\theta) = \eta[1 - \rho\theta^2/(0.25 + \theta^2)].$$

The fixed points are then given by

$$\theta[1 - \rho\theta^2/(0.25 + \theta^2)] = (d/D)\sin\theta \qquad (3.2.18)$$

Fig. 3.23

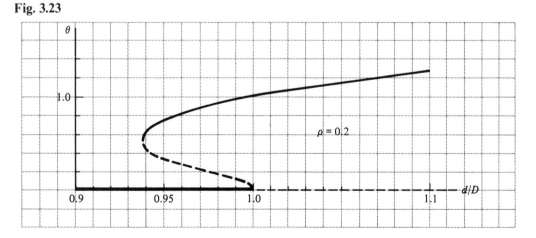

instead of (3.2.15). The solution set, when $\rho = 0.2$, is shown in the Fig. 3.23 (only positive θs are shown). This is very different from the solution set of (3.2.15), for there are now two bifurcation points, one associated with a simple turning point ($d/D \simeq 0.94$), the other with the double point ($d/D = 1$). This is an example of a *bistable system*, which has two (nontrivial) stable states for a range of control parameter values. The physically important repercussion of this type of solution set is that such systems can exhibit a *hysteresis effect* when the control parameter is increased and then decreased across both bifurcation points. Note that any change which is made in a control parameter must be 'slow' or '*adiabatic*' with respect to the equation of motion (3.1.1), for otherwise the equations must be treated as nonautonomous. The resulting behavior of x is illustrated in Fig. 3.24; When c is increased past c_d, the state $x = 0$ becomes unstable and any outside perturbation will cause the system to 'jump' to one of the distant

Fig. 3.24

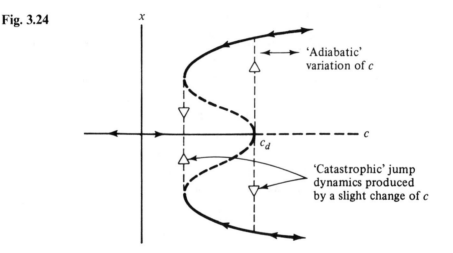

available stable branches. When c is then lowered below c_d the system may remain in the stable state $x \neq 0$ until c reaches the turning point, then it must jump back to the stable state, $x = 0$, when c is decreased further. These jumps are *catastrophic changes* which are brought about by small changes in c. A dramatic and heuristic example of such a catastrophic jump is an earthquake, which results from a monotonic increase in a strain (corresponding to c increasing past c_d). Somewhat below the critical strain, the system can again be viewed as being bistable.

Hysteresis effects can, of course, also occur without implying that there are double points present. An example is a discharge tube (e.g. containing neon gas) which has a slowly increasing current as the voltage is increased ($V < V_c$), due to increasing ionization of the gas (Fig. 3.25). When the voltage reaches a critical voltage a transition is made to a 'self-sustained discharge', which sustains a large current even when $V < V_c$. We could make up a rough model of this 'solution set', as indicated, without any

Fig. 3.25

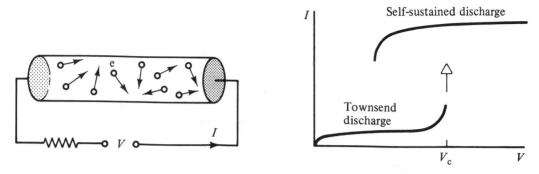

suggestion of a double point bifurcation, but there is a strong hysteresis. The physics here cannot be put in that mold. See the discussion of optical bistability for another form of hysteresis effect.

The above solution sets are two examples of the following group of double point bifurcations (with or without a turning point bifurcation) which are frequently encountered. The terminologies *supercritical, subcritical* and *transcritical* bifurcations are due to Benjamin (1976), and all occur in fluid flows (Stuart, 1977) (Fig. 3.26).

Fig. 3.26

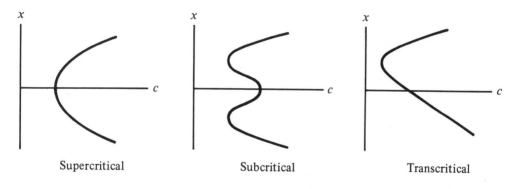

Supercritical   Subcritical   Transcritical

Whether or not a hysteresis effect occurs in the subcritical or transcritical case depends on the level of fluctuations which are present to perturb the system (these perturbations are not in the dynamic model). If the fluctuations are energetic enough, then the system will always be in the lowest energy branch of the bifurcation diagram (for each value of c), and there will be no hysteresis.

As discussed in Section 3.1, in gradient systems the term 'lowest energy' can be reasonably borrowed from mechanics to simply mean the region where the 'potential', $V(x)$, is a minimum (because, in a damped mechanical system, there is no kinetic energy contribution as $t \rightarrow + \infty$). In some fields of research (e.g., catastrophe theory) this case of migration to the lowest energy branches is referred to as a *Maxwell situation* (see

Fig. 3.27

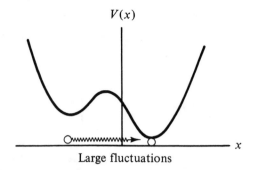

Large fluctuations

Fig. 3.27). If the fluctuations are 'very small', the system remains in its metastable state as c is varied, regardless of the energy relative to other stable states, until that metastable state vanishes – at which point it jumps to the other stable state. This leads to a *hysteresis* effect, and is referred to as a *delay situation*.

Exercise 3.11. Determine the bifurcation points and the stable branches of $F(x, c) = (x + c)(x^2 - 2x + c)$. Sketch the solution set in the control-phase space.

A classic example of a supercritical type of double point is the *Curie point bifurcation* in magnetic systems. Usually the temperature, T, is taken as the control parameter, so that the equilibrium magnetization M vs. T is the control-phase space. The typical solution set is shown in Fig. 3.28 (note that M can be positive or negative for T less than the Curie Temperature T_c).

Fig. 3.28

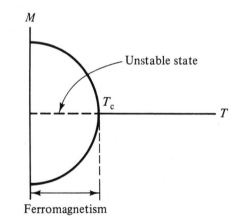

Unstable state

T_c

T

Ferromagnetism

Exercise 3.12. Note that many of the above results rest on the implicit function theorem and the differentiability of $F(x, c)$ with respect to both variables. This need not, however, be the case. For example, let

$$\dot{x} = - x(x - a(c))(x - b(c)),$$

where $a(c) = \alpha + \beta(c - c_1)^{1/2}$, $b(c) = \alpha - \lambda(c - c_1)^{1/2}$, $\lambda \neq \beta$, and α, β, λ are positive. Therefore this $F(x, c)$ is not differentiable in c at c_1. Show, however, that the solution set is (somewhat) like the above discharge tube case.

Up to now, in discussing bifurcations, we have largely concentrated on the branching of fixed (equilibrium) points as a control parameter is varied. That is very important, particularly in first order systems, because in this case all trajectories either tend toward or away from such fixed points. This is not true in higher dimensions (see Fig. 3.29) where the variable x might, for example, refer to the radial distance from the

Fig. 3.29

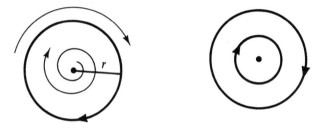

origin ($x \equiv r$). Then a fixed point of x would physically signify a periodic orbit. This is the case in Landau's equation (3.1.7) where $r = |A|$. Similarly, two fixed points means two periodic orbits (one enclosing the other). Such stable (unstable) periodic orbits are referred to as stable (unstable) limit cycles. This, of course, is getting ahead of our present story, since this dynamics occur in dimensions higher than one. Nonetheless it is instructive to review our present results in the light of this interpretation (e.g., what are the periodic dynamics associated with (Fig. 3.30). This will be discussed in Chapter 5.

Fig. 3.30

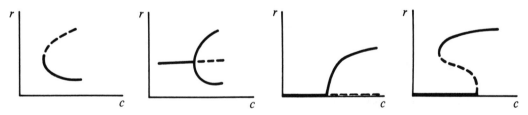

As noted before, in all the above diagrams of solution sets we could have also included the trajectory which corresponds to each value of c, which gives a complete picture of the dynamics in these cases (see Fig. 3.31).

It is possible, however, to have dynamic bifurcations (nonequivalent changes in the phase portrait) which do not correspond to any of these examples, because the higher order derivatives also vanish at the bifurcation point (e.g., all second order derivatives

Fig. 3.31

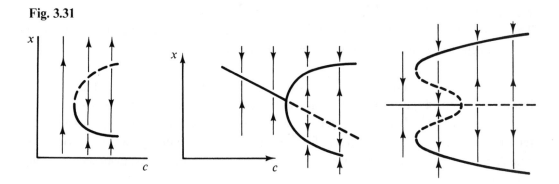

in (3.2.9)). A very simple example of this is the system

$$\dot{x} = c(x - x_0) \equiv F(x, c) \tag{3.2.19}$$

which has the solution

$$x = x_0 + A \exp(ct).$$

If $c < 0$ then $x \to x_0$ as $t \to \infty$ (all solutions tend to x_0; it is an *attractor*), whereas if $c > 0$ then all solutions (with $A \neq 0$) are unbounded ($|x| \to \infty$, as $t \to \infty$). We note that $\partial F(x, c)/\partial x = c$, so that $c = 0$ is a possible bifurcation point. Clearly, from the fact that the phase portraits shown in Fig. 3.32 are not equivalent it follows that $c = 0$ is indeed a

Fig. 3.32 $\xrightarrow{\hspace{1cm}} \bullet \xleftarrow{\hspace{1cm}} \;\; x$ $\xleftarrow{\hspace{1cm}} \bullet \xrightarrow{\hspace{1cm}} \;\; x$

                 $c < 0$ $c > 0$

bifurcation point (for any x!). We again have a type of 'exchange' of stability at the bifurcation point. But note that all derivatives of $F(x, c)$ vanish at $c = 0$, so that the previous discussion does not cover this case. We can also have less extreme examples, where some higher order derivatives do not vanish, and these must be likewise examined in more generality. As we will see shortly, these cases are in some sense not typical (not 'generic') because they are subject to major dynamical changes when only a small change is made in the function $F(x, c)$. This is illustrated by the following exercise.

Exercise 3.13. The equation $\dot{x} = cx - ax^2$ ($a \neq 0$) is easily integrated for arbitrary initial states, $x(0) = A_0$, and arbitrary constants (c, a). Determine the bifurcation point (c_0, x_0). Sketch the phase portraits for $c < c_0$ and $c > c_0$, noting all attractors and *repellors* in the phase space. (Solutions move away from a repellor as $t \to +\infty$.) Determine whether the constant a is also a control parameter which can produce a bifurcation.

The situation changes dramatically from the above examples if the phase space is restricted by physical requirements. Thus consider a simple laser model in which

$$n(t) = \text{number of photons} \ (\geqslant 0) \tag{3.2.20}$$

and

$$\dot{n} = GNn - n/\tau = \text{gain} - \text{loss}, \tag{3.2.21}$$

so the gain (G) is proportional to the number of excited atoms (N) and the number of photons (producing stimulated emission). The constant τ represents the lifetime of the photon in the laser. The number of excited atoms N equal the number N_{p} maintained by an external pump minus the number returned to the ground state by the photons, αn, so $N = N_{\mathrm{p}} - \alpha n$. Substituting this into (3.2.21) yields

$$\dot{n} = -c_1 n - c_2 n^2 \quad (n \geqslant 0), \tag{3.2.22}$$

where

$$c_1 = \tau^{-1} - GN_{\mathrm{p}}; \quad c_2 = \alpha G > 0.$$

Now, in contrast with the above problem and example, the phase space is restricted to $n \geqslant 0$, so that the migration of a fixed point into the region $n \geqslant 0$ now does represent a bifurcation. When, c_1 goes from $c_1 > 0$ to $c_1 < 0$, the physical solution bifurcates, giving rise to the sustained laser, namely

$$n \to -c_1/c_2 \text{ as } t \to \infty \quad \text{if } -c_1 \equiv GN_{\mathrm{p}} - \tau^{-1} > 0$$

(threshold condition). A plot of this flow in the combined phase control space looks that shown in Fig. 3.33. We note that (3.2.22) is yet another example of a *Riccati equation*, which is identical in form to the logistic equation (3.1.6) and the Landau equation (3.1.7), but models a very different physical system.

Fig. 3.33

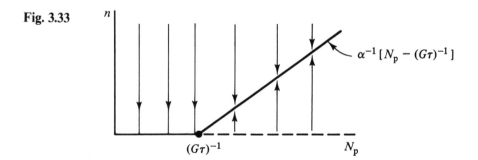

Exercise 3.14. For a nonpolynomial case, consider a charged particle (m, q) in a highly viscous mechanism, (μ), and acted upon by both a constant (E_0) and periodic electric field $E_{\mathrm{p}}(x)$. In suitable units we obtain the gradient approximation

$$\dot{x} = C - A\cos(\pi x) \equiv F(x, c).$$

Fig. 3.34

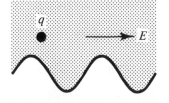

How many bifurcation points does this system have? Sketch a portion of the solution set and the flow in the (x, c) plane (see Fig. 3.31). Indicate the stable and unstable branches of the solution set. How are C and A related to (m, q, μ, E_0, E_p)?

3.3 Structural stability, gradient systems and elementary catastrophe sets

At the end of the last century Henri Poincaré initiated qualitative approaches to the study of differential equations which led to (among other things) topology, differential geometry, and further developed the concept of bifurcation (which had also been studied in special cases by C. Jacobi, 1834, and Lord Rayleigh, 1877). Some aspects of simple bifurcations were discussed in the last section.

Closely associated with the concept of bifurcations is that of *structural stability* of a system of equations, $\dot{x} = F(x, c)$, $(x \in R^n, c \in R^k)$, introduced by Andronov and Pontriagin (1935). The basic idea here is that, since the exact form of the functions $F(x, c)$ can never be known in physical situations (e.g., the precise values of the control parameters c), it should be a basic requirement of any physical model, based on the system $\dot{x} = F(x, c)$, that a slight change in F, $F \to F + \delta F$ should not (usually) result in any 'essential' change in the solutions of this system. The first obvious physical question is: what in general constitutes an 'essential change' in a physical system? One possibility is to associate the 'essential' physical change with a topological change in the phase portrait of $\dot{x} = F(x, c)$, discussed in the last chapter. That is, if the general character of the entire family of solutions of $\dot{x} = F(x, c)$ is (topologically) changed, then this presumably represents a major modifications in the system's behavior. This gives one very appealing possibility, because the precise mathematical notion of *topological orbital equivalence* (TOE) can then be applied. Of course there may be other important physical changes which may occur when two systems are TOE (e.g., their periods can change, they may become Lyapunov unstable, but remain Poincaré stable, etc.). Nonetheless it appears that some form of topological equivalence might be a useful way of identifying a 'significant' physical change, and so we consider Andronov's and Pontriagin's concept.

A system $\dot{x} = F(x)$ is said to be *structurally stable* if, for any sufficiently small smooth change, δF, in the function F, $F \to F + \delta F$ (e.g., $\delta F \in C^r$ and $|\delta F(x)| < \delta$ for all $x \in R^n$), the

Fig. 3.35

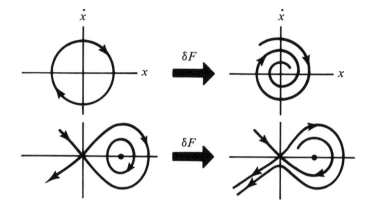

phase portrait of $\dot{x} = F(x)$ and $\dot{x} = F(x) + \delta F$ are topologically orbitally equivalent. Andronov, Vitt, and Khaiken referred to such systems as *'coarse systems'*. The diagrams in Fig. 3.35 illustrate structurally unstable systems on the left, since a small change in the vector field, δF, produces a phase portrait which is not TOE.

To illustrate this, consider the simple case considered before,

$$\dot{x} = cx \equiv F(x).$$

No sufficiently small change of F will change the character of the solutions, except if $c = 0$, which is a bifurcation point. Thus this equation is structurally stable unless $c = 0$.

The system $\dot{x} = 0$, however, is structurally unstable since changing it slightly to $\dot{x} = \varepsilon x \exp(-x^2)$ produces a phase portrait which is not equivalent, regardless of how small ε may be (the $\exp(-x^2)$ term was inserted just to keep δF bounded for all x).

Exercise 3.15. Show that $\dot{x} = cx^2$ is not structurally stable, and determine the region of phase space ($x \in R$) where its flow is most easily altered.

Exercise 3.16. Determine if the harmonic oscillator $\dot{x}_1 = x_2$, $\dot{x}_2 = -x_1$, is structurally stable.

This concept of structural stability for general F is apparently too restrictive to be useful. For example, while it has been proved that most F are structurally stable for $\dim(x) \equiv n \leqslant 2$ by Debaggis (1952), it has been proved that this is not so if $n \geqslant 4$ by Smale (1966), and not so if $n = 3$ (Smale and Williams, 1970). Therefore, in most dimensions, structural stability does not hold for most functions, but only to a limited class of functions. Thom (1975, p. 29) has raised the interesting idea that *structural stability* and *computability* may in fact be incompatible requirements – a point we will return to later.

Exercise 3.17. To understand why most systems (in a mathematical sense) are structurally stable for $n = 2$, it is instructive to think of why some flows in phase space are uncommon, and structurally unstable. In the two following cases, flows on a sphere S^2, and torus T^2, what do you think would be the more common (structurally stable) flows, rather than the indicated (structurally unstable) flows? To do this, visualize what would happen when some small continuous vector field is added to the one shown in Fig. 3.36. Referring to Chapter 2, what physical systems have the dynamics with configuration manifolds S^2 and T^2, and with velocities as indicated? What do the structurally stable vector fields represent physically?

Fig. 3.36

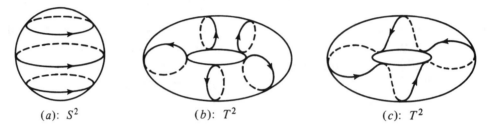

(a): S^2 (b): T^2 (c): T^2

If structural stability of general systems is too restrictive, Thom suggested (1969) that one should focus attention on those systems which tend to have fixed points which are asymptotically stable ('*attractors*') in the sense of Lyapunov. The idea here is that the observed physical states (at least the stationary states) are related to the asymptotic states of $x(t)$ (as $t \to +\infty$). This is of course more interesting in higher dimensional systems, as schematically illustrated in Fig. 3.37. Thom suggested that, for these

Fig. 3.37

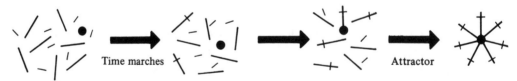

Time marches   Attractor

systems, what is important is the stability of these attractors – once again in a structural sense. Thus, his considerations are limited to systems which tend to 'damp' asymptotically to time independent states (attractors which are fixed points). While this is an important class of systems, it does not include any undamped systems. However, it should not be concluded that his theory is ultimately concerned only with these fixed points, for that is not the case. Both the standard dynamics of $x(t)$, as well as 'adiabatic' dynamics of control parameters can give these systems considerable richness, as will be seen.

Following Thom, we consider now only *gradient systems*, for which F has the special form

$$F = -\nabla V(x, c) \quad (x \in R^n). \tag{3.3.1}$$

Thus the n functions $F_j(x)$ are replaced by only one function, $V(x, c)$, which is, moreover, assumed to be 'smooth' (differentiable a sufficient number of times). The restriction of F to the gradient form (3.3.1) represents a major restriction. Hamiltonian systems

$$\dot{p} = -\partial H/\partial q, \quad \dot{q} = \partial H/\partial p \quad (p, q \in R^n),$$

of course, are similarly restrictive and are, furthermore, restricted to phase spaces of even dimensions.

Exercise 3.18. Show that, corresponding to every Hamiltonian system, in any R^2, there is a gradient system such that its flow is everywhere orthogonal to the Hamiltonian flow (Fig. 3.38). What is the 'potential', $V(p, q)$, of this gradient system?

Fig. 3.38

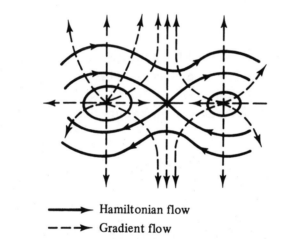

⟶ Hamiltonian flow
--→ Gradient flow

For gradient systems the fixed points x^*,

$$F(x^*, c) = 0 \tag{3.3.2}$$

are asymptotically stable points (attractors) provided that

$$L(x) \equiv V(x) - V(x^*) > 0$$

in some neighborhood of x^*. This follows from the fact that $L(x)$ is a strict Lyapunov function, because

$$F \cdot \nabla L = (-\nabla V) \cdot \nabla V = -(\nabla V)^2 < 0.$$

Thom then calls these attractors structurally stable, if a perturbation $F \to F + \delta F$

generates a corresponding perturbation in the fixed point $x^* \to x^* + \delta x^*$ (i.e., the magnitude of the shift δx^* is bounded by the magnitude of δF). The condition for this is that $F(x) + \delta F(x)$, evaluated at $x^* + \delta x^*$, should be zero for small δx^*. From the Taylor expansion for small δx^* we have

$$F_j(x^*) + \delta x^* \left(\frac{\partial F_j}{\partial x} \right)_{x^*} + \delta F_j(x^*) = 0 \quad (j = 1, \dots, n)$$

The first term vanished by (3.3.2), and these equations do give a solution for the δx_j^* if the determinant of their coefficients does not vanish

$$\det \left| \frac{\partial F_j}{\partial x_i} \right| = \det \left| \frac{\partial^2 V}{\partial x_i \partial x_j} \right| \neq 0 \quad (\text{where } x = x^*). \tag{3.3.3}$$

Hence, if (3.3.3) is satisfied the fixed point x^* is *structurally stable*.

This is closely related to functions called *Morse functions*. A Morse function $M(x)$: $R^n \to R$, is any function such that, if $(\partial M / \partial x_i) = 0$ at some point for all $i = 1, \dots, n$, then the Hessian does not also vanish at this point, i.e.

$$\text{if } \nabla M = 0 \quad \text{then } \det |\partial^2 M / \partial x_i \partial x_j| \neq 0. \tag{3.3.4}$$

Two examples of two dimensional Morse functions, for which $\nabla M = 0$ at $(0,0)$, are illustrated in Fig. 3.39 (a) and (b). The other two possibilities are $-x_1^2 - x_2^2$ and $-x_1^2 + x_2^2$. By way of contrast, a simple non-Morse (NM) function, for which $\nabla NM = 0$ also at $(0, 0)$, is illustrated in (c). You should convince yourself that (3.3.4) is not satisfied for this function.

Fig. 3.39

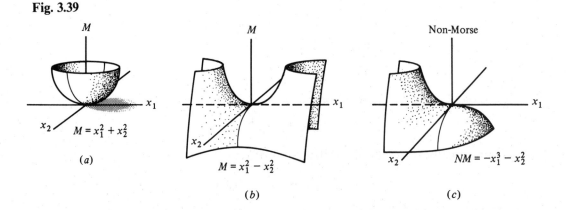

$M = x_1^2 + x_2^2$

(a)

$M = x_1^2 - x_2^2$

(b)

$NM = -x_1^3 - x_2^2$

(c)

Such a function is structurally stable in the sense that, if $\delta M(x)$ is a small perturbation of $M(x)$, there is a diffeomorphism $h(x)$ such that

$$M(h(x)) + \delta M(h(x)) = M(x) + \alpha, \tag{3.3.5}$$

where α is some constant. Morse functions are equivalent to the functions $\pm x_1^2 \pm x_2^2 \cdots \pm x_n^2$. Such functions are the common ('generic') case near extrema, because extrema are usually quadratic in form.

Exercise 3.19. Show that $M = x_1^2 + x_2^2$ is a Morse function, and determine one possible $h(x)$ and α which satisfies (3.3.5), if $\delta M = \varepsilon x_1 x_2$. In other words, x_1 and x_2 are replaced by $h_1(x_1, x_2)$ and $h_2(x_1, x_2)$ on the left side of (3.3.5).

The important point is that *most functions near extrema, $\nabla V = 0$ are Morse functions,* and hence *most fixed points of gradient systems are structurally stable.*

The next step is to introduce the control parameters c, so that fixed points $(F(x, c) = 0)$ are surfaces in the phase-control space used in the last section. Also some of the variations of F can then be represented by a change in the control parameters

$$\delta F = F(x, c + \delta c) - F(x, c).$$

From the implicit function theorem we know that if

$$F(x, c) = 0 \qquad (3.3.6)$$

has a solution for $x \in R$, $c \in D$, and if the Jacobian of F does not vanish,

$$\det \neq |\partial F_i / \partial x_j| \neq 0 \quad (x \in R, \ c \in D), \qquad (3.3.7)$$

then there is a subregion of D such that there is a unique solution $x = g(c)$ (in R). This uniqueness condition is the same as the condition that the fixed point of gradient systems, $\dot{x} = F = -\nabla V$, be structurally stable, (3.3.4). Since most $V(x, c)$ are Morse functions, which are structurally stable, the set of c values for which (3.3.6) and (3.3.7) fail to hold must be of measure zero in the control space $c \in R^k$. Thom called this set, for which

$$\nabla V(x, c) = 0 \quad \text{and} \quad \det|\partial^2 V / \partial x_i \partial x_j| = 0, \qquad (3.3.8)$$

the *elementary catastrophe set, K,* in the control space. The catastrophe set represents those values of c where fixed points become degenerate (i.e., two fixed points $x_1(c), x_2(c)$ become equal as $c \to K$). By referring back to the diagrams of the last section, one can see that this degeneracy occurs at a bifurcation point in c-space. Thus *the catastrophe set K is a set of bifurcation points,* and it is a manifold of dimension $(k - 1)$ in the control space, $c \in R^k$ (hence has measure zero). (In the last section, where $k = 1$, the sets K consisted of a finite number of points, except in Exercise 3.14. These sets all have dimension (capacity) zero.)

The functions $V(x, c)$ are non-Morse functions (i.e., atypical functions) when $c \in K$. Thus both the functions $V(x, c)$ for $c \in K$, and the set K itself, are not typical. Nonetheless Thom has argued that some sets K, and their associated $V(x, c)$, are of great physical importance, for they are associated in some way with the dramatic (near-discontinuous) changes in 'forms', 'structures' (morphology), or behavior in complex systems. He had biological systems particularly, but not solely, in mind. Certainly, in some cases this idea is clearly correct, as illustrated by the hysteresis example of the last section. However, there has also been proposed a hodgepodge of biological, sociological, economic, and mechanical 'applications' of catastrophe theory. Many of these applications have been roundly criticized, with considerable justification in many cases, (e.g., Sussman (1977), Kolata (1977), Guckenheimer (1978), Arnold (1984)), and have been called 'the height of scientific irresponsibility' by M. Kac. For a specific and authoritative critique of Zeeman's applications, see Smale (1978) as well as Thom (pg. 633ff in Zeeman (1977)). Despite many possible misapplications, it would certainly be unwise to summarily dismiss Thom's program, as ill-defined as it appears to be at present ('The truth is that CT is not a mathematical theory, but a body of ideas, I dare say a state of mind': Thom, 1977). Applications to many clearly defined situations such as crystal structures, many phase transitions (lattice, magnetic, etc.), dissipative mechanical systems and optical phenomena, seem fairly likely in the future. Even more intriguing is its possible role in describing many hydrodynamic situations involving not only bifurcations but discontinuities (breaking waves, and shocks which are outside the realm of present methods), and biological applications, which are presently mostly heuristic in nature. We will return to these ideas later. Also see Section 3.6 for some of Thom's perspectives.

What gives a mathematical structure to Thom's idea, and forms the backbone of the present day *elementary catastrophe theory* is Thom's further restriction to catastrophe sets which are themselves structurally stable. This is based again on the physical observation that 'forms', 'structures', or 'behavior', which are presumably related to the K sets, also have a great deal of stability to modifications in the syste. It should be noted, however, that Thom views *catastrophe theory* as being much more general than our present concerns with *elementary catastrophe theory* (e.g., Zeeman, 1977, p. 633).

The *catastrophe set K* of *V* is called *structurally stable*, if the set K^* associated with $V^*(x, c) = V(x, c) + \delta V$ can be mapped into K by a homeomorphism, $h(c): K^* \to K$, for arbitrary small δV. We next turn to the use of this criterion in constructing Thom's elementary catastrophe sets, which have this required stability.

3.4 Thom's 'universal unfolding' and general theorem

We are now concerned about the possible forms of non-Morse (i.e., atypical) functions, $V(x, c)$, and in particular about the nature of their catastrophe (singular) set $c \in K$, given by the solution of

$$\nabla V(x, c) = 0 \quad \text{and} \quad \det|\partial^2 V/\partial x_i \partial x_j| = 0. \tag{3.4.1}$$

For $n = 1$, there is only one variable x, and this becomes simply

$$\left(\frac{\partial V}{\partial x}\right)_{x^*} = 0 \quad \text{and} \quad \left(\frac{\partial^2 V}{\partial x^2}\right)_{x^*} = 0 \tag{3.4.2}$$

So (once again) the set K is composed of those values of c which give degenerate fixed (singular) points x^*, for the equation $\dot{x} = -\nabla V$.

Now consider a particular fixed point, which we can take to be $x^* = 0$, and a particular $c \in K$, which we can also take to be $c = 0$. Then, if $V(x)$ is analytic at $x = 0$ (e.g., the 'infinitely flat' potential $\exp(-1/x^2)$ is excluded), then (3.4.2) requires that

$$V(x) = a_3 x^3 + a_4 x^4 + a_5 x^5 + \cdots \quad \text{(at } c = 0) \tag{3.4.3}$$

in the neighborhood of $x = 0$. What we do now is to consider (in turn) the cases (i) $a_3 \neq 0$; (ii) $a_3 = 0$, $a_4 \neq 0$; (iii) $a_3 = 0$, $a_4 = 0$, $a_5 \neq 0$; etc. Note that we are dealing only with the region near the fixed point, not a global investigation.

Assume, first, that $a_3 \neq 0$, in which case the higher order terms can be neglected if x is sufficiently small (this becomes trickier if $n > 1$). By scaling x, and possibly introducing a reflection $(x \to -x)$, we can take $a_3 = +1$, so $V(x) \simeq x^3$ near the fixed point if $c = 0$ also. Thom calls this function an *'organizing center'*. The idea now is to determine the catastrophe set associated with a function $V(x, c)$, when $c \neq 0$, which reduces to this particular organizing center when $c = 0$ (i.e., $V(x, c)$ satisfies (3.4.2) for $x^* = 0$, and $V(x, c = 0) = x^3$). This development of the catastrophe set K in the control space, away from the origin $c = 0$, is called by Thom the *'unfolding'* of the K set.

Next, the only variations of V which can be important, for small x, are

$$V(x, c) = c_1 x + c_2 x^2 + x^3, \tag{3.4.4}$$

where any irrelevant constant term is ignored. By a shift in the x origin this can always be put into the canonical form

$$V(x, c) = x^3 + cx. \tag{3.4.5}$$

Note again that additive constants can be ignored. The expression (3.4.5) is referred to as the *universal unfolding* of the organizing center $V = x^3$.

Exercise 3.20. Prove that (3.4.4) cannot generally be transformed into $V(x) = x^3 + cx^2$ (if $x, c \in R$). Determine the shift, $x = \bar{x} + x_0$, which transforms (3.4.4) into $\bar{x}^3 + c\bar{x}$, and obtain $c(c_1, c_2)$.

In this case there is only one control parameter of (topological) significance, and Thom calls this *codimension one*. Note that any variation δV of V, which is of significance near $x = 0$, will again produce (3.4.4), and this is homeomorphic to (3.4.5). Thus the structurally stable catastrophe set is simply given by

$$\frac{\partial V}{\partial x} = 3x^2 + c = 0 \quad \text{and} \quad \frac{\partial^2 V}{\partial x^2} = 6x = 0$$

or, in other words, by the point $c = 0$. In general, the dimension of the K set is one less than the codimension (i.e., if $c \in R^k$, $\dim(K) = k - 1$). Hence in the present case $\dim(K) = 1 - 1 = 0$ (a point).

The present case is familiar from the last section. The only new feature is the introduction of the K set. To remind you, if $c > 0$ there are no fixed points, whereas for each $c < 0$ there are two fixed points (one stable, one unstable). The flow in the phase-control space looks like that shown in Fig. 3.40, which follows from $\dot{x} = -\partial V/\partial x = -3x^2 - c$. This structure in the phase-control space, is the origin of its name: the *fold catastrophe* (even though the K set is a single point).

Fig. 3.40

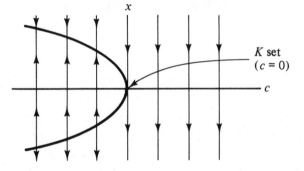

Recalling again that the gradient dynamics can be viewed as the inertialess limit $(m \to 0)$ of

$$m\ddot{x} = -\dot{x} - \partial V/\partial x,$$

then V acts as a potential in this overdamped dynamics. The above flow pattern then can be related to the structure of $V(x, c)$ for $c > 0$ and $c < 0$, as shown in Fig. 3.41. As c goes from $c < 0$ to $c > 0$, the metastable state vanishes at $c = 0$ and then $x \to -\infty$ as $t \to \infty$.

Fig. 3.41

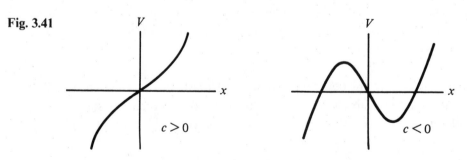

Next we consider the case $a_3 = 0$, but $a_4 \neq 0$ in (3.4.3). Then, near $x = 0$, only the term $a_4 x^4$ is significant, and by rescaling x this can be reduced to $\pm x^4$. Because of the even power it is not possible to alter the sign by a reflection ($x \rightarrow -x$), so there are now two organizing centers. Again, for $c \neq 0$, the most general alteration of $V = \pm x^4$ near $x = 0$ is of the form

$$V = \pm (x^4 + c_3' x^3 + c_2' x^2 + c_1' x).$$

By a shift in x this can always be put in the canonical form

$$V = \pm (x^4 + c_2 x^2 + c_1 x). \tag{3.4.6}$$

which is now the universal unfolding of the organizing center $\pm x^4$. An important new feature has now been added. There are two essential control parameters (codimension two).

The catastrophe set K is determined by (3.4.2),

$$c_1 + 2c_2 x + 4x^3 = 0; \quad c_2 + 6x^2 = 0,$$

so that $c_1 + \frac{4}{3} c_2 x = 0$ or, finally,

$$8c_2^3 + 27c_1^2 = 0 \quad (K \text{ set}). \tag{3.4.7}$$

If this K set is plotted in the control space, as shown in Fig. 3.42, it has the form of a cusp. For this reason this is known as the *cusp catastrophe* (now the cusp is in the control space).

Fig. 3.42

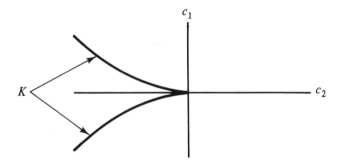

This single K set holds for both signs in (3.4.6), and they are differentiated as the cusp $(+)$ and *dual cusp* $(-)$ catastrophes. Typically the 'potential' $V(x, c)$ in the two cases (and for $c_2 < 0, c_1 < 0$) looks like that shown in Fig. 3.43, which are clearly very different in their stability properties. The fixed point surface in the phase-control, given by $4x^3 + 2c_2 x + c_1 = 0$, has the structure of a pleat beginning at the origin (see Fig. 3.44). The form of the potential at the various points in the control space, indicated in Fig. 3.45, are illustrated. The catastrophe occurs in the 'delay situation' in going from $3 \rightarrow 2$ or $4 \rightarrow 5$, where a system in a metastable state (MS) drops to the state of lower

Fig. 3.43

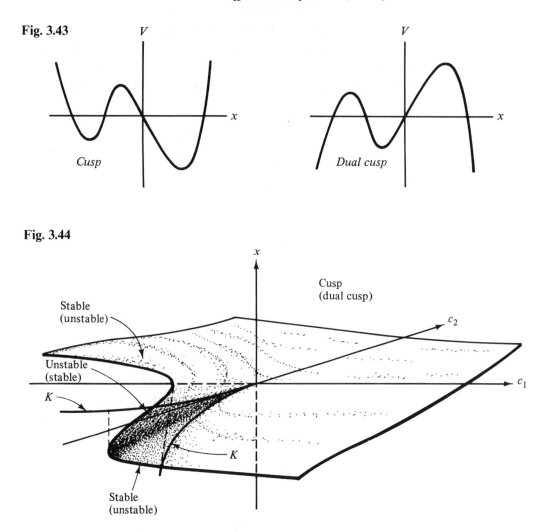

Fig. 3.44

potential which is absolutely stable (AS). If there are strong external fluctuations acting on the system, the 'Maxwell situation', then the transitions will not take place at the catastrophe set K but at what Thom called the 'shock set' S, which is the line $c_1 = 0$, $c_2 < 0$ in the control space. At this line the two potential minima are equal, so across this line the metastable (MS) and absolutely stable (AS) states exchange from the upper to lower sheet, as indicated in Figs. 3.45 and 3.46.

It is perhaps worth noting the similarities and differences between the fixed point surface in this cusp situation and the equilibrium surface of the van der Waal's equation $(p + a/v^2)(v - b) = RT$, illustrated in Fig. 3.47. (For R. Thom's viewpoint on this subject see his 'Mathematical concepts in the theory of ordered media', pp. 385–420, in *Physics of Defects* (Session XXXV, Les Houches, Ecole d'été de Physique Théorique, 1980, North-Holland Publishing Company, 1981). The correlation between (p, v, T)

Fig. 3.45

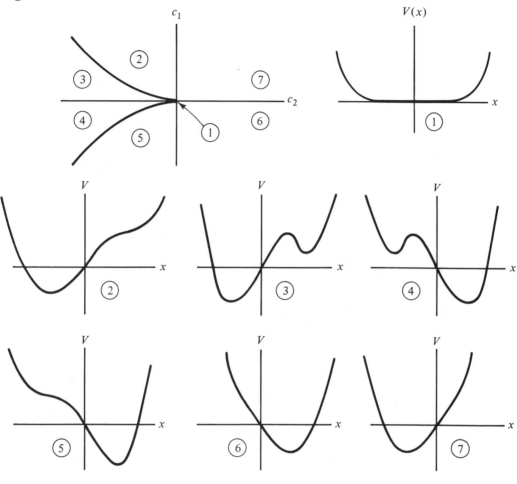

Fig. 3.46

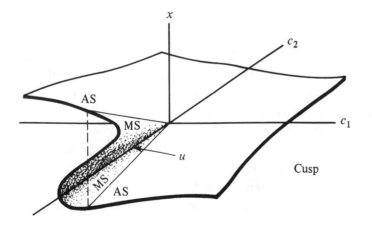

Cusp

Fig. 3.47

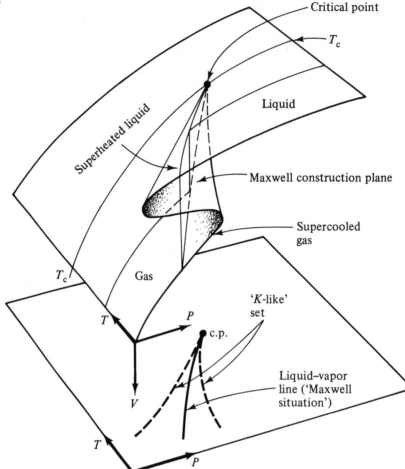

and (x_1, c_1, c_2) on these surfaces is obvious. Even the metastable and unstable portions agree with observations. However, this in no sense implies that some dynamics in the physical system bears any relationship to gradient dynamics (see also point 6 in the following summary). Indeed, the variables (p, T) are equilibrium concepts, and therefore some significant generalization would be needed to bring them into any dynamic model (presumably also requiring other variables for a deterministic system). Most conspicuously missing from this cusp model, and van der Waal's equation, are the physical states on Maxwell's construction plane ('Maxwell's rule'). Since this plane of equilibrium states is not in this model, this clearly indicates the necessity of more refined ideas concerning the collective behavior of molecules. Thus, while the catastrophe surfaces may bear some similarities to physical forms or observations, care should be exercised in jumping to simplistic conclusions. As Thom has warned (Zeeman, 1977, p. 636): 'When presenting catastrophe theory to people, one should never state that, due to such and such a theorem, such and such a morphology is going unavoidably to appear. In no case has mathematics any right to dictate anything to reality. ... If reality does not obey the theorem – that may happen – this proves that some unexpected constraints cause some lack of transversality, which makes the situation all the more interesting.'

Thom's next universal unfolding is

$$V = x^5 + c_3 x^3 + c_2 x^2 + c_1 x_1, \tag{3.4.8}$$

which has codimension three, so we can no longer draw the fixed point surface in the phase-control space, which now has dimension four. The catastrophe set of (3.4.8) is now a surface in the control space $C \subset R^3$. It looks as shown in Fig. 3.48. This is referred to as the *swallow's tail* catastrophe.

Fig. 3.48

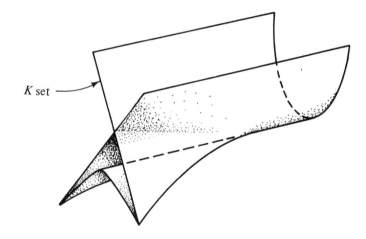

K set

Exercise 3.21. Obtain the equation $G(c_1, c_2, c_3) = 0$ for the swallow's tail catastrophe set.

Surfaces of this type arise in wave phenomena (e.g., Courant and Hilbert, 1953, Vol. II, p. 617, Fig. 54), in connection with Feynman diagrams, in beams from electron microscopes and caustics from water droplets, as illustrated in Section 3.1. For higher order catastrophes see the references. Sections of four dimensional 'butterfly' catastrophes have also been observed in light caustics, Figs. 3.3, 3.5 (Berry, 1976).

Thom's general theorem concerning structurally stable functions is not limited to one x variable or three control parameters, but is considerably more general. To give some appreciation of this generality, we will simply quote the theorem without further comment. One should, however, be able to recognize the limited area, which was discussed above, contained within this theorem.

Thom's Theorem. Most smooth (infinitely differentiable) functions $V(x, c)(x \in R^n, c \in R^k$ with $k \leq 5$) are structurally stable. For this family $(V : R^n \times R^k \rightarrow R)$ and any point $(x, c) \in R^n \times R^k$ there is a choice of coordinates for c in R^k and for x in R^n, such that the x_i vary smoothly with c, in terms of which the function $V(x, c)$ has one of the following local forms: a constant plus

x_1	(not a fixed point)
$x_1^2 + x_2^2 + \cdots + x_r^2 - x_{r+1}^2 - \cdots - x_n^2$	(nondegenerate fixed point; Morse function)
$x_1^3 + c_1 x_1 + (M)$	('fold' catastrophe set)
$\pm (x_1^4 + c_2 x_1^2 + c_1 x_1) + (M)$	('cusp' (+); 'dual cusp' (−))
$x_1^5 + \sum_{k=1}^{3} c_k x^k + (M)$	(swallowtail)
$\pm \left(x_1^6 + \sum_{k=1}^{4} c_k x^k \right) + (M)$	(butterfly and dual butterfly)
$x_1^7 + \sum_{k=1}^{5} c_k x^k + (M)$	(wigwam)

$x_1^2 x_2 \pm x_2^3 + c_3 x_1^2 + c_2 x_2 + c_1 x_1 + (N)$ (hyperbolic (+), elliptic (−) umbilic)

$\pm (x_1^2 x_2 + x_2^4 + c_4 x_2^2 + c_3 x_1^2 + c_2 x_2 + c_1 x_1) + (N)$ (parabolic (and dual) umbilic)

$x_1^2 x_2 \pm x_2^5 + c_5 x_2^3 + c_4 x_2^2 + c_3 x_1^2 + c_2 x_2 + c_1 x_1 + (N)$

  (second hyperbolic (+) and second elliptic (−) umbilic)

$\pm (x_1^3 + x_2^4 + c_5 x_1 x_2^2 + c_4 x_2^2 + c_3 x_1 x_2 + c_2 x_2 + c_1 x_1)$ (symbolic (dual) umbilic)

where

$$(M) = x_2^2 + \cdots + x_r^2 - x_{r+1}^2 - \cdots - x_n^2 \quad \text{and} \quad (N) = x_3^2 + \cdots + x_r^2 - x_{r+1}^2 - \cdots - x_n^2$$

This yields eleven elementary catastrophe sets (not counting the duals). Thom's original list did not contain those involving c_5. For $k > 5$ the number of forms is 'infinite' – that is most functions are not structurally stable (for more details see Poston & Stewart: *Taylor Expansions and Catastrophes*).

Casting caution aside, we end this section with a brief summary of some motivations and concepts:

Ideas and questions	Comments and answers
1. Physical systems have *some* type of stability in their 'form', 'structure', or 'behavior'. Are the models $\dot{x} =$	If $x \in R^n$, then most systems $\dot{x} = F(x)$ are *not* structurally stable if $n > 2$. Either the mathematical concept is irrelevant to the

Ideas and questions	Comments and answers
$F(x)$ structurally stable?	physical idea, or the models are too general.
2. Consider instead gradient systems, $\dot{x} = -\nabla V$, and assume that observables are fixed points (attractors). Are the fixed points structurally stable?	They are if and only if $\lvert \partial^2 V / \partial x_i \partial x_j \rvert \neq 0$ at the fixed point. Then V is a Morse function, $M(x)$. *Most* functions *are* Morse functions, but most $F \neq -\nabla V$, so that the number of relevant models is limited (but see 5).
3. Consider changes of systems induced by control parameters, c. If $V = V(x, c)$, what fixed points are *not* structurally stable?	A set of measure zero in $c \in R^k$, called the elementary catastrophe set K (Thom). For $c \in K$, $V(x, c)$ is *not* a Morse function.
4. If the topological structure of K is somehow to be related to physical 'forms', 'structures' or 'behaviors', is the set K structurally stable?	Need to consider the atypical class of non-Morse functions. Assume that $V(x, c)$ is analytic near an atypical fixed point, taken to be $(x, c) = (0, 0)$. From these 'organizing centers' (i.e., differing leading terms in the Taylor series in R^n of this class, $V(x, 0)$) generate the *typical* (*atypical!*) $V(x, c)$.
5. Can these results be applied to dynamic systems other than gradient systems?	Yes. If $F(x, c) = 0$ implies that $\nabla G(x, c) = 0$, for some G, then Thom's theorem can be applied to $G(x, c)$. It is not required that $F = -\nabla V$ everywhere in the phase space (see Steward (1982)), e.g., $F_x = x^3 + xy^2 + x$, $F_y = y$ so $F = 0$ only at $x = 0$, $y = 0$. Set $G = x^2 + y^2$ so $\nabla G = 0$ only if $F = 0$.
6. How does the classic study of simple bifurcations, (SB, section (2)) differ from ECT (Elementary catastrophe theory)?	The former must avoid the degenerate singularities of the organizing centers, because the implicit function theorem breaks down at these points. The classic theory considers 'stringing together' such SB (see Fig. 3.7), but ECT is non-local in the *complete* control-phase space, required for the degenerate singularities.

3.5 Catastrophe machines

Simple 'machines' which illustrate catastrophes are useful to construct and examine. A few such machines will be discussed briefly in this section.

The Poston machine (see Poston and Steward, 1976) consists of a uniform wheel of mass m and radius r, with a mass M attached a distance $c \leqslant r$ from its axis (on a radial spoke) (Fig. 3.49). The regular bifurcation, associated with a change in c, has already been discussed in Section 3.2.

Fig. 3.49

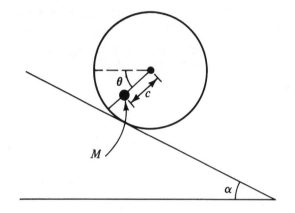

A very nice demonstration of hysteresis can be made with *Benjamin's 'machine'*, which consists of a coiled bicycle brake cable (in a plastic sheath), with one end fixed in a board and the other and free to slide through a hole in the board (see Fig. 3.50). Starting

Fig. 3.50

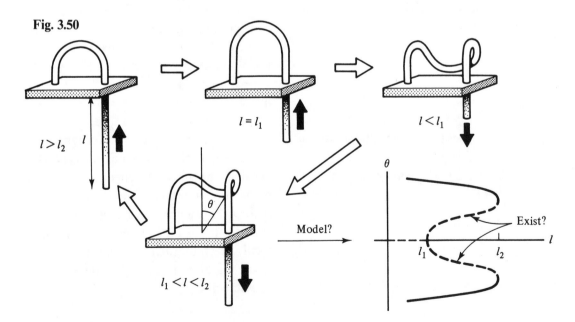

with a short arch (a long length, l, beneath the board) the length is decreased, increasing the height of the arch. The cable will remain upright until some length, l_1, is reached, at which point it flops over to one side (possibly going below the level of the board). If we now increase l, so that it is greater than l_1, it will remain in a 'flopped over' configuration, which one can characterize by the maximum angle which it makes with the vertical, θ. Finally, when l is increased past some value l_2 which is considerably larger than l_1, the cable will pop into an upright position. The hysteresis effect is very pronounced in this system (as contrasted with the plastic strip model discussed above), but the modeling of this system is not very obvious. While the 'solution set' indicated in the diagram is suggestive of what is observed, the unstable branches associated with this set would appear to be totally fanciful in the case of a cable (whereas they might have a reality for a more rigid strip). A similar 'missing unstable branch' occurs in the self-sustained discharge discussed above. In many such cases these simple solution sets are suggestive, but the systems may be considerably more complicated than these models indicate.

A 'machine' which exhibits interesting global properties is Zeeman's catastrophe machine. This consists of a disk of radius r which is pivoted at its center, where there is friction (it can be viewed as an overdamped system, obeying 'gradient dynamics'). Two rubber bands, each with an unstretched length l_0, are attached to a perpendicular peg at the edge of the disk. The other end of one band is attached to the point $(x, y) = (-d, 0)$ in the plane of the disk, where $d > l_0 + r$. This ensures that, if $\theta = \pi$, this rubber band is under tension. The end of the other rubber band is put at some arbitrary points in this plane. This is the *control point*, and there are two control parameters (x, y) or (r_c, θ_c), as illustrated in Fig. 3.51. The lengths of the rubber bands are l and l_c respectively.

Fig. 3.51

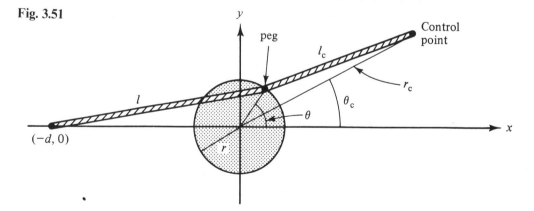

The dynamic variable of the system is θ. If, for simplicity, we treat the disk dynamics as highly damped, then its equation of motion is

$$\mu\dot{\theta} = -\partial V(\theta)/\partial\theta, \tag{3.5.1}$$

where μ is a 'coefficient of friction' (sliding = static!), and $V(\theta)$ is the potential of the rubber bands. If they are elastic, with an elastic constant k, the potential energy is

$$V(\theta) = \tfrac{1}{2}k\,(l(\theta) - l_0)^2 + \tfrac{1}{2}k\,(l_c(\theta) - l_0)^2, \tag{3.5.2}$$

where it is assumed that $l(\theta) > l_0$, $l_c(\theta) > l_0$, and

$$\begin{aligned} l^2 &= d^2 + r^2 - 2rd\cos(\pi - \theta) \\ &= d^2 + r^2 + 2rd\cos(\theta) \\ l_c^2 &= r^2 + r_c^2 - 2r_c r\cos(\theta - \theta_c). \end{aligned} \tag{3.5.3}$$

The equilibrium (fixed point) surface of (3.5.1), $\theta(r_c, \theta_c)$, is given by the solutions of $\partial V/\partial\theta = 0$. This is generally quite complicated, as can be seen from (3.5.2) and (3.5.3). Indeed, there can be several values of θ which satisfy $\partial V/\partial\theta = 0$, for given (r_c, θ_c). Which θ-value occurs experimentally depends on how we approach the point (r_c, θ_c). In other words, the system has bistable states and exhibits hysteresis effects. It is strongly recommended that you build your own machine, to better appreciate the following global phenomena.

To understand this system, first consider 'large' values of r_c. This causes the peg to nearly lie on the radius line r_c, so $\theta \simeq \theta_c$. If θ_c is varied, θ varies smoothly, with $\theta \approx \theta_c$. In other words the equilibrium surface $\theta = \theta(r_c, \theta_c)$ is smooth and single-valued.

We next note that, if $\theta_c = 0$ (so $\theta = 0$), and we decrease r_c from some large value, when l_c becomes less than $l\,(= d + r)$, the equilibrium of the system is sustained by an outward radial force of the disk. Therefore, if $l_c \equiv r_c - r$ (at $\theta_c \equiv 0$) is much less than $d + r$, we might expect that 'dramatic' effects will occur as θ crosses the value $\theta = 0$.

Indeed, a little experimentation shows that as r_c is decreased, and θ_c is moved from positive to negative values, and back again, the value of θ exhibits a 'jump' phenomena (Fig. 3.52). The values of (r_c, θ_c) where it jumps lies on two different curves, depending

Fig. 3.52

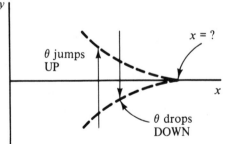

on whether θ_c is increasing or decreasing (Fig. 3.53). Moreover, these curves intersect in a cusp, so locally the equilibrium surface is a cusp catastrophe surface, as illustrated. The global picture of this equilibrium surface is, however, much more sophisticated, because there are four such cusps for $0 < \theta \leqslant 2\pi$ (Fig. 3.55).

Fig. 3.53

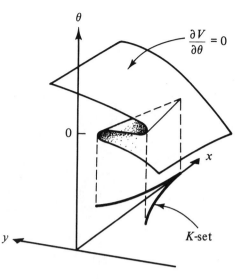

The equilibrium surface, using (3.5.2) and (3.5.3), is given by

$$\frac{\partial V}{\partial \theta} = - k(l(\theta) - l_0)(r \, d \sin \theta / l(\theta))$$

$$+ k(l_c(\theta) - l_0)(r_c r \sin (\theta - \theta_c)/l_c(\theta)) = 0. \qquad (3.5.4)$$

The factors $l_0/l(\theta)$ and $l_0/l_c(\theta)$ make this rather complicated. However, one cannot simply set $l_0 = 0$ without losing the interesting effects. On the other hand, as noted by Poston and Woodcock (1973), the analysis is greatly simplified if we only set $l_0/l_c(\theta) = 0$ in (3.5.4). This means that we use two different types of rubber bands, with the unstretched length of the control band being negligible. In this case, (3.5.4) reduces to

$$(1 - (l_0/l(\theta))) \, d \sin \theta - x \sin \theta + y \cos \theta = 0 \qquad (3.5.5)$$

where $x = r_c \cos \theta_c$, $y = r_c \sin \theta_c$. This is still a complicated function of θ, but it defines (implicitly) the equilibrium surface $\theta = S(x, y)$.

We note that, if θ is fixed, the solutions of (3.5.5) are straight lines in the control space. These are illustrated in Fig. 3.54 for $r = 1, d = 10, l_0 = 5$, and use values of θ which differ by 0.1. In (a) $-\pi/2 < \theta < \pi/2$, whereas $-\pi < \theta < 0$ in the (b). The locus of these curves yields the local catastrophe set, K, where both $\partial V/\partial \theta = 0$ and $\partial^2 V/\partial \theta^2 = 0$.

When all values of θ are used, the resulting diagram is illustrated (now with $\Delta\theta = 0.2$) (Fig. 3.55). The global catastrophe set is a 'cusped-diamond'. The global nature of the equilibrium surface is illustrated below (based on a diagram by Poston and Stewart, 1978). It is something like a spiral slide, except near the central axis there are jumps.

Exercise 3.22. The catastrophe set is given by $\partial^2 V/\partial \theta^2 = 0$ on the surface $\partial V/\partial \theta = 0$. This is generally quite complicated, but it is fairly easy to find the location of the cusps

Fig. 3.54

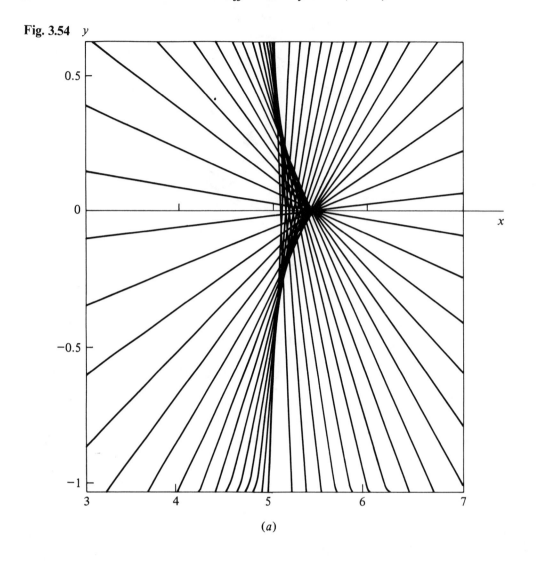

(a)

on the x-axis in the case $\partial V/\partial\theta = 0$ is given by (3.5.5). Obtain these cusp points, $(x(\theta = 0),$ $y = 0)$ and $(x(\theta = \pi), y = 0)$, in terms of d, l_0, and r (Fig. 3.56).

The equilibrium surface does not indicate what portion is stable, unstable, or metastable (only locally stable). To understand this we need to look at the form of the potential energy

$$V(\theta) = \tfrac{1}{2}k(l(\theta) - l_0)^2 + \tfrac{1}{2}k\,l_c(\theta)^2.$$

This is illustrated in the following diagram, in which eleven values of r_c are used for each of two values of θ_c. Since the above surface, $\partial V/\partial\theta = 0$, represents the extrema of $V(\theta)$, the stability, instability and metastability can easily be determined from the diagram. We begin by decreasing (or increasing) r_c along the x axis ($\theta_c = 0$). Here $V(\theta)$ looks as

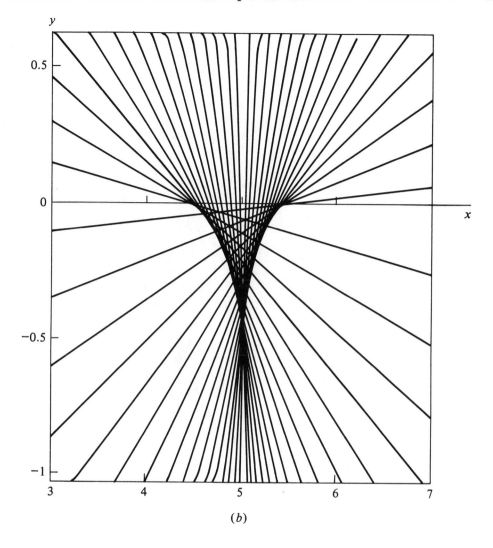

(b)

shown in Fig. 3.57. Then $\theta_c = -0.02$ (radians), is set slightly off the x-axis and r_c is again varied. The resulting $V(\theta)$ are shown in Fig. 3.58. The remaining challenge is to relate this diagram with the Fig. 3.58, which simply takes a little concentration!

An important generalization of Zeeman's machine was described by Woodcock and Poston (1976). It gives a rare mechanical example of a *butterfly catastrophe set*. The modification of Zeeman's machine consists of using two identical fixed rubberbands, symmetrically attached about the x axis, as illustrated in Fig. 3.59. In addition the control band is made double, to balance the added restoring force of the second fixed band. In this case the potential energy, (3.5.2) is replaced by

$$V(\theta) = k(l_c(\theta) - l_0)^2 + \sum_{i=1}^{2} \tfrac{1}{2} k(l(\theta - \phi_i) - l_0)^2, \qquad (3.5.6)$$

Fig. 3.55

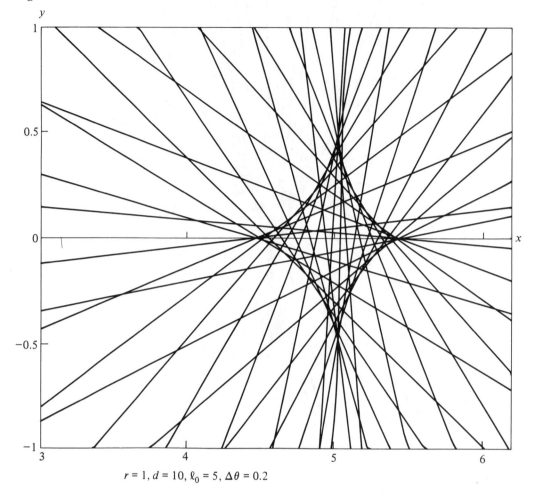

$$r = 1, d = 10, \ell_0 = 5, \Delta\theta = 0.2$$

where $\phi_1 = \phi$ and $\phi_2 = -\phi$ (see the Fig. 3.60, and $l(\theta)$ is still given by (3.5.3). Again for simplicity, setting $l_0 = 0$ only in the first factor of (3.5.6), the equilibrium surface, $\partial V/\partial\theta = 0$, now becomes

$$x = y\cot(\theta) + \frac{d}{2}\sum_{i=1}^{2}(1 - (l_0/l(\theta - \phi_i)))\sin(\theta - \phi_i)/\sin\theta \qquad (3.5.7)$$

which reduces to (3.5.5) if $\phi = 0$.

If we set $\phi = 0.7$ (radians), $d = 4$, and $l_0 = 2$, the resulting (double) *butterfly catastrophe set* can be seen in Fig. 3.60.

Exercise 3.23. Use physical reasoning to determine in what direction the wheel lies with respect to this diagram (i.e., where is the origin in the previous figure?).

Fig. 3.56

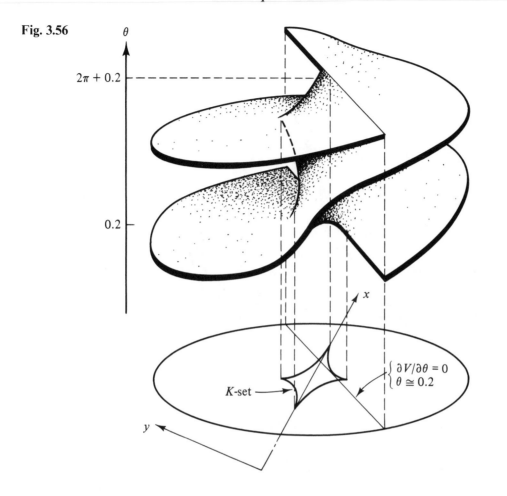

An interesting question (suggested by G. Francis) is to determine the character of the bifurcation between the cusp and the butterflies as a function of ϕ. It might seem reasonable that the two butterflies simply develop smoothly from the cusp as ϕ is increased from zero, as schematically illustrated in Fig. 3.61. It turns out, however, that no butterflies appear until ϕ reaches a finite value.

To show this, we first note that when there is a simple cusp, the lines (3.5.7) intersect the x-axis ($y = 0$) at decreasing values of x, as θ goes from $\pi/2$ to π. When there is a butterfly structure, this intersect, given by

$$x(\theta) = \tfrac{1}{2}d \sum_{i=1}^{2} (1 - (l_0/l(\theta - \phi_i))) \sin(\theta - \phi_i)/\sin\theta, \tag{3.5.8}$$

first decreases and then increases as θ passes through some angle θ_b. This angle is where the catastrophe set crosses the x-axis, with $\pi/2 \leqslant \theta_b < \pi$. It is given by the angle where $dx(\theta)/d\theta = 0$, in this range $(\pi/2, \pi)$. Using (3.5.8), (3.5.3), and the fact that $\phi_1 = -\phi_2 \equiv \phi$,

Fig. 3.57

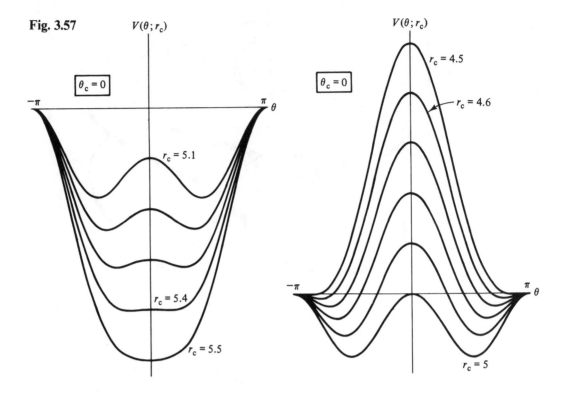

we readily obtain the implicit equation for $\theta_b(\phi)$,

$$\sum_{i=1}^{2} \{l^2(\theta_b - \phi_i)\sin\phi_i + rd\sin^2(\theta_b - \theta_i)\sin\theta_b\}/l^2(\theta_b - \phi_i) = 0. \qquad (3.5.9)$$

We see that this butterfly angle, $\theta_b(\phi, d/r)$, is not a function of $l_0\,(\neq 0)$. If we again set $r = 1$, $d = 4$, and determine $\theta_b(\phi)$ from (3.5.9), the graph shown in Fig. 3.62 is obtained. We see that the butterfly catastrophe set only exists (i.e., $\theta_b < \pi$) only if $\phi > \phi_c = 0.53$ radians ($\simeq 30$ degrees). Not surprisingly, we also find that the critical angle, $\phi_c(d)$, increases with d (e.g., $\phi_c(6) \simeq 0.61$). However, $\phi_c(d)$ remains nonzero even when $d = l_0 + r$, the minimum d which insures tension in the rubberbands (assumed in (3.5.6)).

Exercise 3.24. Determine $\phi_c(d = 3)$ in the above system, $r = 1$.

Catastrophe surfaces are all the more interesting when they are directly observable. This is what makes the observation of *light caustics* a very pleasurable and esthetic experience. It is, moreover, a prime example of where the concept of catastrophe surfaces can be applied without any association with dissipative dynamics or gradient dynamics (in the usual sense), thereby illustrating its greater range of applicability. A

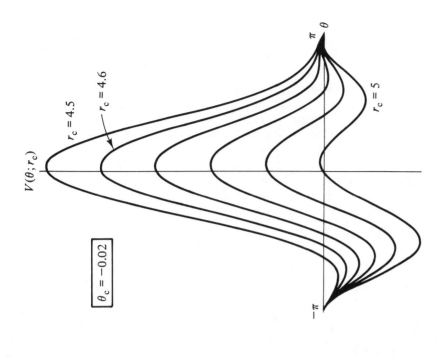

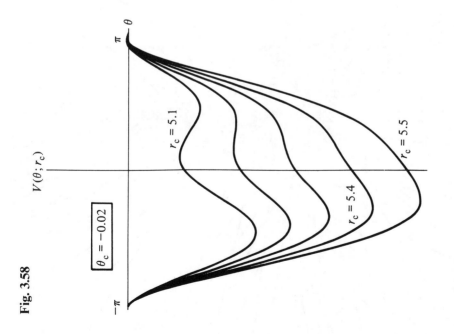

Fig. 3.58

Fig. 3.59

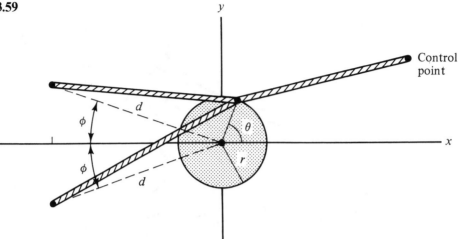

Fig. 3.60

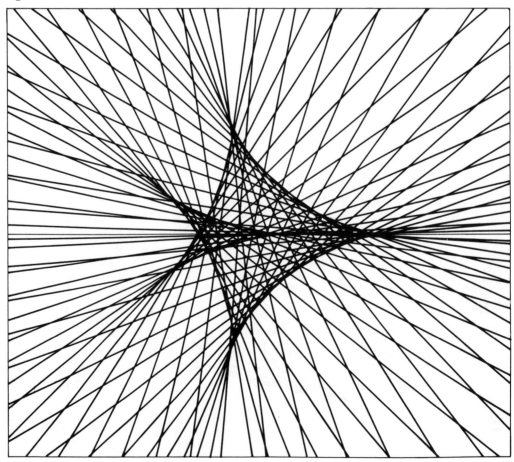

Fig. 3.61

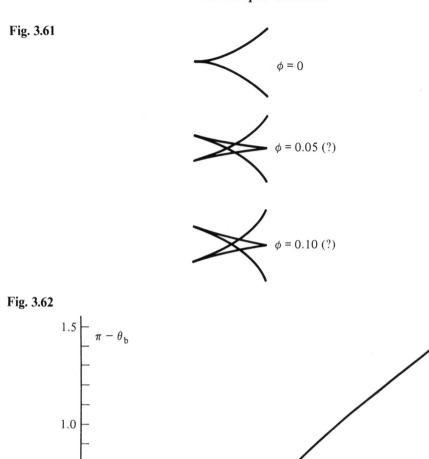

$\phi = 0$

$\phi = 0.05$ (?)

$\phi = 0.10$ (?)

Fig. 3.62

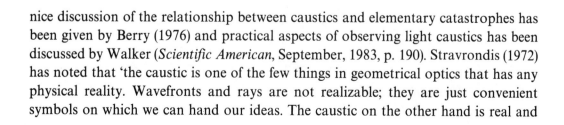

nice discussion of the relationship between caustics and elementary catastrophes has been given by Berry (1976) and practical aspects of observing light caustics has been discussed by Walker (*Scientific American*, September, 1983, p. 190). Stravrondis (1972) has noted that 'the caustic is one of the few things in geometrical optics that has any physical reality. Wavefronts and rays are not realizable; they are just convenient symbols on which we can hand our ideas. The caustic on the other hand is real and becomes visible by blowing a cloud of smoke in the region of the focus of a lens.'

An extensive discussion of wavefront dislocations, caustics as catastrophes and diffraction catastrophes is given by Berry (1981).

3.6 The optical bistability cusp catastrophe set

While we have seen some relationships of catastrophe sets to the fixed points of dynamic gradient systems, these sets may also arise in very different physical contexts. One interesting case involves *optical bistability* (for extensive references, see Wolf, 1984).

Optical bistability is a phenomena which is produced in part by the nonlinear behavior of the electromagnetic absorptivity of an excitable medium – namely, the absorptivity decreases when the radiation intensity become large. In its simplest form, we consider the envelope amplitude of the radiant electric field, $E(x)$, as it travels through this excitable medium. Its spatial variation is determined by

$$\frac{dE}{dx} = -\chi(|E|^2)E, \tag{3.6.1}$$

where $\chi = \chi_a + i\chi_d$ is the dielectric susceptibility. We ignore the complex (dispersive) part, setting $\chi_d = 0$, and keep only the absorptive part, χ_a. When the medium becomes saturated by a large intensity, $|E|^2$, its absorptivity decreases approximately as

$$\chi = \chi_a = \frac{\alpha}{1 + (E^2/I_s)}, \tag{3.6.2}$$

where α is the linear absorptivity. I_s is called the saturation intensity. If E^2 becomes comparable with I_s the medium begins to saturate, and χ_a decreases significantly. A simple physical system which exhibits this saturation effect is a medium containing atoms with two electronic states whose energy difference is in resonance with the monochromatic radiation. When the atoms have their electron excited to the upper state by the intense radiation more rapidly than they can decay to the lower state, they can no longer absorb radiation, so χ_a decreases. This saturable medium is located between $x = 0$ and $x = L$ (see Fig. 3.64).

Fig. 3.63

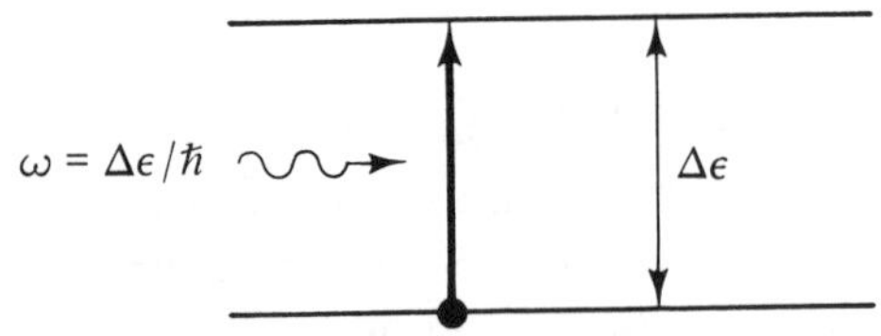

$\omega = \Delta\epsilon/\hbar$ $\Delta\epsilon$

The second element needed for the bistable effect, is for some of the radiation which exists in the medium, $E(L)$, to be fed back and become part of the radiation entering the

Fig. 3.64

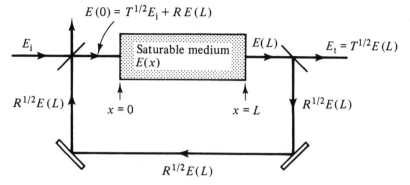

$$E(0) = T^{1/2}E_i + RE(L)$$

E_i

Saturable medium
$E(x)$

$E(L)$ $E_t = T^{1/2}E(L)$

$R^{1/2}E(L)$ $x = 0$ $x = L$ $R^{1/2}E(L)$

$R^{1/2}E(L)$

medium, $E(0)$. This is accomplished with several mirrors, as illustrated in Fig. 3.64. What is experimentally controlled is the incoming monochromatic field, E_i, incident upon a mirror of reflectivity R and transmissivity $T = 1 - R$. Thus the transmitted field through this mirror is $E_t = T^{1/2}E_i$. (Note: the square root comes from the fact that T and R refer to the fraction of $|E|^2$ transmitted and reflected.) The field exiting the medium, $E(L)$, is partially transmitted through an identical second mirror, so that the transmitted field is $E_t = T^{1/2}E(L)$, whereas the reflected field is $R^{1/2}E(L)$. This reflected field is fed back, with the help of two perfectly reflecting mirrors, to add to the field $T^{1/2}E_i$ at the left (incident) mirror (see diagram). At this mirror, the amount of the returned field, $R^{1/2}E(L)$, which is reflected is $R^{1/2}(R^{1/2}E(L)) = RE(L)$. Hence the field incident upon the saturable medium is $E(0) = T^{1/2}E_i + RE(L)$. (Note that the return path length is such that these two waves are in phase, and hence additive.) Therefore the fields at $x = 0$ and $x = L$ are related by the (given) incident field E_i.

We next introduce the dimensionless fields

$$\varepsilon(x) = E(x)/I_s^{1/2}, \quad \varepsilon_t = E_t/(TI_s)^{1/2}, \quad \varepsilon_i = E_i/(TI_s)^{1/2} \qquad (3.6.3)$$

Then the above discussion shows that

$$\varepsilon(0) = T\varepsilon_i + R\varepsilon(L); \quad \varepsilon_t = \varepsilon(L) \qquad (3.6.4)$$

and, from (3.6.1), (3.6.2) and (3.6.3), $\varepsilon(x)$ satisfies

$$\frac{d\varepsilon}{dx} = -\alpha \frac{\varepsilon}{1 + \varepsilon^2}. \qquad (3.6.5)$$

Thus

$$\left(\frac{1}{\varepsilon} + \varepsilon \right) d\varepsilon = -\alpha\, dx,$$

and using the boundary condition for $\varepsilon(0)$, (3.6.4), this yields the solution of (3.6.5)

$$\ln(\varepsilon) + \tfrac{1}{2}\varepsilon^2 = -\alpha x + \ln(T\varepsilon_i + R\varepsilon_t) + \tfrac{1}{2}(T\varepsilon_i + R\varepsilon_t)^2.$$

Finally, setting $x = L$ and using $\varepsilon(L) = \varepsilon_t$, we obtain the equation for the transmitted radiation

$$\alpha L = \ln\left(T\frac{\varepsilon_i}{\varepsilon_t} + R \right) + \tfrac{1}{2}(T\varepsilon_i + R\varepsilon_t)^2 - \tfrac{1}{2}\varepsilon_t^2, \tag{3.6.6}$$

and $R = 1 - T$.

There are three control parameters for this system, αL, T, and ε_i, which determine the transmitted field, ε_t, using (3.6.6). We can simplify the analysis by considering the limit of small transmissivity, $T \to 0$. This limit simplifies the logarithmic term in (3.6.6), and thereby reduces the number of control parameters. In this limit we have

$$\ln\left(T\frac{\varepsilon_i}{\varepsilon_t} + R \right) \equiv \ln\left(1 + T\left(\frac{\varepsilon_i}{\varepsilon_t} - 1\right) \right) = T\left(\frac{\varepsilon_i}{\varepsilon_t} - 1\right) + O(T^2)$$

and (3.6.6) reduces to

$$\frac{\alpha L}{T} = \left(\frac{\varepsilon_i}{\varepsilon_t} - 1\right) + \varepsilon_t(\varepsilon_i - \varepsilon_t) + O(T). \tag{3.6.7}$$

Thus we have a new control parameter,

$$A \equiv \alpha L/T \tag{3.6.8}$$

Fig. 3.65

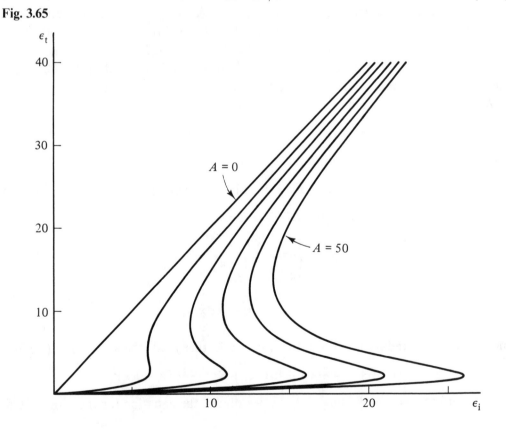

which replaces αL and T. If we ignore terms of $O(T)$, (3.6.7) yields

$$\varepsilon_i = \varepsilon_t\left[1 + \frac{A}{(1 + \varepsilon_t^2)}\right].\qquad(3.6.9)$$

This is a cubic equation for $\varepsilon_t(A, \varepsilon_i)$, in terms of A, and ε_i. There are three real roots for ε_t provided that A is larger than a critical value, A_c.

The curves $\varepsilon_t(\varepsilon_i)$, for various values of A are illustrated in Fig. 3.65. It is easy to show from (3.6.9), that $\partial\varepsilon_i/\partial\varepsilon_t = 0$ when

$$\varepsilon_t^2 = \tfrac{1}{2}(A - 2) \pm \tfrac{1}{2}(A^2 - 8A)^{1/2}.\qquad(3.6.10)$$

If $A > A_c = 8$, then (3.6.10) has two real roots for ε_t, where $\partial\varepsilon_i/\partial\varepsilon_t = 0$. Thus, when $A > A_c$ there are three real roots of (3.6.9) for ε_t, for some range of values of ε_i. A more complete (dynamic) analysis can establish that the central root represents a state which is unstable to perturbations, whereas the outer two states are stable. Hence the terminology 'bistability'.

Since there are two control parameters (ε_i, A) determining ε_t, it is more symmetric and revealing to plot the surface $\varepsilon_t(\varepsilon_i, A)$, given by (3.6.9), over the (ε_i, A) plane. This yields the cusp catastrophe surface, illustrated in Fig. 3.66. The projection onto the (ε_i, A) plane of the singularities of this surface, $|\partial\varepsilon_t/\partial\varepsilon_i| = \infty$, are obtained by substituting

Fig. 3.66

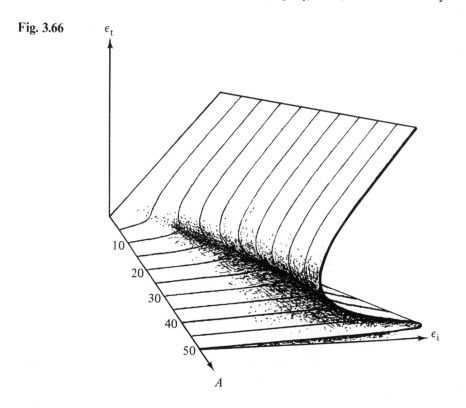

(3.6.10) into (3.6.9). The resulting catastrophe set is illustrated in the lower diagram.

This analysis, of course can be refined by using (3.6.6) for finite T, rather than the approximation (3.6.7). However, this simpler analysis illustrates essential features (Fig. 3.67).

Fig. 3.67

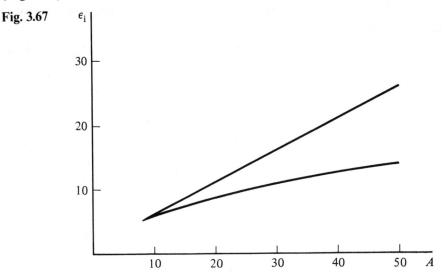

As we will see in Chapter 5, the existence of this catastrophe form of bistability in a system, affords the possibility of very dramatic dynamical effects once we make one of the control parameters time dependent. It indeed does not require any great insight to anticipate that if, for example, $\varepsilon_i(t)$ is made 'appropriately' periodic, some strange dynamics might result (Fig. 3.68). We will return to these details in Chapter 5.

Fig. 3.68

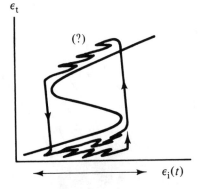

3.7 Some of René Thom's perspectives

Reading René Thom is a unique experience – one which I recommend for its visions, if not for its precision. As Zeeman (1977) put it: '...I must confess that I often find his [Thoms's] writing obscure and difficult to understand, and occasionally I have to fill in

99 lines of my own between each 2 lines of his before I am convinced.' Arnold (1984) likewise noted: 'A particular aspect of Thom's work on catastrophe theory is his original style: he established a fashion in not giving even sketchy formulations of results, let alone proofs.'

Nonetheless, here are a few of his more perspicuous quotes, which give some flavor of his perceptions, motivations, and scientific imagination!

> 'from a macroscopic examination of the morphogenesis of a process and local and global study of its singularities, we can try to reconstruct the dynamic that generates it.'

> '... the intuitive notion is that even bifurcations and catastrophes may occur in a structurally stable way, according to a fixed algebraic model.'

> '... a deterministic system may exhibit, in a 'structurally stable way', a complete indeterminancy in the qualitative prediction of the final outcome of its evolution.'

Fig. 3.69

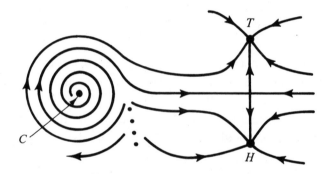

A deterministic, structurally stable, 'coin toss' is illustrated in Fig. 3.69. Starting from neighborhood of C [coin toss], no matter how small, there is equal chance of ending up at Heads or Tails. Actually this is not a good model of a real coin toss, which is much more controllable.

> '... we can at least ask one question: many phenomena of common experience, in themselves trivial... – for example, the cracks in an old wall, the shape of a cloud, the path of a falling leaf, or the froth on a pint of beer – are very difficult to formalize, but is it not possible that a mathematical theory launched for such homely phenomena might, in the end, be more profitable for science?' [Some of these examples remind one of Mandelbrot's studies of fractals.]

> '1. Every object or physical form, can be represented as an attractor C of a dynamical system on a space M of internal variables.
> 2. Such an object is stable, and so can be recognized, only when the corresponding attractor is structurally stable.

3. All creation or destruction of forms, or morphogenesis, can be described by the disappearance of the attractors representing the initial forms, and their replacement by capture by the attractors representing the final forms. This process, called catastrophe, can be described on a space P of external variables.
4. Every structurally stable morphological process is described by a structurally stable catastrophe, or a system of structurally stable catastrophes on P.
5. Every natural process decomposes into structurally stable islands, the chreods. The set of chreods and the multi-dimensional syntax controlling their positions constitute the semantic model.'
(We didn't get to the semantic model – for this you have to read Thom!)

'Are these models subject to experimental control? Can they give experimentally verifiable predictions? At the risk of disappointing the reader, I must answer in the negative: this is an inherent defect of all qualitative models, as compared with classical quantitative models', (but)'... there are two reasons that might commend them to the scientist... every quantitative model first requires a qualitative isolation from reality... (and) our ignorance of the limits of quantitative models.'

'So what I am offering here is not a scientific theory, but rather a method; the first step in the construction of a model is to describe the dynamical models compatible with an empirically given morphology, and this is also the first step in understanding the phenomena under consideration. It is from this point of view that these methods, too indeterminate in themselves, lead not to a once-and-for-all explicit standard technique, but rather to an art of the models.'

'Physicists, if they want one day to obtain information about very small processes at subquantic level, will need the intermediary of the interaction of a highly controlled process with an enormous degree of amplification. Such processes, in which an infinitesimal perturbation may cause very large variations in the outcome, are typically catastrophes.'

'A mathematician cannot enter on subjects that seem so far removed from his usual preoccupations without some bad conscience. Many of my assertions depend on pure speculation and may be treated as day-dreams, and I accept this qualification – is not day-dream the virtual catastrophe in which knowledge is initiated? At a time when so many scholars in the world are calculating, is it not desirable that some, who can, dream?'

Comments on exercises

(3.1) (3.1.3) has two independent solutions, u_1, u_2, and these are related to solutions of (3.1.2b) by $x_i = -(1/a_2)\,\mathrm{d}(\ln u_i)/\mathrm{d}t$. Then $u_i = \exp\{-\int^t a_2(t')x_i(t')\,\mathrm{d}t'\}$. The general solution of (3.1.3) is $u = \alpha u_1 + \beta u_2$ and this must yield the general solution of (3.1.2b) through the relation $x = -a^{-1}\,d(\ln u)/dt = -a_2^{-1}(t)(d/dt)\ln[\alpha\exp(-\int^t a_2(t')x_1(t')\,dt') + \beta\exp(-\int^t a_2(t')x_2^2(t')\,dt')]$.

(3.2) If $s^2 \equiv a_1^2 - 4a_0 a_2 \neq 0$ there are either two fixed points ($s^2 > 0$) or none ($s^2 < 0$) on the real axis. The general solution of the Riccati equation is (using Exercise 3.1.1) $x = -1/2a_2[\alpha(a_1 + s)\exp(st) + \beta(a_1 - s)\exp(-st)]/(\exp(st) + \beta\exp(-st))$. If $s = i\omega$, then we must take $\alpha = \beta^*$. If $s^2 > 0$, the solution either tends to a fixed point or infinity in a finite time ($< 2\pi/\omega$). If $s = 0$, the fixed point is degenerate, $\dot{x} = a_2(x + a_1/2a_2)^2$, so $s = -a_1/2a_2 - \beta/a_2(\alpha + \beta t)$. One solution goes to the fixed point in an infinite time, the other to infinity in an infinite time (both very slowly, $\sim 1/t$). [Note: The Riccati equation $dz/dx = E - V(x) - z^2$ is equivalent to the time independent Schrödinger equation for a particle in the potential $V(x)$, $d^2\psi/dx^2 + V(x)\psi = E\psi$, where $z = (d\psi/dx)/\psi$.]

(3.3) Periodic solution: $\dot{r} = 0$. The equation for $\dot{r}$ does not satisfy the Lipschitz condition where $\dot{r} = 0$. This corresponds to the inside envelope of the tangents to the family (same E) of backscattered orbits (Fig. 3.70). Note that the members of

Fig. 3.70

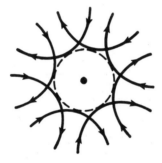

this family only differ in an ignorable coordinate (θ), hence this 'caustic' shows up as a solution. If $a < 0$, there are two roots for $\dot{r} = 0$, $r = (a/2E) \pm (1/2E)[a^2 + (2M^2/m)E]^{1/2}$ if $E < 0$. These are the apsidal distances of the elliptic orbits. At these distances, $\ddot{r} \neq 0$, so they are not physical solutions, unless they are identical. How about other $V(r) = ar^{-n}$?

(3.4) $F(x, c) = 0$ at $x = 0$ or $c + x - x^2 + \frac{1}{3}x^3 = 0$. If $x = 0$, $\partial F/\partial x = 0$ only if $c = 0$, so this is a possible bifurcation point. Moreover $(\partial^2 F/\partial x^2)_0 = 2$ so this is a bifurcation point. In the second case above, we use $\partial F/\partial x = c + 2x - 3x^2 + \frac{4}{3}x^3 = 0$ and eliminating c yields $x - 2x^2 + x^3 = 0$. If ($x \neq 0$) then $x = 1$, and so $c = \frac{1}{3}$ is another possibility. Moreover $(\partial^2 F/\partial x^2)_0 = (2 - 6x + 4x^2)_{x=1} = 0$, and we are back to an inflection point.

(3.5) $F = 0$ and $\partial F/\partial x = 0$ if $(\frac{1}{2}c_2)^2 - \frac{1}{2}c_2^2 - c_1 = 0$, or $c_1 = -\frac{1}{4}c_2^2$. This has a structure shown in Fig. 3.71.

Fig. 3.71

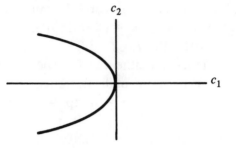

(3.6) The added terms are smaller than x^2 near $x = 0$, hence do not modify the results of (3.2.2) in this region. Globally, however, something new has (possibly) been added. In the case (a) $\partial F/\partial x = (4x^2 + 2)x$ which can only vanish at $x = 0$, so there is nothing new here globally. In the case (b), $\partial F/\partial x = (3x + 2)x$, so $x = -\frac{2}{3}$ and $c = \frac{4}{27}$ is a possible new bifurcation point, and is so; $(\partial^2 F/\partial x^2) = -2$.

(3.7) A block which slides down a frictional inclined surface, with a height $y(x) = A - x(x^2/3 - c)$, so its potential energy is $mgy(x)$ (Fig. 3.72). This has a local maximum if $C > 0$.

Fig. 3.72

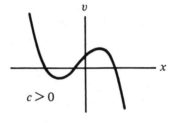

$c > 0$

Fig. 3.73

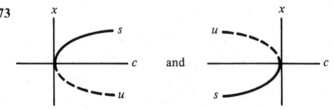

and

(3.8) To determine the stability, look at F vs. x (see Fig. 3.73).

(3.9) $F(x, c) = 0$ cannot have three roots, for a given c, such that $(\partial F/\partial x)$ has the same sign at all of these roots. Hence they cannot all have the same stability property. Note also the example following this exercise.

(3.10) (A) → (c) for $c \simeq 0$, and also (a) near $c = 1$;
     (B) → (e) for c near zero, and also (b) near $c = -1$;
     (C) → (b);
     (D) → (f).

(3.11) Solution set: $-x = c$ and $x = 1 \pm (1-c)^{1/2}$; $\partial F/\partial x = 0$ at $3x^2 + 2(c-2)x - c = 0$. Bifurcation points $(x = 0, c = 0)$, $(x = 1, c = 1)$ and $(x = 3, c = -3)$ (Fig. 3.74).

Fig. 3.74

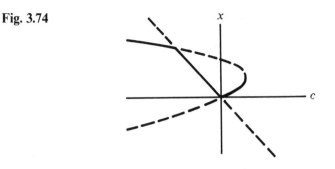

(3.12)

(3.13) Then there are two fixed points $x = 0$, $x = c/a$, and $\partial F/\partial x = 0$ at these fixed points

Fig. 3.75

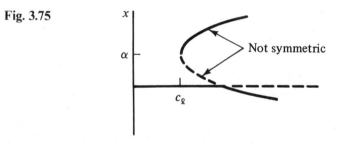

Not symmetric

only if $c = 0$. Hence the bifurcation point is $(x_0, c_0) = (0, 0)$ (Fig. 3.75).

$c < 0$ $x = 0$: attractor; $x = c/a$: repeller

$c > 0$ $x = 0$: repeller; $x = c/a$: attractor

Note: these two portraits are TOE, but at $c = 0$ the flow is nonequivalent. This is also true of the cases (a), (b), (f), and (g) in the Fig. 3.16. The constant a does not produce a bifurcation since it can be scaled out of the equation (set $\bar{x} = ax$) (Fig. 3.14).

Fig. 3.76

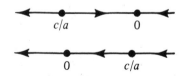

(3.14) There are an infinite number of bifurcation points, $(x = n, c = -A)$, $n = 0, \pm 2$, $\pm 4\ldots$, and $(x = m, c = A)$, $m = \pm 1, \pm 3, \ldots$. $c = qE_0/\mu$, $A = qE_p/\mu$ (Fig. 3.77).

(3.15) The fixed point at $x = 0$ breaks into two fixed points when (for example) any $\varepsilon x \exp(-x^2)$ is added to the right side; so that $\dot{x} \simeq \varepsilon x + cx^2$ (near $x = 0$). Thus the origin is the structurally unstable region of R^1.

Fig. 3.77

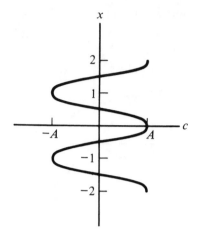

(3.16) It is not structurally stable. The addition of any new function on the right causes the closed orbits to become spiral inward or outward. Try it.

Fig. 3.78

(3.17) (a) Two fixed points (say north and south poles) with a spiral out of one and into the other; (b) and (c) orbits which spiral around the torus but do not close.

(3.18) $V(p, q) = H(p, q)$. Then $F = -\nabla V$ gives a vector field orthogonal to $(-\partial H/\partial q, \partial H/\partial p)$. That is, $(\partial V/\partial p, \partial V/\partial p)\cdot(-\partial H/\partial q, \partial H/\partial p) \equiv 0$.

(3.19) Try $h_1 = ax_1$, $h_2 = x_2 + bx_1$; show that $\alpha = 0$, $a^2 = (1 - (\varepsilon/2^2)^{1/2}$, and find b.

(3.20) The shift $x = \bar{x} + x_0$ eliminates the term $\bar{x}^2$ from (3.4.4) if $3x_0 + c_2 = 0$, in which case $c \equiv c_1 - \frac{1}{3}c_2^2$. The linear term in $\bar{x}$ cannot be eliminated by a real x_0 if $3c_1 > c_2^2$.

(3.21) (3.3.8) yields $5x^4 + 3c_3x^2 + 2c_2x + c_1 = 0$ and $10x^3 + 3c_3x + c_2 = 0$. This yields $3c_3x^2 + 3c_2x + 2c_1 = 0$, and using this to eliminate x^3 from the previous equation produces $30c_2x^2 + (20c_1 - 9c_3^2)x - 3c_2c_3 = 0$. We can proceed to

Fig. 3.79

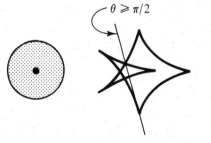

eliminate x^2, and finally obtain $x(c_1, c_2, c_3)$ from $(20c_1c_3 - 9c_3^3 - 30c_2^2)x - 3c_2c_3^2$ $- 20c_1c_2 = 0$. This x can then be eliminated from the simplest quadratic above to give $G(c_1, c_2, c_3) = 0$.

(3.22) Cusp points; $x(0) = [1 - (l_0/(d + r))]d$, $x(\pi) = [1 - (l_0/(d - r))]d$; note what happens if $l_0 = 0$. These cusps are easy because (3.5.5) is trivially satisfied. This is not true for the other cusps.

(3.23) The influence of the angular separation, ϕ, is greatest near the wheel (Fig. 3.79).

(3.24) $\phi_c(d/r = 3) \simeq 0.455$.

4

Models based on first order difference equations

4.1 General considerations

In this chapter we will consider some of the new dynamic features that occur in models based on first order difference equations

$$\Delta x \equiv x_{n+1} - x_n = G(x_n, c),$$

which can be put in the form

$$x_{n+1} = F(x_n, c) \tag{4.1.1}$$

where, again, c are control parameters ($c \in R^k$). In this form, the dynamics can be equally well viewed as a sequence of *mappings*, $F: x_n \rightarrow x_{n+1}$.

The function $F(x, c)$ in (4.1.1) is single-valued, which is required of all maps by definition, but may otherwise have a variety of properties all of which represent quite different dynamical behavior, and are related to very different physical systems. Most commonly we take $F(x, c)$ to be a continuous function of x, based on the physical assumption that 'neighboring states' (represented by contiguous points, x) should have essentially the same change in one 'step', and hence essentially the same value of $F(x, c)$. This is not always the case (consider what happens to physical states near catastrophes), but it is the most common case (Fig. 4.1).

Fig. 4.1

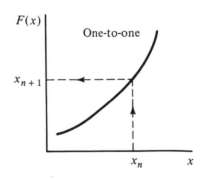

A more common difference which occurs between the dynamics of systems is whether or not they have a 'conservative' or 'dissipative' character to their motion. A typical class of conservative differential systems are Hamiltonian systems, in which the volume elements in the phase space move in such a way as to 'conserve' their volume (the Liouville theorem). Many other systems can also be conservative, as discussed in Chapter 2, depending on the existence of other conserved (invariant) measures $d\mu = \rho(x)\,dx > 0$. The analogous systems in the case of maps involve functions $F(x, c)$, which 'keep track' of each point in the phase space, by mapping each point to a unique point – in other words, a one-to-one map together with the continuity requirements. Such a map is called a *homeomorphism* (appendix A). While conservative systems must be homeomorphisms, the converse is not always true, as will be illustrated shortly.

The second main category of maps involves those which map two or more phase points onto another phase point – that is, a *many-to-one-map* (Fig. 4.2). It is clear that

Fig. 4.2

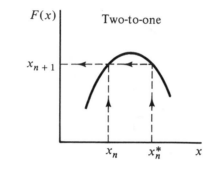

this has a 'compressional' effect on volume elements in the phase space and is, in that sense, 'dissipative' in character. In such systems many states tend to 'settle down' to some select group of states. An example which immediately comes to mind is a pendulum with friction, so that nearly all initial states finally end up at rest, with the pendulum hanging down (except for the single inverted initial state, $\theta = \pi$!). More interestingly, this 'attracting' set of states in the phase space need not simply be at rest (fixed points), but may be a dynamic set. Thus 'dissipation' in the present sense does not necessarily imply that all states 'die', but rather that there is an attraction toward some final, generally dynamic, set of states, called an *attractor*.

The dynamics of equation (4.1.1) is much richer than a first order ODE because it is free from the continuity restrictions of differential equations. x_n can 'jump around' on the real axis, provided that the map is many-to-one, whereas $x(t)$ can only pass a point once, if $\dot{x} = F(x)$ (Fig. 4.3). This freedom makes it possible for (4.1.1) to exhibit several

Fig. 4.3

Differential motion ; Difference motion

interesting types of bifurcation sequences, leading to various forms of coherent and 'chaotic' behavior, depending on the magnitude of a control parameter c. To obtain a comparable degree of dynamic richness from physically motivated differential equations requires at least a third order ODE system (see Section 10).

One origin of one-dimensional maps, but certainly not the only possibility (see the logistic map below), is that they are generated by a Poincaré map of a flow in a limited region of a two-dimensional manifold. For example, consider a flow on an annulus, $1 \leqslant x^2 + y^2 \leqslant 4$, with the inside points rotating outward, as illustrated. If we take the surface $(y = 0, \dot{y} < 0)$ in the annulus as a Poincaré surface, the map for successive x_n points might look as illustrated in Fig. 4.4. The points $x = 1, 2$ are fixed points of the

Fig. 4.4

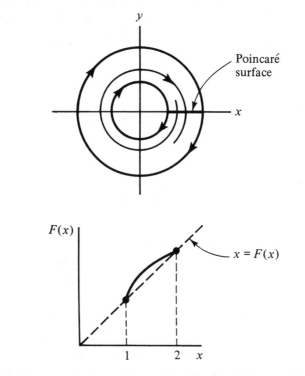

map, corresponding to periodic orbits bounding the annulus. Note that $F(x) > x$, except at the boundaries, so that $x_{n+1} > x_n$, which is the case for outward motion. Also note that, although this map is a homeomorphism, the only 'conserved' (invariant) set of points are the points $x = 1$ or 2. Thus there is no function $\rho(x) > 0$, for all $1 \geqslant x \geqslant 2$, which remains constant under this map, so the system is not conservative in this sense.

Exercise 4.1.

(a) What flow on the above annulus could yield the Poincaré map shown in Fig. 4.5(a)?

(b) What flow, and on what two-dimensional manifold, could yield the Poincaré map in Fig. 4.5(b)?

Fig. 4.5

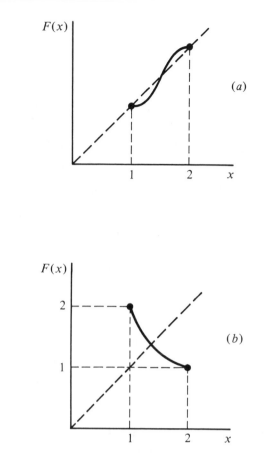

However, the Poincaré map of autonomous second order systems is of limited interest because such maps must be one-to-one, due to the uniqueness of flows in the R^2 phase space. To the contrary, as we will see, a series of bifurcations which leads to 'chaotic' behavior depends on the nonuniqueness of the inverse of the map (4.1.1). In other words the occurrence of chaotic motion depends on the fact that at least two different initial points x_n, x'_n can map to the same point x_{n+1}. Such many-to-one maps can only occur from some '*projection*' process (see Chapter 1) which reduces a higher order system with unique solutions to a lower order model with nonunique dynamics. This is schematically illustrated in Fig. 4.6, where the unique dynamics in three dimensions is projected onto nonunique dynamics in the (x, y) plane. However, the 'map' of this dynamics along the dashed line need not be of the form (4.1.1). In other words, there is no basic reason for there to be any causal relationship between the values x_n and x_{n+1}, so the function $F(x, c)$ in (4.1.1) need not exist. Put another way, there is generally no reason for a variable, x, to be exactly *self-deterministc*. To take a familiar example: the change in the area of the shadows of tree leaves on the ground is neither unique nor self-deterministic, even for deterministic motion of the leaves. Therefore, usually such maps

Fig. 4.6

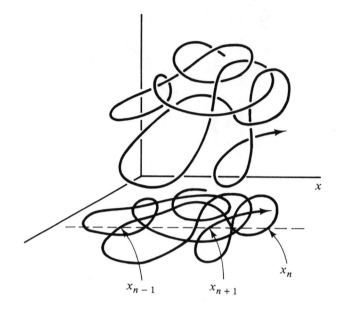

(4.1.1) are rather crude approximations of projections, based on loose arguments and optimistic expectations. This is particularly obvious when the low order maps are intended to represent such complicated systems as population dynamics in biological systems, various economic market dynamics, neurological networks, or even such relatively simple systems as periodically driven nonlinear oscillators.

The fact that complicated systems can not always be reduced to a few interrelated variables is well-illustrated by Fig. 4.7 (after Gardner, 1981, 1986) from the field of economics. The curve on the left (The Laffer Curve) purports to relate the taxation rate to the government's revenue. If there is too much taxation, people will not work as hard for a salary (they will look for other non-cash benefits), so the revenue will drop. Some supply-side economists believe that this curve accurately represents a relationship between taxation and revenue. However, the 'technosnarl' which occurs in the 'neo-Laffer' curve on the right appears to represent much more closely the best available data for the U.S. economy over the past 50 years. Of course, even the neo-Laffer curve is metaphorical. The point here is that there is no map Tax Rate → Government Revenue (except near the trivial end points), because there is no causal connection. Human nature is too complicated for such a connection! What is indeed fortunate, and often amazing, is that some complicated systems appear to have dynamics which can be approximated by simple first order models. While various 'generic' arguments are frequently proposed, we often do not have any profound physical understanding of this good fortune. These points will be discussed in more detail later.

We will find in the following sections that, in addition to the occurrence of various

Fig. 4.7

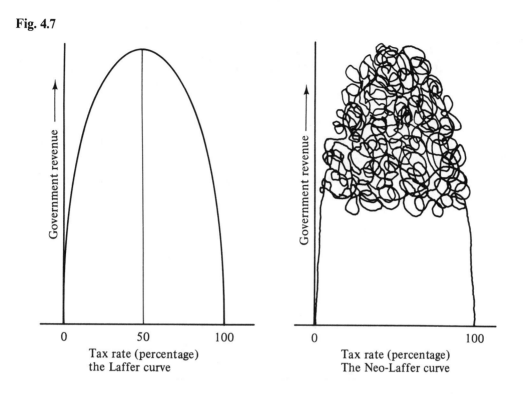

Tax rate (percentage)
the Laffer curve

Tax rate (percentage)
The Neo-Laffer curve

forms of coherent and chaotic motions, other very important results have been discovered about the dynamics of (4.1.1). One result involves asymptotic ($n \to \infty$) quantitative results (numbers!) which depend only on certain general qualitative features of the function $F(x, c)$. Put another way, quantitative values can also have a form of *structural stability*, since these values remain unchanged when $F(x, c)$ is smoothly varied. One group of asymptotic quantitative features, which has been identified for a large class of functions, $F(x, c)$, has been called *'quantitative universality'* by the discoverer, Feigenbaum (1978, 1979).

This discovery was historically preceded by the observation of *qualitative 'universal sequences'* generated by an even larger class of functions, $F(x, c)$, and noted by Metropolis, Stein and Stein (1973). They observed that a large class of maps, $F(x, c)$, generate similar qualitative dynamic patterns. An example of such a qualitative pattern is obtained by following the map of the special point x_m, where max $F(x, c) = F(x_m, c)$. A pattern of Ls and Rs is then generated by the mapping sequence of x_m to the L (left) and R (right) of the maximum of $F(x, c)$. Moreover, these qualitative (L, R) patterns changed (bifurcated) in a common 'universal sequence' for a wide class of $F(x, c)$, as the control parameter c is increased. These 'universal' features', which have also been observed in a number of experimental situations, will be discussed in Section 3.

4.2 Two-to-one maps: the logistic map

A classic example of (4.1.1) is the so-called *logistic equation*, the discrete form of the relatively unexciting logistic differential equation noted before. The logistic map is

$$x_{n+1} = cx_n(1 - x_n) \quad (0 \leqslant x_n \leqslant 1, 0 < c \leqslant 4). \tag{4.2.1}$$

In biological examples of such a map, x_n represents the population of some species in the n'th generation, normalized by the maximum population which the species can achieve due to mutual toxicity, or to competition for nutrients.

The essential feature of the present class of models, of which (4.2.1) is an example, is that $F(x, c)$ has the (x, c) dependence indicated in Fig. 4.8. More specifically, $F(x)$ has

Fig. 4.8

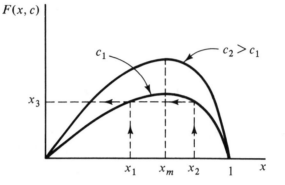

only one maximum $F(x_m) \leqslant 1$ and vanishes at the end points $F(0) = F(1) = 0$. Thus this generally maps the interval $[0, 1]$ into a smaller region, unless $F(x_m) = 1$. $F(x)$ also maps two points into one point (e.g., x_1 and x_2 map into x_3), so it is a two-to-one map. If $F(x)$ has certain properties of smoothness (differentiability, etc.), then the variety of dynamics is correspondingly limited.

One important class of maps involves maps of an interval into itself, $F:[0, 1]$, where $F(x)$ has only one maximum, x_m, and has everywhere a negative '*Schwarzian derivative*' (see Appendix D). This requires that $|\partial F/\partial x|^{-1/2}$ has a positive second derivative on

Fig. 4.9

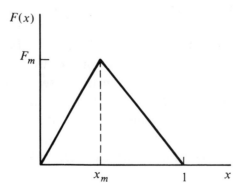

both sides of the maximum (not in intuitive condition!). We will call maps with everywhere negative Schwarzian derivatives *S-maps*. In this case, Singer (1978) has proved that, for each value of c, there is at most one stable periodic solution of (4.1.1) and, if it exists, the iterates of x_m tend toward this periodic solution (i.e., x_m is in the 'basin of attraction' of the periodic solution). The logistic map, (4.2.1), is an example of a S-map (Appendix D), and Singer's result will simplify our analysis of this dynamics.

There are other two-to-one maps (Fig. 4.9) which are frequently considered, and which are not S-maps. An important example is the so-called 'tent map' (sometimes a leaning tent!

$$F(x) = a(1 - x_m)x \quad (x \leqslant x_m),$$

$$F(x) = ax_m(1 - x) \quad (x_m \leqslant x \leqslant 1), \tag{4.2.2}$$

and $ax_m(1 - x_m) \leqslant 1$. This has a discontinuous derivative at $x = x_m$, and is not an S-map. It may have more than one stable periodic solution. Guckenheimer (1979) has obtained the conditions for which (4.2.1) is topologically conjugate to some tent map (4.2.2) (see Exercise 4.8).

Before considering any details it should be emphasized that (4.2.1) is a mapping of the interval $0 \leqslant x \leqslant 1$ into itself, which is two-to-one, and this gives rise to many of the interesting features noted in Section 1. As discussed in that section, while the mapping dynamics is deterministic, it is not unique because it is a two-to-one mapping. Hopefully the example of the biological origin of the logistic map makes it reasonable that both determinism and non-uniqueness may occur (approximately) in some observables, even if there is an underlying unique dynamics in all the variables. It is the self-determinism of this nonunique dynamics which is the most profound aspect of this 'projection'. Generally, of course, this simple one-dimensional determinism does not occur, as has been pointed out in Section 1.

We will now consider some of the properties of the general two-to-one maps frequently illustrating these properties with the logistic map (4.2.1). For a very nice (now classic) introductory article on this subject, see May (1976). We first introduce the notation for a sequence of mappings of some point x,

$$F^0(x) = x, \quad F^{n+1}(x) = F(F^n(x)). \tag{4.2.3}$$

A point is said to be a *fixed point* of a map if

$$F(x) = x \quad \text{(fixed point)}$$

and it is called a *period-n point* $(n = 0, 1, 2, \ldots)$, if

$$F^n(x) = x \quad \text{and} \quad F^k(x) \neq x \quad \text{(for } k < n). \tag{4.2.4}$$

A fixed point is simply a period-one point. If there is one period-n point, then there are n of them.

Exercise 4.2. Prove the last statement.

The sequence of values of x_n which are generated by the map can be obtained in a simple fashion by the following graphical construction. On a graph of $F(x)$ vs. x we also plot the straight line $x_{n+1} = x_n$. Starting at some point x_0, a vertical path is taken to obtain $F(x_0)$, and then a horizontal until it intersects the straight line. This intersection point occurs at x_1. The second map (of x_0) is then obtained by repeating this procedure, beginning at x_1, which then yields x_2. This is illustrated in Fig. 4.10.

Fig. 4.10 $x_{n+1} = F(x_n)$

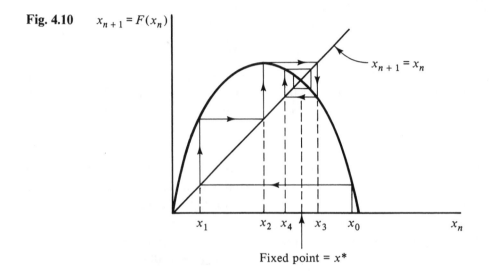

Fixed point $= x^*$

This figure also illustrates a sequence of maps converging to a fixed point, x^*. Note that all *fixed points* of a map are graphically represented by the intersection of the straight line with the function $F(x)$.

Exercise 4.3. This method can be used to establish a special case of *Brouwer's fixed point theorem*, which holds in R^n. Prove that, for any continuous map, $F(x)$, which takes the interval $[0, 1]$ into itself (so $0 \leqslant F(x) \leqslant 1$) there must be at least one fixed point $x^* = F(x^*)$.

In the above illustration the fixed point is stable, since nearby points converge towards it. A fixed point, x^*, is stable if

$$|dF(x)/dx|_{x^*} < 1 \quad \text{(stable)}. \tag{4.2.5}$$

This can be shown by taking an initial point nearby, $x_0 = x^* + \Delta$, and expanding $F(x_0)$ for small Δ,

$$F(x^* + \Delta) = F(x^*) + \Delta \quad (dF/dx)_{x^*} = x^* + \Delta \quad (dF/dx)_{x^*} = x_1,$$

where we have assumed that $(dF/dx)_{x^*} \neq 0$ (see superstable fixed points, below). Therefore the distance to the fixed point is decreased by the map, $|x_1 - x^*| < |x_0 - x^*|$, if (4.2.5) is satisfied. In the case of the logistic map (4.2.1), the fixed point is given by

$$x^* = cx^*(1 - x^*) \quad \text{or} \quad x^* = 1 - c^{-1} \quad (\text{if } c > 1), \tag{4.2.6}$$

ignoring the trivial fixed point $x^* = 0$, which is the only one if $c < 1$. Using the condition (4.2.5), we find that the fixed point (4.2.6) is stable provided that

$$|c(1 - 2x^*)| < 1 \quad \text{or} \quad 1 < c < 3 \equiv c_1. \tag{4.2.7}$$

A curious feature of maps is that there can be points which are *eventually periodic*. The non-periodic point x_e is said to be eventually periodic if $x = F^m(x_e)$ is a periodic point for some finite m. These points can exist because $F(x)$ is not one-to-one. For example, for the logistic map, with $c = 3$, $x_e = \frac{1}{3}$ is eventually periodic when $m = 1$, and $x_e = \frac{1}{2} \pm (5^{1/2}/6)$ are both eventually periodic for $m = 2$.

Exercise 4.4. What are the eventually periodic points for the logistic map ($c = 3$) with $m = 3$? Beware of complex numbers!

Exercise 4.5. Another strange feature of these maps is that there are an infinite number of points that map into an unstable fixed point (try to show this using the inverse of the above graphical method).

As c is increased past the value $c_1 = 3$ the fixed point becomes unstable and points are then mapped towards two other points, x_1^* and x_2^*, as shown in Fig. 4.11(*a*). These two

Fig. 4.11

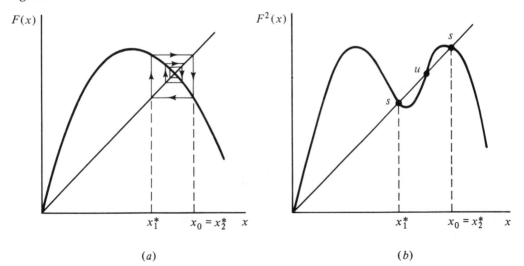

(*a*)   (*b*)

points are period-two points of $F(x)$, that is

$$F(x_1^*) = x_2^* \quad \text{and} \quad F(x_2^*) = x_1^*.$$

This means that after two maps their values are repeated, or

$$F^2(x_k^*) = F(F(x_k^*)) = x_k^* \quad (k = 1, 2). \tag{4.2.8}$$

Therefore, when c passes through the value c_1, the asymptotic limit $(n \to \infty)$ of most points changes from a period-one state $(\lim_{n \to \infty} x_n = x^*)$ to a period-two state $(\lim_{n \to \infty} x_n \in \{x_1^*, x_2^*\})$. This topological change in the asymptotic states therefore represents a *bifurcation* at $c = c_1$ (extending the concept of bifurcations to maps in a natural way). It can be seen from (4.2.8) that period-two points can be viewed as fixed points of the map $F^2(x)$. The graphical representation of this is shown in Fig. 4.11(*b*). All of the intersections of $F^2(x)$ with the straight line are fixed points, but the central point is unstable, whereas the outer fixed points are stable provided that

$$|dF^2(x)/dx|_{x_k^*} < 1 \quad (k = 1, 2). \tag{4.2.9}$$

This follows in the same way that (4.2.5) was established.

It is important now to establish the fact that if one of these fixed points goes unstable when c is increased, then the other fixed point also goes unstable. Indeed, what we will show is that the derivative of F^2 is the same at both fixed points,

$$(dF^2(x)/dx)_{x_1^*} = (dF^2(x)/dx)_{x_2^*} \tag{4.2.10}$$

To show this we simply need to use the chain rule of differentiation. Consider any point x_0. Then

$$F(x_0) = x_1 \quad \text{and} \quad F(x_1) = x_2, \quad \text{or} \quad F^2(x_0) = x_2.$$

Therefore

$$(dF^2/dx_0) = (dx_2/dx_0) = (dF(x_1)/dx_1)(dx_1/dx_0)$$
$$= (dF(x_1)/dx_1)(dF(x_0)/dx_0)$$

It is obvious that this can be generalized for any n to read

$$dF^n(x_0)/dx_0 = \prod_{k=0}^{n-1} dF(x_k)/dx_k. \tag{4.2.11}$$

Now we specialize this result to the case where the set of points $\{x_0, x_1, \ldots, x_{n-1}\}$ are period-n points of $F(x)$ or, in other words, fixed points of $F^n(x)$. A little thought makes it clear that, regardless of which fixed point is used in the left side of (4.2.11) (the one we called 'x_0'), the derivatives on the right are always taken at the same set of points, namely the period-n points of $F(x)$. Since the derivatives on the right are always taken at the same set of points, the values on the left are all equal. Thus we have proved that, for any two period-n points of F^n, x_k^* and x_l^*,

$$(\mathrm{d}F^n/\mathrm{d}x)_{x_k^*} = (\mathrm{d}F^n/\mathrm{d}x)_{x_l^*} \qquad (4.2.12)$$

Since the fixed points of F^n are stable if and only if

$$|\mathrm{d}F^n/\mathrm{d}x|_{x^*} < 1 \quad \text{(stable fixed point)} \qquad (4.2.13)$$

(by using the same reasoning as in (4.2.9)), the result (4.2.12) shows that this set of n fixed points are either all stable or all unstable. Note that F^n has other fixed points, but they do not belong to the period-n set of $F(x)$. This result is of basic importance in establishing the character of a sequence of bifurcations which are exhibited by such maps, as we will now discover.

We now return to F^2 and consider what happens as c is further increased. The function $F^2(x)$ develops larger maxima and a smaller minimum, and its slope at the fixed points thereby increases. Finally a value of c is reached (call it c_2) where the condition (4.2.9) is no longer satisfied at the fixed points, and they become unstable. This is illustrated in Fig. 4.12(a). Now F^2 has three fixed points, but they are unstable. On the other hand, the two fixed points which become unstable at $c = c_2$ 'give birth' to a stable cycle of period $2^2 = 4$. This is made fairly clear by considering the graph of $F^4(x)$

Fig. 4.12

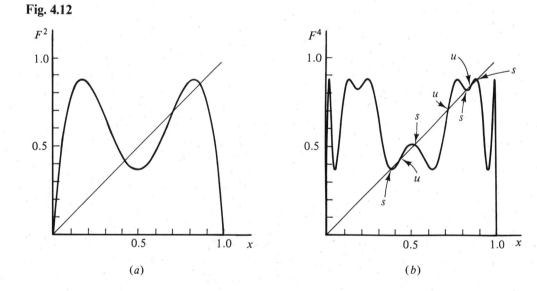

(a) $\qquad\qquad\qquad\qquad\qquad$ (b)

(Fig. 4.12(b)). The two outside unstable fixed points have each 'bifurcated' two stable fixed points of F^4 (that is, the condition (4.2.13) is satisfied at each of the points numbered on the x-axis). In the case of the logistic map, this bifurcation occurs at the value $c_2 = 1 + 6^{1/2} \simeq 3.45$. In the control-phase space these bifurcations look like Fig. 4.13, where the dashed portions are unstable fixed points of F^4. We recognize here

Fig. 4.13

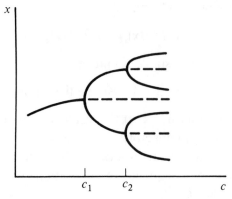

the '*pitch-fork*' (*double point*) bifurcation structure and the *exchange of stability feature* of F^n, found in the last chapter in a different context. Here, however, we find a much more interesting, indeed exciting, bifurcation phenomena, which Thom (1969) viewed as an example of what he called a '*generalized catastrophe*'. As c increases, the above period-four set becomes unstable, giving birth to a stable period-eight, and this process of period doubling repeats indefinitely. That is, there is a set, $\{c_n\}$, such that

$$\text{if } c_{n+1} > c \geqslant c_n, \text{ there is a stable period-}2^n \text{ cycle} \qquad (4.2.14)$$

and, moreover, it turns out that

$$\lim_{n \to \infty} c_n \equiv c_\infty \text{ (finite).} \qquad (4.2.15)$$

In the control-phase space this *infinite* set of *period doubling bifurcations* looks something like that shown in Fig. 4.14. We have sequentially labeled the branches $\pm$,

Fig. 4.14

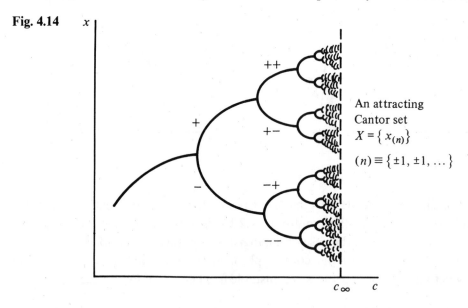

An attracting
Cantor set
$X = \{ x_{(n)} \}$
$(n) \equiv \{ \pm 1, \pm 1, \dots \}$

so that each bifurcation point, $x_{(n)}$ can be associated with its branch by a '$\pm$' subscript (e.g., $(n) = \{+, +, -, +, -, -\}$).

The infinite set, $X = \{x_{(n)}\}$, stable period-2^n ($n \to \infty$) points obtained as $c \to c_\infty$ is interesting to study, since it is an *attractor* of most of the points in the interval $(0, 1)$. It, of course, does not attract the unstable periodic points. The set X is nowhere dense because between each two stable points must be an unstable point. Therefore X is nowhere a continuum, and in fact it is a Cantor set (see Chapter 2, Section 6, and Appendix A). We might ask if it has a dimension (capacity or information) greater than zero. A nice method of obtaining the dimension of such sets has been devised by Chang and McCowan (1984), and will be discussed in Section 7.

The values of c_n, where the period-2^n cycle first occurs (equation (4.2.14)) depend, of course, on the function $F(x, c)$, as does the limiting value c_∞ (4.2.15). The following values of c_n can be obtained for the logistic map.

n	c_n	n	c_n
1	3	5	3.568 759
2	3.449 499	6	3.569 692
3	3.544 090	7	3.569 891
4	3.564 407	8	3.569 934
		∞	3.569 946

These bifurcation values of c are not easy to obtain by any means except numerical methods. There is, however, another set of c values, which we will call $\{c_k^*\}$, that are more accessible and contain the same topological information about the map $x_{n+1} = F(x_n, c)$ as the set of bifurcation values $\{c_n\}$.

Fig. 4.15

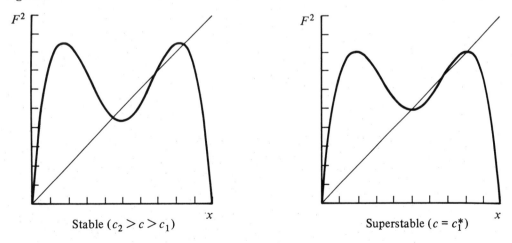

Stable $(c_2 > c > c_1)$ Superstable $(c = c_1^*)$

We need first to introduce the concept of a *superstable* 2^n-cycle. Fig. 4.15 illustrates the distinction between a stable 2-cycle and a superstable 2-cycle. Here we will assume that $F(x)$ is smooth (see Section 3 for some generalizations). In the case of a superstable 2-cycle the stable fixed points of F^2 occur at the maxima, and minima of F^2. It should also be obvious that $c_2 > c_2^* > c_1$. We know from the general result (4.2.12) that if one of these fixed points (corresponding to period-2 points of F) occurs at an extrema of F^2, then so must the other fixed point (i.e., $(dF^2/dx)_{x_1^*} = 0 = (dF^2/dx)_{x_2^*}$). The difference in the sequence of points generated by $F^2(x)$ in the two cases is illustrated above. In the superstable case a point near the fixed point, $x_0 = x^* + \Delta$, maps to $x_1 = x^* + (1/2)(\Delta)^2(d^2F^2/dx^2)_{x^*}$, which is now *quadratically* close to x^*, since it differs by Δ^2. This is a much more rapid approach to x^* than a linear dependence, Δ. The fact that the superstable fixed points occur at the extrema of F^2 (or, more generally, F^{2^n}) makes it easier to determine their associated values of c. We therefore define these values, c_n^*, by this condition

$$c = c_n^* \quad \text{if} \quad (dF^{2^n}/dx)_{x^*} = 0 \quad (x^*: \text{fixed point of } F^{2^n}). \tag{4.2.16}$$

Let us see how this works using the logistic map,

$$F(x) = cx(1 - x),$$

It follows from (4.2.11) that x_m is also an extremum of $F^{2^n}(x)$ by taking $x_0 \equiv x_m$. If c is such that x_m is also a fixed point of $F^{2^n}(x)$ then, according to the definition (4.2.16), this value of $c = c_n^*$ corresponds to a superstable period-2^n cycle. We conclude that if x_m, (4.2.17), is a period-2^n point

$$F^{2^n}(x_m, c) = x_m, \tag{4.2.18}$$

then $c = c_n^*$, so this is a superstable period-2^n.

Let us see how this works using the logistic map,

$$F(x) = cx(1 - x),$$

Fig. 4.16

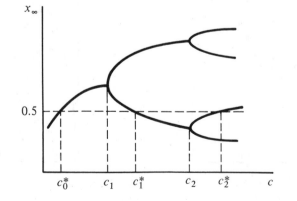

for which $x_m = \frac{1}{2}$. This is also a fixed point of F provided that

$$c\tfrac{1}{2}(1 - \tfrac{1}{2}) = \tfrac{1}{2} \quad \text{or} \quad c_0^* = 2.$$

Next, we have

$$F^2(x) = c[cx(1 - x)](1 - [cx(1 - x)]), \tag{4.2.19}$$

and $F^2(x_m) = x_m$ if

$$c^2(4 - c) = 8$$

or $c_1^* \simeq 3.2361$ (Fig. 4.16). One can proceed in this fashion and obtain the following values for the period 2^n superstable control parameters.

Fig. 4.17

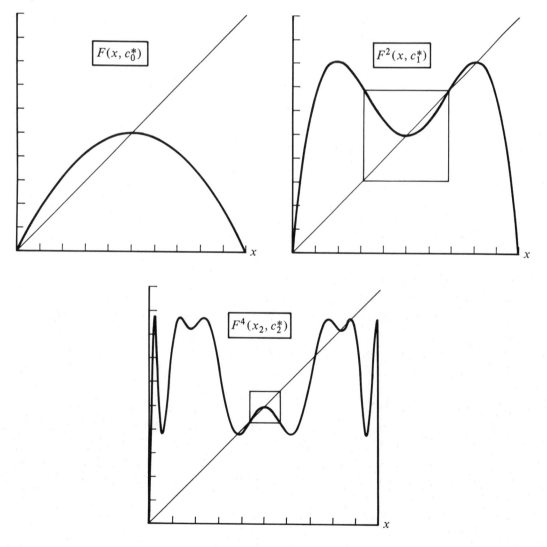

n	c_n^*	n	c_n^*
0	2.00	5	3.569 244
1	3.236 068	6	3.569 793
2	3.498 562	7	3.569 913
3	3.554 641	8	3.569 939
4	3.566 667	∞	3.569 946

The first few functions $F^{2^n}(x, c_n^*)$ are shown in Fig. 4.17 (for $n = 0, 1, 2$). We note that the functions $F^{2^n}(x_1, c_n^*)$ all look similar in the regions around $x = 0.5$, except for a change in scales (and inversions). This is illustrated by the boxes in Fig. 4.17. Moreover, the asymptotic features $(n \to \infty)$ in these boxes would not be changed even if $F(x, c)$ looked quite different away from the point $x = 0.5$. This leads to the consideration of possible 'universal features' and scaling properties.

4.3 Universal sequences and scalings

Metropolis, Stein and Stein (1973) were the first to consider superstable c values and discovered that a wide class of two-to-one maps of an interval onto itself all had a *qualitative feature* in common, which they called *U-sequences* (universal). They considered maps

$$x_{n+1} = c f(x_n) \equiv F(x_n, c) \quad (0 \leqslant c \leqslant c_m) \tag{4.3.1}$$

where $f(x)$ is continuous and continuously differentiable, with a single maximum at $x = x_m$. Note that the control parameter is multiplicative in (4.3.1) with some bound c_m. It has been known for a long time that, if $f(z)$ is analytic (in the complex plane), then any periodic solution of this map contains the point x_m within its region of attraction. Metropolis, Stein and Stein (MSS) found this to also be true for much more general functions $f(x)$ and, in particular, more general than the S-maps discussed above. They considered those values of c such that the point x_m is a periodic point of $F^p(x, c)$. In other words, they considered the superstable $c = \bar{c}_p$ defined by

$$x_m = F^p(x_m, \bar{c}_p), \tag{4.3.2}$$

where now p need not be 2^n. Note that c_n^* of (4.2.18) equals $\bar{c}_{2^n}$. The *number of* $\bar{c}_p$ values, for different values of p, they found to be 1 (for $p = 2$), 1(3), 2(4), 3(5), 5(6), 9(7), 16(8), 28(9), etc.(!), for a total of 2370 $\bar{c}_p$ values for $p \leqslant 15$. These results were obtained from extensive computer studies. They considered the set of points generated by one of these $(\bar{c}_p)$ maps, $x_1 = F(x_m, \bar{c}_p), x_2 = F^2(x_m, \bar{c}_p), \ldots$, starting from x_m, and considered whether $x_k > x_m$ (a 'R' point) or $x_k < x_m$ (a 'L' point). Then any periodic set is characterized by some pattern, such as

$$x_m \to R \to L \to L \to L \to R \to R \to x_m \quad (p = 7) \tag{4.3.3}$$

which contains $(p - 1)$ terms (R, L). They represented such a pattern with the obvious notation RL^3R^2. What they discovered was that the sequence of these patterns (and hence the ordering of the periods p), as c is increased through the values contained in the set $\{\bar{c}_p\}$, is independent of the details of the function $f(x)$, within a large class of functions. The values of the $\bar{c}_p$ of course do depend on the function $f(x)$, but not the sequence of these patterns. This sequence they called the *U-sequence*. For periods of seven or less, the first half of this sequence is

$$R, RLR, RLR^3, RLR^4, RLR^2, RLR^2LR,$$
$$RL, RL^2RL, RL^2RLR, RL^2R, \dots (+ \text{ eleven more}), \tag{4.3.4}$$

where $p = (1 + \text{sum of the exponents}) \leqslant 7$. For the logistic equation $(x_m = \frac{1}{2})$, the corresponding $\bar{c}_p$ values for the patterns (4.3.4) are

$$\bar{c}_p = 3.236, 3.499, 3.628, 3.702, 3.739, 3.774, 3.832, 3.845, 3.886, 3.906, \dots . \tag{4.3.5}$$

For the first three patterns above, the actual ordering of the iterates along the x-axis for the logistic map (taking $x_1 = 0.5$) is

$$x_1 < x_2; \quad x_3 < x_1 < x_4 < x_2; \quad x_3 < x_1 < x_5 < x_4 < x_6 < x_2.$$

A given period can occur with different patterns, at different values of c. Thus, for $p = 5$, the only possible patterns are RLR^2, RL^2R, and RL^3. For a more exhaustive (and exhausting) list of all the 209 patterns through period eleven, see MSS.

Such patterns, and short sequences of such patterns, have now been observed in a number of experimental investigations. Many of these observations are related to aperiodic dynamics in chemical and solid state systems, and the two states (R and L above) may refer, for example, to large and small amplitude outputs from the systems. Then, as some control parameter is varied (e.g., concentrations, flow rates, voltages, etc.) a limited sequence of such patterns is observed.

As an example, Fig. 4.18 shows a 'map' which is experimentally related to the dynamics of the bromide concentration in a well-stirred Belousov–Zhabotinskii reaction (Simoyi, Wolf and Swinney 1982). This is a discussed further in Chapter 8. We simply note here that, as they varied the flow rate of the chemical reactants, the sequence of periodic states which they observed was found to be:

Observed:	R;	RLR;	RLR³LR;	RLR³LRLR;	RLR³;	RLR²
MSS index:	1;	2;	3;	4;	5;	16,
Observed:	RL;	RL²RL;	RL²RLR²L;		RL²R;	RL²; RL³RL
MSS index:	29;	30;	31;		51;	94; 103

Fig. 4.18

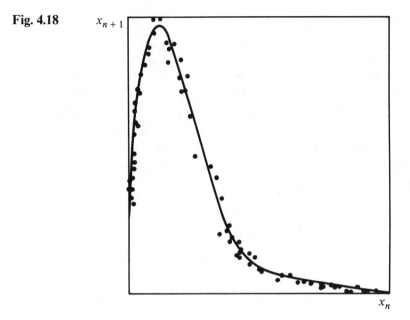

where R and L are right and left of the maximum of the above 'map'. Here the MSS 'index' refers to the placement of the periodic patterns in the MSS listing, given in an appendix of their study. Note that, while many members of the U-sequence were not detected (for a variety of experimental reasons), none of the observed patterns occurred out of the order given by MSS.

This 'universal' *qualitative* discovery by MSS was followed by a 'universal' *quantitative* discovery by Feigenbaum (1978), again with the aid of computer calculations. He considered various subsets of the set $\{\bar{c}_p\}$ introduced by MSS, one of which is (4.2.18), that is $p = 2^n$. What is particularly useful about the set $\{c_n^*\}$ is that it makes a scaling property quite obvious. As can be seen from Fig. 4.19, if one concentrates on the region around x_m (here $x_m = \frac{1}{2}$), the sequence of functions $F(x, c_0^*)$, $F^2(x, c_1^*)$, $F^4(x, c_2^*)$ (etc.) all look similar except for an inversion process and a *scale reduction*. The scale reduction occurs both in the range of x and also in the magnitude of the variation of $F^{2^n}(x, c_n^*)$ over that (boxed) range of x. To represent this scaling, it is useful to center the origin of these figures at $(x = \frac{1}{2}, F = \frac{1}{2})$, so we set

$$y = x - 1/2 \tag{4.3.6}$$
$$G(y, c) \equiv F(y + 1/2, c) - \tfrac{1}{2}$$

We can see from (Fig. 4.20) that (roughly, at first)

$$G(y, c_0^*) \simeq -2.5\, G^2\left(-\frac{y}{2.5}, c_1^*\right)$$

$$G^2(y, c_1^*) \simeq -(2.5)^2 G^4\left(-\frac{y}{(2.5)^2}, c_2^*\right)$$

Fig. 4.19

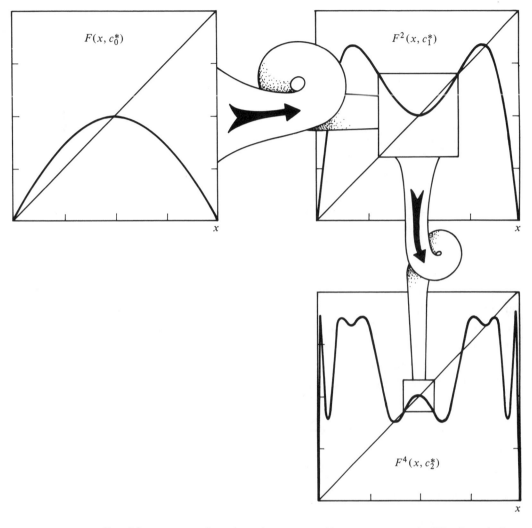

or, more generally, this suggests that there is some scaling parameter (call it α) such that

$$(-\alpha)^n G^{2^n}(y(-\alpha)^{-n}, c_n^*)$$

goes to some universal function of y as $n \to \infty$, call it $g_0(y)$, if α is selected correctly. Feigenbaum found that, for

$$\alpha = 2.502\,907\,875\,095\,892\,848\,5\ldots \tag{4.3.7}$$

this indeed happens,

$$\lim_{n \to \infty} (-\alpha)^n G^{2^n}(y(-\alpha)^{-n}, c_n^*) \equiv g_0(y). \tag{4.3.8}$$

More important than all the significant figures in (4.3.7), Feigenbaum found that both α

Fig. 4.20

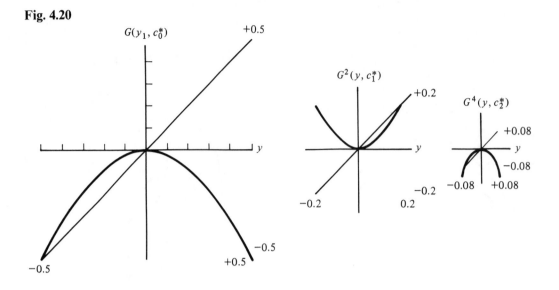

and the function $g_0(y)$ are independent of the precise form of $G(y)$, provided only that it has a single quadratic maximum at $y = 0$. For example, $F(x, c) = c \sin(\pi x)$, $(0 \leqslant x \leqslant 1)$, or $G(y, c) = c \sin(\pi(y + \frac{1}{2})) - \frac{1}{2}$ has the same asymptotic scaling. In other words α *is a universal scaling*, and $g_0(y)$ *is a universal function*. [See Feigenbaum for a discussion of a family of universal functions.]

Feigenbaum also found from his computer studies that

$$\lim_{n \to \infty} \frac{c_{n+1}^* - c_n^*}{c_{n+2}^* - c_{n+1}^*} = \delta \tag{4.3.9}$$

where

$$\delta = 4.669\,201\,609\,102\,9\ldots \text{(etc!)} \tag{4.3.10}$$

is again a *universal constant*. Equation (4.3.9) can also be written

$$c_n^* = c_\infty^* - A^* \delta^{-n} \quad (n \to \infty) \tag{4.3.11}$$

The constants (c_∞^*, A^*) do depend on $G(y)$, and hence are not universal constants (Fig. 4.21). The result (4.3.11) is not limited to the superstable values of c, but also holds for example, for the *period-doubling bifurcations* (at c_n)

$$c_n = c_\infty - A\delta^{-n} \quad (n \to \infty). \tag{4.3.12}$$

An important lesson that can be learned from this discovery is that, in addition to qualitatively similar dynamic systems (e.g., topologically equivalent) which have some form of structural stability, there are also quantitatively similar bifurcation sequences which presumably also have some form of structural stability as indicated by the universal numbers (α, δ). It is known that there are other 'universal numbers' (α_k, δ_k)

Fig. 4.21

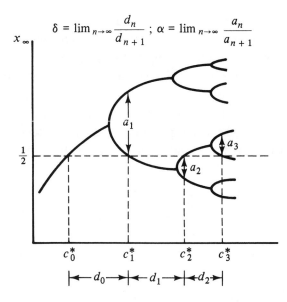

Fig. 4.21

$$\delta = \lim_{n \to \infty} \frac{d_n}{d_{n+1}} \; ; \; \alpha = \lim_{n \to \infty} \frac{a_n}{a_{n+1}}$$

associated with more complicated maps which have other bifurcation sequences. This suggests (in a truly nebulous fashion) that there may exist a form of 'generalized' catastrophe theory which can relate these quantitative families, the universal set of 'universal numbers' $\{\alpha_k, \delta_k\}$. Whether all of these 'universal numbers' have physical significance, remains to be studied.

Feigenbaum has suggested that these universal constants should be observable in bifurcation sequences of much more complex systems than the present 'simple' maps. These issues, however, are not well understood at present (e.g., for a cautious view, see Wilson, 1983). Feigenbaum (1980) found that such period doubling bifurcations occur in higher order differential equations (e.g., the forced Duffing equation), and that they appear to yield the same bifurcation sequence, (4.3.9), with the numerical constant (4.3.10). This will be discussed further in Chapter 5, Section 14.

Other more complex systems also have at least several period-two bifurcations,

Fig. 4.22

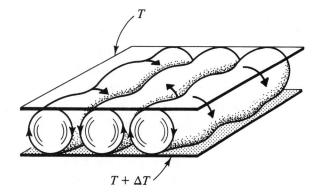

whose successive amplitudes appear to be related to (4.3.7). This is illustrated by Libschaber and Maurer's (1979) study of the Rayleigh–Bénard convection of fluid between two plates held at different temperatures (see Fig. 4.22). As the temperature difference ΔT is increased, the convective vortical rolls of fluid become oscillatory (see the review by Busse, 1985). Feigenbaum's analysis (1979) of Libschaber and Maurer's data is illustrated, indicating four period-doublings of the initial frequency, f_1. He estimated the amplitude of the temperature frequency spectrum, as illustrated in Fig. 4.23 by the horizontal lines, and related their values to α, (4.3.7). That theory

Fig. 4.23

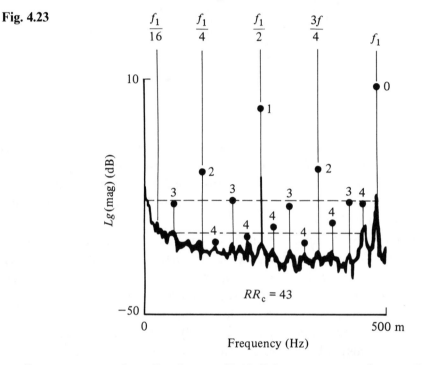

predicts an asymptotic scaling factor of 8.18 db between successive amplitudes, whereas the experimental ratios were 8.4 ± 0.5 between $n = 3$ and $n = 2$ and 8.3 ± 0.4 between $n = 3$ and $n = 4$ (after discarding the highest frequency amplitude). The ratio between $n = 1$ and 2 was not sufficiently 'asymptotic' to fit the theory.

For $c > c_\infty$, the logistic map can, in addition to the infinite number of unstable 2^n period solutions, generate both new stable periodic solutions (e.g., 3×2^n, 5×2^n, etc.) and also aperiodic (nonrepeating) solutions depending on the value of c. We will consider first these *aperiodic solutions*.

An aperiodic solution involves a point which is mapped sequentially through 2^n 'bands' $\{l_k \leqslant x \leqslant u_k; k = 1, \ldots, 2^n\}$, instead of onto 2^n points, as in the case of periodic solutions. In other words, the points map periodically through 2^n disjoint bands of x values, but are not periodic (or eventually periodic) points. This is illustrated in the figure below. For this class of aperiodic sequences Lorenz (1980) has introduced the terminology '*semiperiodic*' sequences, which he also found to occur in Poincaré maps of solutions of his third order differential model, which will be discussed in another chapter.

To be more precise, let $[x_k]_N$ be the range of the infinite set of point, $(F^N)^j$ $(j = 1, 2, \ldots)$, generated by F^N, beginning with $x_k = F^k(x_0)$. In other words $[x_k]_N = \max (x_{k+jN}) - \min (x_{k+jN})$, for all j (see Fig. 4.24). The mapping is called *semi-*

Fig. 4.24 $\underline{N = 4}$

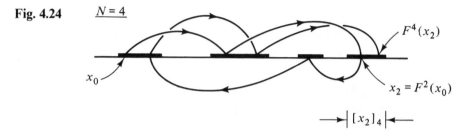

periodic if the ranges $[x_k]_N$ are disjoint for $0 \leqslant k \leqslant N$, but the ranges $[x_k]_M$ overlap for $0 \leqslant k < M$, if $M > N$.

When c is above c_∞ it is frequently found that points do indeed map periodically through such a collection of 2^n 'bands', or disjoint x-regions. The number of such bands goes to infinity as c *decreases* toward c_∞. In other words, the number of bands, N, decreases as c increases and this is referred to as *reverse bifurcation*. This is loosely illustrated in Fig. 4.25, showing a magnification of a small region of x near $c_\infty (c > c_\infty)$.

Fig. 4.25

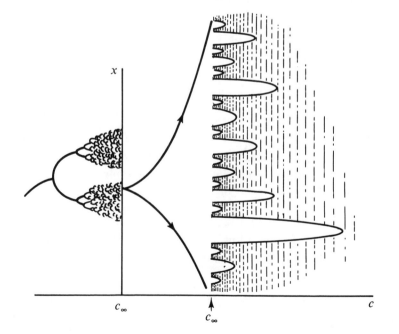

As the shaded region illustrates, these bands widen and then overlap (in pairs) as c is increased. Chang and Wright (1981) found that this series of semiperiodic bifurcations also satisfy Feigenbaum's asymptotic relationship, now with

$$\hat{c}_m \sim 3.569\,945\,671\ldots + \hat{A}\hat{\delta}^{-m} \quad (m \to \infty)$$

where $\hat{A} = 0.494\,454\ldots$ and $\hat{\delta} = 4.669\,201\ldots$. The band $N = 0$ is at $c = 3.678\,57\ldots$, beyond which odd periodic solutions occur. The shaded region in the above figure is meant to convey the fact that the dynamics of x_n, while periodic in the bands, 'skips around' in an erratic fashion each time it returns to a band. It's motion, while deterministic, is quite different for the motion of another initially nearby point. We will call such motion '*chaotic*'. One of our objectives will be to obtain various methods of characterizing *chaos*, both quantitatively and qualitatively.

On the larger scale of the following figure, many of these details are lost, and we can see at most only four of the semiperiodic bands. The density of points is not uniform, as the shading illustrates, and this leads to the rather enticing 'overlapping veils' appearance in this increasingly complex region. Contained in this region are an infinite number of '*windows*', so called because the chaotic 'veil' disappears in this region, and we can 'see through a window'. The veil is replaced by a simple periodic structure at some value of c, and then the map goes through a series of bifurcations much the same as occurs for c below c_∞, until another 'chaotic' region is reached. Only three of these windows are illustrated in the following figure (namely those that begin with period six, with period five, and with period three). The remaining windows have been illustrated in some detail by Collet and Eckmann (1980). We will examine them further in the following sections.

The nature of the periodic solutions (and only one, at most, can be stable for S-maps) is given by the following important and remarkable theorem:

Theorem (Sharkovsky, A.N., *Ukranian Math. J.*, **16**(1), 61 (1964))

> *Let T be the ordered set* $\{3 < 5 < 7 < \cdots < 2 \times 3 < 2 \times 5 < 2 \times 7 < \cdots < 2^2 \times 3 < 2^2 \times 5 < 2^2 \times 7 < \cdots < 8 < 4 < 2 < 1\}$. *Let* $f : I \to I = [0, 1]$, *be a smooth map such that* $f(0) = f(1) = 0$ *which has a single critical point. If* $m < n$ *relative to the order in the set T, and f has a periodic point with prime (i.e., shortest) period m, then f has a periodic point of period n.*[†]

Thus for the logistic example, in Fig. 4.26 where $c = 3.627\ldots$ there is a period six solution, the beginning of a 'window'. According to the above theorem there are periodic (unstable) solutions with periods 2×5, 2×7, and so forth, following to the

---

[†]For a further discussion of this theorem, see Guckenheimer (1977), and the simplified proof by Osikawa and Oono (1981).

Fig. 4.26

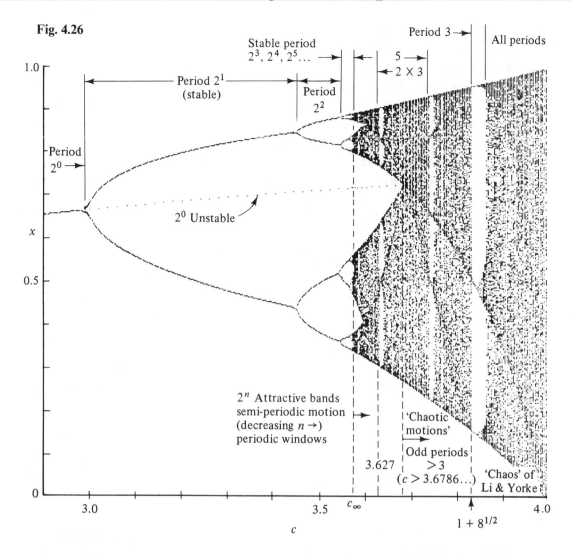

right through the set T. Note that the odd periods have not yet occurred (they begin at $c = 3.6786...$, with a new group of infinite periods!). As c is increased the periodic solutions have periods moving to the left through the T set. When $c = 1 + 8^{1/2}$, the period-3 window is reached, and we are at the first member of the T set (all periods exist).

Some appreciation of why periodic solutions of a general continuous map must obey some ordering, as given by Sharkovsky's T set, can be obtained without much difficulty. The maps produced by a continuous $f(x)$, which has a periodic point $p = f^k(p) (p \neq f^i(p)$ if $i < k)$, can be analyzed quite readily by using a *directed graph* (or '*digraph*') method suggested by Straffin (1978), which is discussed in an Appendix E.

For $c > c_\infty$ the difficulty in representing the periodic vs. aperiodic regions of c is related to the following features. The values of c for which the logistic map has stable periodic solutions fall into continuous bands. Two values (c_1, c_2) which have no stable periodic solutions are apparently separated by such a continuous band of c values, associated with stable periodic solutions. Thus the set of nonperiodic c values is apparently nowhere dense. This does *not* imply, however, that the set of nonperiodic c values has zero measure. These features were first conjectured by Lorenz (1964), who gave an example of a nowhere dense set of points which nonetheless has a finite measure, thereby bearing some resemblance to the present nonperiodic c set. We have already seen one example of such sets, the fat Cantor sets discussed in Chapter 2. We next consider Lorenz's example.

Consider the points $\alpha_{mn} = (2m - 1)/2^n$, restricted to the interval $0 < \alpha < 1$, which form an enumerable set. A few points are illustrated in Fig. 4.27. Around each of these points

Fig. 4.27

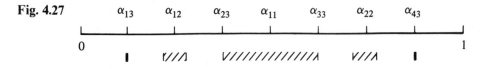

construct an interval of length $(1/2)^{2^n}$, which rapidly decrease in length as n increases. For $n > 3$ the intervals are essentially points on the above scale (as illustrated). Since there are 2^{n-1} intervals for each n, the total length contained in all intervals (counting their overlaps) is

$$\sum_{n=1}^{\infty} 2^{n-1} 2^{-2^n} = 0.390\,747\ldots$$

Moreover, since some of these intervals overlap, the intervals clearly have a measure less than 0.4 (say). Hence the set of points outside of these continuum intervals have a measure greater than 0.6. However, since the α_{mn} are a dense set of points, the points outside their enclosing intervals are nowhere dense. Thus the set of points outside these intervals is nowhere dense but has a finite measure (>0.6!). The largeness of this measure is a result of the rapidly decreasing length of the intervals as n increases – a feature which is also true of the periodic c intervals, as the period increases. Whether this decrease is rapid enough to leave a finite measure for the nonperiodic c set appears to be unknown.

As fascinating as much of this 'fine structure' may be from a mathematical point of view, it is important to keep in mind that many of these details cannot be observed in experiments. Thus, for example, unstable periodic solutions are experimentally unobservable, there is no observational distinction between semiperiodic motion and periodic motion with very long periods, or with semiperiodic motion with large periods which is observationally aperiodic. Similarly, while Cantor sets are great fun mathematically, they do not exist in physical observations. The same is true of fractals.

Only finite approximations can be observed. However, aside from quantum mechanical restrictions (e.g., the uncertainty principle), the ultimate observational restriction can vary greatly in different experiments, which perhaps justifies considering various asymptotic temporal features, and the entire Cantor set. In any case, we will continue to consider these features that are based on continua. This distinction between mathematical and physical continuity was discussed long ago by Poincaré, for example in his *Mathematics and Science: Last Essays*, Chapter III (Dover Pub., 1963). More recent considerations of this basic issue can be found in Section 10 of this chapter (also see Chapter 10, Section 16).

4.4 Tangent bifurcations, intermittencies

The appearance of a period-three solution was believed at one time to represent a type of 'benchmark' in this bifurcation process. It is on the one hand the first member of Sharkovsky's T set, so that all periodic solutions are now present. Secondly, the type of bifurcation which generates these odd periodic solutions are quite different from the 'pitchfork' period 2^n bifurcations. The nature of these bifurcations can be most readily appreciated by considering the graph of $F^3(x, c)$ for the two values of c shown (Fig. 4.28). Note, however, that this is the last such odd bifurcation as c is increased, which then completes Sharkovsky's T set. As c increases from 3.7 to 3.9 the second, fourth, and seventh extrema of F^3 become tangent, and then pass through the 45° line, all at the same value of c. Thus, in addition to a continuing unstable fixed point, three new fixed points of F^3 are 'born out of the blue', and then bifurcate into six fixed points. Only one x-value of each pair satisfies $|dF^3/dx| < 1$, and hence is stable. This *tangent bifurcation* contrasts with the pitchfork type, in which an existing fixed point splits into two stable and one unstable fixed point (recall double point bifurcations, and their exchange of stability). In the logistic case the period three bifurcation occurs at $c = 1 + 8^{1/2}$, and is illustrated in Fig. 4.28 in the control-phase space, showing the stable and unstable branches. Note that the symmetry of the stable–unstable 'horseshoes' is broken by the continuing unstable fixed point. Also note that unstable fixed points of F^3 can be next to each other (contiguous) for the same value of c, whereas this cannot occur for flows (Chapter 3).

Another feature of tangency bifurcations is that, just before the bifurcation, the dynamics of the map produces an effect called *intermittency* which is responsible for the 'folded veil' appearance in Fig. 4.26. Fig. 4.29 shows F^3 for $c = 3.828$, which is slightly lower than the period-three bifurcation point ($c = 1 + 8^{1/2} = 3.828\,427\ldots$). The slight opening between F^3 and the straight line can not be seen on this scale. When the mapped points come into the vicinity of this near-tangency region, the subsequent maps are 'held up' (delayed) in this region (with period three, of course). In other words the maps of $F(x)$ stay sequentially close to the three near-tangency regions, behaving

Fig. 4.28

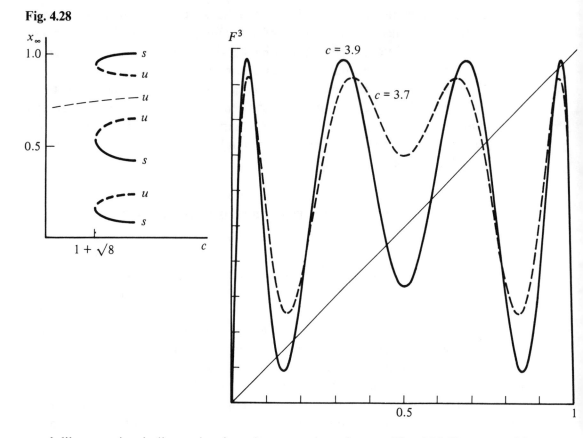

much like a semiperiodic motion for a large number of maps. Fig. 4.30 illustrates this behavior in the present case, for the logistic map. It will be noted that it takes 42 iterations to 'break out' of this semiperiodic behavior. Moreover, this map returns to $x = 0.5065$ after 165 iterations, and hence returns to the semiperiodic behavior for another forty or so iterations. In other words the dynamics spends one quarter of its time in this semiperiodic state. This percentage increases dramatically as c is increased. If $c = 3.8284$ it takes 200 iterates of $F(x)$ before this semiperiodic behavior ends, and after 85 more iterates it returns to the semiperiodic dynamics, which now represents 70% of its behavior.

Intermittencies of various types are rather common features of 'turbulent' states. Thus in steady state turbulence in fluids it is not uncommon to see orderly ('semiperiodic') motion over a period of time in various regions of space, which then breaks up into much more chaotic behavior, only to become orderly at another space-time region. The temptation, of course, is to draw some parallels between these different forms of intermittency, but a solid connection (if it exists) awaits further research.

As noted above, these tangent bifurcations give birth to the 'windows' in the chaotic dynamics. The bifurcation structure within a window is essentially a microcosm of the

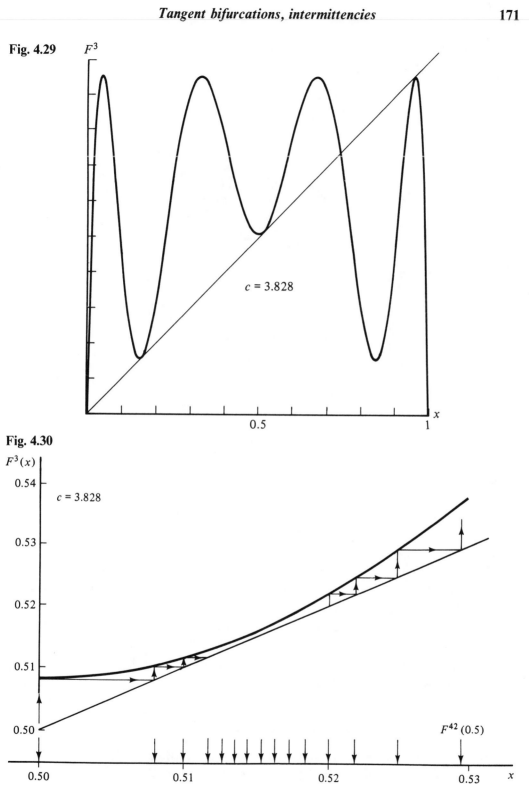

Fig. 4.29

F^3

$c = 3.828$

x

0.5   1

Fig. 4.30

$F^3(x)$

0.54

$c = 3.828$

0.53

0.52

0.51

0.50

$F^{42}(0.5)$

0.50   0.51   0.52   0.53 x

scheme which occurs for $1 \leqslant c \leqslant 4$, but multiplied by the periodicity of the window. This is schematically illustrated in Fig. 4.31 for the period-three window, where there

Fig. 4.31

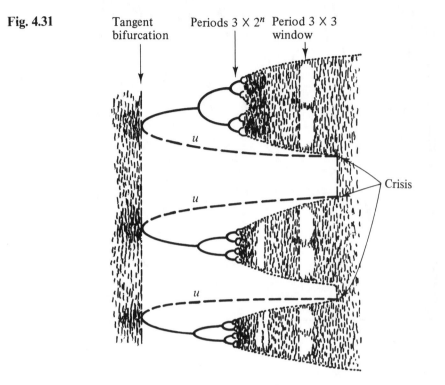

are three microcosms. Thus the former period-2^n bifurcations now become period-3×2^n cycles. This is followed by three chaotic attracting regions, which in turn have their periodic windows. A period-nine (3×3) window is illustrated (and, of course, in that window is ...!). The end of a window is characterized by a bifurcation in which the chaotic attractor discontinuously increases in size to form a continuous attractor (rather than three regions). This occurs at the value of c where the unstable branches of the tangent bifurcation intersects the chaotic attractors. This intersection bifurcation has been termed an interior *crisis* (Grebogi *et al.*, 1982), and represents a distinctly different type of 'bifurcation' (note that the term 'bifurcation' has become generalized to include modification of chaotic sets).

4.5 Characterizing 'deterministic chaos'

The occurrence of windows, and in particular the period-three window which is easiest to observe, presents the first opportunity to give some detailed meaning to the term '*deterministic chaos*'. The adjective 'deterministic' will be implied in all our discussions, but will frequently be omitted for brevity.

Quite generally, we will see that chaos has many 'faces', describing different features of this multifaceted concept. Broadly speaking, we can divide chaotic aspects into:

Qualitative features	Quantitative features
Correspondence with coin tossing (Bernoulli sequences); Topological entropy	(uses measure/distance concepts) Dynamic limiting properties; Lyapunov exponents; Mixing; ergodicity; Kolmogorov–Sinai entropy. Power spectrum; Correlation functions.

We will discuss several of these features in this chapter, beginning with a qualitative description of the dynamics.

To characterize discrete dynamics, such as maps, in a qualitative fashion, it is useful to replace the continuum of numbers (i.e., R^1) by some finite set of symbols. The letters of the alphabet, or any other finite set of symbols, is acceptable. Such a characterization of the dynamics might be referred to as a dynamics of symbols, or more concisely, *symbolic dynamics*. Unfortunately the latter description has been used in a more restrictive sense by mathematicians (see Section 4.11). In any case, the patterns involving R and L, used by Metropolis, Stein and Stein in this U-sequences (Section 4.3), is an example of a qualitative description of periodic dynamics in terms of symbols. Their description is a special case of a much more general method of qualitatively describing dynamics symbolically.

To do this, we first divide the phase space into cells of our choosing. We assume that either the phase space is finite (say, the interval [0, 1]), or the dynamics remain bounded (e.g., on some bounded integral manifold). Then we can divide the space into a finite number of cells. This is referred to as *partitioning the phase space*. A partition, $\{A_i\}$, is any finite set of regions, which are disjoint and cover the dynamic region of interest; call it X.

$$\text{Partition} = \left\{ A_i | i = 1, \ldots, n; \ A_i \cap A_j = \phi, \text{ if } i \neq j; \ \bigcup_{i=1}^{n} A_i = X \right\}. \quad (4.5.1)$$

Such a partitioning of X is illustrated in Fig. 4.32. Metropolis, Stein and Stein used the partition $\{R, L\}$ of the space [0, 1]. More generally, we can describe the dynamics of a mapping by the sequence of symbols corresponding to the partitions visited by a

Fig. 4.32

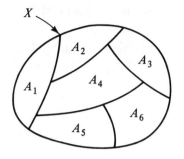

solution,

$$A_2, A_6, A_3, A_3, A_3, A_1, A_5, A_1, A_4, \ldots. \tag{4.5.2}$$

We will now use this partitioning method to describe the dynamics in the period-three window. Specifically, we will take $c = 3.83$, which has a stable period-three orbit, as illustrated. We choose to introduce the partition, involving four regions, L^-, L, R, R^+ (Fig. 4.33), and note that the mapping of the region L^- covers both L^- and L. We

Fig. 4.33

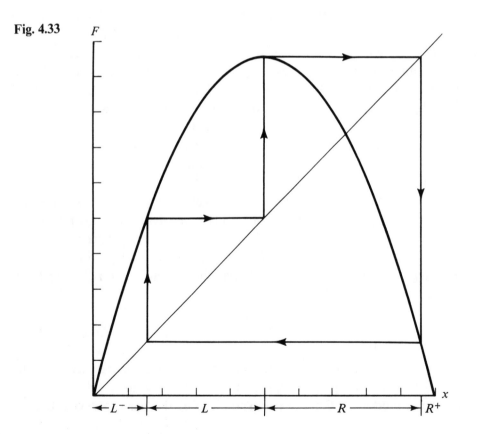

write this as $FL^- = L^- \cup L$. Similarly, for the other regions,

$$FL = R; \quad FR = R \cup L; \quad FR^+ = L^-.$$

That is, all points in R^+ map into L^-, and in fact FR^+ covers L^-. Similarly points in L^- either map to L^- or L. After enough iterations, any initial point of L^- enters L (Fig. 4.34). In other words the region $L \cup R$ is an attractor. All points in $L \cup R$ stay in this region.

Fig. 4.34

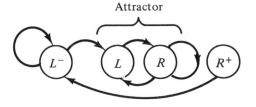

Therefore we will focus our attention on the dynamics between L and R (only). We note, however, that this dynamics is not symmetric. Points in L always map to R, whereas those in R may map to R or L. Because we want to compare the dynamics with the process of flipping a coin, which is symmetric, we will consider the map $G \equiv F^2$, rather than F. This is illustrated in Fig. 4.35, and now we use the symbols (H, T) for the two regions of interest, obviously to relate them to a coin. We note that the dynamics is now symmetric, since

$$GH = H \cup T \quad \text{and} \quad GT = H \cup T.$$

Now we return to deterministic chaos. What is one characterization of this concept? It seems reasonable to say that a system behaves chaotically if there are dynamic solutions which behave like the flipping of a coin (Fig. 4.36). A series of coin flips is characterized by some infinite sequence of symbols

$$H, H, T, H, T, T, T, H, T, \ldots. \tag{4.5.3}$$

Since the probability of H or T turning up in the next flip is 1/2, this is called a *Bernoulli sequence*. In our deterministic system, there is, of course, no probability involved – it is simply a question of comparing sequences. This leads to one definition of deterministic chaos:

> *A deterministic system is* **chaotic** *if we can find, for any prescribed Bernoulli sequence, an initial state which will dynamically produce the same sequence of symbols, as it moves through some fixed partition of the phase space. In that case, we will say that initial state corresponds to the prescribed Bernoulli sequence.*

Fig. 4.35

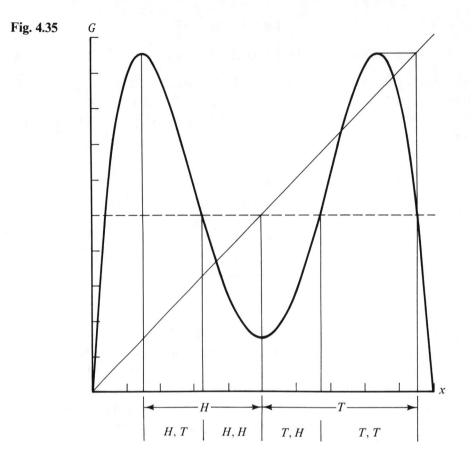

Fig. 4.36

To see that this correspondence can be made in the period-three region, consider the figure for the map G. If the initial state is in the x-region labeled (H, T), then this mapping will take it from the H region to the T-region. In other words, an initial state in the (H, T)-region produces the sequence H, T in its first step. Similarly, beginning in (T, H) will produce the sequence T, H in one step.

For two steps, consider G^2. Now, for example, any initial state in the region HTT will begin in H, map to T, then again map to T. We see from Fig. 4.37 that there are initial

Fig. 4.37

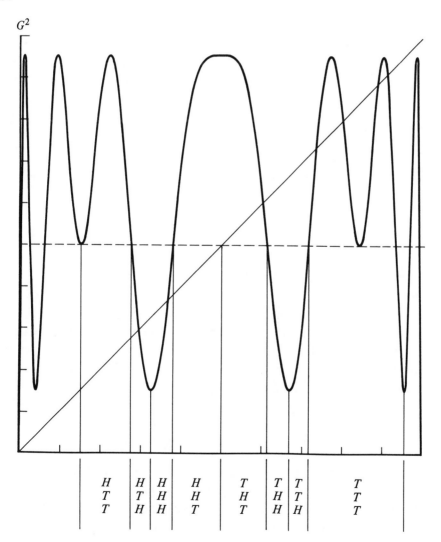

H	H	H	H	T	T	T		T
T	T	H	H	H	H	T		T
T	H	H	T	T	H	H		T

conditions which will produce any of the 2^3 sequences of length 3,

$H, T, T;$ $H, T, H;$ $H, H, H;$ $H, H, T;$ $T, H, T;$ $T, H, H;$ $T, T, H;$ $T, T, T.$

Similarly, by examining G^3, you can see that there are initial conditions which will generate any of the 2^4 sequences of length 4. We conclude that the dynamics of the system can produce an orbit which corresponds to any Bernoulli sequence – so the system is chaotic, in that sense.

The fact that the appearance of a period-three mapping is sufficient (but it turns out not to be necessary) to produce one form of chaos (to be defined shortly) was recognized first by Li and Yorke (1975). The theorem of Li and Yorke deals with the general equation

$$x_{n+1} = F(x_n)$$

where $F(x)$ gives a continuous mapping of a closed interval (say $[0, 1]$) into itself. It says that if F maps some point a into a point $b(= F(a))$, and b into $c (= F(b))$, and c into d $(= F(c))$, all satisfying the inequality shown in Fig. 4.38, then the points in the interval

Fig. 4.38

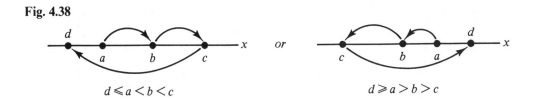

$$d \leqslant a < b < c \qquad\qquad\qquad\qquad d \geqslant a > b > c$$

$[0, 1]$ will necessarily map in a well-defined chaotic fashion. In particular the above condition is satisfied if there is a point with period three (i.e., $d = a$), hence their title, '*Period Three Implies Chaos.*'; this only holds in one dimension (see Section 6.4).

One of the chaotic features in this case is that there are points which have any desired period n (Sharkovsky's theorem), and there is an uncountable number of points which are not even asymptotically periodic. We will call this set S. If q is a periodic point, p is called an asymptotically periodic point if

$$\lim_{n \to \infty} |F^n(p) - F^n(q)| \to 0. \tag{4.5.5}$$

Hence this result says that there is a uncountable set of points, S, which do not even tend toward a periodic solution (nor are any points in S periodic).

Perhaps what is of even greater interest are two other results due to Li and Yorke, which give us another (non-Bernoullian) way to characterize '*chaotic motion*'. The formal statements are that, for every $p, q \in S$ $(p \neq q)$

$$\limsup_{n \to \infty} |F^n(p) - F^n(q)| > 0, \qquad \text{(I)}$$

and

$$\liminf_{n \to \infty} |F^n(p) - F^n(q)| = 0. \qquad \text{(II)}$$

We now need to define and interpret these results.

First of all recall that any infinite sequence of points, $\{x_n\}$, on a closed (compact) interval has at least one subsequence $\{x_{n_j}\} \in \{x_n\}$ which possesses a limit point (the Bolzano–Weierstrass theorem); that is $\lim_{n_j \to \infty} x_{n_j} \equiv l_{\{n_j\}}$ (dependent, of course, on the subsequence). The definition of lim sup in (I) is

$$\limsup_{n \to \infty} x_n = \max_{\{n_j\}} l_{\{n_j\}} \equiv l_{\max}. \qquad (4.5.6)$$

that is, it is the maximum of all limits obtained from all subsequences of $\{x_n\}$. Similarly, in (II),

$$\liminf_{n \to \infty} x_n = \min_{\{n_j\}} l_{\{n_j\}} = l_{\min}, \qquad (4.5.7)$$

the minimum of all limits obtained from all subsequences. Applied to (II) this says that the sequence of values, $|F^n(p) - F^n(q)|$, obtained from any two initial points (p, q) in S, has a subsequence of mappings such that

$$\lim_{n_j \to \infty} |F^{n_j}(p) - F^{n_j}(q)| = 0$$

In other words we can find maps F^{n_j} of p and q which are arbitrarily close to each other as $n_j \to \infty$.

On the other hand (I) says that we can also find a sequence of mappings such that

$$\lim_{n_j \to \infty} F^{n_j}(p) = p_0, \quad \lim_{n_j \to \infty} F^{n_j}(q) = q_0 \quad \text{and} \quad |p_0 - q_0| = l_{\max} > 0.$$

Therefore what (I) and (II) state is that the mappings, F^n, of p and q remain erratic as $n \to \infty$ in the sense that they both become arbitrarily close for some steps, n_j, and remain a finite distance apart ($l_{\max}$) for another subset of these maps. This is schematically illustrated in Fig. 4.39.

This is an interesting characterization of a chaotic mapping, but it has the disadvantage that it yields no quantitative measure of this chaotic motion. In particular it gives no measure of how rapidly mixing occurs for the mapping of adjacent points. This is a characteristic which is measured by the Kolmogorov–Sinai entropy, to be discussed later, and the Lyapunov exponents (Section 6). Another feature not clear from this theorem is that, while the set S is uncountable, it may have (Lebesgue) measure zero. It should always be kept in mind, however, that 'measure' comparisons

Fig. 4.39

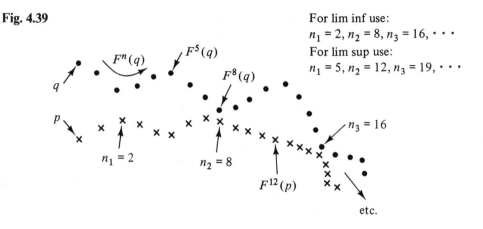

For lim inf use:
$n_1 = 2, n_2 = 8, n_3 = 16, \cdots$
For lim sup use:
$n_1 = 5, n_2 = 12, n_3 = 19, \cdots$

are strictly mathematical concepts, and need not represent the physical importance of the quantities.

The chaos defined by Li and Yorke is by no means the only possible definition of chaos, and the condition of period three is not necessary even for their form of chaos. Various definitions of chaos and sufficient conditions have been discussed by Oono and Osikawa (1980).

Other forms of chaos will arise in the study of higher order systems, but at the heart of all chaotic dynamics is the *fold-over* ('*kneading*') nature of the flows or maps. This is illustrated in Fig. 4.40, where it is shown how a map (e.g., the logistic map, with $c = 4$)

Fig. 4.40

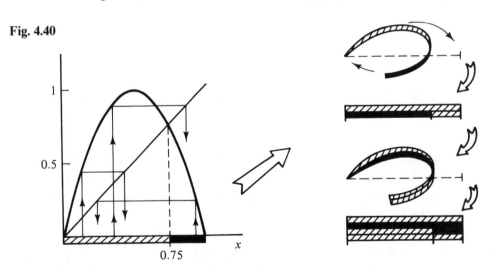

can take the interval $[0, 1]$ and fold it double onto the same interval. Note that the dark and stripped regions are separated by the fixed point.

It should be noted that the logistic map, with $c = 4$ (i.e., $F(x) = 4x(1 - x)$), has no attracting periodic orbits, as illustrated in the exercise below. At the same time, the periodic orbits are dense in the interval $(0, 1)$ (i.e., any region in the interval contains a point which is a periodic orbit). Moreover, the dynamics is *ergodic* over $[0, 1]$ for this value, $c = 4$. This means that the only invariant subregions have either Lebesgue measure zero or one. This will be discussed further in Section 8.

Exercise 4.6. (a) If period three implies chaos, why doesn't the period-three window of the logistic map appear chaotic? Think about what is required to observe a chaotic set. (b) Using ideas from the above fold-over nature of the $c = 4$ logistic map, together with *Brouwer's fixed point theorem*, can you speculate on the distribution of periodic points in $[0, 1]$?

Exercise 4.7. Show that $x_n = \sin^2(2^n \pi \theta)$ is a solution of $x_{n+1} = 4x_n(1 - x_n)$ for all θ, and comment on the distribution of both the periodic and the nonperiodic solutions (also see S.M. Ulam, *A Collection of Mathematical Problems* (Interscience Pub., 1960) pp. 150).

In Chapter 4 we discussed the topological similarities of the continuous flows produced by ODE. This led to the concept of TOE. An analogous property can be defined for two maps, $F(x)$ and $G(x)$, if their respective dynamics can be related by any homeomorphism, $h(x)$. If we can find $h(x)$ such that $F(x)$ and $G(x)$ can be related by $G \cdot h = h \cdot F$ (as operators on the points of R^n), then F and G are said to be *topologically conjugate* (TC). These sequential operations are illustrated schematically in Fig. 4.41. Written in terms of functions, rather than as operators, $F(x)$ and $G(x)$ are TC if

$$G(h(x)) = h(F(x)) \tag{4.5.8}$$

Fig. 4.41

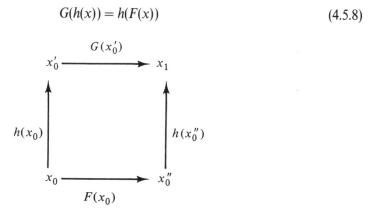

for all x, and some homeomorphism $h(x)$. In particular if P_1 and P_2 are two *Poincaré maps* of TOE flows, then P_1 and P_2 are *topologically conjugate*.

Exercise 4.8. Show that the (differentiable) map $F(x) = 4x(1 - x)$ $(1 \geqslant x \geqslant 0)$ and the *tent map* $G(x) = 2x$, $(\frac{1}{2} \geqslant x \geqslant 0)$, $G(x) = 2(1 - x)$, $(1 \geqslant x \geqslant \frac{1}{2})$ are related by the homeomorphism $h(x) = (2/\pi) \arcsin x^{1/2}$ (noted first by S.M. Ulam and J. von

Neumann; see P. Stein and S. Ulam, 'Nonlinear transformation studies on electronic computers', *Rozprawy Metamatyczne*, **39**, 401–84 (1964)). In other words, $G \cdot h = h \cdot F$ (as operators). Determine the stability of the periodic orbits of F, by determining the stability of the periodic orbits of G.

4.6 Lyapunov exponents

One measure of 'chaotic' motion can be related to the sensitivity of the dynamical behavior of a system when its initial state is changed by a small amount. If the system's behavior is very *sensitive to initial conditions*, so that nearby points in the phase space separate 'fast' (say exponentially with time, or iterates of a map) over most of the phase space, then the system (not just a particular solution) can reasonably be described as being dynamically unstable. This may not be easy to establish because we may not be able to study most orbits of a system, but have to be satisfied with (hopefully) some 'representative' orbits. We are then effectively investigating the Lyapunov stability of an orbit, γ, but limiting our concern to exponential instabilities.

The same question can obviously also be raised about the properties of a map

$$x_{n+1} = f(x_n).$$

Let (x_0, y_0) be two nearby initial points in the phase space (R^1) and

$$x_n = f^n(x_0), \quad y_n = f^n(y_0).$$

If these points separate exponentially with n,

$$|y_n - x_n| = A \exp(\lambda n) \quad (\lambda > 0)$$

where $A = |y_0 - x_0|$. For large n.

$$(1/n) \ln |y_n - x_n| \to \lambda \quad (\text{large } n).$$

However, in the case of motion in a bounded region such exponential separation cannot occur for very large n, unless the initial points (x_0, y_0) are very close. Therefore, before taking the limit $n \to \infty$, we must take the limit $|x_0 - y_0| \to 0$. This then defines a constant

$$\lambda = \lim_{n \to \infty} \frac{1}{n} \lim_{|x_0 - y_0| \to 0} \ln \left| \frac{x_n - y_n}{x_0 - y_0} \right|$$

or

$$\lambda = \lim_{n \to \infty} \frac{1}{n} \lim_{|x_0 - y_0| \to 0} \ln \left| \frac{f^n(x_0) - f^n(y_0)}{x_0 - y_0} \right|$$

$$= \lim_{n \to \infty} \frac{1}{n} \ln \left| \frac{d f^n(x_0)}{dx} \right|$$

Now, using (4.2.11), we obtain

$$\lambda_\gamma = \lim_{n \to \infty} \frac{1}{n} \sum_{k=0}^{n-1} \ln \left| \frac{\mathrm{d}f(x_k)}{\mathrm{d}x_k} \right| \tag{4.6.1}$$

which is called the *Lyapunov exponent*, for the orbit $\gamma \equiv \{x_n = f^n(x_0); n = 0, 1, \ldots\}$. If a group of solutions (i.e., with different initial x_0) are asymptotically attracted to the same ergodic subset of the phase space, then their Lyapunov exponents will all be the same, and the subscript γ can be dropped. This is true because the limiting set of points, large k, determine the value of λ_γ in (4.6.1).

More generally, if we know the probability distribution function, $P(x)$, of the asymptotic $(n \to \infty)$ set of points for all equally probably initial conditions, then an average Lyapunov exponent (for all initial states) can be obtained by replacing the sum ('time average') in (4.5.1) by the phase average for all initial states

$$\bar\lambda = \int \ln |\mathrm{d}f/\mathrm{d}x| \, P(x) \, \mathrm{d}x. \tag{4.6.2}$$

In practice, it frequently occurs that most initial states (in a measure sense) are attracted to some ergodic subset and therefore we can drop both the subscript on λ_γ and the bar on $\bar\lambda$, and discuss only the unique Lyapunov exponent of the system. However, the above provisions should be kept in mind for more general situations.

Since $P(x)\,\mathrm{d}x$ is the probability of finding a mapped point, x_n, in the region $\mathrm{d}x$ of x as $n \to \infty$, this distribution must be an invariant function of this map. That is

$$P(y) = \int \delta(y - f(x)) \, P(x) \, \mathrm{d}x. \tag{4.6.3}$$

Exercise 4.9. A useful example of these ideas is the case of the logistic map with $c = 4$, $f(x) = 4x(1 - x)$, which was originally studied by M. Kac (*Ann. Math.*, **47**, 33–49 (1946)). Show first that $P(x) = [\pi(x(1 - x))^{1/2}]^{-1}$ is a normalized probability distribution on $[0, 1]$. Next, show that it satisfies the invariance property (4.6.3) for this map. Finally, show that the Lyapunov exponent for this map is $\lambda = \ln 2$.

Exercise 4.10. Using (4.6.2), show that any stable (unstable) periodic orbit has a negative (positive) Lyapunov exponent.

In some simple cases the invariant distribution can be obtained explicitly, and the Lyapunov exponent can be computed, using (4.6.2). For example:

Exercise 4.11. Show that for the 'tent map' (see Fig. 4.42) $f(x) = x/x_0$ (for $0 \leqslant x \leqslant x_0$), $f(x) = (1 - x)/(1 - x_0)$ (for $x_0 \leqslant x \leqslant 1$) the invariant distribution is $P(x) = 1$, and obtain $\bar\lambda$ as a function of x_0. What's the most unstable map in this family, $\{0 < x_0 < 1\}$?

Fig. 4.42

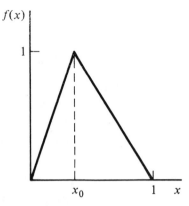

On very rare occasions, we can determine $\bar{\lambda}$ if $P(x)$ is not known. Thus, if $f(x)$ consists of any number of linear sections with slopes of equal magnitude, $df/dx = \pm n$, then $\bar{\lambda} = \int P(x) \ln n \, dx = \ln n$, regardless of the details concerning $P(x)$. This is clearly an exceptional situation!

The Lyapunov exponent can, of course, be obtained directly by evaluating (4.6.1), without obtaining $P(x)$. Shaw (1981) used $\log_2$ to obtain Fig. 4.43 for λ vs c, for

Fig. 4.43

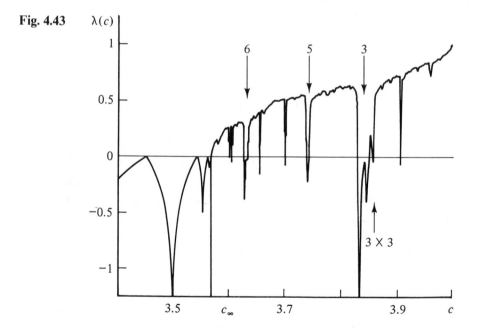

the logistic map, using 300 values of c, separated by 0.002. Each value of λ was obtained by using 100 000 maps in (4.6.1). The negative regions correspond to attractive periodic orbits which have the relatively wide parameter windows. The windows corresponding to periods 6, 5, and 3, and even the 'microcosm' 3×3 window (Section 4), are labeled in

Fig. 4.43. The infinite number of other attractive orbits cannot be seen on this scale. We should also emphasize several facts which are included in the above figure:

c = 4

> This is the most chaotic situation in the sense that λ is the largest, so the set of points have the greatest sensitivity to initial conditions. On the other hand, essentially all initial states move continually over the entire interval $(0, 1)$. Therefore this set cannot be regarded as an attractor.

c = c$_\infty^{(k)}$

> These are the Cantor sets which occur at the end of the bifurcation sequences $k2^n$, in the period k window. Our previous c_∞ is $c_\infty^{(1)}$ in the present notation. Some of these are illustrated in the above figure at the upward crossing of the c axis, so that $\lambda = 0$. Therefore these sets are not sensitive to initial conditions. On the other hand they are attractors of most initial states.

We will see later that there are dynamic systems which combine the features of these two types of sets. Namely, there are sets which on the one hand are *attractors* and yet

Fig. 4.44

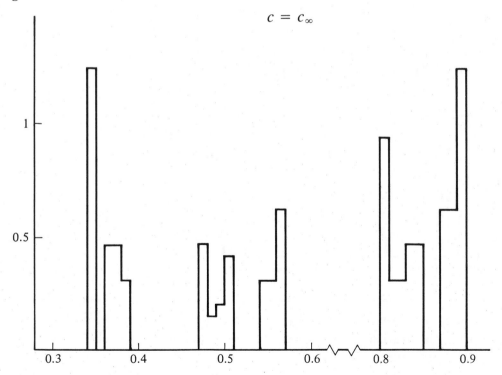

$$c = c_\infty$$

have positive Lyapunov exponents, $\lambda > 0$, so that they are *sensitive to initial conditions*. Such a dynamic set is sometimes referred to as a *strange attractor*, although there is presently no generally accepted precise definition of a strange attractor. We will encounter such sets in Chapter 6.

The determination of the invariant distribution, $P(x)$, for general maps can be obtained to whatever accuracy that one is willing to invest computer time (and money!). The situation is simplest if all initial points (except a set of measure zero) go asymptotically to some ergodic region, as discussed above. In that case, $P(x)$ can be approximated by a histogram generated by the mappings of any representative initial state, provided that enough iterates are taken to approximate the asymptotic (invariant) set. As an example, a coarse histogram (bin size $\Delta x = 0.01$) for the logistic map at c_∞ is shown in Fig. 4.44. We see that all points of the nonenumerable set, X (see Fig. 4.14), are contained within six disjoint regions on this scale (note the broken x scale). If the bin size is reduced to $\Delta x = 10^{-3}$, then the histogram looks like the following. There are now twenty disjoint regions which are discernible (Fig. 4.45). More

Fig. 4.45

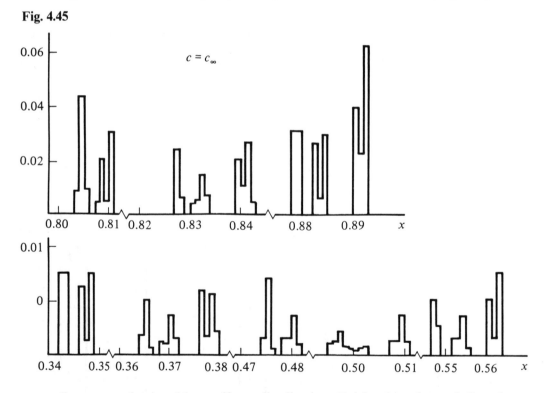

generally, one can begin with a uniform distribution, $P_0(x) = 1$, and map it for a large number of x values, yielding $P_1(x) = F(P_0(x))$. Continuing this process will asymptotically yield the invariant $P(x)$ (see, e.g., Hoppensteadt and Hyman, 1977, Shaw, 1981, and Chang and Wright, 1981).

4.7 The dimensions of 'near self-similar' cantor sets

The limit set of stable 2^n periodic points of the logistic map, obtained as $c \to c_\infty$, has the approximate distribution illustrated in Section 4.6. From this information we can obtain an aproximate value for either the capacity or information dimension of this set. These dimensions are discussed in Section 2.6, which may need to be reviewed.

Exercise 4.12. From Fig. 4.45, estimate d_c. Note that $\varepsilon = \Delta x = 10^{-3}$ in this figure. The answer is not very accurate, as we will see below. Note also that the estimate of d_1 involves a considerable amount of additional labor.

While, in principle, the values of d_c and d_1 can be obtained in this fashion by decreasing ε, a large number of cells, $N(\varepsilon)$, are required for significant accuracy. Chang has proposed a much more rapidly converging method of obtaining the dimensions of a class of such Cantor sets, which is based on their near self-similar structure (recall the general definition of Cantor sets, Appendix A, does not imply self-similarity).

We begin with the set, X (see (4.2.15)ff), obtained in the limit of the bifurcations of stable 2^n cycles, as $c \to c_\infty$,

$$x_n = F^n(1/2, c_\infty) \quad (n = 1, 2, \ldots). \tag{4.7.1}$$

Recall that, for an S-map, the maximum of $F(x)$ tends to any stable periodic orbit. Next, we use the iterates to make a geometrical construction of the attracting set, X, based on the sequential removal of open regions of x which clearly do not contain the attractor. This sequential removal of open regions is, of course, precisely the way Cantor sets are constructed, as discussed in Chapter 2. The clever aspect of this method is to recognize and use the near self-similarity which occurs in these sequential removals, in order to terminate the iterates (4.7.1) at a low order, and still obtain an accurate estimate of d_c.

Consider first the points x_1 and x_2, obtained from (4.7.1), which are illustrated in Fig. 4.46. It is not difficult to see that all $0 < x < 1$ will eventually map into the region $x_2 < x \leqslant x_1$, or $(2, 1)$ for brevity. Therefore the attractor X is contained in $(2, 1)$. Next consider the iterates x_3 and x_4, also illustrated in the figure. We can readily see that all points will eventually map into the union of $(2, 4)$ and $(3, 1)$. That is, we have removed the open interval, $x_4 < x < x_3$, from $(2, 1)$ in the process of constructing the attractor. Now we can proceed in two ways, one being exact, the other an approximation based on a self-similar approximation. The exact method involves obtaining the next two iterates of (4.7.1) in each of the regions $(2, 4)$ and $(3, 1)$. This involves the points x_5, x_6, x_7, and x_8. These points are illustrated in Fig. 4.47. We conclude, as before, that the attractor must lie in the union of $(2, 6), (8, 4), (3, 7)$, and $(5, 1)$. In other words, we have removed the open intervals $x_6 < x < x_8$ and $x_7 < x < x_5$ from $(2, 4)$ and $(3, 1)$ respectively. The result is exact, but we are still a long way from obtaining d_c if we follows this procedure.

Fig. 4.46

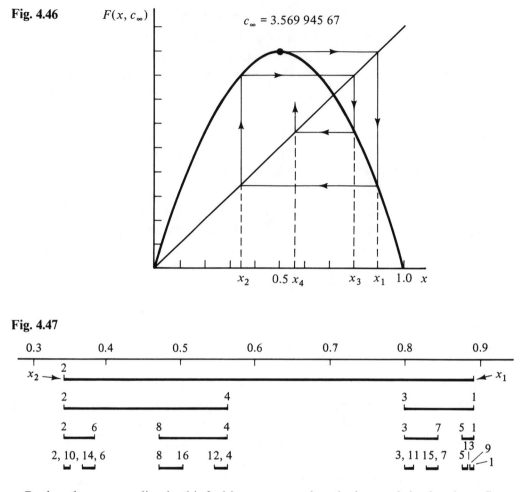

Fig. 4.47

Rather than proceeding in this fashion, we note that the intervals in the above figure have nearly a self-similar structure (allowing for a trivial interchange in the ordering of the long and short intervals). If $L(i, j)$ is the length of the interval (i, j), then we find for the logistic map, with $c = c_\infty = 3.569\,945\,67\ldots$, the following values: $L(1, 2) = 0.5499$; $L(1, 3) = 0.0885$, $L(2, 4) = 0.2200$; $L(1, 5) = 0.0140$, $L(3, 7) = 0.0380$, $L(4, 8) = 0.0876$, and $L(2, 6) = 0.0385$. If these were exactly similar then, for example, we would find the following equalities on the left, compared with the actual values on the right:

$$L(4, 8)/L(2, 4) = L(2, 4)/L(1, 2), \quad \text{whereas } 0.3980 \simeq 0.4001$$
$$L(2, 6)/L(2, 4) = L(1, 3)/L(1, 2), \quad \text{whereas } 0.1749 \simeq 0.1609$$
$$L(1, 5)/L(1, 3) = L(1, 3)/L(1, 2), \quad \text{whereas } 0.1580 \simeq 0.1609$$
$$L(7, 3)/L(1, 3) = L(2, 4)/L(1, 2), \quad \text{whereas } 0.4292 \simeq 0.4001.$$

We see that these intervals are only approximately similar, so the Cantor set being produced is not exactly self-similar on this scale. Moreover, as n is increased, the successive groups of 2^{2n} intervals do not become increasingly similar to the group of 2^n intervals, relative to the original interval $(1, 2)$. Nevertheless this approximate self-similarity can be used to obtain increasing accurate estimates of d_c, as n is increased.

Exercise 4.13. If you would like to try some computing, set up the ratios of lengths which would have to hold if the group of sixteen intervals, obtained in the next step beyond the above figure, is similar to the four intervals in the figure. Then compute these ratios, and see how similar are the groups. Note that we are now examining whether the Cantor set is self-similar to these four intervals. Note also that adjacent points, producing the intervals (i, j), satisfy $|i - j| = 16$ in the sixteen intervals.

If the Cantor set were self-similar in the two intervals $(1, 3)$ and $(4, 2)$, then the dimension of this set is, according to the discussion given in Exercise 2.29, given by

$$L(4, 2)^{d_c} + L(3, 1)^{d_c} = L(1, 2)^{d_c}. \tag{4.7.2}$$

This gives the lowest approximation of d_c. In the next approximation, we can use the iterated intervals $(2, 6)$, $(8, 4)$, $(3, 7)$, and $(5, 1)$, and assume that the Cantor set is self-similar on this scale. This yields for d_c

$$L(2, 6)^{d_c} + L(8, 4)^{d_c} + L(3, 7)^{d_c} + L(5, 1)^{d_c} = L(1, 2)^{d_c}. \tag{4.7.3}$$

Proceeding in this fashion we can obtain higher approximations for d_c, based on Cantor sets which are self-similar in the $2^n \equiv N$ intervals generated by the 2^{n+1} iterates of $x = 1/2$. The result of this, based on the equations (4.7.2), (4.7.3), and obvious generalizations, are tabulated:

Self-similar Cantor set, based on N intervals	d_c
2	0.5261
4	0.5329
8	0.5346
16	0.5354
32	0.5359

Chang and McCowan (1984) obtained a more rapidly convergent result, based on the use of the universal function (satisfying $\alpha f^{*2}(x/\alpha) = f^*(\alpha)$, $\alpha = -2.5029$) to generate the intervals, and a suitable extrapolation scheme to obtain d_c. They obtained $d_c = 0.5379$, so the above result is accurate to 0.4%. It is clearly much more accurate than the result in Exercise 4.12. They also applied this method to other near self-similar

Cantor sets, generated by bifurcation sequences which are quite different from the common $k2^N$ periodic sequences. They considered the sequence of periodic solutions consisting of periods $N, N^2, N^3, \ldots$, and obtained the dimension of the limiting Cantor sets. For extensive numerical data on both d_c, d_1, and the associated Feigenbaum exponents, see their paper.

4.8 Invariant measures, mixing and ergodicity

From an experimental viewpoint it is important to known something about the likelihood of observing a dynamical feature or characteristic of a system. This likelihood, or probability, is presumably related to some 'measure' of those states in the phase space which have the dynamical feature of interest.

By a *measure*, $\mu(A)$, of a set A in R^n we will mean a set function (i.e., a function which, for any set, gives a real number, $\mu: A \to R$) with the properties: (a) $0 \leqslant \mu(A) \leqslant \infty$, (b) $\mu(A \cup B) = \mu(A) + \mu(B)$ for any disjoint sets $A \cap B = \varnothing$ (the empty set), and (c) $\mu(\varnothing) = 0$. In particular, the measure of all set (e.g., all phase space) does not need to be bounded. If it is bounded, then μ can be normalized to unity.

A classic example where a measure is introduced is the Liouville theorem, where the interest is in the behavior of 'most' solutions rather than the possibly exceptional behavior of a set of states 'of measure zero'. Such a set makes no contribution to the integral invariant. In the general Liouville theorem the local measure, $d\mu = \rho(x) \, dx (x \in R^n)$, weights the region dx with a density, $\rho(x) > 0$, so as to make the total measure of any propagating region, $\Omega(t)$, a constant in time. More succinctly, this is simply called an *invariant measure*. If the dynamics is viewed as a diffeomorphism f^t, discussed in Chapter 2, and A is some set of points in the phase space ($A = \Omega(t)$ in the Liouville theorem), then this invariance of the measure of A can be written

$$\mu(A) = \mu(f^t A) \quad \text{(all } t\text{)}.$$

In other words the measure of the set A (which equals $\int_\Omega \rho(x) \, dx$ in the Liouville theorem) is equal to the measure of the set of points which are obtained from the solutions of $\dot{x} = F(x)$, with the states in A as initial conditions.

However, in the case of maps which are not one-to-one, it is necessary to generalize this concept of an invariant measure. Assume that f is a two-to-one map in some regions of the phase space. For example, assume that the sets A and B both map onto the same region, C, so $f(A) = C, f(B) = C$ (Fig. 4.48). In this case we say that the inverse map of C, denoted by $f^{-1}(C)$, is the union of A and B.

$$f^{-1}(C) \equiv A \cup B. \tag{4.7.1}$$

This definition holds whether or not A and B are disjoint. In other words the inverse map of a set of points, C, is simply the set of all points which map into C. Using this

Fig. 4.48

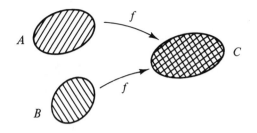

definition we can now generalize the definition of an invariant measure to include the case of maps which are not one-to-one (not homeomorphisms). The measure is said to be an *invariant* of the map $f:R^n \to R^n$ if, for any set C

$$\mu(C) = \mu(f^{-1}C) \tag{4.8.2}$$

where $f^{-1}C$ are all the points that map onto C.

It is useful to consider some explicit examples of such *invariant measures*. One such example has already been obtained for the dynamics on an integral manifold, namely

$$d\mu = \Delta(x_1, \ldots, x_{n-1}) \, dx_1 \cdots dx_{n-1} \text{ in (2.4.5), where } \Delta \text{ is given by (2.4.9):}$$

$$\Delta(K_0, x_1, \ldots, x_{n-1}) = \rho(x_1, \ldots, x_n^0(K_0, x_1, \ldots, x_{n-1}))/(\partial K/\partial x_n)_{x_n = x_n^0}.$$

Note, again, that this measure is defined on a manifold of dimension $(n-1)$ in R^n.

We now consider a second example that involves a two-to-one map, namely

$$x_{n+1} = 4x_n(1 - x_n) \equiv f(x_n)$$

This is the logistic map, with $c = 4$, which is examined also in above exercises. Now consider the set

$$C = \{x : a \leqslant x \leqslant b\},$$

illustrated in Fig. 4.49. Since $x_n = \frac{1}{2} \pm \frac{1}{2}(1 - x_{n+1})^{1/2}$, the two regions which map into C

Fig. 4.49

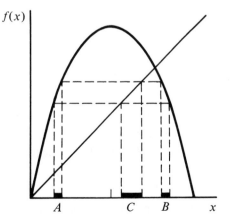

are

$$A = \{x : \tfrac{1}{2} - \tfrac{1}{2}(1-a)^{1/2} \leqslant x \leqslant \tfrac{1}{2} - \tfrac{1}{2}(1-b)^{1/2}\}$$
$$B = \{x : \tfrac{1}{2} + \tfrac{1}{2}(1-b)^{1/2} \leqslant x \leqslant \tfrac{1}{2} + \tfrac{1}{2}(1-a)^{1/2}\}.$$

For $d\mu = \rho dx$ to be an invariant measure requires

$$\int_a^b \rho(x)\, dx = \int_{\frac{1}{2} - \frac{1}{2}(1-a)^{1/2}}^{\frac{1}{2} - \frac{1}{2}(1-b)^{1/2}} \rho(x)\, dx + \int_{\frac{1}{2} + \frac{1}{2}(1-b)^{1/2}}^{\frac{1}{2} + \frac{1}{2}(1-a)^{1/2}} \rho(x)\, dx.$$

Since this must hold for any region, we can differentiate with respect to b, obtaining the condition

$$\rho(b) = \tfrac{1}{4}(1-b)^{-1/2}\rho(\tfrac{1}{2} - \tfrac{1}{2}(1-b)^{1/2}) + \tfrac{1}{4}(1-b)^{-1/2}\rho(\tfrac{1}{2} + \tfrac{1}{2}(1-b)^{1/2})$$

which must hold for all $0 < b < 1$. From the symmetry of the map about $x = \tfrac{1}{2}$ it is clear that the two factors on the right are equal, so that what is required is a solution of

$$\rho(x) = \tfrac{1}{2}(1-x)^{-1/2}\rho(\tfrac{1}{2} - \tfrac{1}{2}(1-x)^{1/2}).$$

The solution of this is (obviously?)

$$\rho(x) = \alpha[x(1-x)]^{-1/2} \qquad (4.8.3)$$

where α is an arbitrary constant, which may be selected to normalize the measure ($\int_0^1 \rho(x)\, dx = 1$, if $\alpha = \pi^{-1}$). The invariant density, $\rho(x)$, is illustrated in Fig. 4.50.

Fig. 4.50

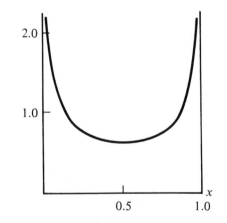

With the use of such invariant measures it is possible to give other descriptions of 'chaotic motion' which take into account the probability that this motion can actually be observed. One such indication of 'scrambled' motion is called *mixing*. A frequent example of mixing dynamics involves stirring a mixed drink, consisting of 20% rum and 80% Coca Cola (for Arnold and Avez, p. 19), or 90% gin and 10% vermouth (for P.R. Halmos, 'Lectures on ergodic theory', *Math. Soc. Japan*, 1956 pp. 36, 37). The drink is really mixed if, in any fixed spatial region, we find that the fraction of rum in this region is 20%, or 10% vermouth in the stronger drink. This example is a dynamical case involving a one-to-one map (e.g., stroboscopic map of this flow), whereas we need to also include the case of maps which are not one-to-one.

Let μ be an invariant measure of a many-to-one map $f : R^n \to R^n$, and let A be an *arbitrary* fixed region (see Fig. 4.51). Let B be another arbitrary, but dynamic set, where

Fig. 4.51

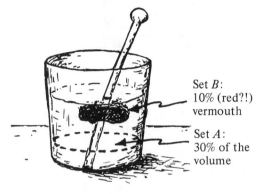

Set B:
10% (red?!)
vermouth

Set A:
30% of the
volume

both A and B have nonzero measure. The then dynamics of B that is generated by this map, $f^k(B)$ (all k), is said to be *mixing* if

$$\lim_{k \to \infty} \mu[f^{-k}(B) \cap A] = \mu(A)\mu(B). \qquad (4.8.4)$$

Here the notation $A \cap B$ represents those points which are common to the sets A and B (their intersection). A dynamic system is said to be a *mixing system* provided that (4.8.4) holds for all sets A and B. If the map is one-to-one this can be 'run forwards' to produce the condition

$$\lim_{k \to \infty} \mu[f^k(B) \cap A] = \mu(A)\mu(B) \qquad (4.8.5)$$

which is somewhat more transparent in its meaning. Thus in the case of Halmos' drink (Figs. 4.51 and 4.52), and for the region $\mu(A) = 0.3$, the mixing dynamics yields

$$\lim_{t \to \infty} \mu(A \cap f^t B) = 0.1 \times 0.3 = .03.$$

Note that these are normalized measures ($\mu(A) \leqslant 1$).

The property of mixing is a strong characterization of chaos. In particular, if a system is mixing then it is chaotic in the sense of Li and Yorke (Section 4.5), but the converse is not necessarily true (e.g., see Oono and Osikawa, 1980). Moreover, if a system is mixing it is necessarily ergodic. A system is *ergodic* under a map f if the measure of every *invariant set*, $A = f(A)$, is either $\mu(A) = 1$ or $\mu(A) = 0$. In other words the only invariant sets either have measure zero, or else represent all states except for a set of measure zero. The importance of ergodicity is presently unclear, but it is at least doubtful whether this

Fig. 4.52

property is either necessary or sufficient for many properties of statistical equilibrium or irreversible properties of systems. It is also unclear whether the strong property of mixing (if, in fact, it exists for real systems) is necessary for the existence of these statistical properties, but it appears to be more relevant. In any case it is a widely investigated property of systems.

To illustrate some of these ideas with a simple dynamic system, we consider a flow (rather than a map) in phase space, given by

$$x = a \sin \theta, \quad \dot{x} = a^2 \cos \theta \quad \text{(all } a \geqslant 0\text{)}, \qquad (4.8.6)$$

where $\dot{a} = 0$, and $\dot{\theta} = a$ is an arbitrary constant satisfying $a \geqslant 0$. This system is a nonlinear oscillator, with a frequency dependent on the amplitude.

Exercise 4.14. Show that a time independent constant of the motion for this oscillator is

$$K(x, \dot{x}) = x^2 + [x^4 + 4\dot{x}^2]^{1/2}.$$

On the manifold $K(x, \dot{x}) = K_0(a)$, what is the function $K_0(a)$? The manifolds $(a = 1.5)$ are illustrated in Fig. 4.53. The flow (4.8.6) satisfies $\ddot{x} = -a^2 x$. Why isn't the system a harmonic system? What is the proper second order equation of motion for this system?

On an integral manifold, $K(x, \dot{x}) = K_0$, the arclength $\int_{\theta_1(t)}^{\theta_2(t)} d\theta$ is invariant in time, so $d\mu = d\theta/2\pi$ is an invariant (normalized) measure on this manifold. Let A be a fixed segment $(\theta_1 \leqslant \theta \leqslant \theta_2)$ on $K(x, \dot{x}) = K_0$, and B be a dynamic segment $(\theta_3(t) \leqslant \theta \leqslant \theta_4(t))$. How does $\mu[A \cap f^t B]$ behave as a function of time in this case? It is illustrated in Fig. 4.54, which you should justify for yourself (note that the arclengths are large in this case). Clearly, in this case, this function has no limiting value as t increases, since it is periodic in time, so (4.8.5) is not satisfied. Hence the system is not mixing on integral manifolds (not surprising, since this is R^1). Note, however, that this system is (trivially) *ergodic* on its integral manifolds. On the other hand, the measures of these segments in

Fig. 4.53

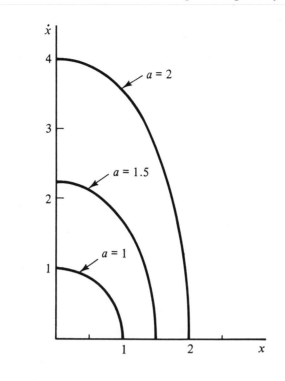

Fig. 4.54

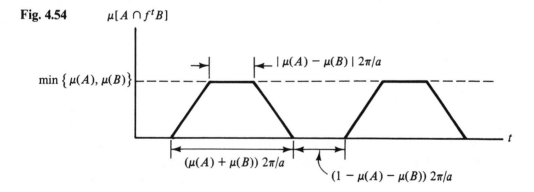

$\mu[A \cap f^t B]$

the phase space (R^2) are both zero, so the above result is not relevant to the question of whether (4.8.6) is a mixing system in R^2. These results are, however, useful in what follows.

Next, let us consider some two dimensional regions in the phase space. We note that the measure $d\mu = d\theta da/2\pi(a_2 - a_1)$ is invariant (here $a_1 \leqslant a \leqslant a_2$ is the normalizing range).

Exercise 4.15. Obtain this invariant phase plane measure of (4.8.6) in the form $d\mu = \rho(x, \dot{x}) \, dx d\dot{x}$ (i.e., involving only, $x, \dot{x}$, so that $\rho(x, \dot{x}$ is a phase function).

Let A and B be the sets

$$A = \{x, \dot{x} | a_1 \leqslant a \leqslant a_2, \theta_1 \leqslant \theta \leqslant \theta_2\}$$

$$f^t B(0) \equiv \{x, \dot{x} | a_1 \leqslant a \leqslant a_2, \theta_3(t, a) \leqslant \theta \leqslant \theta_4(y, a)\}.$$

Now what happens is much more interesting. Because the arclengths at different values of a rotate at different rates, an initially wedge-shaped region, $B(t = 0)$, will become transformed into a spiral region between the manifolds $K(x, \dot{x}) = K_0(a_1)$ and $K(x, \dot{x}) = K_0(a_2)$. This is illustrated schematically in Fig. 4.55. A little thought will hopefully

Fig. 4.55

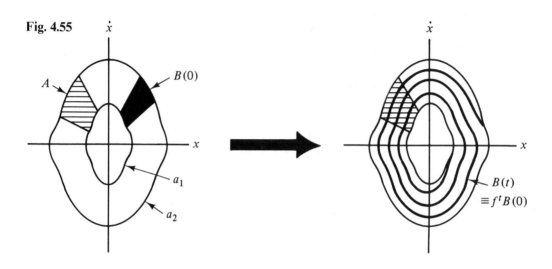

convince you that now the function $\mu[A \cap f^t B]$ will settle down to a limiting value as $t \to \infty$. Moreover, it should be clear that this limit is $\mu(A)\mu(B)$, so that (4.8.5) is satisfied.

The rate at which the mixing occurs (that is, the time scale required for (4.8.5) to become nearly satisfied) depends on the difference in the two frequencies, a_1, a_2. Typically one expects this relaxation time to be of the order of $(a_2 - a_1)^{-1}$. In the limit $a_2 \to a_1$, this relaxation time goes to infinity, so there is no relaxation, which was our first example above.

Now, although the last two regions A and $B(t)$ do satisfy (4.8.5), this does not mean that the system (4.8.6) is a mixing system in R^2. Indeed the system is not a mixing system, because it does not satisfy (4.8.5) for all A and $B(t)$ in the phase space. This is so because the integral of the motion, $K(x, \dot{x})$, is an isolating integral in R^2, for it is a smooth function of the phase variables. For example, if $B(t)$ is the region

$$B(t) = \{x, \dot{x} | a_2 < a < a_3, \theta_3(t, a) \leqslant \theta \leqslant \theta_4(t, a)\}$$

then $A \cap B(t) = 0$ for all t (see figure), and obviously (4.8.5) is not satisfied. For the system to be mixing in R^2, the region $B(t)$ must become distributed over all of the

Fig. 4.56

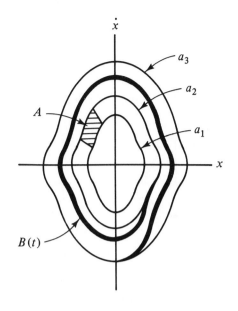

phase space (Fig. 4.56). Therefore the system (4.8.6) is neither mixing on a manifold $K(x, \dot{x}) = K_0$ (i.e., in R^1), nor in the full space R^2.

In higher dimensional phase spaces, it is possible to have isolating integrals, and still have mixing on their integral manifolds. The above system has a phase space with too low a dimension for this mixing to be possible on the integral manifolds.

Classic examples of mixing maps in R^2 are given in Chapter 6.

4.9 The circle map

In this section we will consider a map which has an entirely different physical origin from the logistic map, and which exhibits correspondingly different effects. We are interested here in the influence of one ('driver') oscillator when coupled to a second oscillator, and will attempt to model this effect by a simple one-dimensional map. The dynamics will be governed by two control parameters, the frequency ratio of the (uncoupled) oscillators, Ω, and their coupling strength, K. Moreover the map will be one-to-one for $K < 1$, but is two-to-one if $K > 1$, in which case it has many of the chaotic properties of the logistic map. Since we have already discussed these chaotic properties in some detail, we will largely concentrate here on the one-to-one mapping region, $K < 1$. For more details, see Lanford (1987).

Let ϕ and ψ be the phase angles of two oscillators. If they are decoupled, then their motion can be represented on a torus, T^2, in 'our R^3', as discussed in Chapter 2. Taking a cross section of this torus, we can obtain a Poincaré map for the phase angle of one of these oscillators. This map occurs on a (topological) circle, or one-torus T^1.

We now distinguish between the two oscillators, treating the ψ-oscillator as a prescribed 'driver' of the other, and proceed to study the Poincaré map of the 'influenced' ϕ-oscillator. The physical nature of the ϕ-oscillator is not specified in the present study, but is implicit in the following models. This will be discussed briefly at the end of this section.

Let $(P(\phi), P(\psi))$ be the periods of the uncoupled (ϕ, ψ) oscillators (Fig. 4.57). The

Fig. 4.57

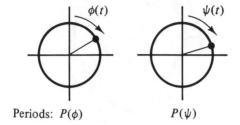

Periods: $P(\phi)$ $P(\psi)$

successive points of the Poincaré map are $\phi_n = \phi(t = nP(\psi))$, and when the oscillators are not coupled, $\phi_{n+1} - \phi_n = 2\pi P(\psi)/P(\phi)$ (Fig. 4.58), and the Poincaré map is simply

$$T\phi \to \phi + \Phi$$

or

$$\phi_{n+1} = \phi_n + \Phi, \quad \Phi = 2\pi P(\psi)/P(\phi). \tag{4.9.1}$$

This is illustrated in Fig. 4.58. When there is coupling between the oscillators, this map

Fig. 4.58

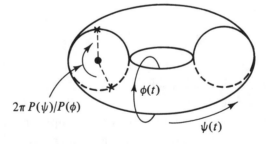

$2\pi P(\psi)/P(\phi)$ $\phi(t)$ $\psi(t)$

is modified to

$$\phi_{n+1} = \phi_n + \Phi + F(\phi_n) \equiv \phi_n + t(\phi_n), \tag{4.9.2}$$

where the function $F(\phi)$ must satisfy $F(\phi + 2\pi) = F(\phi)$ for all ϕ. This simply expresses the fact that the effect of the ψ-oscillator on the same phase point of the ϕ-oscillator must be the same (the notion is on T^2). Note that, while ϕ and $\phi + 2\pi$ are the same phase points, the *cumulative* values of the ϕ_n ($n = 1, 2, \ldots$) are generated by

$$T\phi \to \phi + \Phi + F(\phi) \equiv \phi + t(\phi). \tag{4.9.3}$$

In other words $\phi \in R^1$, and therefore can tend to infinity.

Note also that the shift $t(\phi)$ in (4.9.1) is generally only defined up to a multiple of 2π. However, once it is fixed for some ϕ, it remains unique by continuity. The origin of this ambiguity in $t(\phi)$ reflects a physical ambiguity inherent in this Poincaré map; namely, this map does not indicate how many full periods of the ϕ-oscillator occur in the time $P(\psi)$. This physical information must be implicitly contained in the function $t(\phi)$ – because this function reflects the influence of the ψ-oscillator on the ϕ-oscillator for some specific rotational motion of the latter oscillator. This physical fact is not explicit in the present models of coupled oscillators.

To illustrate a simple example of the coupled map, (4.9.2), consider the case of a ϕ-oscillator which tends to rapidly (relative to $P(\psi)$) return to its unperturbed orbit after it is perturbed by an outside source. Such an oscillator is called a relaxation oscillator, and will be described in more detail in the next chapter (we are ahead of the differential equations story!). Next, assume that the ψ-oscillator acts on the ϕ-oscillator impulsively (instantaneously) every $P(\psi)$ seconds, changing its velocity. Moreover, assume that $P(\phi) > P(\psi)$. The resulting motion of the ϕ-oscillator then looks something like shown in Fig. 4.59. If the unperturbed motion is circular in the phase space, $(x, \dot{x})$,

Fig. 4.59

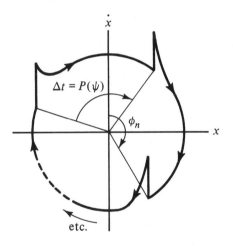

then 'spikes' are added, because of the impulsive interaction (changing $\dot{x}$, but not x). Only three impulses are shown in the figure, for simplicity. The phase shifts caused by the impulsive at ϕ_n is $F(\phi_n)$ in (4.9.2), which can clearly be positive or negative, depending on whether it advances or retards ϕ relative to its unperturbed motion (in the figure, two impulses might produce $F < 0$, and one might yield $F > 0$).

A frequently used form of the map (4.9.2), which may bear some approximate relationship to the above coupled oscillators, is

$$\phi_{n+1} = \phi_n + \Phi - K \sin(\phi_n). \tag{4.9.4}$$

Sometimes the substitution

$$\phi_n = 2\pi\theta_n, \quad \Phi = 2\pi\Omega \tag{4.9.5}$$

is used, so that

$$\theta_{n+1} = \theta_n + \Omega - \left(\frac{K}{2\pi}\right)\sin(2\pi\theta_n). \tag{4.9.6}$$

Ω is not, of course, an angular frequency (despite its notation), but a pure number

$$\Omega \equiv P(\psi)/P(\phi), \tag{4.9.7}$$

the ratio of the periods of the two uncoupled oscillators. The maps (4.9.4) and (4.9.6) are known as *circle maps*. These maps are one-to-one provided that $d(T\phi)/d\phi > 0$ for all ϕ, and since $T\phi = \phi + \Phi - K\sin\phi$, this requires

$$1 > K \quad \text{(one-to-one)}. \tag{4.9.8}$$

Examples of oscillators to which this model has been applied are the heartbeat (Glass and Perez, 1982) and the Rayleigh–Bénard vortical motion in fluids (Fein, Heutmaker, and Gollub 1985), and optical devices. Applications of the circle map to other systems (Josephson junctions, charge density waves) have also been proposed (Bohr, Bak and Jensen, 1984).

Exercise 4.16. Refer to the last figure and the circle map (4.9.4). How is K related to the figure? What physical conditions are required for the relaxation properties of the ϕ-oscillator, and for the value of Ω, in order for (4.8.4) to be a (possibly) reasonable approximation?

We use, as usual, the notation

$$T^{n+1}\phi = T(T^n\phi) = \phi_{n+1} \tag{4.9.9}$$

for the iterates of the map. Poincaré introduced the concept of a *rotation number*, defined by

$$2\pi\rho \equiv \lim_{n\to\infty} \frac{T^n\phi - \phi}{n} \tag{4.9.10}$$

which is independent of ϕ. In terms of the variable θ, (4.8.5), the related concept is called the *winding number*

$$W(K,\Omega) = \lim_{n\to\infty} \frac{\theta_n - \theta_0}{n}, \tag{4.9.11}$$

which is simply related to ρ, (4.9.10) by

$$W(K,\Omega) \equiv \rho(K, \Phi = 2\pi\Omega). \tag{4.9.12}$$

The rotation number, $2\pi\rho$, is the time-averaged angular rate of rotation of the ϕ-oscillator. If $K = 0$, then $\phi_n = \phi_0 + n\Phi$, and $2\pi\rho = \Phi$, by (4.9.10). Therefore, in this case the rotation number is simply the angular step at each iteration, and $\rho = \Omega = W$. When $K \neq 0$ the rotation rate differs from Φ. This is most readily illustrated at the other extreme, when $K > \Phi$ (Fig. 4.60). From (4.9.4) it can be seen that, for some values of ϕ_n, the

Fig. 4.60

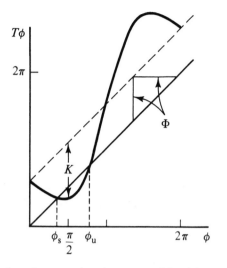

map yields $\phi_{n+1} < \phi_n$, so that the rotation is reversed in this region of ϕ. Also, if we plot $T\phi$ vs. ϕ, it is clear that $T\phi$ has two fixed points, ϕ_s (stable) and ϕ_u (unstable). The stability properties follow from

$$\frac{d}{d\phi}(T\phi) = 1 - K \cos \phi$$

which is < 1 (> 1) if $\phi < \pi/2$ (respectively, $\phi \gtrsim \pi/2$). Since

$$\lim_{n \to \infty} \phi_n = \phi_s \quad (\text{if } \phi_0 \neq \phi_u),$$

it follows that

$$2\pi\rho = \lim_{n \to \infty} \frac{\phi_n - \phi_0}{n} = 0$$

for all ϕ_0. In other words, asymptotically in time, the points of the Poincaré map do not rotate about the center of the circle (they go to fixed points). Thus, if

$$1 > K > \Phi, \qquad \text{then } \rho = 0$$

or

$$1 > K/2\pi > \Omega, \quad \text{then } W = 0 \tag{4.9.13}$$

When $K = \Phi$, it can be seen from Fig. 4.60 that $\phi_s = \phi_u = \pi/2$. The mappings in the neighborhood of this fixed point all move 'forward' ($\phi_{n+1} > \phi_n$), and Arnold referred to such points in any cyclic orbit as (forward) *semistable points*. They clearly occur at bifurcation points (the catastrophes set of K, Φ), Fig. 4.61.

Fig. 4.61

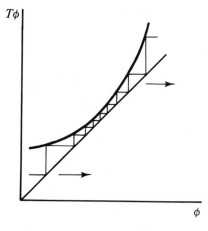

In general, Poincaré proved that if $\rho = m/n$, where m and n are relatively primed integers (ρ is a rational number), then there exists solutions of

$$T^n \phi = \phi + 2\pi m. \qquad (4.9.14)$$

Indeed there are usually $2n$ such solutions, which alternate between the n stable values $(\phi_1^s, \ldots, \phi_n^s)$ and the unstable set $(\phi_1^u, \ldots, \phi_n^u)$, where $T\phi_j = \phi_{j+1}$. These sets are called cycles of T. A rotation number $\rho = m/n$ will only occur for a particular set of maps $T(K, \Phi)$, which Arnold referred to as the *level set* of $\rho(T)$. The level set of $\rho = 0$, which we just determined, (4.9.13), is illustrated in Fig. 4.62. Poincaré stated [for a proof, see

Fig. 4.62

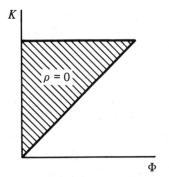

Arnold (1965)] that $\rho(T)$ is a continuous function of T, so $\rho(K, \Phi)$ is a continuous function in the special case of the circle map. Thus the level sets are all continuous regions in the (K, Φ) plane, and have simple boundaries.

From a physical point of view, a level set may be considered in several different ways. Consider, for example, increasing the frequency of the ψ-oscillator, while maintaining a constant coupling, K, to the ϕ-oscillator (Fig. 4.63). If $\rho = m/n$ over some range of $P(\psi)$, this means that the ϕ-oscillator tends to a cyclic mode which is '*locked in*' to the driving oscillator's period (ρ remains constant as $P(\psi)$ is varied). The phenomena is called the

Fig. 4.63

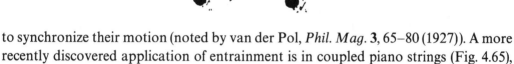

entrainment of the ϕ-oscillator by the ψ-oscillator, and is discussed further in Chapter 5. A classic (17th century) example of this entrainment was Christian Huygens' observation that when two clocks are hung back to back on a wall (Fig. 4.64), they tend

Fig. 4.64

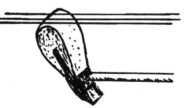

to synchronize their motion (noted by van der Pol, *Phil. Mag.* **3**, 65–80 (1927)). A more recently discovered application of entrainment is in coupled piano strings (Fig. 4.65),

Fig. 4.65

which are intentionally mistuned, to produce a sustained aftersound – but we will save this for Chapter 5, where other physiological examples will also be noted.

While it is less common in practice, we might also take two oscillators, with fixed physical properties (specified by Φ), and increase the coupling constant between them. Once again, we may obtain entrainment at a given value of ρ only for a limited range of K, and another entrainment value may occur for larger K.

Exercise 4.17. It is useful at this point to give a physical interpretation of the level set of $\rho = 0$. Specifically, consider the $P(\phi) \gg P(\psi)$, with $K > \Phi$, and interpret the meaning of

$\rho = 0$. Make the physical interpretation again, if $P(\psi) = P(\phi)$ (see the discussion following (4.8.32)). Prove that generally,

$$\rho(\Phi, K) + l = \rho(\Phi + 2\pi l, K) \quad (l\text{:any integer}).$$

The boundaries of the level sets occur where the stable and unstable solutions become semistable points, ϕ^{ss}. This occurs when the slope of $T^n\phi$ equals one at ϕ^{ss} (see our previous tangency condition). Hence the boundary of the level set $\rho = m/n$ is given by

$$T^n\phi = \phi + 2\pi m, \qquad (4.9.14)$$

together with the tangency requirement

$$[\mathrm{d}(T^n\phi)/\mathrm{d}\phi]_\phi = 1,$$

which is the same as (recall (4.2.11)),

$$\prod_{j=0}^{n-1} (\mathrm{d}T/\mathrm{d}\phi)_\phi = 1, \qquad (4.9.15)$$

where $\phi_j = T^j\phi$ are the semistable points.

These level sets are separated by irrational ρ values. In any neighborhood of an irrational value of ρ is a level set, because the rational values of ρ are dense. Thus the function $\rho(K, \Omega)$ has a very strange property, as a function of Φ, which is schematically illustrated in Fig. 4.66. ρ is a continuous function, which is constant in some finite

Fig. 4.66

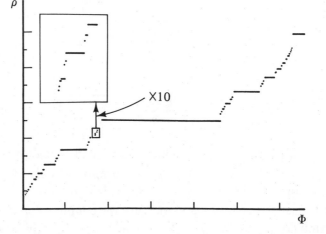

interval of Φ, whenever ρ takes on a rational value, $\rho = m/n$ (read the sentence again!). That is indeed a strange function, which is an example of functions known as *Cantor–Lebesgue functions*. The schematic figure of ρ (it can't really be drawn!) is known, quite appropriately, as the '*devil's staircase*'.

To illustrate these level sets, for $\rho = \frac{1}{2}$ the equations (4.9.14) and (4.9.15) yield

$$2\pi = 2\Phi - K \sin\phi - K \sin(\phi + \Phi - K \sin\phi)$$

$$1 = (1 - K \cos\phi)(1 - K \cos(\phi + \Phi - K \sin\phi)).$$

Expanding Φ and $(\sin\phi, \cos\phi)$ in powers of K, one finds, after some labor, the relationship

$$\Phi = \pi \pm (K^2/4) + O(K^4). \tag{4.9.16}$$

For $K = 1$, the actual boundary value is $\Phi = \pi \pm 0.232\,37\ldots$ (Arnold), compared with this result of $\pi \pm 0.25$. Arnold reported that a similar calculation for the boundaries of the level set $\rho = \frac{1}{3}$ yields

$$\Phi = \frac{2\pi}{3} + \frac{3^{1/2}}{12} K^2 \pm \frac{7^{1/2}}{24} K^3 + O(K^4). \tag{4.9.17}$$

Generally, for $\rho = m/n$

$$\Phi = 2\pi m/n + O(K^{n-1})$$

but the leading term may vanish, as in (4.9.16). In any case the level sets are very narrow for larger values of n. It is because of the rapid decrease in the range of the level sets with increasing n that there remains a finite measure for the irrational values of ρ. Their structure is illustrated in Fig. 4.67, which is adopted from Arnold. These structures

Fig. 4.67

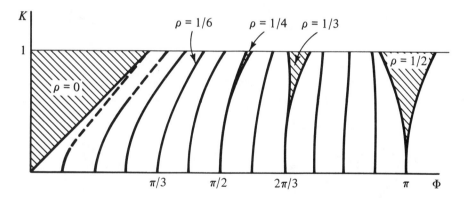

have since become known as 'Arnold tongues'. Note that Arnold's figure may be extended using (see Exercise 4.16)

$$\rho(\Phi, K) + l = \rho(\Phi + 2\pi l, K).$$

When $K > 1$, the circle map becomes two-to-one in regions of ϕ, and period doubling bifurcation series, together with many other bifurcations can occur. These studies go beyond our present concerns, but can be found discussed in the literature (e.g., Ostlund *et al.*, 1983, Aronson *et al.*, 1982, Bélair and Glass, 1985).

4.10 The 'suspension' of a tent map

In the introduction of this chapter, we considered briefly the possible relationship between maps and the continuous dynamics (flows) in the phase space associated with a system of ODE. The reason that this may be of interest is that physical systems are frequently modeled by ODE. It is nice, therefore, to have some insight into the possible relationshp (map → flow) in order to better understand the possible relationship between a map and physical systems. In particular we can determine how complicated the system must be (e.g., the dimensions of the phase space R^n).

When the map is a diffeomorphism (one-to-one, etc.) then it can always be regarded as the first-return map of a flow in some manifold, by a very simple operation, called a *suspension* (Nitecki, 1971). The idea is that, if the diffeomorphism, f, maps the manifold M onto itself, $f: M \to M$ then we construct a new manifold, $\langle M \rangle$, which is the product space of M and the unit interval $I = (0, 1)$, $\langle M \rangle = M \times I$, and satisfying the relationship $(x, 0) = (f(x), 1)$. Thus, if $M = I$ (the unit interval), then $\langle M \rangle$ is either an annulus or a Mobius strip (Fig. 4.68), depending on whether the map $f(x)$ is orientation preserving or

Fig. 4.68

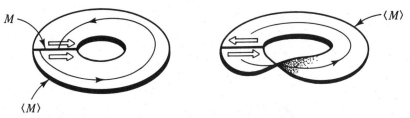

not. Similarly, if $M = S^1$ (a circle), the new manifold, $\langle M \rangle$, is either a two-torus, T^2, or else a Klein bottle (Fig. 4.69), again depending on whether or not $f(x)$ is orientation

Fig. 4.69

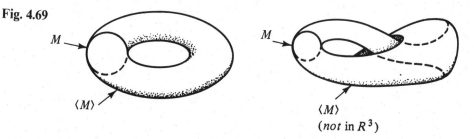

preserving. While this construction of a flow is nice from a mathematical point of view, the resulting flow on the manifold $\langle M \rangle$ may not be clearly related to any physical system of interest. That, of course, requires a separate investigation.

When a map $f(x)$ is continuous, but two-to-one, it is much more difficult to find a flow in some manifold M^* for which $f(x)$ can be related to (but is not) a Poincaré map. To illustrate the problems which are involved, we follow Haken's study (1985) of this question for a tent map. Following the reasons discussed in the introduction of this chapter, it is clear that a two-to-one map cannot be a first-return map of a flow in some other manifold. This is simply because flows have unique orbits through each point. Therefore a two-to-one map can only be related to one coordinate of a two or more dimensional surface of section, rather than the entire surface of section. Therefore chaotic motion can only result from a flow in three or more dimensions.

Let us construct a rectangular region, with one side parallel to the x-coordinate, which is mapped according to some tent map (4.2.2). The other coordinate we will denote as z. The question is, can this rectangular region have a continuous one-to-one first-return map

$$x' = f^*(x, z), \quad z' = g^*(x, z) \quad (0 \leqslant x \leqslant 1, 0 \leqslant z \leqslant 1) \tag{4.10.1}$$

which is produced by some flow in the R^3 phase space (x, y, z), and is such that x maps according to the tent map,

$$f(x) = f^*(x, z) = \begin{matrix} 2x \,(\text{if } 0 \leqslant x \leqslant \tfrac{1}{2}) \\ 2(1-x)\,(\text{if } \tfrac{1}{2} \leqslant x \leqslant 1) \end{matrix} \tag{4.10.2}$$

for all $0 \leqslant z \leqslant 1$? Here we take the tent map which maps $[0, 1]$ onto itself (as in the case of the logistic map with $c = 4$), Fig. 4.70.

Fig. 4.70

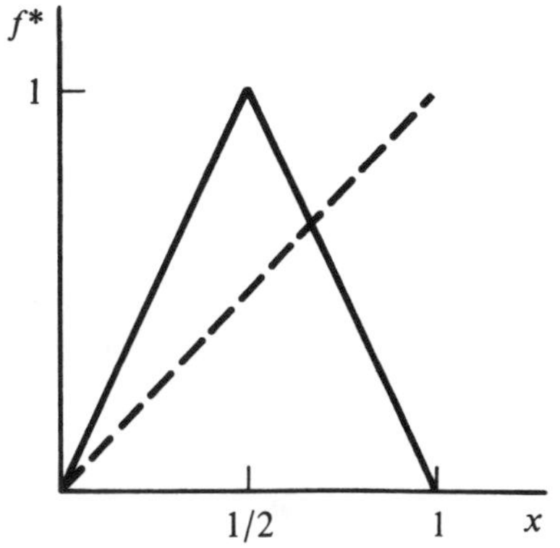

The flow is required to take all points $(0 \leqslant x \leqslant \tfrac{1}{2},\ z)$ onto the interval $0 \leqslant x \leqslant 1$, suitably changing z so as to leave room for the other half of the rectangular region, $x \geqslant \tfrac{1}{2}$ (Fig. 4.71). A simple (area preserving) way of accomplishing this is to cut the rectangle along $x = \tfrac{1}{2}$, and put a 'hinge' at $x = \tfrac{1}{2}, z = 1$. As the points flow around the z axis the hinge moves to $(x = 1, z = \tfrac{1}{2})$ and the initial region $(\tfrac{1}{2} \leqslant x \leqslant 1, 0 \leqslant z \leqslant 1)$, rotates around the hinge, so that it is mapped onto $1 \geqslant z > \tfrac{1}{2}$. The z distances are

Fig. 4.71

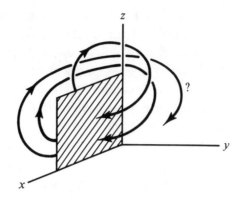

Fig. 4.72

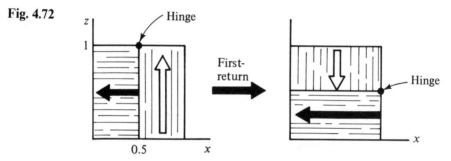

compressed by a factor $\frac{1}{2}$, as the x distances increasing by a factor 2 (Fig. 4.72).

This proposed flow is, of course, not unique. There might be (for example) any number of complete rotations of the present entire flow about any periodic 'hinge' circling the z axis (Fig. 4.73).

Fig. 473

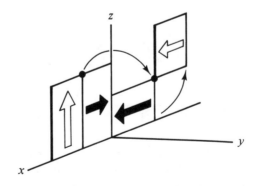

The above proposed flow does not, however, satisfy continuity nor closure in R^3. The first difficulty is there cannot be a cut associated with a flow that depends continuously on the initial conditions. That is, Fig. 4.73 must look more like shown in Fig. 4.74

Fig. 4.74

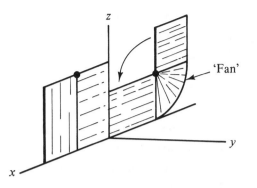

(containing some 'fan' which preserves continuity). A second problem is that the two boundaries which are coming together (see Fig. 4.74) cannot both be closed, for this would lead to a two-to-one map of the points on these boundaries, and that cannot occur for unique flows. To overcome these difficulties for a flow in R^3, appears to be possible only if the Poincaré surface of section has a fractal boundary structure, such as the one schematically illustrated in Fig. 4.75. Part of the fractal structure arises from the

Fig. 4.75

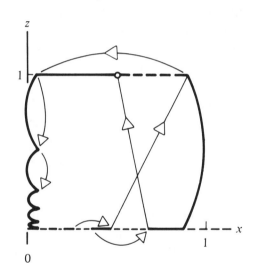

'continuity fan', which produces a self-similar fractal structure on the left side. The other fractal aspect is the closed (solid line) and open (dashed line) boundaries, which is essentially self-similar on the lower 'edge'. The lower left hand corner of this boundary must apparently be a sophisticated 'intersection' of two types of fractals.

These results indicate, that in order to obtain a physically meaningful system, which has a Poincaré surface of section yielding a tent-like two-to-one map, requires that the flow occur in R^n, with $n \geqslant 4$. In other words, physically realistic systems which have the dynamic complexity of a logistic map, apparently must have a dynamical dimension greater than, or equal to, four. (Also see Sverdlove (1977), Mayer-Kress and Haken (1987)).

4.11 Mathematics, computations, and empirical sciences

Even at this early stage in our exploration of nonlinear dynamics, we encounter an issue which may be one of the most basic lessons, which will come from the study of nonlinear dynamics. The general issue concerns the interface between mathematics, computations, and empirical sciences. The present specific issue is related to the complicated, 'chaotic' dynamics which we have encountered in the simple two-to-one maps of this chapter. Many features of this dynamics were uncovered because we had the help of computers to perform the necessary iterations. In certain select cases, such as the logistic map with $c = 4$, we could mathematically establish that chaotic motion does occur, by determining that the Lyapunov exponent is positive, and actually obtain its (precise) value. Such cases are rare indeed, but that is not the issue we will consider here.

What we are concerned with at present is the fact that both empirical sciences and computations are fundamentally restricted to observations/computations involving a finite amount of information. We can only write down so many numbers in 'our lifetime'! In other words, these occupations deal with THE FINITE – some finite 'precision' (rational numbers), manipulated for some finite amount of time. By way of contrast, as Leibniz expressed it, mathematics is the science of THE INFINITE, which contains the concept of the continuum, and such related concepts as limit point, derivatives, irrational numbers, and Cantor sets. These concepts belong to the realm of THE INFINITE, and of mathematics, not to THE FINITE, which is the realm of computations and empirical sciences.

The idea that computations and empirical sciences should be intimately related, has been concisely expressed by Fredkin (1982) in the form of the postulate: 'There is a one-to-one mapping between what is possble in the real world, and what is theoretically possible in the digital simulation world,' and the corollary 'That which cannot, in principle, be simulated on a computer, cannot be part of physics', where 'physics' presumably means any physical system. Such statements can easily stimulate long discussions, some of which we will return to in Chapter 10.

The point we will discuss here is that the 'chaotic' motion, discovered in this chapter, brings THE INFINITE and THE FINITE face to face (Fig. 4.76)! It does so because

Fig. 4.76

the various forms of 'chaos' which we have defined mathematically were naturally based on concepts from THE INFINITE, whereas our computations and observations fall within the realm of THE FINITE. In most dynamics, which involves regular forms of motion, this distinction is not so clearly apparent because the dynamics never explores the 'fine structures' contained in THE INFINITE, particularly when only finite times are considered. Thus, even for regular motion, infinitesimal distances can be explored, as in the Poincaré recurrence theorem, but this involves arbitrarily long intervals of time. What the chaotic motion of THE INFINITE does is to force us to consider whether, within a very finite time, our actions in THE FINITE have any relationship to these theoretical ideas from THE INFINITE. It is clearly a very basic and important issue, about which there is presently only a very limited understanding. However, since this issue may ultimatey influence our understanding of nature itself, it is certainly worth a brief discussion, despite the present very incomplete insight.

To be much less philosophical, let us consider some really finite dynamics, say a system with only five possible states. We would probably all agree that the deterministic dynamics of such a system cannot be chaotic, in any 'real sense', because after no more than five 'steps' the motion must repeat. In other words, all motion is ultimately periodic, possibly ergodic (left example), or containing several 'short' periods (middle), or several basins of attraction (right) (Fig. 4.77). But what if the system has

Fig. 4.77

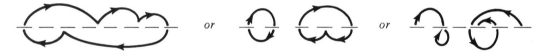

100 states, or 100,000 states, or 10^7 states? Does it make empirical sense to talk about a system with 10^{100} states, even if it is finite? (A.N. Kolmogorov, 1984, indeed suggested that all numbers may be meaningfully, even if not precisely divided into small, medium, large, and extra-large numbers.) Which of these systems can have 'chaotic' or 'complex' forms of dynamics, or can we not make such a distinction in the realm of THE FINITE? The question is rhetorical, of course, since we presently have no generally accepted definitions for 'chaos' or 'complexity' within THE FINITE.

There are a variety of topics that can be raised in this area, such as:

(A) Introduce concepts which deal with questions concerning the possibility of computing numbers (and in particular, dynamical orbits) with finite algorithms (programs). This gets into such topics as Turing's universal computer, computable functions, computable irrational numbers, Kolmogorov's concept of complexity, and questions of polynomial vs. nonpolynomial computing time. These important topics are beyond the scope of our present study, and bear at least questionable relevance to the understanding of the dynamics in THE FINITE (e.g., see Bennett, 1985, 1986).

(B) Try to establish relationships between computed properties and associated properties of the solutions in THE INFINITE. We will examine briefly several examples of such enquiries.

Consider first the computation of an orbit using a computer with its necessarily limited accuracy. If there is an exponential separation of nearby solutions, small differences in the calculation of an orbit rapidly produces totally different results. This, of course, means that the computed result no longer has anything to do with THE INFINITE, but is a consequence of the possible finite numbers of some particular computer. Fig. 4.78 illustrates this fact for the logistic map, with $c = 4$. One computer

Fig. 4.78

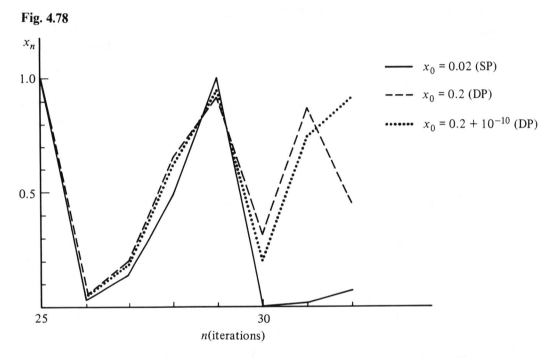

was used, with either single precision (SP: 7 digits) or double precision (DP: 17 digits) accuracy, for the 'same' initial condition $x_0 = 0.2$. In addition a DP iteration was done for an initial condition which differs by 1×10^{-10}.

The SP value of x_n deviates noticably for $n > 26$, and the different initial states are clear for $n > 28$. When $n \geqslant 30$ there is no correlation between the SP and DP solution with the same initial condition. Also, a difference of 10^{-10} in the initial condition produces 'uncorrelated' solutions for $n \geqslant 32$. Thus, roughly speaking, THE FINITE dominates the computed (hence known) solutions after about thirty iterations for the map

$$x_{n+1} = 4x_n(1 - x_n) \quad (0 \leqslant x_0 \leqslant 1). \qquad (4.11.1)$$

The nature of the chaos in the iterates of (4.11.1) can be better understood once we use the Ulam–von Neumann transformation.

$$x_n = \sin^2(\pi y_n), \tag{4.11.2}$$

where y is now only defined modulo 1 (i.e., y is on a unit circle, S^1). Using $4\sin^2(\pi y_n)$ $\cos^2(\pi y_n) = \sin^2(2\pi y_n)$, it is clear that (4.11.1) transforms to the map

$$y_{n+1} = 2y_n \bmod(1), \tag{4.11.3}$$

and y_n is uniquely defined by requiring $1 > y_0 \geqslant 0$. The mod (1) simply means that we are to discard the integer part of y_{n+1} at each step. Using this mod(1) notation, we can readily write down the formal solution to (4.11.3), namely

$$y_n = 2^n y_0 \bmod(1).$$

However, this does not help to clarify much, since the numbers y_n must still be computed.

A much more useful representation of the dynamics of (4.11.3) is obtained by using *binary notation*. This is, of course, also the natural language of computers, with their 'on' and 'off' switches. Since $0 \leqslant y < 1$, we are interested in the binary fractions $2^{-n}(n = 1, 2, 3, \ldots)$, and the binary representation of a number y is given by first making the identification

$$\sum_{n=1}^{\infty} a_n 2^{-n} \quad (a_n = 0 \text{ or } 1), \tag{4.11.4}$$

and then recording these coefficients, a_n. We do this by writing

$$y = .a_1 a_2 a_3 \ldots, \tag{4.11.5}$$

with the understanding that the value of y is given by (4.11.4). A sequence of such symbols, $(a_1, a_2, a_3 \ldots)$, is sometimes referred to as a 'string'.

The binary representation (4.11.5) has the great advantage that if

$$y = \cdot a_1 a_2 a_3 \ldots$$

and

$$y' = 2y \bmod(1) \equiv \cdot a_1' a_2' a_3' \ldots \tag{4.11.6}$$

then $a_k' = a_{k+1}$. In other words, the operation $2y \bmod(1)$ simplify shifts the numbers a_k one place to the left, and discards a_1. For example, if $y = 0.100\,101\,0\ldots$, then $2y \bmod(1) = 0.001\,010\ldots$. Thus the dynamics (4.11.3), in binary notation (4.11.5), simply amounts to shifting the a_k one place to the left, discarding the previous a_1 because the mod (1) operation. [If $y_{n+1} = 2y_n$, without the mod (1) operation, so that y is any real number, then the shift of all a_k in the binary representation, $y = \sum_{n=-\infty}^{\infty} a_n 2^n$, is an example of the so-called *Bernoulli shift*, $a_n' = a_{n+1}$. Also see Smale's horseshoe map, Appendix K.]

Because THE INFINITE dynamics of the particular logistic map (4.11.1) (i.e., $c = 4$) amounts to the shift operation (4.11.6), and this discards the previous a_1 at each iteration, we clearly lose information about the previous state after each iteration. Thus to get a chaotic solution out of the dynamics (4.11.3) requires an initial y_0 which is an infinite, random, binary digit string, (4.11.5). Most $y_0 \in R$ (in a Lebesgue measure sense) are in fact truely random and cannot be computed by any deterministic finite program. Thus, within the realm of THE INFINITE, we see that the chaos of (4.11.3), and hence (4.11.1), resides in this randomness in the real numbers, which are 'used' (in principle) for the initial conditions. Thus, as Ford (1986) has emphasized, the randomness of the dynamics of (4.11.3) and (4.11.1), can be viewed as a 'transcription' of this randomness which is intrinsic to most numbers. But as he and others (e.g., Prigogine and Stengers, 1984) have also emphasized, we do not deal with the real number in computations on experiments. There we must use finite amounts of information.

What happens in the realm of THE FINITE? Because (4.11.3) discard the previous a_1 information from the binary string (4.11.5) at each interation, it is clear that all information in the initial condition

$$y_0 = \cdot a_1 a_2 \cdots a_N \tag{4.11.7}$$

is discarded after N iterations. Thus, putting (4.11.7) on a computer (assuming it can hold the N bits of information) will simply yield $y_k = 0$ for $k \geqslant N + 1$, which is hardly chaotic. Thus there is not anything that looks even vaguely chaotic for (4.11.3) in THE FINITE.

On the other hand, the calculations at the beginning of this section illustrate that calculations (in THE FINITE) of the dynamics of (4.11.1) continue indefinitely in what may appear to be a reasonably 'chaotic', if not altogether 'accurate' fashion (from the point of view of THE INFINITE). It should be noted that this great differences between the calculated behavior of (4.11.1) and (4.11.3), is due to the fact that THE FINITE is very different in the two systems. This is because they are only related within THE INFINITE, namely through (4.11.2), which has no meaning in THE FINITE. Thus there is no reason for their calculated behavior to be related after N steps, given (4.11.7).

This leaves open the question as to the possible relationship between the computed orbits of (4.11.1) and the mathematical orbits, in THE INFINITE – including, perhaps, the question 'Who Cares?'. After all, the only orbits we know about are the computed orbits. So a scientist might take the point of view that she will let the mathematicians worry about THE INFINITE. However, sometimes certain properties of the dynamics in THE INFINITE can be established even though detailed orbits are unknown. Thus one may be able to determine the Lyanpunov exponent, or the invariant measure, or the dimensions of an attractor, or whether the system is ergodic, and so forth. In that case, it would be nice to know if these properties also appear in THE FINITE – and, more specifically, which FINITE? That is, how much information do we need to manipulate to 'approximate' THE INFINITE properties? This is a very large subject, which presently has very few answers, but let us look at a few ideas.

Following Benettin *et al.* (1978) we call a computed orbit (in THE FINITE) a *pseudo-orbit*, and ask 'To what extent does a pseudo-orbit have the unpredictable properties of the exact mathematical orbit (in THE INFINITE), in a deterministically chaotic system?' One response to similar questions has been to introduce the concept of a *shadowing orbit* (The term 'shadowing' was introduced by Bowen, 1978).

To introduce this concept, let

$$x_{n+1} = F(x_n) \quad (0 < x < 1) \tag{4.11.8}$$

yield an exact 'orbit', associated with an initial state x_0. We illustrate this as a continuous curve (for clarity) in Fig. 4.79. A *δ-pseudo-orbit* $(y_0, y_1, y_2, \ldots)$ is one such

Fig. 4.79

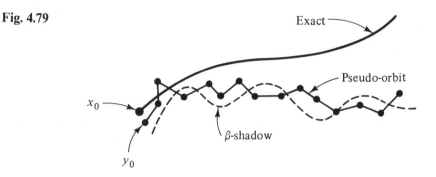

that $|F(y_n) - y_{n+1}| \leqslant \delta$ for all n. It allows for the introduction of a numerical error, or noise, at each step of the iteration. A simple example is a roundoff to some accuracy δ (e.g., $\delta = 10^{-6}$)

$$y_{n+1} = \delta \operatorname{Int} [F(y_n)/\delta] \tag{4.11.9}$$

where $\operatorname{Int}(z)$ equals the integer part of z. In this case $|F(y_n) - y_{n+1}| = |F(y_n) - \delta \operatorname{Int}[F(y_n/\delta)]| \leqslant \delta$, as can be seen by dividing by δ. Since the dynamics of (4.11.9) occurs on a set of rational numbers (a 'lattice') we illustrate the pseudo-orbit as a series of straight line segments (again for clarity) connecting these lattice points (Fig. 4.79). If the dynamics of (4.11.8) is bounded (e.g., $0 \leqslant x \leqslant 1$), then the number of lattice points associated with the pseudo-orbit is finite. (e.g., δ^{-1} in the case (4.11.9)), in which case all pseudo-orbits are eventually periodic. That is y_0 is either a periodic point or y_n is periodic for some n.

In recent years there have been studies (e.g., Bowen, 1978; Guckenheimer and Holmes, 1985; McCauley and Palmore, 1986; Coven, Kan, and Yorke, 1986) to understand under what conditions there exists exact orbits, $\bar{x}_n = F(\bar{x}_n)$, which can 'shadow' pseudo-orbits. More specifically, $F(x)$ is said to have the *shadowing property* if, for every $\beta > 0$ there is a $\delta(\beta) > 0$ such that every δ-pseudo-orbit, y_n, can be β-shadowed by some exact solution $\bar{x}_n$, as defined by

$$|\bar{x}_n - y_n| \leqslant \beta \quad \text{(all } n \geqslant 0\text{)}. \tag{4.11.10}$$

If this holds, then $\bar{x}_n$ is said to β-shadow the pseudo-orbit y_n. Put another way, even though the δ-pseudo-orbit diverges from the orbit (starting at x_0) that it was intended to describe, the pseudo-orbit may be shadowed within a distance β by another exact solution, provided $\delta(\beta)$ is sufficiently small. Notice that this property may not be of much use since, given β, we may not have a computer with the necessary accuracy, $\delta(\beta)$, in which case that computer need not have β-shadowing pseudo-orbits.

If the β-shadowing orbits exist, and are in some sense 'typical' of the entire family of exact orbits (4.11.8), then the computed pseudo-orbits will reflect this same 'typical' behavior, and this feature of the dynamics will not be lost by the finite numerical accuracy. At present it is not known that 'typical properties' of the entire family of exact orbits, if any, are retained by the β-family – (provided they exist!) and hence our original question remains unanswered at present.

Exercise 4.18. (McCauley and Palmore 1986) To explicitly illustrate a β-shadowing orbit, consider the Bernouli shift dynamics $x_{n+1} = 2x_n \bmod (1)$. Assume that some computation method makes a constant error, e (a rational number), in the calculation of this shift, so that the pseudo-orbit obeys $y_{n+1} = 2y_n + e \bmod (1)$. This is a heuristic rather than a realistic example, since no acceptable computation method would make such a systematic error – but let us consider it anyway. Show that, if $x_0 = y_0 + e$, then x_n β-shadows y_n. Determine the minimum β, and also determine a fundamental property of the β-shadowing orbit. Is this property a typical property of the orbits of $x_{n+1} = 2x_n \bmod (1)$?

THE FINITE can also greatly alter the number of basins of attraction of a map, as well as its bifurcation sequences. To be more specific, we consider again the logistic map, but now consider the pseudo-orbits which are in a dynamical space with some finite 'lattice' of values, as in (4.11.9). That is, we divide the interval into a set of rational numbers,

$$x(j) = j/(N+1) \quad (j = 1, 2, \ldots, N), \tag{4.11.11}$$

so that the dynamics takes place on N points (the 'lattice'). It is certainly appropriate at this point to recall the biological origin of the logistic map, in which x represented the (normalized) population of a species in succeeding generations. Clearly this number is rational, even if the maximum population (N in (4.11.11)) is very large. Thus the mathematical logistic map, in THE INFINITE, is quite an extrapolation from its empirical origin in THE FINITE. The dynamics will now be defined so that it is 'similar' to the logistic map, but satisfies the lattice restriction (4.11.11). We will use the

N	P(N)
500	2;7;10
1000	10;10
100 000	81
200 000	5;355
400 000	96;508
800 000	159;192
1 600 000	304

form (4.11.4), now with $\delta = 1/(N+1)$, to define this lattice map $x \to x'$ by

$$x' = \text{Int}\,[(N+1)cx(1-x)]/(N+1) \qquad (4.11.12)$$

where $\text{Int}\,(z)$ is the integer part of z.

As noted at the beginning of this section, for any finite lattice, all initial states must tend to periodic orbits. That is, all x_0 are eventually periodic. Fig. 4.80 illustrates this

Fig. 4.80

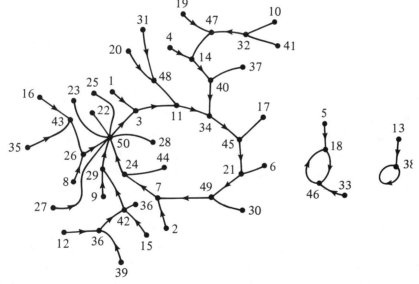

for the case $N = 50, c = 4$. It shows the 'flow' of the iterates of all 50 initial states, and the fact that there are three basins of attraction to periodic orbits with periods $P = 1, 2$, and 9. This is obviously not a very large value of N, and hence not necessarily representative of what happens for larger N.

It is presently unclear in what sense chaotic motion is approximated within THE FINITE, for large values of N. The periods of the asymptotic motion, $P(N)$ apparently do not become proportional to N, nor does the number of such basins increase. This is indicated by the table, obtained from limited samples of initial conditions (for $c = 4$).

Fig. 4.81

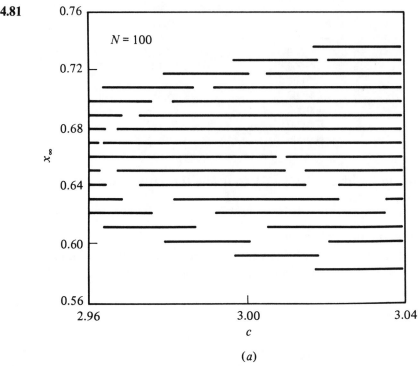

(*a*)

Thus most states are not periodic. However, this alone is not an indication of chaos (e.g., all states could, in one step, go to a fixed point). It is not the asymptotic motion which can characterize 'chaos' but rather some form of 'initial' scrambling of the orbits (e.g., discrete and finite-time forms of Lyapunov exponents). These details are yet to be understood. For two-dimensional examples, see Section 6.14.

In addition to the chaotic dynamics at $c = 4$, it is interesting to see the difference which THE FINITE makes on the dynamics in the region of THE INFINITE bifurcation points. This can be explored at all of THE INFINITE bifurcation points, but we illustrate the modified bifurcation (new definition?), by considering the first period-two bifurcation region, $c \simeq 3$. Fig. 4.81 illustrate all of the periodic points (for all initial conditions) in the attracting periodic cycles, for different numbers of lattice points. Essentially all points come in pairs (period two cycles). Using $x_{n+1} = $ NINT$(cx_n(1 - x_n)N)/N$, with NINT(z) being the nearest integer of z (differing slightly from (4.11.12)), the number of pairs at $c = 3$ is 7 when $N = 100$, 30 when $N = 1000$, 44 when $N = 2000$, and 63 when $N \sim 3{,}300$. Note that the graphic dots have been reduced in size going from the cases $N \leqslant 1000$ to $N \geqslant 2000$. The differences between these structures and the single period-one, ($c \leqslant 3$), or period-two ($c > 3$) attractor of the logistic map in THE INFINITE is very striking. The bifurcation features for large values of c are also dramatically altered, due to the discrete nature of the Int(z) function. It appears to require quite large values of N to recover the

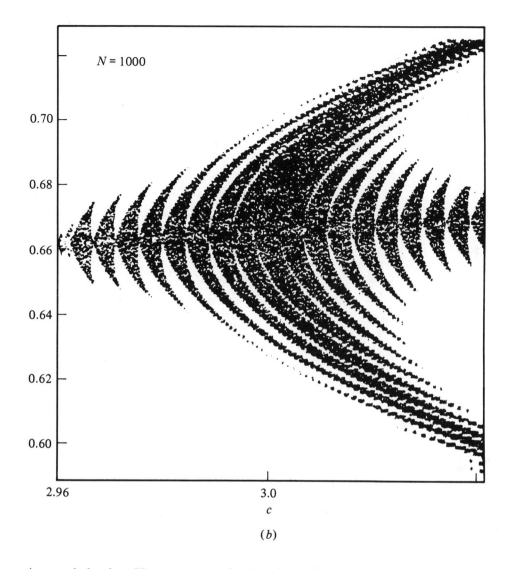

$N = 1000$

c

(b)

continuum behavior. However, note in the above figures the reduction in both the range and the gaps, as N is increased from $N = 2000$ to $N \simeq 3300$. Of course, the higher order bifurcations, the Cantor set attractor, and periodic windows are largely lost in THE FINITE, unless N is very large.

Comments on exercises

(4.1) Also see Section 10 of this chapter.

(4.2) If $F^n(x_0) = x_0$, then $F^{m+n}(x_0) = F^m(x_0)$, so $F(x_m) = x_m$ $(m = 0, \ldots, n-1)$. There are n periodic points.

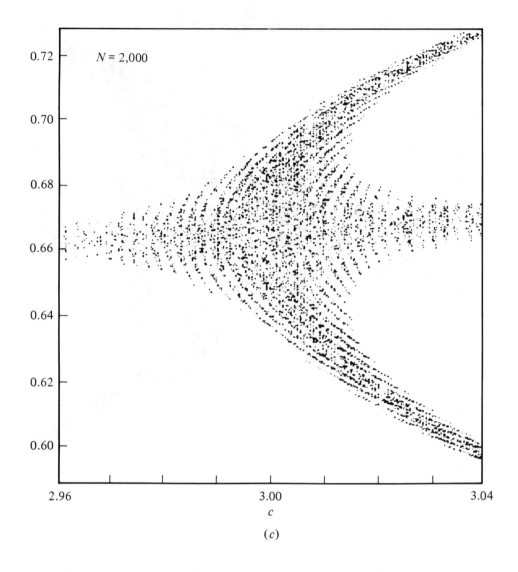

(c)

(4.3) Continuity is the essential element here. Any curve which joins the vertical at $x_n = 0$ to the one at $x_n = 1$ (defining the map), and such that $0 \leqslant F(x_n) \leqslant 1$, must (because of continuity) intersect the diagonal, $x_{n+1} = x_n$, at least at one point (Fig. 4.83).

(4.4) For $c = 3$, $x_0 = \frac{2}{3}$ is a fixed point (period one) of $F = cx(1-x)$. If $F^2(x_e) = F(F(x_e)) = \frac{2}{3}$, then $F(x_1) = \frac{2}{3}$ where $x_1 = F(x_e)$. We first obtain x_1, $3x_1(1-x_1) = \frac{2}{3}$, so $x_1 = \frac{1}{3}$, and $F(x_e) = \frac{1}{3}$; $x_e^2 - x_e + \frac{1}{9} = 0$, $x_e = \frac{1}{2} \pm 5^{1/2}/6$. Similarly, if $F^3(x_3) = \frac{2}{3}$, then $F^2(F(x_e)) = \frac{2}{3}$, so we need $F(x_e) = \frac{1}{2} \pm \frac{5^{1/2}}{6}$, or $x_e = \frac{1}{2}$

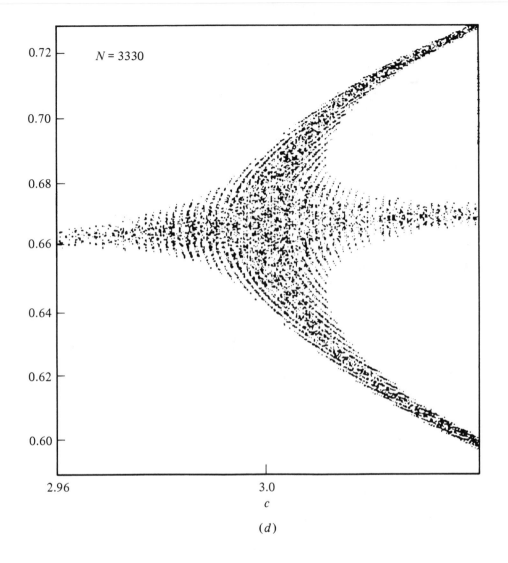

$N = 3330$

2.96   3.0

c

(*d*)

$+ \frac{1}{6}(3 + 2(5)^{1/2})^{1/2}$. Note that $F(x_e) = \frac{1}{2} + \frac{5^{1/2}}{6}$ yields a complex x_e. (Also see Exercise 4.5.)

(4.5)

(4.6) (a) It has a chaotic set, but it cannot be observed because it is unstable. In other words, not all mathematically chaotic sets are of physical (observational) importance.

(b) By continually folding-over the interval $[0, 1]$, it may seem clear that any initial interval, A, will ultimately map to intersect A (i.e., $F^n A \cap A \neq \phi$). This

Fig. 4.82

Unstable periodic
orbit

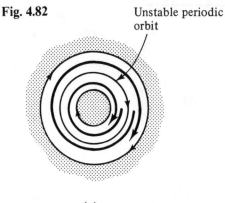

The flow is on a Möbius strip
(see Section 4.9), with a
'central' periodic orbit which
is unstable (see (4.2.5) and
the following discussion)

(a)   (b)

Fig. 4.83 $x_{n+1} = F(x_n)$

Fig. 4.84

x_{n+1}

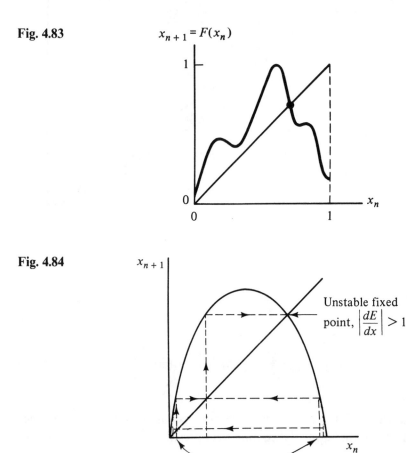

Unstable fixed
point, $\left|\dfrac{dE}{dx}\right| > 1$

An infinite number of points
in the 'wings', which map into
the unstable fixed point

intersect must have a fixed point of F^n, by Brouwer's theorem, and hence contain a period-n point.

(4.7) $x_{n+1} = \sin^2(2^{n+1}\pi\theta) = [2\sin(2^n\pi\theta)\cos(2^n\pi\theta)]^2 = 4x_n[1 - x_n]$; if θ is irrational the solution is nonperiodic. Hence the nonperiodic solutions are a set of Lebesgue measure 1 on the interval $0 < \theta < 1$, but there is also a dense set of periodic solutions (namely all the rationals, $\theta = l2^{-n}/(1 + 2^k)$, which have period k). Note also that we have here an 'analytic solution (??)' of chaotic motion!

(4.8) We want to show that $G \cdot h = h \cdot F$, or:

For $0 < x < 1/2$, $4/\pi \arcsin x^{1/2} = 2/\pi \arcsin(4x(1 - x))^{1/2}$

For $1 > x > \frac{1}{2}$, $2[1 - 2/\pi \arcsin x^{1/2}] = 2/\pi \arcsin(4x(1 - x)^{1/2}$

(a) if $x < \frac{1}{2}$, multiply by $\pi/2$, and take the sine. We need to show $\sin$ $\cdot(2 \arcsin x^{1/2}) = (4x(1 - x))^{1/2}$, but $\sin(2 \arcsin x^{1/2}) = 2 \sin$ $(\arcsin x^{1/2})$ $\cdot[1 - \sin^2(\arcsin x^{1/2})]^{1/2}$ (qed)

(b) if $x > \frac{1}{2}$, similarly we need to show $\sin[\pi - 2 \arcsin x^{1/2}] = (4x(1 - x))^{1/2}$ but $\sin[\pi - 2 \arcsin x^{1/2}] = \sin\pi\cos(2\arcsin x^{1/2}) - \cos\pi\sin$ $\cdot(2 \arcsin x^{1/4} \sin x^{1/2}) = \sin(2 \arcsin x^{1/2}) = $ same as in (a).

All of the fixed points of G^n are unstable ($|\partial G^n/\partial x| > 1$), so all fixed points of F^n (period-n points of F) are unstable.

(4.9) $\int_0^1 [dx/(\pi(x(1 - x))^{1/2})] = \int_0^{\pi/2} [d(\sin^2\theta)/(\pi\sin\theta\cos\theta)] = \int_0^{\pi/2}(2d\theta/\pi) = 1$. $P(y) = \int_0^1 \delta(y - 4x(1 - x)) [\pi(x(1 - x))^{1/2}]^{-1} dx$; the δ function contains an $f(x)$ which gives the same y for two values of x, so split it up, and set $z = 4x(1 - x)$; $P(y) = \int_0^{1/2} \delta(y - z)[dx(\pi(z/4)^{1/2})] + \int_{1/2}^1 \delta(y - z)[dx/(\pi(z/4)^{1/2})]$ and $dx = dz/(4 - 8x)$ and $x = \frac{1}{2} - \frac{1}{2}(1 - z)^{1/2}$ in the first integral and $x = \frac{1}{2} + \frac{1}{2}(1 - z)^{1/2}$ in the second. So $P(y) = \int_0^1 \delta(y - z)[1/(\pi(z/4)^{1/2})][dz/(4(1 - z)^{1/2})] + \int_1^0 \delta(y - z)[1/(\pi(z/4)^{1/2})][dz/(-4(1 - z)^{1/2})] = \int_0^1 \delta(y - z)[dz/(\pi(z)^{1/2}(1 - z)^{1/2})] = [\pi(y(1 - y))^{1/2}]^{-1}$. The Lyapunov exponent can be obtained from (4.6.2) $\lambda = \int_0^1 [\ln|4 - 8x|/(\pi(x(1 - x))^{1/2})] dx = \int_0^1 [\ln 2^2 + \ln|1 - 2x|/(\pi(x(1 - x))^{1/2})] dx = 2\ln 2 + 2\int_0^{1/2} [\ln(1 - 2x)/(\pi(x(1 - x))^{1/2})] dx$ since $d(-\sin^{-1}(1 - 2x)) = dx/(x(1 - x))^{1/2}$; set $(1 - 2x) = \sin\theta$, then $\lambda = 2\ln 2 + 2/\pi \int_{\pi/2}^0 \ln(\sin\theta)(-d\theta)$; now $\int_0^{\pi/2} \ln(\sin\theta) d\theta = -\pi/2\ln 2$ so $\lambda = 2\ln 2 - \ln 2 = \ln 2$.

(4.10) For a periodic orbit (period m), x only 'exists' on the points x_n, such that $f^m(x_n) = x_n (n = 1, \ldots, m)$, so the probability (note $\int_0^1 P(x) dx = 1$) is $P(x) = (1/m)\sum_{n=1}^m \delta(x - x_n)$. Hence $\lambda = \int \ln|df/dx|P(x) dx = (1/m)\sum_{n=1}^m \ln|df/dx|_{x_m} = (1/m)\ln|df^m/dx|_{x_j}$ where x_j is any period-m point. The last equality is due to (4.2.11). Thus the sign of λ is directly related to whether $|df^m/dx|_{x_j} > 1$ (unstable) or < 1 (stable).

(4.11) $P(x) = 1$ is invariant, according to (4.6.3), if $1 = \int_0^1 \delta(y - f(x)) dx$. Substituting $f(x) = z$ in each interval yields $x_0 \int_0^1 \delta(y - z) dz + \int_1^0 \delta(y - z)(1 - x_0)(-dz)$, $x_0 + (1 - x_0) = 1$, so it is indeed invariant. (4.6.2) yields $\bar{\lambda} = -x_0 \ln(x_0) -$

$(1 - x_0) \ln (1 - x_0)$. The most unstable map is given by $d\bar{\lambda}/dx_0 = 0$, which yields $x_0 = \frac{1}{2}$, so $\bar{\lambda}_{max} = \ln 2$.

(4.12) The number of occupied cells is $N(\varepsilon) = 69$, $\varepsilon = 10^{-3}$, so $d_c \simeq \ln (69)/3 \ln 10 = .613$.

(4.13) If you work this exercise out in detail, the general answer will become clear. First of all, 2^{N+1} iterations of the point $x_0 = \frac{1}{2}$ produces 2^N intervals. For a Cantor set which is self-similar to 2^N intervals in $(1, 2)$, the $(2^N)^2$ intervals obtained in the next level of similarity construction must satisfy the following equalities

$$L(K + I2^N, 2^{2N} + K + I2^N)/L(K, K + 2^N) = L(1 + I, 1 + I + 2^N)/L(1, 2)$$

where $K = 1, 2, \ldots, 2^N$, and $I = 0, \ldots (2^N - 1)$. Here

$$L(i, j) = |x_i - x_j|, \quad x_n = F^n(1/2, c_\infty).$$

This result is not particularly obvious, but it is correct. The exercise corresponds to the case $N = 2$. From this, one finds that the Cantor set generated by the iterations of $x_0 = \frac{1}{2}$, is not more self-similar to 2^N intervals than it is to 2^{N-1} intervals (i.e., the percentage deviation from the above equalities does not decrease as N increases). Nonetheless, the estimate of d_c is improved, simply because it is based on the more accurate starting point of 2^N intervals.

(4.14) Eliminate the θ dependence between the equations of (4.8.6) to obtain $a^4 - a^2 x^2 - \dot{x}^2 = 0$. We then obtain $K(x, \dot{x}) = 2a^2$. The constant a is arbitrary, so $\ddot{x} = -a^2 x$ is not a harmonic oscillator. The proper equation is $\ddot{x} = -(x/2)(x^2 + (x^4 + 4\dot{x}^2)^{1/2})$. Is this a Lipschitz system?

(4.15) $\rho(x, \dot{x}) = [2\pi(a_2 - a_1)(x^4 + 4\dot{x}^2)^{1/2}]^{-1}$.

(4.16) K is proportional to the height of the 'spikes' in the figure, provided that the relaxation of the ϕ-oscillator is fast relative to $P(\phi)$. Otherwise the response to the disturbance would not simply depend on the single point ϕ_n (i.e., $K \sin (\phi_n)$). Also the relaxation must be rapid compared to $P(\psi)$, for otherwise there would be an overlap effect, not accounted for in (4.9.4). If the ϕ-oscillator has a relaxation which is some fraction of $P(\phi)$, then the last condition means that Ω cannot be too large. Presumably a 'safe' value would always be $\Omega \lesssim 1$.

(4.17) The set $\rho = 0$ occurs at $\Phi = 0$ when there is no coupling, $K = 0$, $\Phi = 0$ implies that $P(\psi) = 0$ so that the driving oscillator is very fast. If now $P(\psi)$ is very small, the rotation number, (4.9.10) is the average change of ϕ over this short period. In the case (4.9.13) the coupling is strong enough to cause the 'naturally slow' ϕ-oscillator to retain a zero (time-averaged) phase shift relative to the fast ψ-oscillator, i.e., when viewed at each period of the ψ-oscillator.

For the proof, let $\Phi' = \Phi + 2\pi l$. $T_{\Phi'}\phi = T_\Phi\phi + 2\pi l$ so

$$\rho(\Phi') = \lim_{n \to \infty} \frac{T_{\Phi'}^n - \phi}{2\pi n} = l + \rho(\Phi).$$

(4.18) We find that $x_1 = 2y_0 + 2e \bmod (1)$ whereas $y_1 = 2y_0 + e \bmod (1)$. More generally $x_n = y_n + e$, so $|x_n - y_n| = e \leqslant \beta$, if $\beta = e$. All y-solutions are eventually periodic, whereas only a set of measure zero of the x-solutions are periodic. Hence, in this example, the β-shadowing trajectories are not typical of most x-solutions.

5

<hr>

Second order differential systems (n = 2)

5.1 The phase plane

Certainly the most studied nonlinear systems are those described by second order ordinary differential equations. This is due in large part to the fact that such systems include all nonlinear oscillators, and secondly, to the fact that the phase space is a plane, which has a number of very special and unique properties not found in higher dimensional phase spaces. Thus the dynamics is both much more interesting than that obtained from first order differential systems, and yet is restricted to remain in a plane, which is a significant simplifying constraint.

In keeping with our general scheme, we will consider a pair of coupled first order equations, rather than a second order equation. The first thing we will do is to consider solutions of the autonomous equations

$$\dot{x}_1 = F_1(x_1, x_2), \quad \dot{x}_2 = F_2(x_1, x_2). \tag{5.1.1}$$

The solutions of nonautonomous equations can be much more complicated, and the consideration of their properties will largely be reserved until Sections 10–14.

In what follows, we will suppress the dependence of F on the control parameters. Also, to avoid subscripts, these equations are frequently written in the form

$$\dot{x} = P(x, y); \quad \dot{y} = Q(x, y). \tag{5.1.2}$$

The steps which are frequently useful in analyzing these equations are:

(A) Locate all of the *fixed points* of the dynamics (the equilibrium points, $\dot{x} = 0$, $\dot{y} = 0$). These are also referred to as the *singular points* of (5.1.2).
(B) Determine the character of the flow in the neighborhood of each fixed point. This involves a simple linear analysis (usually) about each fixed point.
(C) Use the equation $dx/dy = P(x, y)/Q(x, y)$, which gives the tangent of the orbit at each point, to obtain the *global* topological character of the *flow*. This step connects the local flows found in (B). Since this procedure is numerical (evaluating P/Q at judiciously selected points), it can also be achieved by local numerical integrations, if desired.

(D) In rather rare cases, you may be able to determine an *integrating factor* which allows you to explicitly integrate $dx/dy = P(x, y)/Q(x, y)$. This would then give you a function, $K(x, y) = K_0(x_0, y_0)$, determining the orbit through each point (x_0, y_0) (whose topological character is already known from (B) and (C)). $K(x, y)$ is the only time-independent integral of the motion of this system.

(E) The final piece of information is the time dependence along these orbits, which may make use of $K(x, y)$ to eliminate one variable from one of the equations of (5.1.2). In many cases, however, this detailed information can only be obtained by numerical integration of (5.1.2), using a computer.

Once again, briefly:

*(A) *Locate all fixed points* (FP)
*(B) *Linear analysis about each* FP
*(C) *Global analysis*
*(D) *Integrating factor*
*(E) *Time dependence*

We will now consider these steps in greater detail.

*(A) *Locate all fixed points*

The fixed (singular) points (x^*, y^*) are given by

$$P(x^*, y^*) = 0 = Q(x^*, y^*) \tag{5.1.3}$$

and without loss of generality we can transform the coordinates so that $(x^*, y^*) = (0, 0)$ for one fixed point. Then (5.1.2) can be expanded in a Taylor series, and written in the form

$$\dot{x} = ax + by + P_2(x, y)$$
$$\dot{y} = cx + dy + Q_2(x, y), \tag{5.1.4}$$

where $a = (\partial P/\partial x)_*$, $b = (\partial P/\partial y)_*$, $c = (\partial Q/\partial x)_*$, $d = (\partial Q/\partial y)_*$. It is assumed, of course, that these derivatives exist, in which case the Lipschitz condition is satisfied at the fixed point. P_2, Q_2 in (5.1.4) are at least second degree in (x, y), so that all linear terms are explicitly shown.

*(B) *Linear analysis about each fixed point*

If we 'linearize' these equations to study the dynamics near this fixed point, we obtain

$$\dot{x} = ax + by, \quad \dot{y} = cx + dy. \tag{5.1.5}$$

then there are special solutions

$$x = x_0 \exp(st), \quad y = y_0 \exp(st),$$

provided that the *characteristic exponents* s are given by

$$s = \tfrac{1}{2} p \pm \tfrac{1}{2} [p^2 - 4q]^{1/2}, \tag{5.1.6}$$

where

$$p = (a + d), \quad q = (ad - bc). \tag{5.1.7}$$

If $p^2 \neq 4q$, so that the exponents are not degenerate, we can obtain the general solution of (5.1.5) by summing over the independent solutions, $(i, j = 1, 2)$

$$x = \sum_{i \neq j} \frac{(a - s_j)x_0 + by_0}{(s_i - s_j)} \exp(s_i t); \quad y = \sum_{i \neq j} \frac{cx_0 + (d - s_j)y_0}{(s_i - s_j)} \exp(s_i t) \tag{5.1.8}$$

where $x(0) \equiv x_0$, $y(0) \equiv y_0$. You should verify for yourself that, if you take $x = \sum_i A_i \exp(s_i t)$ and $y = \sum B_i \exp(s_i t)$, and use (5.1.5) and (5.1.6), then (5.1.8) follows.

Exercise 5.1. What is the time dependence in (5.1.8) in the limit $p^2 \to 4q$ (that is, s_1, $s_2 \to s \equiv p/2$)?

The linearization (5.1.5) reflects the dynamics of (5.1.2) near the fixed point $(0, 0)$ provided that $(ad - bc) \neq 0$, i.e., provided that the determinant $|\partial F_i / \partial x_j| \neq 0$ at the fixed points $(F = 0)$. Otherwise the second degree (i.e., quadratic) factors in (P_2, Q_2) play a role in the dynamics near the fixed point.

Exercise 5.2. To illustrate what can happen if $(ad - bc) = 0$, show that $\dot{x} = x + y + x^2 + xy$, $\dot{y} = -x - y + y^2 + xy$ has a solution only for a finite time for all initial states with $(x_0 + y_0) > 0$.

These linearized solutions are widely discussed in numerous references, so the discussion here will be limited to introducing the names given to the various orbital forms in the phase space, and to emphasize that there are two control parameters in this system of equations rather than the apparent four (a, b, c, d). The control parameters, (c_1, c_2), are (p, q) of (5.1.7).

Exercise 5.3. Since the behavior of the solutions of (5.1.5) only depends on the parameters (p, q) of (5.1.7), it must be possible to transform (5.1.5) into equations which depend only on (p, q). Obtain an example of such equations.

The possible orbital forms are:

Fig.5.1

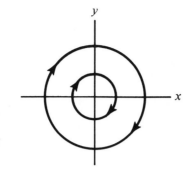

Center. Closed orbits about the fixed point (Fig. 5.1). This point is also referred to as an *elliptic point*. It occurs when $p = 0$ and $q > 0$ in (5.1.6) so that the characteristic exponents are imaginary. For example:

$$\dot{x} = y, \quad \dot{y} = -x$$

Focus. (Spiral) All orbits tend toward (or away from) the fixed point without a limiting direction. This occurs if $4q > p^2 > 0$ in (5.1.6), so that the exponents are complex conjugates. For example: $\dot{x} = ax + y$, $\dot{y} = -x + ay$ $(a \neq 0)$. Let $x = r \cos \theta$, $y = r \sin \theta$ then $\dot{r} = ar$, $\dot{\theta} = -1$. If $a < 0$ it is a *stable focus*; if $a > 0$, an *unstable focus* (as shown in Fig. 5.2).

Fig. 5.2

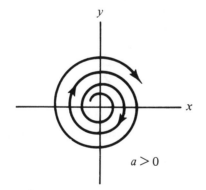

$a > 0$

Node. All orbits enter the fixed point with a limiting direction (several possibilities are illustrated see Fig. 5.3). In this case $p^2 > 4q > 0$, and the exponents are real and of the same sign. For example: $\dot{x} = x$, $\dot{y} = ay$ $(a > 0)$.

Saddle point. Only four orbits connect with the fixed point. This occurs if $q < 0$, so the exponents are real, with opposite signs. For example: $\dot{x} = x$, $\dot{y} = -y$. An orbit which converges to the saddle point as $t \to +\infty$ is called an *incoming separatrix*, whereas if it converges to the saddle point as $t \to -\infty$, it is called an *outgoing separatrix* (Fig. 5.4).

Fig. 5.3

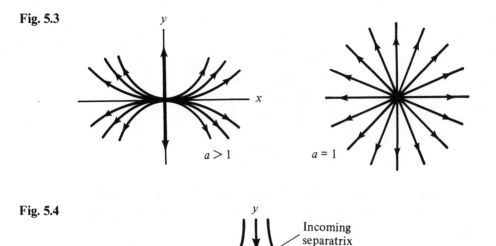

$a > 1$ $a = 1$

Fig. 5.4

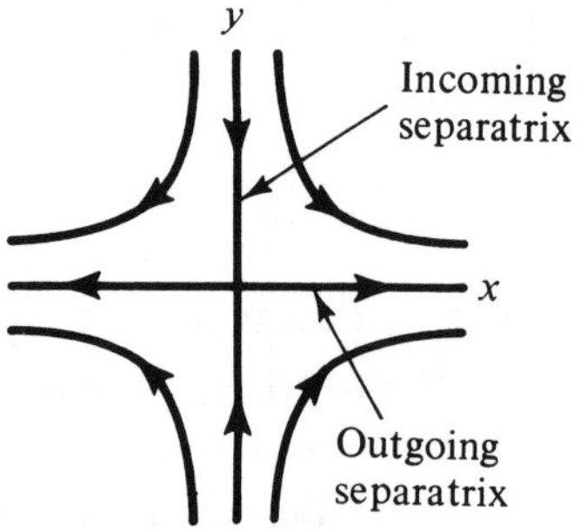

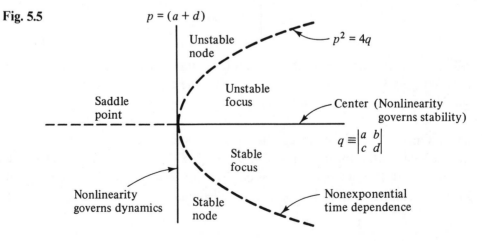

A *hyperbolic point* is defined to be any singular point for which the associated characteristic exponents all have a nonzero real part. Put another way, the eigenvalues of the Jacobian matrix at the fixed point

$$J = \begin{pmatrix} a & b \\ c & d \end{pmatrix} = \begin{pmatrix} \partial P/\partial x & \partial P/\partial y \\ \partial Q/\partial x & \partial Q/\partial y \end{pmatrix}_{(0,0)} \tag{5.1.9}$$

all have a nonzero real part. Thus, only a center is not a hyperbolic fixed point.

These results can be summarized on the control plane (p, q) as shown in Fig. 5.5.

Fig. 5.5

Fig. 5.6

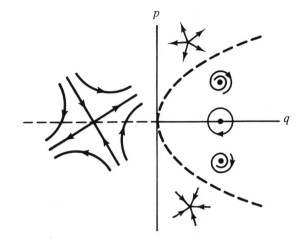

Fig. 5.6 superimposes the phase plane dynamics on the control plane. The solid lines dividing the regions of topologically different orbits, and are the *bifurcation lines* in this control space (yielding 'dynamic phase transitions').

We note that the two control parameters, (5.1.7), governing the topological character of the flow of the linear equations (5.1.5), are the determinant and trace of the matrix (5.1.9)

$$q = \det |J|; \quad p = \mathrm{Tr}\,|J|. \tag{5.1.10}$$

A nonsingular linear transformation to other coordinates

$$\begin{pmatrix} x' \\ y' \end{pmatrix} = A \begin{pmatrix} x \\ y \end{pmatrix} \equiv \begin{pmatrix} \alpha & \beta \\ \gamma & \delta \end{pmatrix} \begin{pmatrix} x \\ y \end{pmatrix}; \quad \det |A| \neq 0 \tag{5.1.11}$$

leaves p and q invariant, so the flows in the (x, y) and (x', y') phase spaces have similar fixed point stability. However the flows are *topologically orbitally equivalent* (TOE) only if $\det |A| > 0$, corresponding to an orientation preserving transformation.

Exercise 5.4. Explicitly show that (p', q'), associated with the $(\dot{x}', \dot{y}')$ equations of the variables (5.1.11), satisfy $p' = p$, and $q' = q$.

An important class of flows are those which represent '*conservative systems*', or *area-preserving* systems. To clarify this feature, consider the following exercise.

Exercise 5.5. First draw some pictures of moving regions, $\Omega(t)$, and decide what type of fixed points of (5.1.5) are possible for area-preserving flows (Fig. 5.7). Next, apply *Liouville's theorem* to (5.1.5), and obtain a precise criterion (determine the value of $s_1 + s_2$).

Fig. 5.7

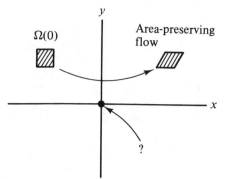

The Lyapunov stability of the fixed points of the non-linear equations (5.1.2) may or may not, be determined by the stability of the linear equations (5.1.5). Thus, for example, we have:

Lyapunov's theorem

> *If the real parts of the characteristic exponents obtained from the linear equations (5.1.5) (of (5.1.2)) are not zero, then the fixed points of (5.1.2) are stable if all real parts are negative, and unstable if at least one real part is positive.*

Thus the *stability of centers* $(p = 0, q > 0)$ *is not determined from the linear equations*, nor the general character of the intrinsically nonlinear case $q = 0$. However all hyperbolic fixed points have local flows similar to those of the linear equations (this is also true in R^n).

Exercise 5.6. To illustrate the problem of determining the stability of a center, consider the following (relatively simple) system

$$\dot{x} = y + x^3; \quad \dot{y} = -x + y^3.$$

Determine whether the origin is stable (and in what sense). If you found that easy, what about the stability of the origin (in the sense of Laplace) for the system

$$\dot{x} = y - x^3; \quad \dot{y} = x - y^3?$$

Exercise 5.7. Determine the character of the bifurcation at $c = 0$ of the system

$$\dot{x} = c + x^2; \quad \dot{y} = -y$$

by drawing the phase portrait for $c > 0$ and $c < 0$.

*(C) Global analysis

Once the local properties of the flow near each fixed point are determined, the next stage of understanding the dynamics concerns the *global flow*. Sometimes this is obvious without further analysis, but sometimes it is not. A simple example is shown in Fig. 5.8(a), with two fixed points. While we can attempt to guess the global topological character of the flow, the answer is not unique, even in this simple case. This is illustrated in the lower part of the figure.

Fig. 5.8

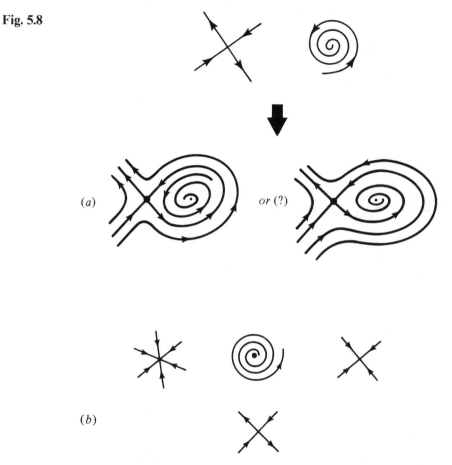

(a)   or (?)

(b)

Also, as food for future thought (see Section 3), we might ask whether it is always possible to find a global flow for a given set of fixed points – for example the set of four fixed points illustrated in Fig. 5.8(b).

We can approach this global problem more systematically by considering the equation

$$\mathrm{d}x/\mathrm{d}y = P(x, y)/Q(x, y). \tag{5.1.12}$$

This gives the tangent of the orbit which passes through (x, y). Before one attempts to integrate (5.1.12), it is easier to put in a few scattered values of (x, y) in the right side and determine the local slope of the orbits at these scattered points. Using this information, together with the previous fixed point information, it is usually possible to guess the global phase portrait. Since this procedure involves the numerical evaluation of $P(x, y)/Q(x, y)$, it can of course also be evaluated by a computer. The point here is to only integrate for the orbits in the regions of interest (uncertainty). This intelligent integration procedure can save considerable time.

These studies reveal structures which are intrinsically global in nature, since they are not determined by the flows around the fixed points. One important example is the *limit cycle*, which is a closed (periodic) orbit, whose neighboring orbits tend asymptotically toward (or away) from it. If they tend toward it as $t \to +\infty$, then it is a stable limit cycle; if they tend toward it as $t \to -\infty$, then it is an unstable limit cycle. Some physical examples are illustrated in Fig. 5.9. In 1947 Minorsky wrote 'Perhaps it is not too great

Fig. 5.9

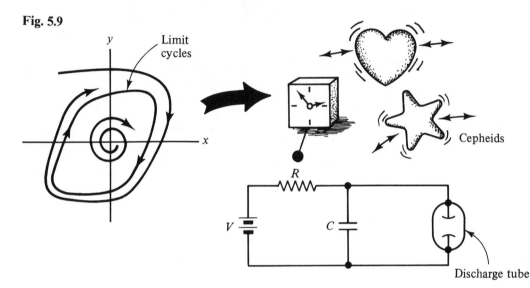

an exaggeration to say that the principal line of endeavor of nonlinear mechanics at present is a search for limit cycles' (pg. 68). While times have changed (and were changing even as he wrote that statement), limit cycles will always remain a very important form of nonlinear dynamics. Indeed, stable nodes, stable focuses, or stable limit cycles represent the simplest forms of so-called *attractors*, about which we will have much more to say later.

A second structure, shown in Fig. 5.10, again involves the *outgoing* and *incoming separatrices* of one saddle point connecting with a second saddle point (respectively as incoming and outgoing separatrices). These two connecting orbits jointly form the boundary of an 'island', containing only periodic solutions, in a 'sea' with nonperiodic solutions. A simple example of this is a frictionless pendulum, to be considered in the next section, or a ball rolling without friction on a curved surface with two maxima of equal height. An orbit, such as these separatrices, which connects two fixed points is called a *heteroclinic orbit*.

Fig. 5.10

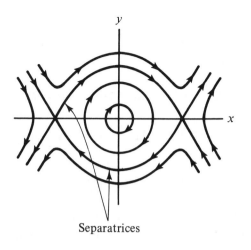

Separatrices

If, on the other hand, a ball rolls on a surface with maxima of different heights, there is a single orbit which converges to the same saddle point both when $t \to +\infty$ and when $t \to -\infty$. Such an orbit is called a *homoclinic orbit*. This situation is illustrated in Fig. 5.11, where the height of the surface is shown in (*a*).

Fig. 5.11

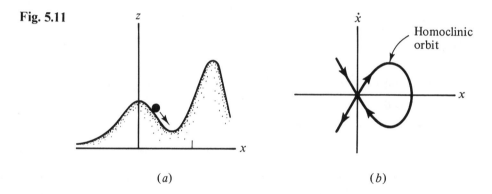

(*a*)   (*b*)

Before proceeding with the general analysis, it is useful to contrast the 'physical content', and global behavior of some special dynamical systems. More specifically, we will contrast the behavior of *Hamiltonian systems, gradient systems*, and standard.

Newtonian systems. These are each special, but physically important examples of the general system (5.1.1). In the present R^2 phase-space, they have the following forms:

Newtonian system

$$m\ddot{x} = F(x, \dot{x}) \quad (F = \text{'Force'}) \tag{5.1.13}$$

or

$$\dot{x} = y, \quad m\dot{y} = F(x, y) \quad (m = \text{'mass'}).$$

linear form, (5.1.5), (5.1.7):

$$p = d = \frac{1}{m}(\partial F/\partial y)_0, \quad q = -c = -\frac{1}{m}(\partial F/\partial x)_0. \tag{5.1.14}$$

Hamiltonian system. Special Newtonian system: $m\ddot{x} = F(x) \equiv -dV(x)/dx$

$$m\dot{x} = y = \frac{\partial}{\partial y}[\tfrac{1}{2}y^2 + mV(x)] \equiv m\frac{\partial H}{\partial y}$$

$$\dot{y} = F(x) = -\frac{\partial}{\partial x}\left[\frac{1}{2m}y^2 + V(x)\right] \equiv -\frac{\partial H}{\partial x}$$

where y is the 'momentum' and $V(x) \equiv -\int^x F(x')dx'$ is the '*potential energy*'. $H(x, y)$ is the Hamiltonian, or the 'total energy'. Hamiltonian systems 'conserve energy', since $H(x, y)$ is a constant of the motion, $dH/dt = (\partial H/\partial x)\dot{x} + (\partial H/\partial y)\dot{y} \equiv 0$.

Linear form, (5.1.5), (5.1.7). Since in (5.1.2)

$$P = \frac{\partial H}{\partial y} \quad \text{and} \quad Q = -\frac{\partial H}{\partial x}, \quad \frac{\partial P}{\partial x} = \frac{-\partial Q}{\partial y}, \tag{5.1.16}$$

We obtain from (5.1.9) $a = -d$, or $p = 0$.

Gradient system
$$\dot{x} = -\partial V/\partial x, \quad \dot{y} = -\partial V/\partial y. \tag{5.1.17}$$

Recall the relationship of these systems with elementary catastrophe theory, Chapter 3. *linear form*, (5.15), (5.17). Since $P = -\partial V\partial x$ and $Q = -\partial V/\partial y$, $\partial P/\partial y = \partial Q/\partial x$, hence $b = c$. Then

$$p^2 - 4q - a^2 + d^2 + 2ad - 4(ad - bc) = (a - d)^2 + 4b^2 \geqslant 0. \tag{5.1.18}$$

Newtonian system (5.1.13) represents the most diverse physical types of motion, such as all damped, or excited nonlinear oscillators. Because p and q, (5.1.14) are arbitrary, Newtonian systems can have fixed points of any type.

Hamiltonian systems are conservative, and therefore do not represent either damped or excited systems (Fig. 5.12). From (5.1.16), $p = 0$ hence the only type of simple fixed

Fig. 5.12

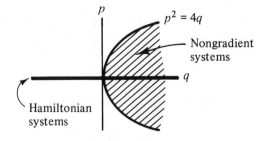

point which they can have are either centers or special saddle points (see Exercise 5.5). These are represented by the horizontal axis in the (p, q) plane – a very limited set of points!

Gradient systems are the antithesis of Hamiltonian systems, both physically and mathematically. First of all, while many Hamiltonian systems have periodic motions, no gradient system has a period solution. For if γ is a periodic trajectory, in the phase space, the line integral $\oint_\gamma (\dot{x}dx + \dot{y}dy)$ cannot vanish (the vector field is everywhere tangent to the curve γ, and has one orientation). However, from (5.1.17), $(\dot{x}dx + \dot{y}dy) = -(\partial V/\partial x)dx - (\partial V/\partial y)dy$. When this is integrated from (x_1, y_1) to (x_2, y_2) it yields $-V(x_2, y_2) + V(x_1, y_1)$. Hence, integrating around a closed curve, we have $(x_1, y_1) = (x_2, y_2)$, so the integral vanishes, and we have a contradiction (Fig. 5.13).

Fig. 5.13

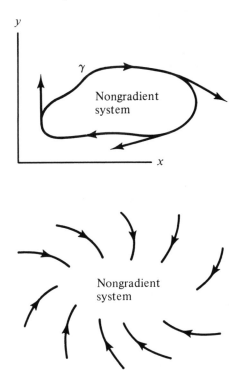

Fig. 5.14

This same analysis in fact shows that a gradient flow cannot even spiral into (or out of) a region of phase space with a constant orientation (say clockwise, Fig. 5.14). As a consequence, at simple fixed points the flow cannot be a focus or a center. This is verified by (5.1.18), which excludes these fixed points (see Fig. 5.13).

As has been noted before in Chapter 3, Section 3, for each Hamiltonian system, (5.1.15), there is a gradient system, (5.1.17), whose vector field is everywhere orthogonal to the Hamiltonian vector field. The 'orthogonal' gradient flow has $V = H$ in (5.1.17) (see Exercise 3.18). This is illustrated in Fig. 5.15 for the undamped pendulum

Fig. 5.15

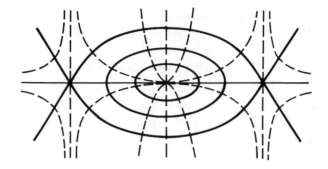

(Hamiltonian). The orthogonal flow of a gradient system, is illustrated by dashed curves in the same figure. Note that both can have the same type of saddle point, but their separatrices are rotated by $\pi/4$.

A final very important physical interpretation of gradient flows can be obtained by considering the Newtonian system (now in R^4)

$$m\ddot{x} = -\mu\dot{x} - \frac{\partial V}{\partial x}; \quad m\ddot{y} = -\mu\dot{y} - \frac{\partial V}{\partial y}. \tag{5.1.19}$$

This represents a system moving in two spatial dimensions (x, y), subjected to a viscous-type damping, $-\mu(\dot{x}, \dot{y})$ and a 'conservative' (potential) force $(-(\partial V \partial x) - (\partial V/\partial y))$. We see that in the limit $m \to 0$ ('inertialess' motion), the system (5.1.19) is essentially identical with the gradient system (5.1.17), once we scale the time, or V, to remove the 'coefficient of viscosity', μ. In other words, if we are not interested in the oscillatory aspects of the dynamics (which is an 'inertial effect'), but only want to know where a system will ultimately go when subjected to damping, we can greatly simplify the analysis by using the 'associated' gradient system (setting $m = 0$ in (5.1.19)). Note that the dimension of the physical (configuration) space is unchanged, but the dimension of the phase space has been reduced from R^4 to R^2, namely from $(\dot{x}, x, \dot{y}, y)$ to (x, y). That is a significant simplification.

Gradient systems afford us the opportunity for some interesting and imaginative modelings of physical systems. Thus, while young people may 'oscillate' back and forth between hot places to swim and cold places to ski, older people and animals move toward their 'ideal' temperature (gradient $(T(x) - T_{\text{ideal}})^2$ dynamics?) (Fig. 5.16). More 'serious', but prosaic, examples would be the motion of interacting charged or magnetic particles in an oil bath, where inertia is dominated by viscosity. Likewise, water flowing downhill through surface soil experiences a very 'viscous' drag, so inertial effects are not important. In the usual ('generic') situation the mountain, mountain passes, and mountain 'bowls' (e.g., old volcano bowls) present a surface which produces a gradient flow with simple singularities, when we look down on the terrain (see the island in Fig. 5.17). The 'tops' of hills or mountains correspond with unstable nodes (UN), the

Fig. 5.16

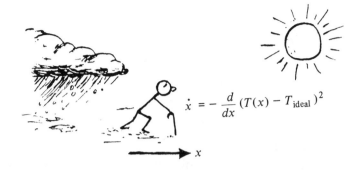

$$\dot{x} = -\frac{d}{dx}(T(x) - T_{\text{ideal}})^2$$

Fig. 5.17

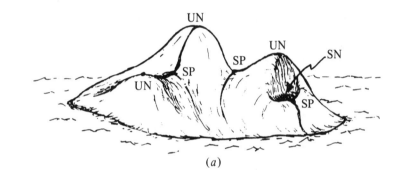

(a)

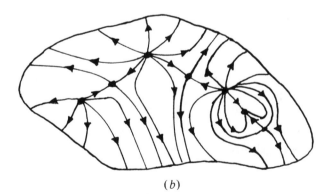

(b)

'passes' with saddle points (SP) and 'bowls' with stable nodes (SN) = which may become lakes. An obvious question is 'What are the possible numbers of UN, SN, and SP on the island?'. We will return to these considerations in Section 5.3.

We next turn to part (D) of the program which was outlined at the beginning of this chapter.

5.2 *(D) Integrating factors

The next level of detailed information about the dynamics comes from integrating

$$dx/dy = P(x, y)/Q(x, y) \tag{5.2.1}$$

to obtain the integral of the motion

$$K(x, y) = K_0 \tag{5.2.2}$$

which defines the orbits in the phase space. As was noted at the beginning of Chapter 3, one approach to integrating (5.2.2) is to obtain the integrating factor, $\mu(x, y)$, which is such that $\mu Q dx - \mu P dy$ is an exact differential of some function $K(x, y)$. Then, from (5.2.1),

$$0 = \mu Q dx - \mu P dy = dK(x, y), \tag{5.2.3}$$

which yields the constant of the motion, (5.2.2). It follows from (5.2.3) that

$$\frac{\partial K}{\partial x} = \mu Q, \quad \frac{\partial K}{\partial y} = -\mu P \tag{5.2.4}$$

and therefore μ must satisfy

$$\frac{\partial}{\partial y}(\mu Q) = -\frac{\partial}{\partial x}(\mu P). \tag{5.2.5}$$

If the original equations of motion (5.1.2) have a unique solution, this implies that unique orbits exist in R^2, so that $K(x, y)$ exists (defining the orbits, by (5.2.1)) and is differentiable (with the possible exception of singular points or curves; see Exercise 5.13). Therefore, by (5.2.4), μ exists. Indeed the function $\mu' = \mu H(K(x, y))$ is an equally good integrating factor, for any function $H(K)$, so there exists infinitely many integrating factors! Isn't it easy, therefore, to find just one? Unfortunately not.

To illustrate some exceptions, when an integrating factor can be found, we consider first a rather trivial case,

$$\dot{x} = Ax^a y^b, \quad \dot{y} = -Bx^c y^d \tag{5.2.6}$$

where A, and B are constants, and $a, b, c, d \geqslant 1$, so that the Lipschitz condition is satisfied at the origin. In this case

$$P = Ax^a y^b, \quad Q = -Bx^c y^d \tag{5.2.7}$$

so that

$$dx/dy = P/Q = -(A/B)x^{a-c} y^{b-d}$$

which yields the differential form

$$Bx^{c-a} dx + Ay^{b-d} dy = 0. \tag{5.2.8}$$

This form is so simple that it only requires the trivial integrating factor $\mu = 1$, because $dK(x, y)$ equals (5.2.8), if

$$K(x, y) = \frac{B}{c-a+1}x^{c-a+1} + \frac{A}{b-d+1}y^{b-d+1} \quad \left(\text{if } \begin{matrix} c-a+1 \neq 0 \\ b-d+1 \neq 0 \end{matrix}\right), \quad (5.2.9)$$

or

$$K(x, y) = \ln x^B + \frac{A}{b-d+1}y^{b-d+1} \quad \left(\text{if } \begin{matrix} c-a+1 = 0 \\ b-d+1 \neq 0 \end{matrix}\right), \quad (5.2.10)$$

and similarly, if $b - d + 1 = 0$. The original differential form (5.2.3) for the equation (5.2.6) is actually

$$Ax^a y^b \, dy + Bx^c y^d \, dx = 0$$

rather than (5.2.8), but the difference is trivial. Now an integrating factor is

$$\mu = x^{-a} y^{-d}, \quad (5.2.11)$$

which again yields (5.2.10). Moreover $x^{-a} y^{-d} H(K(x, y))$ is an equally good integrating factor for any C^1 function $H(z)$.

Exercise 5.8. Draw a (global) phase portrait for the orbits of the system $\dot{x} = -y(a + bx)$, $\dot{y} = cx + \frac{1}{2}by^2$ ($a, b, c > 0$). Follow the analysis of (A), (B), and (C), outlined above, and then fill in the flow pattern using the tangent (dx/dy). Next obtain an integral of the motion, and from this obtain the equations of two heteroclinic orbits.

The situation becomes more interesting if the equations of motion are

$$\dot{x} = \alpha x^{m+1} y^n + A x^{k+1} y^l$$
$$\dot{y} = -\beta x^m y^{n+1} - B x^k y^{l+1} \quad (5.2.12)$$

yielding the differential form

$$(\beta x^m y^{n+1} + B x^k y^{l+1}) \, dx + (\alpha x^{m+1} y^n + A x^{k+1} y^l) \, dy = 0. \quad (5.2.13)$$

If an integrating factor for the portion involving α and β, namely $y^{\alpha-1-n} x^{\beta-1-m} H(x^\beta y^\alpha)$, equals an integrating factor for the portion involving A and B, namely $y^{A-1-l} x^{B-1-k} G(x^B y^A)$, then it can be used as the integrating factor for the entire expression (5.2.13). Therefore we set

$$\mu = y^{\alpha-1-n} x^{\beta-1-m} H(x^\beta y^\alpha) = y^{A-1-l} x^{B-1-k} G(x^B y^A) \quad (5.2.14)$$

where the second equality of course places a restriction on the functions H and G. The simplest selection is $H(z) = z^p$, $G(z) = z^q$, leaving p and q to be determined by the conditions (from the x and y exponents)

$$\alpha - n + p\alpha = A - l + qA; \quad \beta - m + p\beta = B - k + qB,$$

or

$$p(\alpha B - \beta A) = (n - \alpha - l)B - (m - \beta - k)A$$
$$q(\alpha B - \beta A) = (k - B - m)\alpha - (l - A - n)\beta. \tag{5.2.15}$$

Therefore this procedure works provided that $(\alpha B - \beta A) \neq 0$. On the other hand, if $\alpha B = \beta A$, then multiplying (5.2.13) by B yields

$$(\beta x^m y^n + B x^k y^l)(By\,dx + Ax\,dy) = 0$$

and the integrating factor is $\mu = (\beta x^m y^n + B x^k y^l)^{-1} y^{A-1} x^{B-1}$.

Another readily integrable case is when $P(x, y)$ and $Q(x, y)$ are *homogeneous functions* of the same degree n (so $P(x, y) = x^n R(y/x)$, $Q(x, y) = x^n S(y/x)$). The integrability rests on Euler's theorem which states that, for any homogeneous function $\phi(x, y)$ of degree n,

$$x\frac{\partial \phi}{\partial x} + y\frac{\partial \phi}{\partial y} = n\phi. \tag{5.2.16}$$

Then $\mu = (xQ - yP)^{-1}$ is an integrating factor of $Q\,dx - P\,dy = 0$. To prove this requires showing that (5.2.5) holds, or

$$\frac{\partial}{\partial y}\left(\frac{Q}{xQ - yP}\right) = -\frac{\partial}{\partial x}\left(\frac{P}{xQ - yP}\right).$$

This yields

$$P\left(y\frac{\partial Q}{\partial y} + x\frac{\partial Q}{\partial x}\right) = Q\left(x\frac{\partial P}{\partial x} + y\frac{\partial P}{\partial y}\right)$$

which indeed holds because of Euler's theorem, (5.2.16).

Exercise 5.9. Obtain an integrating factor, and integrate (5.2.3), for the nonlinear oscillator $\dot{x} = y[1 + x^2]^{1/2}$, $\dot{y} = -x[1 + y^2]^{1/2}$. Can you obtain an integral expression for the period of this oscillator as a function of the magnitude of K, (5.2.3)?

Exercise 5.10. Show that Lyapunov's theorem does not shed light on the stability of the fixed point of $\dot{x} = -xy$, $\dot{y} = x^2 + y^2$. Obtain an integrating factor, and a constant of the motion $K(x, y)$. Show that this establishes whether the fixed point is stable.

Exercise 5.11. Obtain $K(x, y)$ for $\dot{x} = x^3 - y^3$, $\dot{y} = -x^2 y$. Following the above formal procedure is not always the easiest way to obtain $K(x, y)$ – it is only one sure method.

The integrals for the particular linear examples discussed in Section 1 are readily obtained using the homogeneous integrating factor $\mu = (xQ - yP)^{-1}$. Referring to those specific examples:

Center:
example: $\dot{x} = y$, $\dot{y} = -x$
$K = x^2 + y^2$

Focus:
example: $\dot{x} = ax + y$, $\dot{y} = -x + ay$ $(a \neq 0)$
$K = \ln(x^2 + y^2) + 2a\tan^{-1}(x/y)$

Node: *Saddle point:*
example: $\dot{x} = x, \quad \dot{y} = ay \,(a > 0)$ example: $\dot{x} = x, \quad \dot{y} = -y$
$\qquad K = x^2/y$ $\qquad K = x^2/y$

Note that $\partial K/\partial y$ does not exist on the curve $y = 0$ for two of these cases, but nonetheless $dK/dt = 0$ for all t. Moreover, note that $K = 2\ln(r) - 2a\theta$ is a useful representation in the case of the focus.

Exercise 5.12. Extend these particular results to obtain $K(x, y)$ for the general case, (5.1.5).

Exercise 5.13. *Relativistic dynamics* is very nonlinear, and has a very different type of phase space, because of the requirement that $\beta_k \equiv v_k/c < 1$ $(k = x, y, z)$. Consider a particle of charge q and mass m in a uniform, constant electric and magnetic field, as illustrated in Fig. 5.18. The equations of motion for $\beta \equiv v/c$

Fig. 5.18

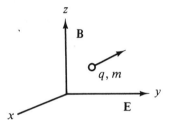

are $(d/dt)(\gamma\beta_x) = \omega_b\beta_y; \quad (d/dt)(\gamma\beta_y) = \omega_e - \omega_b\beta_x; \quad (d/dt)(\gamma\beta_z) = 0$ where $\gamma \equiv [1 - \beta_x^2 - \beta_y^2 - \beta_z^2]^{-1/2}$ and $\omega_e = qE/mc$, $\omega_b = qB/mc$. For simplicity, take $\beta_z(t) \equiv 0$.

(a) Show that $d\gamma/dt = \omega_e\beta_y$, and obtain an integral of the motion.
(b) Draw the (β_x, β_y) phase space, and obtain all the fixed points in the case (i) $\omega_e/\omega_b < 1$ and (ii) $\omega_e/\omega_b > 1$. In each case sketch the corresponding trajectory in (x, y, z) space. Can you explain why these fixed points exist from the point of view of a suitable Lorentz transformation along the x axis?
(c) Obtain an integrating factor for the equations $d\beta_x/dt$ and $d\beta_y/dt$, and then obtain another integral of the motion.
(d) Let $\Omega \equiv \omega_e/\omega_b \equiv E/B$. Draw the phase portrait in the (β_x, β_y) phase space when (i) $\Omega = 0$, (ii) $0 < \Omega < 1$, (iii) $\Omega > 1$. Are the phase portraits of electrons and protons TOE?

5.3 Poincaré's index of a curve in a vector field

One aspect of a flow (vector field) in the phase plane is that a number can be associated with a topological characteristic of the field – which is most interesting near singular points. Consider a simple closed curve, γ, in the plane and a vector field (P, Q) associated with the orbits of

$$\dot{x} = P(x, y) \quad \dot{y} = Q(x, y). \tag{5.3.1}$$

At each point on γ consider the direction of the vector field, as illustrated in Fig. 5.19.

Fig. 5.19

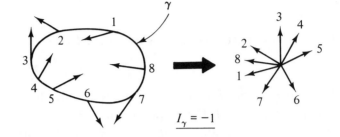

$I_\gamma = -1$

After going around the closed path, γ, either this record of field vectors rotates around the intersection point with γ, or else it does not. The number of rotations of the field vectors about the intersection points with γ is called the *index I_γ of the curve γ*, with $I_\gamma > 0$ $(I_\gamma < 0)$ if the rotation is in the same (opposite) sense as the rotation carried out along γ in the plane.

The index, for the vector field determined by (5.3.1), can be expressed as the line integral

$$I_\gamma = \frac{1}{2\pi} \oint_\gamma d\left(\arctan \frac{Q}{P} \right) = \frac{1}{2\pi} \oint_\gamma \frac{\left(P \frac{\partial Q}{\partial x} - Q \frac{\partial P}{\partial x} \right) dx + \left(P \frac{\partial Q}{\partial y} - Q \frac{\partial P}{\partial y} \right) dy}{P^2 + Q^2} \tag{5.3.2}$$

but usually it is easier to determine I_γ graphically (see Fig. 5.20).

Fig. 5.20

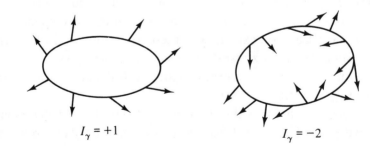

$I_\gamma = +1$ $I_\gamma = -2$

Exercise 5.14. Show that if γ goes around either a (single) center, focus, or node then $I_\gamma = +1$, whereas, if it goes around a saddle point $I_\gamma = -1$. What is the index of the degenerate singular points shown in Fig. 5.21 ((a) is called a *saddlenode*)? Do stable and unstable focii and nodes have the same index?

Fig. 5.21

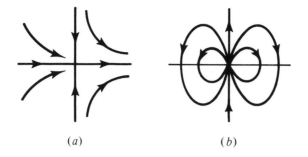

(a)   (b)

Exercise 5.15. Use Green's theorem in the plane, $\iint_\Omega [(\partial P/\partial x) + (\partial Q/\partial y)]\,dxdy = \oint_\gamma (P\,dy - Q\,dx)$, to show that, if $\partial P/\partial x + \partial Q/\partial y$ does not change sign in a simply connected region, Ω, of the phase plane (x, y), then its boundary curve, γ, is not an orbit of $\dot{x} = P, \dot{y} = Q$. Hence, if these equations have a periodic solution, $\partial P/\partial x + \partial Q/\partial y$ must change sign in some region, Ω. This is known as *Bendixson's cirterion*.

The index of a curve has the following *property A*:

> *If a closed curve γ or the vector field is continuously deformed, the index of γ does not change, provided that it does not pass through a singular point of the vector field (i.e., where it vanishes) during this deformation.*

For an elaboration on this property and its proof see, for example, Chinn and Steenrod (1967), or Hartman (1973). From property A it follows:

If $I_y \neq 0$ for γ, then there is at least one singular point in the region enclosed by γ (for otherwise γ could be contracted to a point without crossing a singularity, and end up with the index $I_y = 0$. But this contradicts the assumption that $I_y \neq 0$, using property A).

Because of property A, it is clear that the index I_y of a curve about a single singularity only depends on the nature of the singularity, and not on γ. Using any such a curve, it is then possible to define an *index, I_s, of a singular point*. Exercise (5.14) shows that several singularities can have the same index.

Exercise 5.16. Consider P, Q given by $P + iQ \equiv (x + iy)^n$, where n is an integer and $i = (-1)^{1/2}$. What is the index of the singular point at the origin for arbitrary n (Hint: Note that $z = r\exp(i\phi)$)? Do these singularities belong to the categories discussed in the last section?

One can then prove that I_y, for a simple closed curve, γ, equals the sum of the indices of the enclosed singularities. If W_γ is the region enclosed by γ,

$$I_\gamma = \sum I_s. \quad (s \in W_\gamma) \tag{5.3.3}$$

This follows directly from property A, using the fact that $I_{\gamma'} = 0$ in Fig. 5.22. The minus signs in front of I_1, I_2, and I_3, come from the fact that, in breaking up γ' into γ plus the

Fig. 5.22

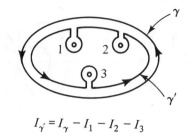

$$I_{\gamma'} = I_{\gamma} - I_1 - I_2 - I_3$$

three loops, the sense of rotation on these loops must be opposite the sense on the original γ'.

An important application of these ideas is to prove Brouwer's *fixed point theorem in* R^2. Consider a compact region W, which is mapped into itself by some continuous maps $F(x): W \to W$. (i.e., $F: x \to x'$, $x_1' = F_1(x)$, $x_2' = F_2(x)$, with $x, x' \in W$). Choose the coordinates so that $x = 0$ is in W. Now consider the vector field $v(x) \equiv F(x) - x$, and the curve γ on the boundary of W. If W is disk-like (convex), so that all points of γ can be connected to the origin by a straight line lying in W, then the vector field $v_t \equiv tF(x) - x$ does not vanish for any $0 \leqslant t \leqslant 1$ (see Fig. 5.23) when x is on γ. The vectors $v_t(x)$ do not

Fig. 5.23

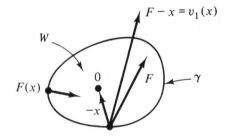

necessarily map γ into W for all t, but that is not important. All we need is that $v_t(x)$ does not vanish on γ, for any $0 \leqslant t \leqslant 1$. Hence, the vector field $F(x) - x$ can be continuously deformed to the field $-x$ (by varying t) without producing a singular point, $v_t(x) = 0$, on γ. Now it is clear that the index of γ in the field $v_0 = -x$ is $+1$. Hence this is the index of the field $v(x) \equiv v_1(x)$, by property A. This means that there must be at least one singular point of $v(x)$ in W. But a singular point of v is a fixed point of F (i.e., if $v(x_0) = 0$, then $x_0 = F(x_0)$). Hence

> *Every continuous map $F: W \to W$ of a convex compact region W into itself has at least one fixed point in W.*

This theorem can be generalized to n dimensions (*Brouwer's fixed point theorem*). If $F(x): S^n \to S^n$ is a continuous map of a compact n-spherical region $S^n = \{x: x_1^2 + x_2^2 + \cdots + x_n^2 \leqslant r^2\}$ into itself, then this map has at least one fixed point in this region.

Fig. 5.24

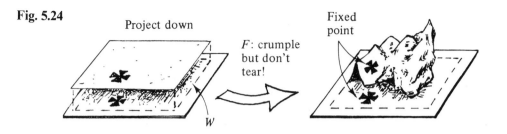

Project down

F: crumple but don't tear!

Fixed point

W

As a simple example, consider a sheet of paper over a plane (Fig. 5.24). The projection of the points on the paper down onto the plane defines a region *W*. Now crumple the paper and place it over the region *W*. The projection now defines a new region *W'* inside *W*. Therefore the process of crumpling and placing defines a map $F: W \to W'$, which is continuous if the paper is not torn. The theorem tells us that there must be some point on the paper whose projection does not change in this process (it ends up over the same point of the plane). This is clearly not necessarily true if the paper is torn (e.g., tear in half, crumple a little, and interchange halves).

Exercise 5.17. Use the above results concerning indices to show that:

Fig. 5.25

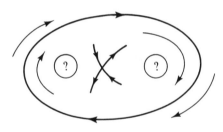

(a) Closed (periodic) orbits cannot exist if there are no singular points (Hint: determine I_γ for a periodic orbit, γ).
(b) If there is only one singular point, and $I_s \neq +1$, then there is no periodic solution (e.g., no limit cycle).
(c) If a stable limit cycle encloses a saddle point, what other singularities must it enclose? See Fig. 5.25.
(d) If $\dot{x} = F$, and $|\partial F_i / \partial x_j| \neq 0$ at all singular points, then any periodic orbit encloses an odd number of singular points. Can you construct a vector field with three such singularities, none of which are stable, all enclosed by a stable limit cycle?

Exercise 5.18.

(a) If a simple node vanishes at a bifurcation point what other simple singular point must also vanish? (See Fig. 5.26.)

Fig. 5.26

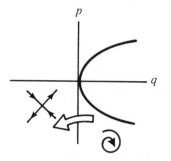

(b) In a bifurcation, if q changes from $q > 0$ to $q < 0$ for a fixed point, what else must occur in this bifurcation?

Exercise 5.19. Show that if $B(x, y)$ is C^1 and such that $\partial(BP)/\partial x + \partial(BQ)/\partial y$ does not change sign in a simply connected region, then there is no periodic solution of $\dot{x} = P(x, y)$, $\dot{y} = Q(x, y)$ in that region. This is known as *Dulac's criterion.*

The above exercises point toward another interesting use of the Poincaré index, which also concerns the question raised in Section 1, namely: What collection of fixed (singular) points are possible in a global vector field (flow)? This might appear to be both a topological and an orientational problem (i.e., involving the distinction between stable and unstable fixed points). However that is fortunately not the case, so that the Poincaré index proves to be adequate, even though it does not distinguish between different orientations (Exercises 5.14, and 5.17(d)). This index can supply the necessary (topological) criterion, if the flow takes place on a compact manifold, M, such as a sphere, S^2, or torus, T^2.

Fig. 5.27 illustrates several flows on a sphere. We can define a Poincaré index of a

Fig. 5.27

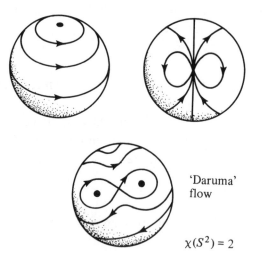

'Daruma' flow

$\chi(S^2) = 2$

singular point by projecting the local flow onto a tangent plane at the singularity, and proceeding as in R^2. In each case, if we add up the Poincaré indices for all singularities, we can see that it equals 2 (the 'Daruma' flow – a Japanese doll – has a center on the back side). This number is called the *Euler characteristic*, $\chi(M)$, of the manifold, so $\chi(S^2) = 2$ (see Hartman, 1973; or Arnold, 1978, for more details). On the other hand, we know that there can be flows on a torus, T^2, which have no fixed points, so the sum of the Poincaré indices is expected to be zero, $\chi(T^2) = 0$ (Fig. 5.28). Indeed, more

Fig. 5.28

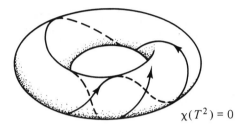

$$\chi(T^2) = 0$$

generally, $\chi(S^n) = 1 + (-1)^n$ and $\chi(T^n) = 0$, and the concept of the Poincaré index can also be generalized to R^n. The sum of these indices, for all singular points of a vector field on a compact manifold, M, equals $\chi(M)$ – which depends only on the manifold, and not the vector field. Thus, if $I(m)$ is the Poincaré index of the singularity m

$$\sum_m I(m) = \chi(M). \tag{5.3.4}$$

The above results on a sphere can be illustrated by several nice examples, due to Tucker and Bailey, Jr. (1950). They noted, among other facts, that:

(1) The wind cannot be blowing everywhere on the Earth at the same time. There must be a windless point.

(2) If the wind is blowing everywhere in the northern hemisphere at a given instant, then there must be a point on the equator where it blows in any prescribed compass direction.

To apply (5.3.4) to the flow in R^2, we need to project R^2 onto S^2, and append the point at ∞ (the north pole of S^2). This frequently eliminates the criterion concerning the finite fixed point, since the appended point can be a fixed point with any (necessary) Poincaré index. Moreover, these results do not depend on the stability properties of the fixed points, as noted above.

Thus, for example, it might seem unlikely that a flow could occur in R^2 which has only two unstable focii, with the same sense of rotation (see Fig. 5.29). Physically we might expect symmetry, in which case there would be a singular line ('stagnation line') between them. However $\sum I(m) = 2$ in this case, so it is possible for such a flow to occur on S^2 (with no additional singularities) and hence also in R^2, with no stagnation line.

Fig. 5.29

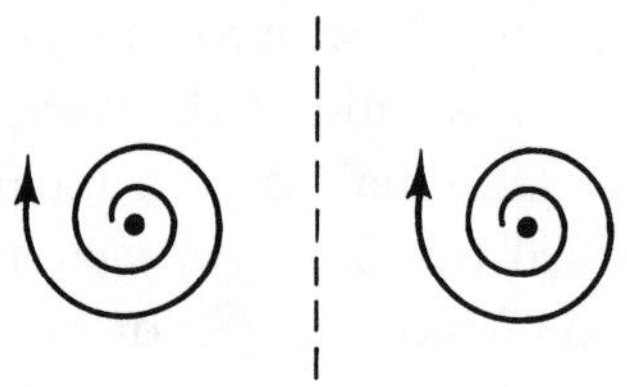

Fig. 5.30

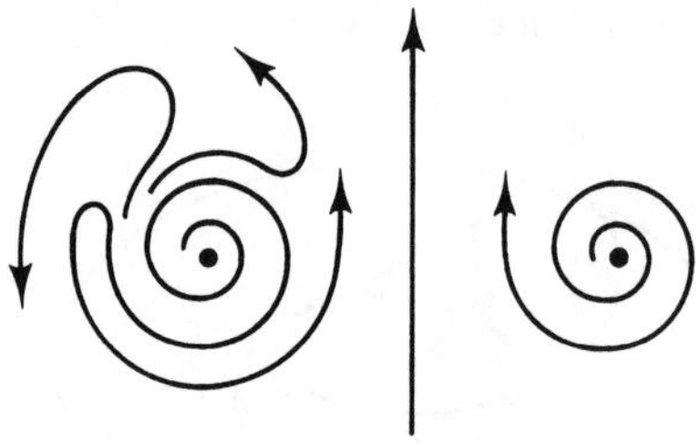

However, the required flow is not symmetric, and looks 'strange' from a physical point of view – but not impossible! (See Fig. 5.30.)

Exercise 5.20. Can you devise equations of motion, (5.1.2), which produce a rotation-reversing flow from an unstable focus, as illustrated in Fig. 5.31?

Fig. 5.31

*********Preview of coming attractions*********

*(E) Time dependence

In order to discuss the time dependence of solutions, we will limit our considerations to the most important systems, namely oscillators. In the following sections we will consider four distinct types of oscillator systems, divided into the following groups:

1. Autonomous systems

(a) *'Passive' oscillators* – characterized by the pendulum and an equivalent polynomial potential oscillator (called the Duffing oscillator). The amplitude and period of these systems can depend on the initial conditions. These systems involve questions concerning the dependency of the period on the amplitude, secular perturbation problems, and the application of various averaging methods (Fig. 5.32).

Fig. 5.32

We will consider one due to Krylov–Bogoliubov–Mitropolsky (KBM).

(b) *Self-exciting (autocatalytic) systems* – characterized by the Rayleigh and van der Pol equations. These systems have constant (time-independent) energy sources and sinks, depending on where the system is in the phase plane (Fig. 5.33). This leads

Fig. 5.33

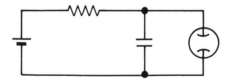

asymptotically ($t \to \infty$) to a periodic motion – the limit cycle. This limit cycle is independent of the initial conditions. Some other basic concepts are the Poincaré–Bendixson theorem, and Hopf bifurcation. For weak excitations we can use the KBM method whereas, for strong excitations, nonanalytic phase plane analysis is required. Also aspects of singular perturbation problems arise, involving 'fast' and 'slow' variables.

2. Nonautonomous systems

(a) *Periodic force applied to a passive oscillator* – such as the forced Duffing oscillator. This yields hysteresis effects (a cusp catastrophe), harmonic and subharmonic generation, and chaotic motion.

Exercise 5.21. Obtain the equations of motion for $\theta(t)$ in these two systems: (A) A rigid, frictionless pendulum of mass m, length l, acted upon by gravity, g, and a vertically oscillating support; (B) the same pendulum pivoted at the edge of a rotating wheel (angular velocity $= \omega$) (Fig. 5.34).

Fig. 5.34

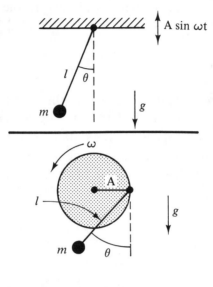

Fig. 5.35

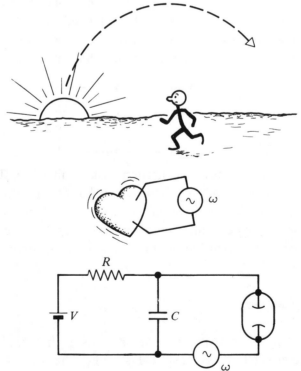

(b) *Periodic force on a self-exciting system* – yields the phenomenon of entrainment, such as occurs in many biological systems. Thus, for example, various circadean rythms in our bodies are modified by the environmental periods (e.g., daylight). More drastic couplings occur in the mechanical pacemakers, used to regulate heartbeats. An electrical example of such coupling is the illustrated system, which is associated with the classic experimental studies of van der Pol and van der Mark, and the theory of Cartwright and Littlewood and of Levinson. They were the first to show that such 'physical' models (ODE) can produce a remarkably 'complicated' *attractor*, associated with fractal structures. These are illustrated in Fig. 5.35.

We will now proceed to consider these topics in more detail.

5.4 The pendulum and polynomial oscillators

One of the classic problems of nonlinear analysis is the determination of the motion of a simple pendulum. While the problem is an 'oldie' it is also a 'goodie', because it gives an explicit example of an amplitude dependent frequency (something not exhibited by a harmonic oscillator). In general this feature requires the complication of introducing new transcendental functions, the elliptic functions, which give analytic expressions for the dynamics, and hence the frequencies. Special cases, involving small (but finite) amplitudes, can then be compared with perturbative analysis of this system. More importantly, these results apply to any nonlinear oscillator with a cubic polynomial nonlinear force ($\ddot{x} = -\omega_0^2 x - \varepsilon x^3$). This is a very common model of nonlinearity, since the potential, $\frac{1}{2}\omega_0 x^2 + \frac{1}{4}\varepsilon x^4$, is the 'natural' symmetric generalization of the harmonic oscillator.

In dimensionless time $(t(g/l)^{1/2})$ (Fig. 5.36), the equation for the angle, θ, of an

Fig. 5.36

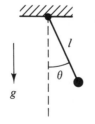

undamped pendulum is

$$\ddot{\theta} + \sin\theta = 0. \tag{5.4.1}$$

The nonlinear force in (5.4.1) is an example of *soft-nonlinearity*, where the restoring force (or torque) is weaker than in the linear system. This is illustrated in Fig. 5.37. The case of hard-nonlinearities will be considered shortly.

Fig. 5.37

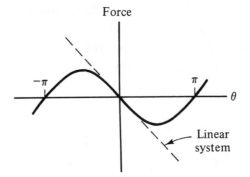

The integral of the motion of (5.4.1) is $\frac{1}{2}\dot{\theta}^2 - \cos\theta = E$ and, if $E \leqslant 1$, it can be written

$$\tfrac{1}{2}\dot{\theta}^2 - \cos\theta = -\cos 2\alpha$$

where α is a constant given by $2\alpha = \max\theta$ (where $\dot{\theta} = 0$). Therefore, since $\cos\theta = 1 - 2\sin^2(\theta/2)$,

$$\dot{\theta}^2 = 2(\cos\theta - \cos 2\alpha) = 4[\sin^2\alpha - \sin^2(\theta/2)].$$

If we introduce the variable ϕ through the definition

$$\sin(\theta/2) = \sin\alpha\sin\phi \quad \text{(note: } \max\phi = \pi/2) \tag{5.4.2}$$

we have

$$\dot{\theta}^2 = 4\sin^2\alpha\cos^2\phi, \quad \text{and also} \quad \tfrac{1}{2}\dot{\theta}\cos(\theta/2) = \sin\alpha\cos\phi\,\dot{\phi}.$$

Hence

$$\tfrac{1}{2}2\sin\alpha\cos\phi\cos\theta/2 = \sin\alpha\cos\phi\,\dot{\phi},$$

or

$$\dot{\phi} = \cos(\theta/2) = [1 - \sin^2\theta/2]^{1/2} = [1 - k^2\sin^2\phi]^{1/2}, \tag{5.4.3}$$

Fig. 5.38

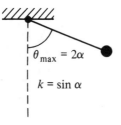

$\theta_{max} = 2\alpha$

$k = \sin\alpha$

where (see Fig. 5.38)

$$k = \sin\alpha; \quad k^2 \leqslant 1. \tag{5.4.4}$$

Integrating (5.4.3) yields

$$t - t_0 = \int_0^\phi [1 - k^2\sin^2\theta]^{-1/2}\,\mathrm{d}\theta. \tag{5.4.5}$$

The function

$$\bar{F}(\phi, k) \equiv \int_0^\phi [1 - k^2 \sin^2 \theta]^{-1/2} \, d\theta \tag{5.4.6}$$

is called the *elliptic integral of the first kind* (see appendix *F*).

Before discussing the properties of this solution, we will derive another equation of motion which is related to (5.4.1), but contains a polynomial force rather than a transcendental force, $\sin \theta$. To make this connection, we define

$$y = \sin \phi \quad (\text{note: max } y = 1). \tag{5.4.7}$$

Using (5.4.3), we find that

$$\dot{y} = \cos \phi \, \dot{\phi} = \cos \phi \, [1 - k^2 \sin^2 \phi]^{1/2} = [(1 - y^2)(1 - k^2 y^2)]^{1/2},$$

and therefore

$$t - t_0 = \int_0^y [(1 - z^2)(1 - k^2 z^2)]^{-1/2} \, dz. \tag{5.4.8}$$

This is, of course, simply another representation of the solution (5.4.5), and the elliptic integral (5.4.6), through the relationship (5.4.7). Since

$$(\dot{y})^2 = (1 - y^2)(1 - k^2 y^2) = 1 - (1 + k^2)y^2 + k^2 y^4$$

the second order equation for y is

$$\ddot{y} = -(1 + k^2)y + 2k^2 y^3 \tag{5.4.9}$$

Thus we have made a connection between an oscillator with a transcendental force, (5.4.1), and one with a polynomial force, (5.4.9).

We now use (5.4.8) to introduce the *Jacobi elliptic function* $sn((t - t_0), k)$ through the definition

$$t - t_0 = \int_0^y [(1 - z^2)(1 - k^2 z^2)]^{-1/2} \, dz \equiv F(y, k) \equiv sn^{-1}(y, k) \tag{5.4.10a}$$

or

$$y = sn((t - t_0), k). \tag{5.4.10b}$$

Fig. 5.39

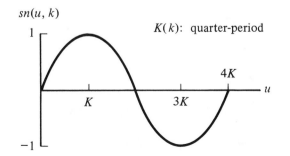

This elliptic function behaves like the trigonometric function, $\sin(t - t_0)$, if $k \ll 1$ (see below, and Eq. (15) of the appendix F). Note that $F(\sin\phi, k) \equiv \bar{F}(\phi, k)$, defined in (5.4.6).

Since $y = \sin\phi = 1/k \sin(\theta/2)$, the solution of the pendulum equation, $\ddot{\theta} + \sin\theta = 0$, is

$$\sin(\theta/2) = k\,sn((t - t_0), k); \quad k \equiv \sin\alpha; \quad \alpha = \max(\theta/2). \tag{5.4.12}$$

The elliptic function $sn((t - t_0), k)$ is a periodic function of t, illustrated in Fig. 5.39, which has a *quarter-period* defined by

$$K(k) \equiv \int_0^1 [(1 - z^2)(1 - k^2 z^2)]^{-1/2}\,dz = \int_0^{\pi/2} [1 - k^2 \sin^2\theta]^{-1/2}\,d\theta. \tag{5.4.11}$$

$K(k)$ is also called the *complete elliptic integral* of the first kind. As can be seen from (5.4.5) or (5.4.10), this is the time required for y to go from 0 to 1, or from θ to go from 0 to θ_{max}. The period of the motion is therefore $4K(k)$, and some values are illustrated in the table.

max θ:	0	20°	40°	60°	80°	140°	178°	180
period$/2\pi$:	1	1.008	1.03	1.07	1.135	1.65	3.46	∞

It will be noted that the period is changed very little for small amplitude oscillations ($\theta_{max} \lesssim 20°$).

The pendulum dynamics, given by (5.4.12),

$$\theta(t) = 2\sin^{-1}[k\,sn((t - t_0), k)]; \quad k = \max(\theta/2),$$

is illustrated in Fig. 5.40, for the cases $\theta_{max} = 40°$, 176°. The lengthening of the period for larger amplitudes is due, of course, to the small velocity and acceleration at large angles ($\theta \simeq 180°$). The small acceleration is a consequence of the *soft-nonlinearity* of this system.

In the case of a polynomial force, the equation of motion is usually not given in the form (5.4.9), but rather in the form

$$\ddot{x} + \omega_0^2 x + \varepsilon x^3 = 0 \quad (\omega_0, \varepsilon: \text{given constants}). \tag{5.4.13}$$

The two equations can be related by scaling. Thus, if we set $x = ay$ and $t = \Omega\tau$ in (5.4.9), we obtain

$$\frac{d^2 x}{d\tau^2} = -(1 + k^2)\Omega^2 x + \frac{2k^2\Omega^2}{a^2} x^3$$

which is of the form (5.4.13), provided that

$$\omega_0^2 = (1 + k^2)\Omega^2 \quad \text{and} \quad \varepsilon = -\frac{2k^2\Omega^2}{a^2}. \tag{5.4.14}$$

Thus the general solution of (5.4.13) can be obtained from (5.4.11), namely (setting $\tau \to t$)

$$x = a\,sn(\Omega(t - t_0), k)(a, t_0: \text{arbitrary constants}) \tag{5.4.15}$$

Fig. 5.40

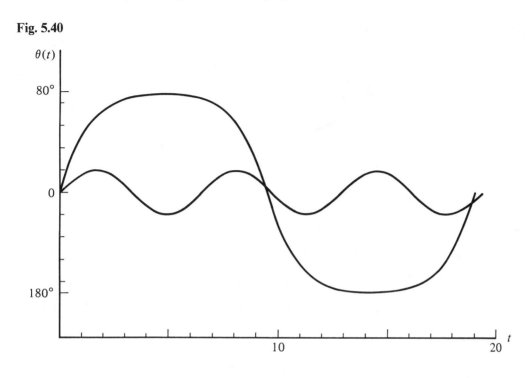

provided that Ω and k satisfy (5.4.14), namely

$$\Omega^2 = \omega_0^2 + \frac{\varepsilon a^2}{2} \quad k^2 = \frac{-a^1\varepsilon}{2\omega_0^2 + \varepsilon a^2}. \tag{5.4.16}$$

The period of this solution, τ, is given by

$$\Omega\tau = 4K(k) = 2\pi[1 + (\tfrac{1}{2})^2 k^2 + (\tfrac{1\cdot3}{2\cdot4})^2 k^4 + \cdots],$$

where the power series in k^2 comes from the expansion of $(1 - k^2 z^2)^{-1/2}$ in (5.4.11). Therefore the angular frequency ω is, by definition,

$$\omega \equiv \frac{2\pi}{\tau} = \frac{\Omega}{[1 + \tfrac{1}{4}k^2 + \tfrac{9}{64}k^4 + \cdots]} = \Omega(1 - \tfrac{1}{4}k^2 + O(k^4))$$

$$= \omega_0\left(1 + \frac{\varepsilon a^2}{4\omega_0^2} + O(\varepsilon^2 a^4)\right)\left(1 + \frac{a^2\varepsilon}{8\omega_0^2} + O(\varepsilon^2 a^4)\right)$$

$$= \omega_0 + \frac{3\varepsilon a^2}{8\omega_0} + O(\varepsilon^2 a^4). \tag{5.4.17}$$

As one would expect, the frequency decreases for soft-nonlinearity ($\varepsilon < 0$), since the restoring force is less than in the linear system (Fig. 5.41).

To treat the case of a hard-nonlinearity polynomial oscillator ($\varepsilon > 0$ in (5.4.13)), we

Fig. 5.41

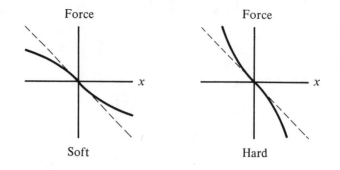

need to introduce the elliptic function

$$cn^{-1}(x, k) \equiv F([1 - x^2]^{1/2}/k^2, k)$$

where $F(y, k)$ is defined by (5.4.10). Now, if

$$x = a\, cn(\Omega(t - t_0), k),$$

one can verify that x satisfies the polynomial oscillator equation of motion (5.4.13), provided that (see Appendix F)

$$\omega_0^2 = \Omega^2(1 - 2k^2), \quad \varepsilon = \frac{2k^2\Omega^2}{a^2}.$$

Note that now $\varepsilon > 0$ (hard nonlinearity). These results can be written

$$\Omega^2 = \omega_0^2 + \varepsilon a^2, \quad k^2 = \varepsilon a^2/2(\omega_0^2 + \varepsilon a^2), \tag{5.4.18}$$

which differs from the soft nonlinearity results, (5.4.16). However, to lowest order, the angular frequency is, from (5.4.18),

$$\omega = \Omega(1 - \tfrac{1}{4}k^2) = \omega_0\left(1 + \frac{\varepsilon a^2}{2\omega_0^2}O(\varepsilon^2 a^4)\right)\left(1 - \frac{\varepsilon a^2}{8\omega_0^2}O(\varepsilon^2 a^4)\right)$$

$$= \omega_0 + \frac{3\varepsilon a^2}{8\omega_0} + O(\varepsilon^2 a^4).$$

Thus the frequency shift is of the same form as (5.4.17) to $O(\varepsilon^2 a^4)$, but now $\varepsilon > 0$, and $\omega > \omega_0$.

While the dynamics of such nonlinear oscillators can be obtained from the tabulated values of the functions $\bar{F}(\phi, k)$ and $F(y, k)$, this procedure is now largely antiquated, due to availability of personal computers. Using a simple program, we can readily compute the dynamics $x(t)$ given by (5.4.13), which circumvents the computation (and inversion) of such functions. A simple fourth order Runge–Kutta integration method is adequate for most equations (see Appendix H). For an extensive collection of computer programs, see Press, Flannery, Teukolsky, and Vetterling (1986).

Equation (5.4.13) can be simplified by scaling, setting $t' = \omega_0 t$, $x' = |\varepsilon|^{1/2} x/\omega_0$ (then dropping the primes) to obtain

$$\ddot{x} + x + \delta x^3 = 0 \quad \begin{pmatrix} \delta = +1: & \text{hard} \\ \delta = 0: & \text{linear} \\ \delta = -1: & \text{soft} \end{pmatrix} \tag{5.4.19}$$

The 'energy', in these units, is

$$E = \tfrac{1}{2}(\dot{x}^2 + x^2) + \delta x^4/4 \equiv \tfrac{1}{2}\dot{x}^2 + V(x). \tag{5.4.20}$$

There may be periodic motion, provided that $\dot{x} = 0$ has at least two real roots for x

$$\tfrac{1}{2}\delta x^4 + x^2 - 2E = 0. \tag{5.4.21}$$

If $\dot{x} \neq 0$ at $x = 0$, so that $E > 0$, there are two real roots if $\delta = +1$, and four real roots when $\delta = -1$, provided that $E < \tfrac{1}{4}$, but no real roots if $E > \tfrac{1}{4}$. This is easily understood by graphing the potential energy, $V(x)$, in (5.4.20) (Fig. 5.42).

Fig. 5.42

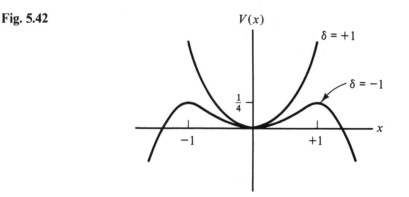

Fig. 5.43(a) illustrates the behavior of the linear oscillator ($\delta = 0$), and of the soft ($\delta = -1$) and hard ($\delta = +1$) nonlinearity, for the common initial conditions $x(0) = 0$, $\dot{x}(0) = 0.7$ ($E = 0.245$). Fig. 5.43(b) shows these same solutions in the phase plane. The orbit of the soft oscillator ($\delta = -1$) is quite close to a separatrix at $E = 0.25$, noted in Section 1.

Exercise 5.22. Sketch the phase plane orbits for the system

$$\ddot{x} - x + \varepsilon x^3 = 0 \quad (\varepsilon > 0).$$

Be specific about the location and types of all fixed points.

This *separatrix structure*, which occurs for an oscillator with soft nonlinearity, is shown in Fig. 5.44. These are orbits obtained from (5.4.19), with $\delta = -1$. The separatrix

Fig. 5.43

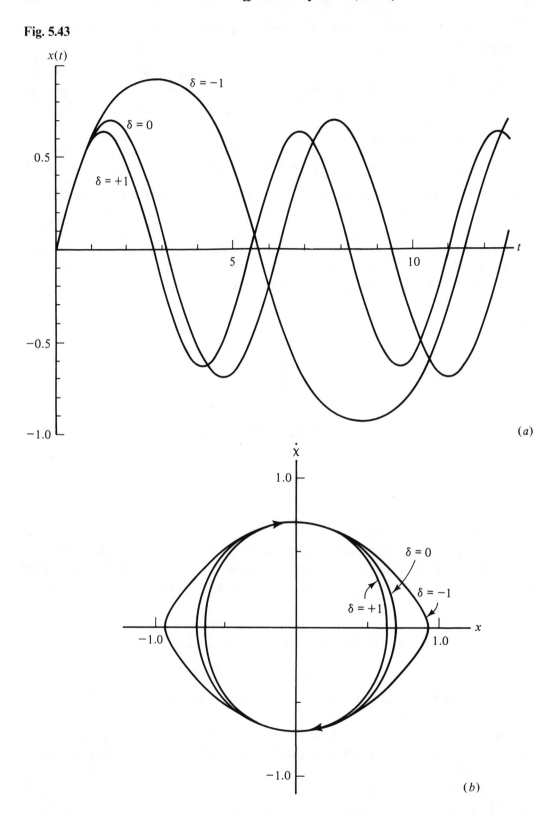

(a)

(b)

Fig. 5.44

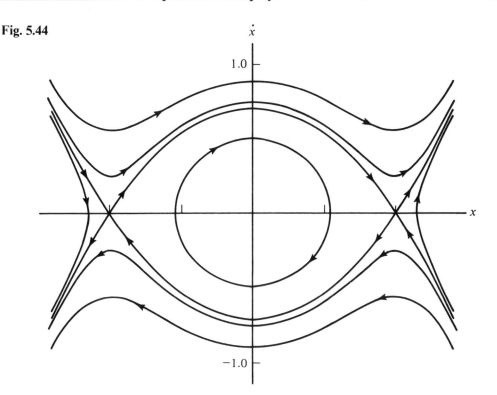

consists of the two orbits which join the two hyperbolic fixed points (saddle points). Any orbit which connects two distinct fixed points is called a *heteroclinic orbit*. Hence a separatrix consists of a pair of heteroclinic orbits.

The restoring force of the pendulum, and in the above polynomial oscillators, is symmetric about the fixed points, at $\theta = 0$ and $x = 0$. In contrast to these cases, the oscillator

$$\ddot{x} + x + x^2 = 0 \tag{5.4.22}$$

has a stronger restoring when $x > 0$ than when $x < 0$ (it is 'positive-hard' and 'negative-soft'). There is now only one hyperbolic point, at $x = -1$, and hence no separatrix. Instead there is an orbit which leaves the hyperbolic point in one direction, and returns in another direction. Such an orbit is called a *homoclinic orbit*. A simple rolling ball example was given in Section 1. The homoclinic orbit and several other orbits are illustrated. The time dependence of $x(t)$ illustrated in Fig. 5.45, both for the periodic orbit and the homoclinic orbit, starting from $x = 0$, $\dot{x} > 0$. Note the symmetry of the periodic orbit (there is no 'quarter-period' in this case). The homoclinic solution takes an infinite time to arrive at $x = -1$.

Exercise 5.23. (a) Use the constant of the motion ('energy') of (5.4.22) to obtain the initial condition, $\dot{x}(0)$, for the homoclinic orbit in Fig. 5.46 (if $x(0) = 0$)? (b) Obtain an integral 'solution', of the character of (5.4.8), for the homoclinic orbit.

Fig. 5.45

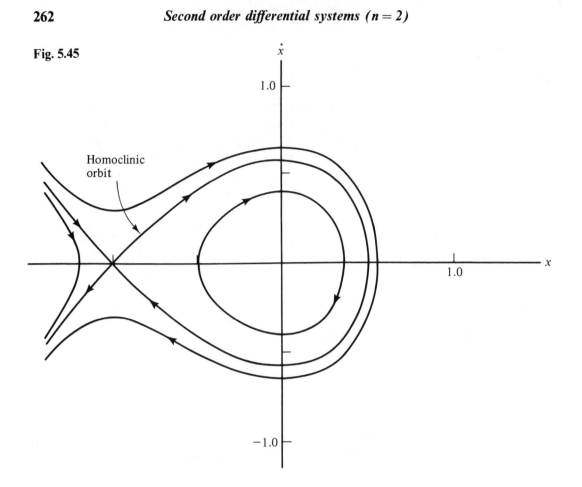

Exercise 5.24. An important physical example of an asymmetric potential is the molecular Lennard–Jones '6–12' potential

$$V(r) = A(r^{-12} - 2r^{-6}),$$

where the distance has been scaled so that the minimum is at $r_0 = 1$ (Fig. 5.47). Set $r = r_0 + x$ and obtain the x-polynomial approximation of $V(r)/A$ from the Taylor expansion about $x = 0$. Obtain all coefficients up to, and including, x^4. See the following exercises for other aspects of such a polynomial potential.

Exercise 5.25. Consider the polynomial oscillator

$$d^2y/d\tau^2 + \omega_0^2 y + Ay^2 + By^3 = 0.$$

(a) Show that this can be scalled from $(\tau, y) \to (t, x)$ such that $\ddot{x} + x + x^2 + \varepsilon x^3 = 0$. What is $\varepsilon(\omega_0, A, B)$?

(b) For what value of ε are there (i) zero, (ii) one or (iii) two hyperbolic points? Sketch the potentials.

Fig. 5.46

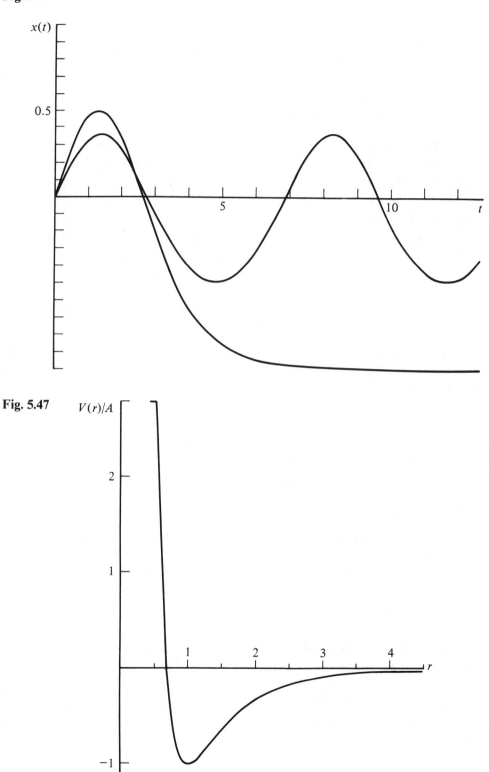

Fig. 5.47

(c) How many homoclinic orbits are there? For what value(s) of ε? What are the values of their energies in terms of ε, $E(\varepsilon)$?

Exercise 5.26. The above figures were obtained by using a Runge–Kutta iteration method to solve the equations of motion. To obtain practice with this method, consider the generalizations of (5.4.19) and (5.4.22),

$$\ddot{x} + x + x^2 + \varepsilon x^3 = 0 \quad \text{or,} \quad \dot{x} = y, \dot{y} = x - x^2 - \varepsilon x^3$$

discussed in the last exercise. Use the Runge–Kutta method (Appendix H) to determine the period, $T(E)$, vs. the energy, E, when $\varepsilon = 0.1$, for the following orbits. Take the initial conditions to be $x(0) = 0$, $y(0) = \dot{x}(0) = (2E)^{1/2}$, and use the time step size $\Delta t \equiv H = 0.05$ (you can use smaller values to check the accuracy). Determine $T(E)$ for $E = 0.01, 0.02, 0.05, 0.10, 0.15, 0.20, 0.25, 0.30, 0.40, 0.50,$ and 1.0. Sketch $T(E)$ vs. E, and explain the 'anomolous' behavior. To obtain $T(E)$, you can put a test in the program so that it stops if $\dot{x} > 0$ and $x(t)x(t - \Delta t) < 0$ (a rough Poincaré map. For Accuracy, see Section 6.10.).

5.5 The averaging method of Krylov–Bogoliubov–Mitropolsky (KBM)

An important group of systems are those which are, in some sense, close to systems which only have simple harmonic periodic motion. The simplest example from this group is the nonlinear oscillator

$$\ddot{x} + x + \varepsilon x^3 = 0 \tag{5.5.1}$$

where the time has been normalized to the unperturbed ω_0^{-1}. ε is a 'sufficiently small' constant to make the nonlinear term a perturbation on the harmonic motion (e.g., $\int |x| \, dt \gg \varepsilon \int |x^3| \, dt$). As we have seen in the last section the exact solution of (5.5.1) can be expressed in terms of elliptic functions. We are interested, however, in using (5.5.1) as a heuristic example before discussing more general systems.

Before discussing an averaging method, of which there are many variations (e.g., see Hagihara, Vol. III, 1974; Sanders and Verhulst, 1985; Nayfeh, 1986), we will first see why a straightforward perturbative method cannot be used on (5.5.1), if one wants to obtain a solution valid for long times ($t \lesssim O(1/\varepsilon)$). If $\varepsilon = 0$ the solution of (5.5.1) is

$$x(t) = a \cos(t + \theta) \equiv x_0(t) \tag{5.5.2}$$

where (a, θ) are arbitrary constants, and we can normalize x so that $a = O(1)$. Suppose we now assume a perturbative series in powers of ε

$$x = a \cos(t + \theta) + \varepsilon x_1(t) + \varepsilon^2 x_2(t) + \cdots \tag{5.5.3}$$

where it is assumed that the x_ns are of order $a = O(1)$. This makes ε a true indication of the magnitude of the successive terms in the series (5.5.3). Then

$$x^3 = a^3 \cos^3 (t + \theta) + \varepsilon \, 3a^2 \cos^2 (t + \theta) x_1(t) + \cdots$$

Substituting this expression in (5.5.1), and setting the coefficient of ε equal to zero (so that (5.5.1) is satisfied to $O(\varepsilon)$), yields

$$\ddot{x}_1 + x_1 + a^2 \cos^3 (t + \theta) = 0.$$

Since $\cos^3 \alpha = \frac{3}{4} \cos \alpha + \frac{1}{4} \cos 3\alpha$, this equals

$$\ddot{x}_1 + x_1 = - \frac{3a^3}{4} \cos (t + \theta) - \frac{a^3}{4} \cos 3(t + \theta)$$

which has the general solution

$$x_1 = a' \cos (t + \theta') - t \frac{3a^3}{8} \sin (t + \theta) - \frac{a^3/4}{(1 - 3^2)} \cos 3(t + \theta). \qquad (5.5.4)$$

Two points should be noted.

(a) The first term, $a' \cos (t + \theta')$, which comes from the solution of the homogeneous equation, is redundant. It already appears in the lowest approximation of (5.5.4). Hence, we want to exclude such terms.
(b) More important, the second term is a *secular term*, being proportional to t and hence becomes unbounded for large t. It is easy to see, however, that the actual solution of the nonlinear equation must be bounded, for $\frac{1}{2}\dot{x}^2 + \frac{1}{2}x^2 + \frac{1}{4}\varepsilon x^4$ is constant and positive definite. In order for $x_1(t) = O(1)$, as was assumed, we see from the second term of (5.5.4) that the time is restricted by

$$t \lesssim O(1). \qquad (5.5.5)$$

This restriction is too severe to make the series (5.5.4) of interest as it stands.

The resolution of this difficulty lies in the fact that the nonlinearity changes the frequency from $\omega_0 = 1$ to $\omega(a)$, which depends on the amplitude a. To take this into account we need to set

$$\omega = 1 + \varepsilon \omega_1 + \varepsilon^2 \omega_2 + \cdots$$

and introduce this unknown frequency into the lowest order term

$$x = a \cos (\omega t + \theta) + \varepsilon x_1(t) + \cdots.$$

Then

$$\begin{aligned}
\ddot{x} &= - a\omega^2 \cos (\omega t + \theta) + \varepsilon \ddot{x}_1 + \varepsilon^2 \, \ddot{x}_2 + \cdots \\
&= - a \cos (\omega t + \theta) + \varepsilon [- 2\omega_1 a \cos (\omega t + \theta) + \ddot{x}_1] \\
&\quad + \varepsilon^2 [\ddot{x}_2 - \omega_1^2 \, a \cos (\omega t + \theta) - 2\omega_2 a \cos (\omega t + \theta)].
\end{aligned}$$

Now we obtain to $O(\varepsilon)$

$$\ddot{x}_1 + x_1 - 2\omega_1 a \cos(\omega t + \theta) = -\frac{3a^3}{4}\cos(\omega t + \theta) - \frac{a^3}{4}\cos 3(\omega t + \theta)$$

The point now is to select ω_1 so that $x_1(t)$ contains no secular terms. This is done by taking $\omega_1 = 3a^2/8$. Then, ignoring the homogeneous solution,

$$x_1 = -\frac{a^3/4}{1 - (3\omega)^2}\cos 3(\omega t + \theta).$$

Also, to this order

$$x = a\cos(\omega t + \theta) - \frac{\varepsilon a^3/4}{1 - (3\omega)^2}\cos 3(\omega t + \theta)$$

$$\omega = 1 + \varepsilon \frac{3a^2}{8}.$$

The secular term found in (5.5.4) comes from the fact that

$$\cos\left(\left[1 + \varepsilon \frac{3a^2}{8}\right]t + \theta\right) \simeq \cos(t + \theta) - \varepsilon \frac{3a^2 t}{8}\sin(t + \theta)$$

only if $\varepsilon t \ll 1$ (i.e., (5.5.5) holds).

Many methods have been used to eliminate these secular terms, including a succession of methods developed by Krylov and Bogogiubov, and later Mitropolsky. This 'averaging method' has been widely used in recent years, but continues to be modified by yet more recent 'rapidly convergent' methods (see Bogoliubov and Mitropolsky, 1961; Hagihara, Vol. III, 1974; Sanders and Verhulst, 1985 for a review of many methods; and Arnold, 1983, for a critical analysis). The following discussion is only a brief introduction to some of these averaging concepts.

We now generalize the discussion to a nonlinear oscillator obeying the equation

$$\ddot{x} + x = \varepsilon f(x, \dot{x}, t) \tag{5.5.6}$$

where ε is again assumed to be small. If $\varepsilon = 0$, $x = a\sin(t + \theta)$, where (a, θ) are constants, and $\dot{x} = a\cos(t + \theta)$. This may be written

$$x = a\sin\phi \tag{5.5.7a}$$

$$\dot{x} = a\cos\phi \tag{5.5.7b}$$

In the phase plane $(x, \dot{x})$, the orbit is circular with radius a, and $\dot{\phi} = 1$, $\dot{a} = 0$.

If $\varepsilon \neq 0$ we can retain the form (5.5.7), but the behavior of $(a(t), \phi(t))$ becomes, of course, much more involved (Fig. 5.48). Let us obtain the equations of motion for (a, ϕ). Differentiating (5.5.7a), and equating to (5.5.7b) yields

$$\frac{\mathrm{d}}{\mathrm{d}t}x = \dot{a}\sin\phi + a\dot{\phi}\cos\phi = a\cos\phi. \tag{5.5.8}$$

Fig. 5.48

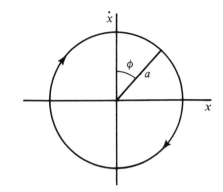

Differentiating (5.5.7b), and using (5.5.6), we obtain

$$\dot{a}\cos\phi - a\dot{\phi}\sin\phi = -a\sin\phi + \varepsilon f(a\sin\phi, a\cos\phi, t). \qquad (5.5.9)$$

Multiplying (5.5.8) by $\sin\phi$, and (5.5.9) by $\cos\phi$, and adding gives

$$\dot{a} = \varepsilon f(a\sin\phi, a\cos\phi, t)\cos\phi \qquad (5.5.10a)$$

Similarly

$$\dot{\phi} = 1 - \frac{\varepsilon}{a} f(a\sin\phi, a\cos\phi, t)\sin\phi. \qquad (5.5.10b)$$

One now has two first order equations rather than one second order equation. The equations (5.5.10) are exact, but are no easier to solve than the original equation (5.5.6). It might be mentioned that the variables (a, ϕ) are closely related to the *action-angle* variables; a being essentially the square of the action, and ϕ the angle.

To proceed we now distinguish between two cases:

$$\text{Autonomous systems: } f(x, \dot{x}, t) = f(x, \dot{x})$$
$$\text{Nonautonomous systems: } \partial f(x, \dot{x}, t)/\partial t \neq 0.$$

For the present we restrict ourselves to autonomous systems, and consider the *averaging method* for obtaining the approximate behavior of (a, ϕ).

Roughly the idea is that, in (5.5.10), $\dot{a}$ and $\dot{\phi} - 1$ are both proportional to a small quantity (ε), hence $a(t)$ is a slowly varying function of time, whereas $\phi = t + O(\varepsilon)$ is a rapidly varying function of time (or $(\phi - t)$ is slowly varying). Thus it should be possible to obtain a 'lowest order' approximation for $\dot{a}$ and $\mathrm{d}(\phi - t)/\mathrm{d}t$, by averaging the term proportional to ε over a period 2π. Since $\mathrm{d}\phi \simeq \mathrm{d}t$, the relevant averages are

$$[\bar{f}_c(a), \bar{f}_s(a)] \equiv \frac{1}{2\pi}\int_0^{2\pi} \mathrm{d}\phi\, f(a, \phi)[\cos\phi, \sin\phi]. \qquad (5.5.11)$$

This gives, for the lowest order approximation

$$\frac{\mathrm{d}a}{\mathrm{d}t} = \frac{\varepsilon}{2\pi} \int_0^{2\pi} f(a \sin \phi, a \cos \phi) \cos \phi \, \mathrm{d}\phi \qquad (5.5.12a)$$

Similarly

$$\frac{\mathrm{d}\phi}{\mathrm{d}t} = 1 - \frac{\varepsilon}{2\pi a} \int_0^{2\pi} f(a \sin \phi, a \cos \phi) \sin \phi \, \mathrm{d}\phi \qquad (5.5.12b)$$

The equations give the average influence of the nonlinearity (averaged over the rapidly varying phase) on (a, ϕ). Therefore (5.5.10) are replaced by (5.5.12), which can be written

$$\dot{a} = \varepsilon \bar{f}_{\mathrm{c}}(a) \quad \dot{\phi} = 1 - \frac{\varepsilon}{a} \bar{f}_{\mathrm{s}}(a). \qquad (5.5.13)$$

The great simplification which has been achieved in this approximation is that $\dot{a}$ and $\dot{\phi}$ depend only on $a(t)$ and not $\phi(t)$. This approximation is frequently quite adequate for obtaining the general behavior of the nonlinear oscillator provided that ε is small (and $a = O(1)$).

To be more specific, let $a = A(t)$ be the solution of (5.5.13) and assume that $1 - (\varepsilon/A)\bar{f}(A) = O(1)$ for all t in question. Then the solution $a(t)$ of (5.5.10) differs from $A(t)$ by less than $\alpha\varepsilon$

$$|a(t) - A(t)| \leqslant \alpha\varepsilon \qquad (5.5.14)$$

for some constant α (independent of ε), at least for all $0 \leqslant t \leqslant 1/\varepsilon$. Thus the perturbation method (5.5.13) has the accuracy (5.5.14) for long times, $O(1/\varepsilon)$. This is referred to as an *asymptotic approximation method*. For proofs of the accuracy of many averaging methods, see Sanders and Verhulst (1985). The above derivation, however, is not suitable for generalization to higher orders in ε, so we will need to consider a more systematic approach.

First, however, consider an example of a damped nonlinear oscillator known as the *Duffing equation*

$$\ddot{x} + x = -\varepsilon(x^3 + 2\mu\dot{x}) \equiv \varepsilon f(x, \dot{x})$$

Using (5.5.12)

$$\frac{\mathrm{d}a}{\mathrm{d}t} = -\frac{\varepsilon}{2\pi} \int_0^{2\pi} \{a^3 \sin^3 \phi + 2\mu a \cos \phi\} \cos \phi \, \mathrm{d}\phi = -\varepsilon\mu a$$

$$\frac{\mathrm{d}\phi}{\mathrm{d}t} = 1 + \frac{\varepsilon}{2\pi a} \int_0^{2\pi} \{a^3 \sin^3 \phi + 2\mu a \cos \phi\} \sin \phi \, \mathrm{d}\phi = 1 + \frac{3\varepsilon a^2}{8},$$

so

$$a(t) = a_0 \exp(-\varepsilon\mu t),$$

and therefore

$$\frac{d\phi}{dt} = 1 + \frac{3\varepsilon}{8}\,a_0^2\exp(-2\varepsilon\mu t).$$

This has the solution

$$\phi = t + \frac{3a_0^2}{16\mu}(1 - \exp(-2\varepsilon\mu t)) + \phi_0.$$

Thus, as the amplitude decreases, so does the (instantaneous) frequency. As $t \to \infty$, $\phi \to t + 3a_0^2/16\mu + \phi_0$, so there has been a finite phase shift (independent of ε). If $\mu = 0$ we recover the frequency shift to $1 + \varepsilon 3a_0^2/8$, obtained in the last section, (5.4.17) which is also the first order frequency shift of the simple pendulum. Since the solution tends toward a fixed point, the condition (5.5.14) is satisfied for all t (in this case).

We now consider the general formulation of Krylov–Bogoliubov–Mitropolsky (KBM) for autonomous systems. Instead of (5.5.7), they set

$$x = a\cos\psi + \varepsilon u_1(a, \psi) + \varepsilon^2 u_2(a, \psi) + \cdots \tag{5.5.15}$$

where $u_k(a, \psi + 2\pi) = u_k(a, \psi)$, and assume that

$$\dot{\psi}(t) = 1 + \varepsilon B_1(a) + \varepsilon^2 B_2(a) + \cdots \tag{5.5.16}$$

$$\dot{a}(t) = \varepsilon A_1(a) + \varepsilon^2 A_2(a) + \cdots. \tag{5.5.17}$$

This retains the feature that $(\dot{a}, \dot{\psi})$ depend only on a. Moreover, to insure that $u_k(a, \psi)$ does not contain $\cos\psi$ or $\sin\psi$, which can be included in $(a\cos\psi)$ by shifting a and ψ, we require

$$\int_0^{2\pi} u_k(a, \psi)\cos\psi\,d\psi = 0; \quad \int_0^{2\pi} u_k(a, \psi)\sin\psi\,d\psi = 0. \tag{5.5.18}$$

Introducing the notation $\partial u_k/\partial a \equiv u_{k,a}$, $\partial u_k/\partial\psi = u_{k,\psi}$, we have

$$\dot{x} = \dot{a}\{\cos\psi + \varepsilon u_{1,a} + \varepsilon^2 u_{2,a} + \cdots\} + \dot{\psi}\{-a\sin\psi + \varepsilon u_{1,\psi} + \varepsilon^2 u_{2,\psi} + \cdots\} \tag{5.5.19}$$

$$\ddot{x} = \ddot{a}\{\cos\psi + \varepsilon u_{1,a} + \cdots\} + 2\dot{a}\dot{\psi}\{-\sin\psi + \varepsilon u_{1,a\psi} + \cdots\} + (\dot{a})^2\{\varepsilon u_{1,aa} + \cdots\}$$

$$+ (\dot{\psi})^2\{-a\cos\psi + \varepsilon u_{1,\psi\psi} + \cdots\} + \ddot{\psi}\{-a\sin\psi + \varepsilon u_{1,\psi} + \cdots\} \tag{5.5.20}$$

The last equation contains the terms $\ddot{a}$ and $\ddot{\psi}$, which, according to (5.5.16) and (5.5.17) are

$$\ddot{\psi} = \{\varepsilon B_{1,a} + \varepsilon^2 B_{2,a} + \cdots\}\dot{a},$$

$$\ddot{a} = \{\varepsilon A_{1,a} + \varepsilon^2 A_{2,a} + \cdots\}\dot{a}. \tag{5.5.21}$$

We shall define the order of approximation simply to be the power of ε retained in (5.5.15–5.5.17), which is then the power of ε to which equation (5.5.6) is satisfied. This ordering differs slightly from that used by KBM, as will be noted below. Then

$$\textit{Zeroth approximation: } x = a \cos \psi, \quad \dot{a} = 0, \quad \dot{\psi} = 1$$

$$\textit{First approximation: } x = a \cos \psi + \varepsilon u_1(a, \psi);$$

$$\dot{\psi} = 1 + \varepsilon B_1(a); \quad \dot{a} = \varepsilon A_1(a) \tag{5.5.22}$$

In the original KBM ordering, the term u_1 is neglected in this order, and included in the second approximation. Using (5.5.22), in equation (5.5.6), yields, with the aid of (5.5.19) and (5.5.20), to $O(\varepsilon)$

$$\varepsilon 2 A_1 \{ - \sin \psi \} + (1 + 2\varepsilon B_1)(- a \cos \psi) + \varepsilon u_{1,\psi\psi}$$
$$+ a \cos \psi + \varepsilon u_1 = \varepsilon f(a \cos \psi, - a \sin \psi). \tag{5.5.23}$$

Now, because of (5.5.18), u_1 and $u_{1,\psi\psi}$ are orthogonal to $\sin \psi$ and $\cos \psi$. Hence, multiplying (5.5.23) by $\sin \psi$ (or $\cos \psi$) and integrating yields the two equations

$$A_1(a) = - (1/2\pi) \int_0^{2\pi} f(a \cos \psi, - a \sin \psi) \sin \psi \, d\psi$$

$$B_1(a) = - (1/2\pi a) \int_0^{2\pi} f(a \cos \psi, - a \sin \psi) \cos \psi \, d\psi \tag{5.5.24}$$

Substituting (5.5.24) back into (5.5.22) gives the equations for $\dot{\psi}$ and $\dot{a}$, which must be solved. Finally $u_1(a, \psi)$ is given by solving the differential equation (5.5.23), namely

$$u_{1,\psi\psi} + u_1 = f(a \cos \psi, - a \sin \psi) + 2 A_1 \sin \psi + 2 a B_1 \cos \psi \tag{5.5.25}$$

and one wants only the particular (i.e., inhomogeneous) solution of (5.5.25), because of the condition (5.5.18).

Aside from the additional function $u_1(a, \psi)$, the present approximation given by (5.5.22), (5.5.24), and (5.5.25), is the same as (5.5.12). Note that $\dot{a}$, $\dot{\psi}$ are not dependent on u_1.

The next order of approximation retains all terms of order ε^2 in (5.5.15)–(5.5.17), so

$$x = a \cos \psi + \varepsilon u_1 + \varepsilon^2 u_2; \quad \dot{\psi} = 1 + \varepsilon B_1(a) + \varepsilon^2 B_2; \quad \dot{a} = \varepsilon A_1 + \varepsilon^2 A_2. \tag{5.5.26}$$

We then find that the terms of $\ddot{x}$, (5.5.20), which are proportional to ε^2 are

$$A_1 A_{1,a} \cos \psi - 2 A_2 \sin \psi - 2 A_1 B_1 \sin \psi + 2 A_1 u_{1,a\psi} - (B_1^2 + 2 B_2) a \cos \psi$$
$$+ 2 B_1 u_{1,\psi\psi} + u_{2,\psi\psi} - B_{1,a} A_1 a \sin \psi.$$

To this we must add u_2 from the term x, and equate this to the first order term from $f(x, \dot{x})$, namely

$$u_1 f_x(a \cos \psi, - a \sin \psi) + (A_1 \cos \psi - a B_1 \sin \psi + u_{1,\psi}) f_{\dot{x}}(a \cos \psi, - a \sin \psi).$$

If we multiply this equation by $\sin \psi$ or $\cos \psi$ and integrate, and use (5.5.18) for both u_1

and u_2, then we obtain the equations for $A_2(a)$ and $B_2(a)$

$$A_2(a) = (-1/2\pi) \int_0^{2\pi} F(a, \psi) \sin \psi \, d\psi;$$

$$B_2(a) = (-1/2\pi a) \int_0^{2\pi} F(a, \psi) \cos \psi \, d\psi \qquad (5.5.27)$$

where

$$F(a, \psi) = u_1 f_x(a \cos \psi, -a \sin \psi) + [A_1 \cos \psi - aB_1 \sin \psi + u_{1,\psi}]$$
$$\times f_{\dot{x}}(a \cos \psi, -a \sin \psi) + (aB_1^2 - A_1 A_{1,a}) \cos \psi$$
$$+ (2A_1 B_1 + aB_{1,a} A_1) \sin \psi$$

The expressions obtained from (5.5.27) together with those from (5.5.24), are then to be substituted into the equations (5.5.15) and (5.5.16), keeping only the factors shown (i.e., $O(\varepsilon^2)$), and solve first for $a(t)$ and then for $\psi(t)$. Obviously nobody goes beyond the second order approximation! Indeed, the series (5.5.15), (5.5.16) are generally not convergent. We will save an application of the second order method for the more interesting case of the van der Pol equation in the next section.

As an indication of the error involved in the first approximation, it has been shown by Bogoliubov and Mitropolsky (1961; Section 13.4) that if (I, ϕ) satisfy the equations

$$\dot{I} = \varepsilon F(I, \phi; \varepsilon), \quad \dot{\phi} = \omega(I, \varepsilon) + \varepsilon G(I, \phi, \varepsilon) \qquad (5.5.28)$$

where F and G are periodic in ϕ (period 2π) and J satisfies the 'average equation'

$$\dot{J} = \varepsilon \bar{F}(J) \equiv \frac{\varepsilon}{2\pi} \int_0^{2\pi} F(J, \phi; 0) \, d\phi \qquad (5.5.29)$$

there is a constant C such that, if $I(0) = J(0)$, then for sufficiently small ε

$$|J(t) - I(t)| < C\varepsilon \quad \text{for all } 0 \leqslant t \leqslant 1/\varepsilon. \qquad (5.5.30)$$

The bound on the error becomes significantly worse for higher order systems. Thus for $n = 4$ (a pair of coupled oscillators) the bounds on $|J(t) - I(t)|$ is (at best) $C\varepsilon^{1/2} (\ln \varepsilon)^2$ for $0 \leqslant t \leqslant 1/\varepsilon$, and generally for $t \simeq 1/\varepsilon$ one has $|J(t) - I(t)| > C\varepsilon^{1/2}$ (note: this is a lower bound). This result is due to Arnold (1965). The reason for this lower bound has to do with resonant coupling between oscillators which is obviously not present for a single oscillator. This will be discussed in more detail later (see Chapter 8, and Appendix L).

5.6 The Rayleigh and van der Pol equations: Andronov–Hopf bifurcation

We next turn to a discussion of '*self-excited*' ('*auto-catalytic*') *oscillators*, which really means that the oscillators have some external source of energy upon which they can draw. Rayleigh, in his paper 'On maintained vibrations' (*Phil. Mag.*, **15**, 229, 1883) briefly considered the equation

$$m\ddot{x} + (-A + B\dot{x}^2)\dot{x} + kx = 0 \quad (A, B > 0). \tag{5.6.1}$$

We will refer to this as *Rayleigh's equation*. Here the common friction term $2\mu\dot{x}$ is replaced by the term $(-A + B\dot{x}^2)\dot{x}$, so that for small $\dot{x}$ there is negative 'friction', $-A\dot{x}$. Energy is thereby injected into the system when $\dot{x}$ is small. This means that the singular point $(x, \dot{x}) = (0, 0)$ is unstable, so that with any perturbation from this point the oscillator is *self-exciting*, and only the (positive) nonlinear term, $B\dot{x}^3$, keeps the motion bounded.

If we differentiate (5.6.1), we obtain

$$m\dddot{x} + (-A\ddot{x} + 3B(\dot{x})^2\ddot{x}) + k\dot{x} = 0,$$

and making the transformation

$$t' = (k/m)^{1/2}t, \quad x' = (3B/A)^{1/2}\dot{x}, \quad \varepsilon = A/(km)^{1/2}$$

the Rayleigh equation reduces to

$$\ddot{x} - \varepsilon(1 - x^2)\dot{x} + x = 0, \tag{5.6.2}$$

where the prime has been dropped on the new variable, x'. Equation (5.6.2) is known as the *van der Pol equation*, due to his extensive research into such systems in the 1920s.

If ε is small, we can obtain an approximate solution by using the KBM averaging method, (5.5.22) and (5.5.24). Setting $f(x, \dot{x}) = (1 - x^2)\dot{x}$, we have

$$f(a\cos\psi, -a\sin\psi) = -(1 - a^2\cos^2\psi)a\sin\psi$$
$$= -(1 - 1/4a^2)a\sin\psi + a^3/4\sin 3\psi.$$

The averaged functions (A_1, B_1), (5.5.24), are

$$A_1 = 1/2\pi \int_0^{2\pi} \left(1 - \frac{a^2}{4}\right)a\sin^2\psi \, d\psi = 1/2a\left(1 - \frac{a^2}{4}\right); \quad B_1 = 0.$$

Therefore $\dot{\psi} = 1$, and $\dot{a} = (\varepsilon/2)a(1 - (a^2/4))$. The equation for a can be written

$$\frac{d}{dt}a^2 = \varepsilon a^2\left(1 - \frac{a^2}{4}\right)$$

and factorizing, $(da^2/a^2) + da^2/(4 - a^2) = \varepsilon dt$, setting $a(0) = a_0$, this gives $\ln(a^2/a_0^2) - \ln[(4 - a^2)/(4 - a_0^2)] = \varepsilon t$. Therefore

$$a = \frac{a_0\exp(\varepsilon t/2)}{\left[1 + \frac{a_0^2}{4}(\exp(\varepsilon t) - 1)\right]^{1/2}}; \quad \psi = t + \psi_0. \tag{5.6.3}$$

Finally, (5.5.25) becomes

$$u_{1,\psi\psi} + u_1 = a^3/4 \sin 3\psi,$$

so

$$u_1 = [a^3/4(1 - 3^2)] \sin 3\psi = -[a^3/32] \sin 3\psi.$$

Then (5.5.22) yields

$$x = a \cos \psi - (\varepsilon a^3/32) \sin 3\psi, \tag{5.6.4}$$

where (a, ψ) are given by (5.6.3).

The very important feature of the van der Pol oscillator is that it is a simple example of a system with a *limit cycle*. Many examples of such cycles in electrical, mechanical, and biological systems can be found in numerous texts. One of the most important limit cycles of our lives is the heartbeat, which van der Pol was very interested in modeling. Another spectacular example of a limit cycle (which ultimately broke down!) was the Tacoma Narrows bridge collapse (Fig. 5.49) on November 7, 1940, where the limit cycle

Fig. 5.49

drew its energy from the wind and involved torsional oscillations of the roadbed of about 70 degrees. The common feature of these limit cycles is that they are 'fed' (e.g., energy) from a constant external source, rather than a periodic action, and then the system 'sheds' this acquired energy at certain locations in phase space. This makes the system formally 'autonomous' despite the outside influence.

In the present case we note that, if $\varepsilon > 0$, according to (5.6.3)

$$\lim_{t \to \infty} a(t) \to 2 \quad (\varepsilon > 0),$$

which is independent of a_0. Thus the asymptotic solution loses its dependence on the initial amplitude, but retains information about its initial phase ψ_0, (5.6.3). If $a_0 < 2$, $a(t)$ increases with time, whereas if $a_0 > 2$, $a(t)$ will decrease with time. In other words, orbits will approach the limit cycle both from inside and outside. If ε is not very large, then the limit cycle is nearly a circle. This is illustrated in the following figure for $\varepsilon = 0.5$.

Exercise 5.27. Obtain the second order KBM approximation of the solution of van der Pol's equation (5.6.2), by evaluating (5.5.27), and then integrating (5.5.26). Compare the actual period 6.39 (if $\varepsilon = 0.5$), 6.65 (if $\varepsilon = 1.0$) with those given by the KBM method.

If $\varepsilon < 0$ (or $A < 0, B < 0$ in the Rayleigh equation), the limit cycle is unstable, whereas the singular point $(x, \dot{x}) = (0, 0)$ becomes stable. Note that $\varepsilon \to -\varepsilon$ is equivalent to $t \to -t$ in (5.6.2). That means that, in terms of Fig. 5.50, the vectors on the orbits are

Fig. 5.50

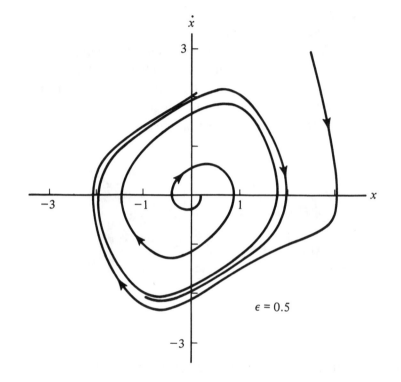

$\epsilon = 0.5$

reversed and the figure is reflected across the x axis ($\dot{x} \equiv -\dot{x}$). Thus, as ε goes through the value $\varepsilon = 0$, there is a *bifurcation* in the solution, because the orbits are not topological orbital equivalent. It is important of course that the *orientation* of the curves be contained in this equivalence, for the trajectories (without the arrow!) are homeomorphic for $\varepsilon > 0$ and $\varepsilon < 0$, as was just noted. The bifurcation just noted is not a common ('generic') situation. A generic bifurcation involving limit cycles is the *Andronov–Hopf bifurcation*, which we will now consider.

The Andronov–Hopf bifurcation, which is frequently simply called the Hopf bifurcation (1940), was first proved in R^2 by Andronov (1930), and had been suggested by Poincaré (1892). The bifurcation usually involves the change in stability of a focus, as some control parameter is varied, together with the 'birth' of a limit cycle, as illustrated in Fig. 5.51.

Fig. 5.51

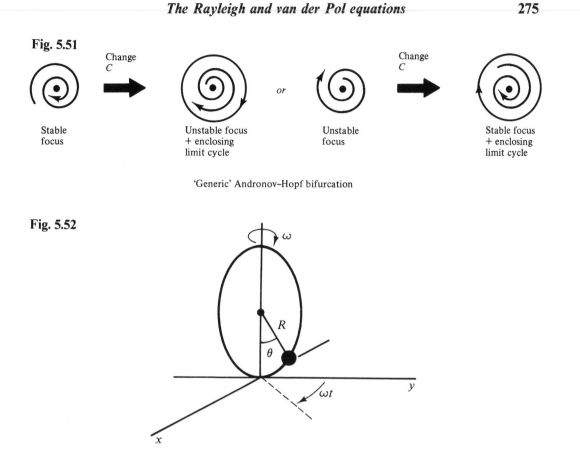

'Generic' Andronov–Hopf bifurcation

Fig. 5.52

A simple physical system which has a Hopf bifurcation (in configuration space) is a bead which slides on a vertical wire hoop of radius R (Fig. 5.52). The hoop is rotated with a constant angular velocity, ω, about its vertical diameter. ω is the control parameter in this system. The position of the bead can be described in terms of the single angular variable θ (see the figure)

$$x = R \sin \theta \sin (\omega t), \quad y = R \sin \theta \cos (\omega t), \quad z = R(1 - \cos \theta).$$

The equation of motion for the bead is found to be (see Exercise 5.21)

$$\ddot{\theta} + \mu \dot{\theta} + (g/R) \sin \theta - \omega^2 \sin \theta \cos \theta = 0,$$

where $\mu \dot{\theta}$ represents the effect due to friction. There is a fixed point at $\theta = 0$ and at $\cos \theta = (g/R\omega^2)$, if $(g/R\omega^2) \leqslant 1$. If $\omega = 0$ and $\mu^2 > 4g/R$, $\theta = 0$ is stable and the θ motion is overdamped (no oscillations). The motion of the bead projected onto the (x, y) plane is unique if $\theta(0) < \pi/2$, and is a stable focus (its rotational velocity in the (x, y) plane is ω). If we increase ω, the characteristic exponent of θ, $s = -(\mu/2) + (1/2) (\mu^2 - 4(g/R) + 4\omega^2)^{1/2}$ goes to zero when $\omega^2 = (g/R)$. If $\omega^2 > (g/R)$, s is positive and $\theta = 0$ is an unstable focus for the (x, y) motion. θ increases until it reaches $\theta = \cos^{-1}(g/R\omega^2)$, and

Fig. 5.53

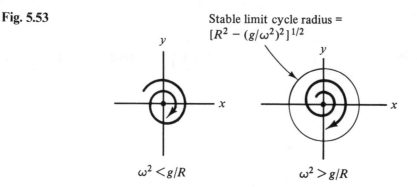

Stable limit cycle radius =
$[R^2 - (g/\omega^2)^2]^{1/2}$

$\omega^2 < g/R$ $\omega^2 > g/R$

the (x, y) motion therefore tends toward a stable limit cycle (see Fig. 5.53). This is an example of one of the two above generic Andronov–Hopf bifurcations.

The above bifurcations can be viewed more completely in the phase-control space, where the bifurcation point is taken to be $c = 0$.

We see that, in this space, there is a surface which contains only periodic solutions of the nonlinear equations (Fig. 5.54). This is an extension of the concept of centers, which

Fig. 5.54

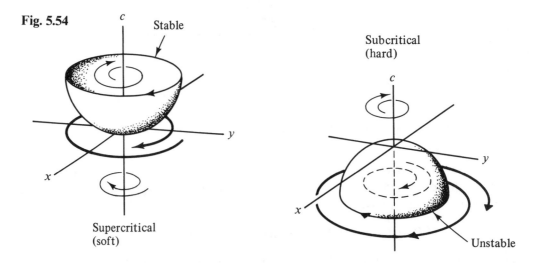

Stable

Subcritical (hard)

Supercritical (soft)

Unstable

we found for linear systems. These periodic solutions are now found to exist at a finite distance from the fixed point $(x = 0, y = 0)$, but usually only if $c \neq 0$. This surface which contains only periodic solutions is called a *center manifold* in the control-phase space.

The periodic solutions are either stable or unstable limit cycles. If the surface consists of stable limit cycles the bifurcation is called *supercritical*, or *soft*. The latter terminology comes from the fact that, if c is only slightly positive (see the left figure), the motion away from the unstable fixed point is very small ('soft') since there is a nearby stable limit cycle. On the other hand, if the center manifold consists of unstable limit cycles, the bifurcation is called *subcritical*, or *hard*. In this case, even if the fixed point is stable $(c \lesssim 0)$, a small (but finite) perturbation can take the state 'outside' the unstable limit cycle, which then will carry the solution to some distant reaches of the phase space (a 'hard' effect). This type of instability is very important because it is not detected by linear analysis, but can produce drastic ('hard') effects, for small but finite perturbations. Such nonlinear instabilities are known in fluid systems, where only 'finite' perturbations of laminar flows produce turbulent states (e.g., Orszag and Kells, 1980; Orszag and Patera, 1983).

To further illustrate these bifurcations, consider the system

$$\dot{x} = y + x(c + a(x^2 + y^2)); \quad \dot{y} = -x + y(c + a(x^2 + y^2)). \tag{5.6.5}$$

If we introduce the (r, θ) variables

$$x = r \sin \theta, \quad y = r \cos \theta,$$

we easily find that

$$\dot{r} = r(c + ar^2), \quad \dot{\theta} = 1.$$

Thus the origin is a stable focus for $c < 0$, and becomes unstable if $c > 0$. We note that, at $c = 0$, the solution is $r = r_0/[2tr_0^2 + 1]^{1/2}$, which is a much slower approach to $r = 0$ than when $c < 0$ ($r \simeq r_0 \exp(ct)$). Note that this bifurcation does not have a center vector field in the entire (x, y) plane, which is expected for a linear system (see Fig. 5.55, from Section 1).

Fig. 5.55

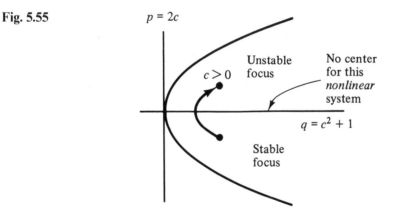

If $a < 0$, there is a stable periodic orbit when $c > 0$, at $r = (-c/a)^{1/2}$. Therefore, if $a < 0$, the system has a supercritical Hopf bifurcation at $c = 0$. If $a > 0$, an unstable periodic orbit occurs for $c < 0$, so the system has a subcritical Hopf bifurcation at $c = 0$. In this 'generic' example, the radius of the limit cycle grows as $|c|^{1/2}$. From the above bead-on-a-hoop example, one can see that this square root dependence depends on the identification of c.

We next note that the essential feature of the system (5.6.5) is that the characteristic exponents of the linearized equations $s(c) = c \pm i$, changes from a stable focus to an unstable focus as c goes from $c < 0$ to $c > 0$. Moreover $\mathrm{Re}\,(\partial s(c)/\partial c) = 1$ does not vanish at $c = 0$. In other words the stability of the fixed point changes at a finite 'speed' as c passes through $c = 0$. These observations are the essential points of the Hopf bifurcation theorem.

One of the versions of the Hopf bifurcation theorem, now generalized to R^n (see Marsden and McCracken, 1976; for a careful discussion, see Arnold, 1983) is the following:

'Hopf' theorem. Assume that

$$\dot{x} = F(x; c) \quad (x \in R^n, c \in R)$$

has an $F(x; c)$ which is analytic in x, and which has a fixed point $x_0(c_0)$, so that $F(x_0(c_0); c_0) = 0$. The linear equations about this fixed point

$$\dot{x}_i = \sum_{j=1}^{n} (\partial F_i / \partial x_j)_0 x_j \quad (i = 1, \ldots, n)$$

are assumed to have solutions $x_j = A_j \exp(st)$, whose characteristic exponents $\{s_i\}$ satisfy the following conditions:

(A) For $c = c_0$, two roots are purely imaginary and conjugate, say
$$s_1(c_0) = -s_2^*(c_0) \quad (*: \text{ complex conjugate})$$

(B) No other purely imaginary roots satisfy $s_j(c_0) = ms_1(c_0)$ (m: integer)
(C) $\text{Re}(\partial s_j / \partial c)_{c_0} \neq 0$ for $j = 1, 2$.

Then the theorem states that there are analytic functions $c(\varepsilon)$ and $T(\varepsilon)$ near $\varepsilon = 0$, such that $c(0) = c_0$ and $T(0) = 2\pi / s_1(c_0)$. Moreover there is a nonconstant solution $x(t; c(\varepsilon))$ which has the period $T(\varepsilon)$ and, in the 'generic' case it contracts to the fixed point $x_0(c_0)$ with a radius $(|c(\varepsilon) - c_0|)^{1/2}$.

The 'generic' (or usual) case is (5.6.5), plus terms of $O(xr^2, yr^2)$. For more details, see Arnold (1983, pg. 267 ff). An interesting nongeneric Hopf bifurcation is exhibited by the system

$$\dot{x} = y + x(c - x); \quad \dot{y} = -x + y(c - x). \tag{5.6.6}$$

Introducing (r, θ), as above, this transforms to

$$\dot{r} = cr - r^2 \sin(\theta); \quad \dot{\theta} = 1.$$

This system satisfies the above therem, but its only periodic solutions are for $c = 0$ (so that its center manifold is the entire (x, y) plane at $c = 0$). This is a structurally unstable system, since the addition of small nonlinear terms will generally produce the previous generic results. Put another way, the generic center manifold is an upright or inverted paraboloid, rather than a plane, in the control-phase space.

This theorem does not tell us whether the bifurcation is subcritical or supercritical, since that depends on the higher derivatives of $F(x)$ at x_0. Moreover the theorem does not indicate whether these periodic solutions are unique or whether other periodic solutions may also occur at $c = c_0$. An example of this is due to Chafee (see Marsden and McCracken 1976). Consider the system

$$\dot{x} = g(x^2 + y^2; c)x + y \quad \dot{y} = g(x^2 + y^2; c)y - x \tag{5.6.7}$$

which reduces, as above, to

$$\dot{r} = rg(r, c), \quad \dot{\theta} = 1.$$

In Chafee's example, take

$$g(r; c) = (r - c^{1/3})^2 (2c^{1/3} - r).$$

The characteristic exponents are $s(c) = 2c \pm i$, and hence satisfy the above theorem, but there are two limit cycles for $c > 0$, at $r = c^{1/3}$ and $r = 2c^{1/3}$ (Fig. 5.56). They both have

Fig. 5.56

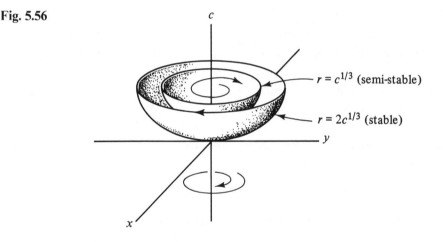

the period 2π. Note that $g(r; c)$ is not differentiable in c at $c = 0$, which permits this nonuniqueness.

If the condition $\text{Re}\,(\partial s_1(c)/\partial c)_{c_0} \neq 0$ in Hopf's theorem is not satisfied, then a great variety of bifurcations are possible. These are sometimes referred to as *degenerate Hopf bifurcations*. Thus, in the system (5.6.7), if

$$g(r, c) = -(c - r)^2$$

the characteristic exponents are $s(c) = -c^2 \pm i$. There is a bifurcation at $c = 0$, but $\text{Re}\,(\partial s/\partial c)_0 = -2c = 0$. The fixed point at $x = y = 0$ does not change stability in this example. The limit cycles at $r = c(c > 0)$ are semi-stable in this system (Fig. 5.57).

Fig. 5.57

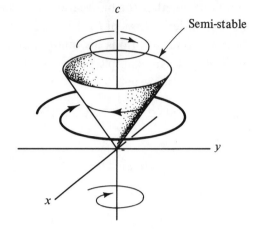

Exercise 5.28. Some other non-Hopfian bifurcations are:

(a) $g(r,c) = c(c-r)^2$ (b) $g(r,c) = (c-r^2)(4c-r^2)^2$
(c) $g(r,c) = -r^2 + c^2$ (d) $g(r,c) = c(c+r^2)^2(c-r)$.

Sketch the phase portraits in the phase-control space (as above) for each of these cases. Specify the structure of all center manifolds, and the stability of their limit cycles or periodic orbits.

The Hopf bifurcation concerns the local 'birth' or 'death' of a limit cycle around a fixed point. It is a localized phenomenon and theorem. A more global method of determining the existence of periodic solutions in R^2 is given by the *Poincaré–Bendixson theorem*, which can be applied to the Hopf bifurcation as well as other situations. It concerns a general system in R^2,

$$\dot{x} = P(x,y), \quad \dot{y} = Q(x,y), \tag{5.6.8}$$

which is assumed to have unique solutions in some region D of the plane. Moreover it is assumed that these solutions exist for all $0 \leqslant t < \infty$. Let K be a compact subset of D, and assume that there is some solution, γ, of (5.6.8) which remains in the region K as time goes to infinity (Fig. 5.58). The theorem states that the region K must contain at least

Fig. 5.58

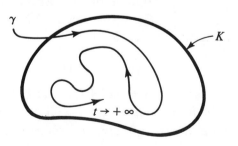

one fixed point, $P = 0$, $Q = 0$, or else a periodic Γ (there may be any number of each, but there must be at least one in K). Moreover the solution γ must tend either to the fixed point or to the periodic solution Γ as $t \to +\infty$, in which case Γ is the limit cycle of the solution, γ.

The theorem is also frequently applied in the following form: If the region K contains no fixed points of (5.6.8), then it contains a periodic solution of (5.6.8) which is the limit cycle of γ. Note that the region K as well as the limit cycle Γ may be large, so there is nothing localized about this theorem.

An immediate concern about the usefulness of such a theorem is how do we know if a solution, γ, stays in some region K? After all, if we know the solution, we don't need the theorem to discover a limit cycle! The way around this difficulty is to look for a region, K, such that the vector field (P, Q) points toward the interior of K at all points on its boundary (Fig. 5.59). In that case, any solution which enters K remains in K, and the Poincaré–Bendixson theorem can be applied.

Fig. 5.59

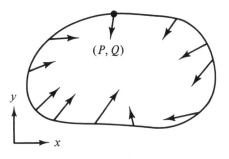

On a casual level, the *Poincaré–Bendixson theorem* is very easy to believe. Assume that a bounded region K contains no singular points of $\dot{x} = F(x)(x \in R^2)$ and γ is an orbit which enters K and remains there as $t \to +\infty$. If there is no fixed point in K, what can γ be like, considering the fact that it is trapped in K and cannot intersect itself? It can do a good deal of meandering, but because $F(x)$ is regular enough to yield unique solutions, it must have the same orientation when two parts of γ are in the same region of phase space. As $t \to \infty$, $x(t)$ begins to run out of space which is not near another position of γ, so it finally must 'settle down' to regular motion. On the other hand there is no singular point in K, by assumption, so the only possibility is that it approaches a periodic solution, Γ, as $t \to \infty$. Since the Poincaré index of a periodic solution is $+1$, we know there must be a singular point inside Γ, so K must have a 'hole' interior to Γ which contains this singular point (the hatched region in Fig. 5.60). In other words, K is an annulus, or possibly has more than one hole. More details concerning this theorem, together with some aspects of Birkhoff's α and ω-limit sets, are given in Appendix G. For an extensive discussion and some generalizations, see Hartman (1973).

Fig. 5.60

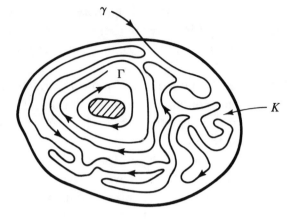

Exercise 5.29. The nonlinear oscillator

$$\ddot{x} + (c_2 + x^2)\dot{x} + c_1 x + x^3 = 0$$

incorporates features of both the van der Pol and Duffing equation. Let $x_1 \equiv x$, $x_2 = \dot{x}$.

(a) Determine all fixed points in the (x_1, x_2) phase plane for all values of (c_1, c_2)

(b) Determine the character of each fixed point for all values of (c_1, c_2), obtain all the bifurcation lines in the (c_1, c_2) plane. Label which fixed points are associated with each bifurcation line.

(c) Sketch the flow pattern over all the phase space which contains all fixed points. Do this for each topologically different region of control space. Most are easy, but the case $c_1 < 0$ and $c_2 < 0$ presents an interesting challenge. In involves a very different type of bifurcation. You may need some computer help.

While we are discussing the Poincaré–Bendixon theorem, it is interesting to consider its possible application in higher dimensions. For example, if there is a bounded region Ω in R^3, and the flow is inwardly directed at every point on the boundary of Ω, does Ω necessarily contain either a fixed point or a periodic orbit? If the flow is sufficiently smooth (i.e., C^N, large N), then it appears very likely that the answer is 'yes', but this has not been proved. Moreover, the fixed point, or periodic orbit need not be the ω-limit set of any inflowing orbit. For example, we could have a flow into a torus which asymptotically tends to an interior torus, which consists of an incommensurable (nonperiodic) orbit. However, inside this torus, it appears that there must be a periodic (e.g., unstable) orbit (Fig. 5.61). In any case, if the flow is only C^1, then the answer to the above question is 'no' (Schwitzer, 1972). Moreover, in R^4, even if the flow is C^∞ the answer is 'no' (Fuller, 1952). These 'no's', of course, are all mathematical responses, and specific physical systems, which are much more dynamically restrictive, may have a Poincaré–Bendixson 'effect', but proofs are more elusive.

Fig. 5.61

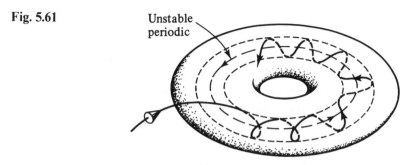

Unstable periodic

5.7 The Lotka–Volterra and chemical reaction equations

Lotka in studying autocatalytic chemical reactions (1910) and the theory of competing species (1920), which was later followed by Volterra (1928, 1956), considered the pair of equations

$$\dot{N}_1 = \alpha_1 N_1 - \lambda_1 N_1 N_2 \quad (N_k \in R_+)$$
$$\dot{N}_2 = -\alpha_2 N_2 + \lambda_2 N_1 N_2 \quad (\alpha_1, \alpha_2, \lambda_1, \lambda_2 > 0). \tag{5.7.1}$$

In these equations N_k represents the number, or concentration of species k, and therefore carries the requirement that $N_k(t) \geqslant 0$ for all t. We will refer to the strictly positive phase space, $N > 0$, as R_+^2. Note that, if $N_k(t)$ obeys any equation of the form

$$\dot{N}_k = N_k F_k(N_1, N_2, \dots) \quad (N \in R_+^2), \tag{5.7.2}$$

then $N_k(t)$ will remain in R_+^2 ($N_k(t) \neq 0$ for finite t) if F_k is Lipschitzian, and also on the axes. (5.7.1) is of this form.

In (5.7.1), N_1 would increase exponentially in number, and N_2 would likewise decrease, if it were not for their interaction, which is proportional to the 'binary collision' rate $N_1 N_2$ (Fig. 5.62). Species 2 in this case is the predator whose survival

Fig. 5.62

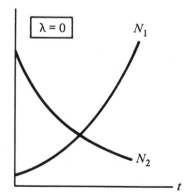

$\lambda = 0$ N_1

N_2

t

depends on destroying members of species 1 (the prey). Such *predator–prey equations* have subsequently been widely studied in a variety of contexts, with numerous variations from the original form (5.7.1). The term $\alpha_1 N_1$ represents *autocatalytic* or *self-reproductive* process, as in the logistic differential equation. A survey of many of these results can be found in Goel, Maitra and Montroll (1971). More recent references can be found in Ebeling and Peschell (1985), and Eigen and Schuster (1979). For a broad introduction to population modeling, see Frauenthal (1979) and Freedman (1982).

Equation (5.7.1) gives oscillatory solutions due entirely to the nonlinear terms (a truly nonlinear oscillator). This is clearer if they are written in the form

$$\dot{N}_1 = N_1(\alpha_1 - \lambda_1 N_2); \quad \dot{N}_2 = -N_2(\alpha_2 - \lambda_2 N_1)$$

or, if

$$M_1 = N_1(\lambda_2/\alpha_2), \quad M_2 = N_2(\lambda_1/\alpha_1) \tag{5.7.3}$$

they become

$$\dot{M}_1 = \alpha_1 M_1(1 - M_2), \quad \dot{M}_2 = -\alpha_2 M_2(1 - M_1). \tag{5.7.4}$$

Thus $M_1 = M_2 = 1$ is the fixed point of this system, and it is a center of the linearized equations. Moreover, even for the nonlinear equations, the solutions are all periodic (the (N_1, N_2) plane is a center manifold.) This is, of course, an exceptional case (i.e., this system is not structurally stable – see Fig. 5.63). It is very different from the limit cycle dynamics of the van der Pol equation.

Since

$$dM_2/dM_1 = -\frac{\alpha_2}{\alpha_1}\frac{M_2(1 - M_1)}{M_1(1 - M_2)} \tag{5.7.5}$$

the shape of the closed orbit (in the (M_1, M_2) plane) depends only on the ratio (α_2/α_1) and looks something like what is shown in the above figure. The Fig. 5.63 is adopted from a paper on this subject.

Fig. 5.63

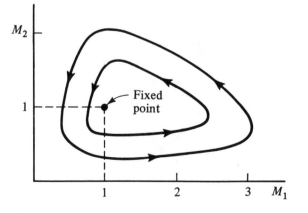

Exercise 5.30. What is wrong with Fig. 5.63? Integrate (5.7.5) to obtain the equation for the curves in. Does this make the integration of (5.7.4) easy?

One example of a predator–prey interaction is illustrated in Fig. 5.64, which is based on the trading records of the Hudson Bay Company. It indicates the fluctuations in the

Fig. 5.64

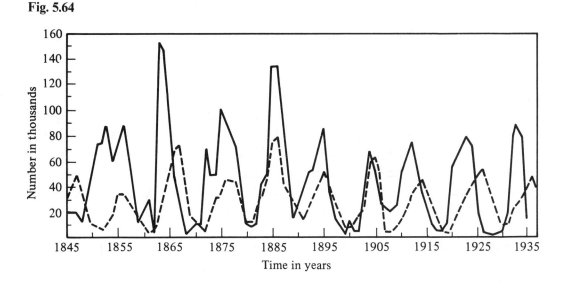

catches by trappers of the Canadian lynx and its prey, the snowshoe hare, over a span of ninety years. Which is the lynx and which is the hare? Are you sure? Look around 1850, 1860, and 1870. Did hares eat lynx? See Gilpin (1973).

Another general area which has dynamic models of the form (5.7.2) involves *bimolecular chemical reactions*. If x and y are the concentrations of two reacting chemicals ($x \in R_+, y \in R_+$), and the only interactions are bimolecular processes (i.e., $A + B \to C$), then Tyson and Light (1975) pointed out that the only possible dynamic equations for x and y are of the form

$$\dot{x} = a_0 \pm a_1 x + a_2 y - a_3 x^2 \pm a_4 xy + a_5 y^2 \tag{5.7.6}$$

$$\dot{y} = b_0 \pm b_1 y + b_2 x - b_3 y^2 \pm b_4 xy + b_5 x^2, \tag{5.7.7}$$

where all $(a_k, b_k) \geqslant 0$. Thus, for example, the reaction $x + x \to 3x$ does not occur, since it does not conserve mass. Therefore x interacting with x can only be of the form $x + x \to$ (something else), causing x to decrease. Another process involving two molecules of x, say (something) $+ 2x \to 3x$ is a trimolecular process, excluded by assumption. Hence, from these considerations, the coefficient of x^2 in (5.7.6) must be negative. Also, a chemical can only disappear at a rate proportional to some power of its concentration, which explains the positive signs of the coefficients a_k and b_k, for $k = 0, 2$, and 5. The remaining coefficients may be positive or negative.

The next question which might be asked about such a system is whether it can sustain oscillations, and possibly a stable limit cycle. Since (5.7.1) is a special case of the system (5.7.6) and (5.7.7) (namely, with $a_k = 0 = b_k$, for $k = 0, 2, 3, 5$), it is clear that it can sustain neutral oscillations – that is, with the phase space being a center manifold. The interesting question, however, is whether (5.7.6)–(5.7.7) can yield a stable limit cycle motion.

The fixed points of this system are given, of course, by the roots $(x_0 \in R_+, y_0 \in R_+)$ of

$$\left. \begin{aligned} a_0 \pm a_1 x_0 + a_2 y_0 - a_3 x_0^2 \pm a_4 x_0 y_0 + a_5 y_0^2 = 0 \\ b_0 \pm b_1 y_0 \pm b_2 x_0 - b_3 y_0^2 \pm b_4 x_0 y_0 + b_5 x_0^2 = 0 \end{aligned} \right\} \qquad (5.7.8)$$

Tyson and Light pointed out that, because the Poincaré's index of any periodic orbit is $I_y = +1$, a limit cycle cannot enclose a saddle point without also enclosing two foci and/or nodes (see Exercise 5.17). That would be quite an extraordinary collection of fixed points, among the four possible solutions of (5.7.8). We, of course, are only interested in the roots $x_0 \neq 0$ and $y_0 \neq 0$ (which is assured if $a_0 \neq 0$, $b_0 \neq 0$ respectively), since the fixed points must be interior to a closed orbit in R_+^2.

Exercise 5.31. Referring to the general linearization method, (5.1.5) and (5.1.5), prove that, if $x_0 \neq 0$, $y_0 \neq 0$, then both $a < 0$ and $d < 0$, so that $q = a + d < 0$ (hence any node or focus is stable). You will need to use (5.7.8) to eliminate terms with ambiguous signs. This is a result obtained by Tyson and Light.

Fig. 5.65

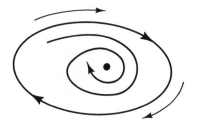

or

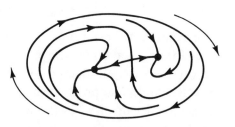

The results of this exericse establishes that (5.7.5)–(5.7.6) cannot have a stable limit cycle, even if the above extraordinary collection of fixed points exist, as illustrated in Fig. 5.65.

Kolmogorov (1936) suggested a number of possible conditions on the functions $F_k(N_1, N_2)$ of (5.7.2), which he felt were reasonable from the viewpoint of biology, and might make it possible to draw a limited number of conclusions concerning the global dynamics of the system. Following Kolmogorov, we think of N_1 as the number of prey, and N_2 as the number of predators (the opposite of (5.7.1)), and we will call F_k the corresponding 'growth rates'. Kolmogorov suggested the following conditions:

(1) $\partial F_1/\partial N_2 < 0$: The growth rate of the prey decreases as the predator population increases.
(2) $\partial F_1/\partial S < 0$ $(S = (N_1^2 + N_2^2)^{1/2})$: The prey growth rate decreases as the total population increases.
(3) $F_1(0,0) > 0$: Preys increase in number for sufficiently small total population.
(4) $F_1(0, N_2^*) = 0$: There is a predator population, N_2^*, such that the prey growth rate is zero when their number is zero.
(5) $F_1(N_1^*, 0) = 0$: There is a prey population, N_1^*, such that its growth rate is zero, even when there are zero predators.
(6) $\partial F_2/\partial N_2 < 0$: The growth rate of predators decreases as their population increases.
(7) $\partial F_2/\partial S > 0$: The growth rate of predators increases with increasing total population.
(8) $F_2(\hat{N}_1, 0) = 0$: If there are $\hat{N}_1$ prey, the predator growth rate is zero when their number is zero.
(9) $N_1^* > \hat{N}_1$, otherwise the prey disappear.

Unfortunately the conditions (6) and (7) are contradictory along the N_2 axis. Later Albrecht, Gatzke, Haddad and Wax (1974) corrected this error, and proved the theorem (which Kolmogorov only stated) under slightly different conditions. They retained the conditions (1), (2) $(\partial F_1/\partial S \equiv (\partial F_1/\partial N_1)N_1 + (\partial F_1/\partial N_2)N_2)$, (7), and established (9), but reduced the other conditions to (all $N_k \in R_+$)

(3′) There exists a $N_1^* > 0$ such that $(N_1 - N_1^*)F_1(N_1, 0) < 0$ for all $N_1 \neq N_1^*$
(4′) There exists a $N_2^* > 0$ such that $(N_2 - N_2^*)F_1(0, N_2) < 0$ for all $N_2 \neq N_2^*$
(5′) There exists a $\hat{N}_1 > 0$ such that $(N_1 - \hat{N}_1)F_2(N_1, 0) > 0$ for all $N_1 \neq \hat{N}_1$
(6′) $\partial F_2/\partial N_2 \leqslant 0$.

The conditions (3′), (4′), (5′) are more general than Kolmogorov's conditions, but, on the other hand, do not have his simple physical interpretations. Albrecht, Gatzke, Haddad and Wax then proved:

Kolmogorov's theorem

There exists a unique fixed point (N_1^0, N_2^0) of (5.7.2) in R_+^2. If it is unstable, then there is at least one periodic orbit. The outer most periodic orbit is semistable from the outside. The inner most periodic orbit is semistable from the inside. This is illustrated in Fig. 5.66. If there is only one periodic orbit, it is simply a stable limit cycle.

Fig. 5.66

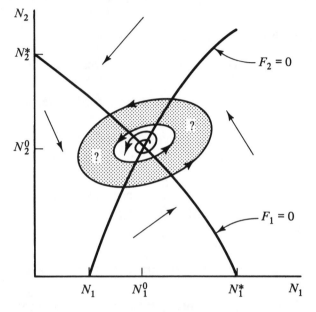

However the theorem does not indicate the conditions for such a unique cycle, nor the number of possible periodic cycles in the shaded region of the figure. Finally, if there is no periodic orbit, then (N_1^0, N_2^0) is a global attractor. Note that this differs from Kolmogorov's claim that, if (N_1^0, N_2^0) is a stable node or focus, it is a global attractor. This need not be the case. It could be enclosed by limit cycles.

We will return to the study of the generalized Lotka–Volterra system,

$$\dot{N}_i = N_i\left(k_i + \sum_{j=1}^{n} a_{ij}N_j \right)(N \in R_+^n),$$

and other modifications in Chapter 8. There we will see the application of such modeling to a variety of systems involving complex organization, competition, and evolutionary questions. But that is getting ahead of our story!

5.8 Relaxation oscillations; singular perturbations

For large values of the parameter k in the van der Pol equation,

$$\ddot{x} - k(1 - x^2)\dot{x} + x = 0, \tag{5.8.1}$$

the limit cycle is very different from the circular form, which we obtained for small k. Even for the moderate value of $k = 3.33$, the limit cycle is quite distorted as illustrated in Fig. 5.67. Note also that the time intervals, indicated on the orbit, exhibit regions of fast

Fig. 5.67

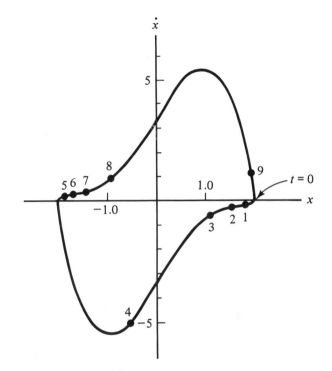

and slow motion in the phase space. These modifications of the uniform circular motion ($k = 0$) are clearly not perturbative. Self-exciting (autocatalytic) periodic systems which have such fast and slowly varying 'configurational' states in their period, are referred to as *relaxation oscillators*. There are numerous physically interesting and important examples of such relaxation oscillations. They are typically characterized by a relatively slow storage of energy, followed by a rapid release of this energy. This type of 'switch-on' and 'switch-off' between two very different dynamic states requires some form of *hysteresis*. In other words the system has two different dynamic states for the same applied conditions (it is dynamically bistable), and which state it presently occupies depends on the past history of the applied conditions (e.g., were they previously larger or smaller?).

An example of a relaxation oscillation is the extension of a violin string caused by its static friction with the bow (Fig. 5.68(*a*)). Multiple 'escapes' of the string from the bow, as it is uniformly drawn across the string, produce the sustained note. The difference between the static and sliding frictions produce the two different dynamic states, which are responsible for the hysteresis (see, e.g., J.C. Schelleng, 'The physics of the bowed string', *Sci. Amer.*, **230** (1), 87, 1974). A simpler example, is a bucket attached to the end of a flexible rod (Fig. 5.68(*b*)), which is continuously filled with water, until it becomes

Fig. 5.68

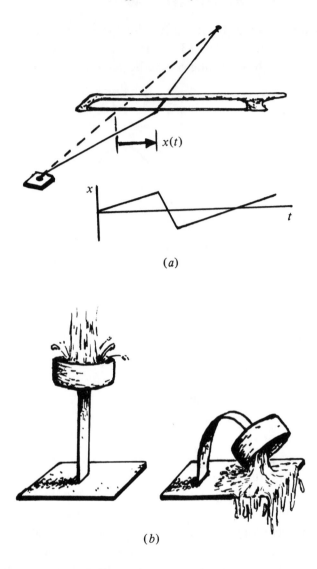

(a)

(b)

unstable and rapidly dumps out the water and returns to be filled. The hysteresis here is the one associated with the bistable states of the rod, discussed in Chapter 3.

A simple electrical relaxation oscillator involves a capacitor to store energy, and a discharge of some form, which rapidly releases this energy. As also noted in Chapter 3, a discharge tube (such as a neon tube) has a current–voltage relationship which can produce the hysteresis effect. To treat this dynamically it is necessary to include a small inductance L which represents both the parasitic inductance and also models inertial effects within the tube. Using the fact that the total voltage drop around any closed loop is zero, the left loop yields

$$V_\mathrm{o} = V_\mathrm{c} + R(i + C\dot{V}_\mathrm{c}) \tag{5.8.2}$$

and the right loop,

$$V_t + L\frac{di}{dt} = V_c. \tag{5.8.3}$$

The tube characteristic relationship, $i = I(V_t)$, is a multi-valued function for $V_1 \geqslant V_t \geqslant V_2$ (see Fig. 5.69). The single-valued relationship $V_t = V(i)$ can be introduced into (5.8.3). Differentiating this equation, and using (5.8.2) to eliminate $\dot{V}_c$, yields

$$Ld^2i/dt^2 = \dot{V}_c - V'di/dt$$
$$= (V_0 - V_c)/RC - i/C - V'\,di/dt \quad (V' \equiv dV/di).$$

Fig. 5.69

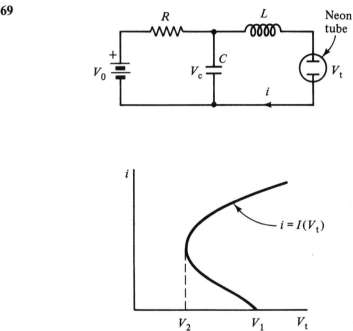

Finally V_c can be eliminated from this equation, using (5.8.3),

$$RCL\frac{d^2i}{dt^2} + (RCV' + L)\frac{di}{dt} + Ri + V = V_0 \tag{5.8.4}$$

The equilibrium state for the current i_0 is is given by $Ri_0 = V_0 - V(i_0)$, illustrated in Fig. 5.70. The value of i_0 falls where shown on the figure provided that R is not too large. Perturbations of this state yield the characteristic exponents, s, determined from

$$RCLs^2 + (RCV'(i_0) + L)s + R + V'(i_0) = 0.$$

Therefore

$$(2RCL)s = -(RCV'(i_0) + L) \pm [(RCV'(i_0) + L)^2 - 4RCL(R + V'(i_0))]^{1/2}, \tag{5.8.5}$$

Fig. 5.70

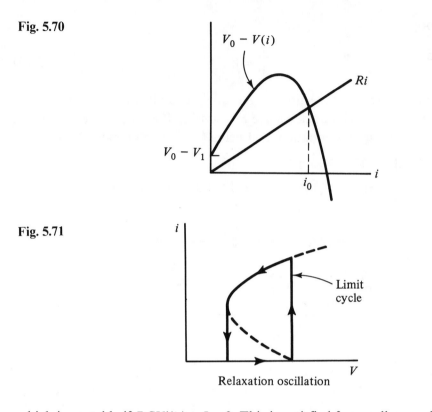

Fig. 5.71

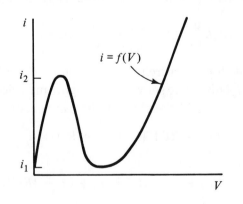

Relaxation oscillation

which is unstable if $RCV'(i_0) + L < 0$. This is satisfied for small enough L if i_0 is in the region where $V'(i_0) < 0$ as indicated in Fig. 5.71. Moreover, if this i_0 is the only fixed point $(V_0 > V_1)$, then all initial states tend toward near-discontinuous relaxation oscillations.

A more modern version of this relaxation circuit is the Fitzhugh–Nagumo model for mimicing the firing of *neurons* in nervous systems. In this case the discharge tube is replaced by a tunnel diode whose current–voltage characteristics are illustrated in Fig. 5.72. The circuit is also illustrated (Fig. 5.73). Again, considering the rate charging

Fig. 5.72

Fig. 5.73

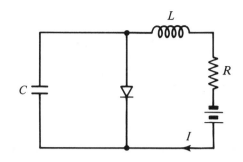

of the capacitor

$$C dV/dt = -I - f(V),$$

and the zero right-loop voltage drop

$$L dI/dt = V - V_o - RI,$$

we can eliminate the current, and obtain

$$LC \frac{d^2V}{dt^2} + (Lf'(V) + RC)dV/dt + Rf(V) + V = V_o. \qquad (5.8.6)$$

This is analogous to (5.8.4), and again yields relaxation oscillations. In the present context, this represents the periodic 'firing' of a neuron when subject to a constant stimulus (V_o).

A number of other systems are known to possess characteristic hysteresis interrelations between variables. Some of these systems are presently referred to as *bistable*. Thus *optical bistability* is where there are two possible transmission intensities for a range of incident radiation intensities (Fig. 5.74). This is discussed in some detail in Chapter 3. Similarly, *nuclear magnetic resonances* have been observed to be bistable (e.g., see discussion by D. Meier, *et al.*, in *Evolution of Order and Chaos* (Haken, ed.; Springer-Verlag, 1982)). When such bistable systems are activated dynamically, relaxation oscillations can obviously occur.

Fig. 5.74

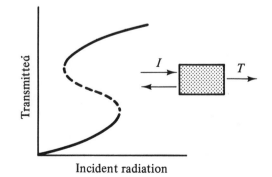

Incident radiation

In the case of relaxation oscillations, the averaging methods discussed in Section 5 are no longer a useful approach, and we need an entirely different line of attack, which is known as *singular perturbation theory*. This subject dates at least from Liouville's research of the Sturm–Liouville equation with a large parameter,

$$\ddot{y} + [\lambda^2 f(x) + g(x)]y = 0 \quad (\lambda \to \infty),$$

which in quantum mechanics is related to the so-called WKBJ approximation. In contrast to these linear differential equations, the subject of singular perturbations of nonlinear equations is much less developed. Early research by Fredrichs and Wasow (1946) who studied systems of the form

$$\dot{x}_k = F_k(x); \quad \varepsilon\dot{x}_n = F_n(x) \quad (k = 1, \dots, n-1)$$

for small ε, and by Levinson (1948), can be found referenced and discussed in Contributions to the theory of nonlinear oscillations (S. Lefschetz, ed., *Annals Math. Studies*, No. 20, Princeton University Press, 1950). Also, in addition to the general references, see Cesari (Section 10.5, *et. seq.*). The discussion here will represent only a simple introduction to this interesting subject, as it applies to the self-exciting oscillator systems. This particular application was originally investigated by Flanders and Stoker (1946) and LaSalle (1949).

The Rayleigh and van der Pol equations are special cases of the more general equations (respectively)

$$\ddot{z} + kF(\dot{z}) + z = 0, \tag{5.8.7}$$

$$\ddot{x} + k\Phi(x)\dot{x} + x = 0. \tag{5.8.8}$$

which are related by $x = \dot{z}$, $\Phi(x) = \mathrm{d}F(x)/\mathrm{d}x$. Many physical systems involving self-excitations can be modeled by these general equations. All that is required for the self-excitation and limit cycle behavior is that $F(\dot{z})$ and $\Phi(x)$ have the general form shown in Fig. 5.75. Referring back to (5.8.7) and (5.8.8), it is clear that functions of this form

Fig. 5.75

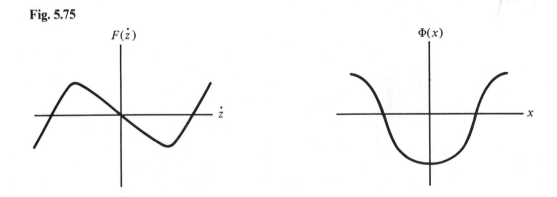

represent negative 'viscosity' near the fixed point $(0, 0)$. At larger values of $\dot{x}$ (or z), these change sign and limit the motion to finite cycles.

If we make the transformation

$$t = kt', \quad k = (\varepsilon)^{-1/2},$$

and then drop the prime, the van der Pol-like equation, (5.8.8), becomes

$$\varepsilon\ddot{x} + \Phi(x)\dot{x} + x = 0. \tag{5.8.9}$$

The limit $k \to \infty$ in (5.8.8) corresponds to $\varepsilon \to 0$ in (5.8.9).

We next note that when k is large in (5.8.1), the limit cycle becomes very distorted in the phase plane $(x, \dot{x})$. Indeed this distortion gets worse, as k is increased. It is useful, therefore to introduce new variables which have a limit cycle that tends to a fixed shape as $k \to \infty$ (or $\varepsilon \to 0$ in (5.8.9)). One possibility is to introduce the so-called *Liénard variable*

$$y = \varepsilon\dot{x} + \int_0^x \Phi(x')\mathrm{d}x' \equiv \varepsilon\dot{x} + F(x). \tag{5.8.10}$$

Then, using (5.8.9), we readily obtain the pair of equations

$$\dot{y} = -x \tag{5.8.11a}$$

$$\varepsilon\dot{x} = y - F(x). \tag{5.8.11b}$$

We will see that, as $\varepsilon \to 0$, the limit cycle goes to a limiting shape in the *Liénard phase plane*, (x, y). This is very useful for graphical purposes. It might also be noted that (5.8.7) can likewise be written

$$\dot{z} = y, \quad \dot{y} = -z - \varepsilon F(y),$$

which is mathematically equivalent to (5.8.11).

In the limit of 'strong relaxation' $\varepsilon \to 0$, the order of the differential equation (5.8.9) decreases, as does (5.8.11b). A reasonable question is: 'How does the solution of (5.8.11) with $\varepsilon = 0$ correspond to the solution obtained in the limit $\varepsilon \to 0$?.' While this might seem to be a reasonable question, it is also ill-defined, because a change of the order of the differential equation also changes the number of allowed initial conditions which are required (or allowed). The analysis for small ε involves a 'singularity' in the highest order derivative term, and therefore requires a different ('singular') perturbation treatment.

For $\varepsilon = 0$, the solution of (5.8.11b) in the Liénard phase plane $(x, \dot{y})$ is simply $y = F(x)$, which defines a curve, Γ. However (5.8.11a) requires that the flow on Γ is in the direction indicated in Fig. 5.76, and since the only fixed point is the origin, we have a contradiction at the extrema of $F(x)$. Thus the solution of (5.8.11) must move away from $y = F(x)$ between the extreme of $F(x)$. We note from (5.8.11b) that, if ε is small and

Fig. 5.76

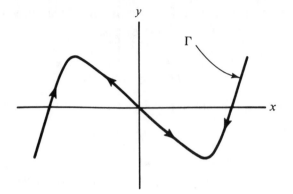

$y = F(x)$ then $\dot{x}$ is large and positive, whereas, if $y < F(x)$ then $\dot{x}$ is large and negative. Thus the limiting ($\varepsilon \to 0$) behavior of the limit cycle in the *Liénard plane* is a very simple symmetric shape, as illustrated. The horizontal portions represent the 'fast switches' between two different configurational 'states' (i.e., distinct ranges of x, the configurational variable of (5.8.9)). These two states represent the *bistable character* of the system (Fig. 5.77).

Fig. 5.77

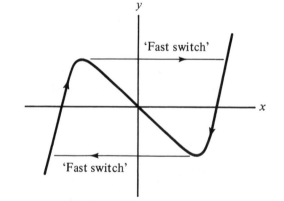

This limiting form of the limit cycle in the Liénard plane is very useful in many contexts. As an example we consider a bifurcation involving the *discontinuous 'birth or death'* of a limit cycle, which is an example of a so-called *blue sky catastrophe* (Abraham and Stewart, 1986; Abraham and Simó, 1985). This contrasts strongly with the Andronov–Hopf bifurcation, which involves a continuous birth and growth of the limit cycle's radius. Here the limit cycle will appear and disappear with a finite size, as a control parameter is varied (Fig. 5.78).

This can be illustrated by modifying the van der Pol equation, to read

$$\varepsilon\ddot{x} + \Phi(x)(\dot{x} - A) + x = 0.$$

This amounts to changing the force from $-x$ to $-x + A\Phi(x)$, which is, of course, a

Fig. 5.78

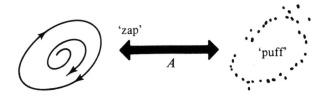

'zap'

'puff'

A

select form of nonlinearity (you might compare this with Exercise 5.29). Now we take $\Phi(x) = x^2 - 1$ (the van der Pol case, if $A = 0$), and obtain the Liénard-variables equations

$$\varepsilon\dot{x} = y - F(x), \quad \dot{y} = A\Phi - x \qquad (5.8.12)$$

where $F(x) = -x + (x^3/3)$. The fixed points of (5.8.12) are

$$y = F(x), \quad x = \frac{1}{2A} \pm \frac{1}{2A}(1 + 4A)^{1/2} \qquad (5.8.13)$$

(Fig. 5.80). If ε is small the limit cycle follows close to the curve $y = F(x)$ of $|x| > 1$, and near $y_c = \pm \frac{2}{3}$ if $|x| < 1$. As A is increased one fixed point tends toward this limit cycle.

Fig. 5.79

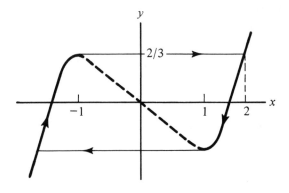

The critical value of x is roughly x_c, given by $\frac{2}{3} = F(x_c)$, or $x_c = 2$. This gives an estimate for the critical value of A, namely $A_c\Phi(2) - 2 = 0$, or $A_c = \frac{2}{3}$. This simple analysis, of course, only suggests that something dramatic may happen if $A_c \gtrsim \frac{2}{3}$ (if ε is small). To determine what occurs requires a computer analysis.

Exercise 5.32. (a) Determine the character of the fixed points (5.8.13), and sketch the vector field when the fixed point $x > x_c$. (b) Set $\varepsilon = 0.1$, and use a Runge–Kutta iteration to determine the critical value of A for which the limit cycle vanishes.

Another advantage of having the limiting shape of the limit cycle in the Liénard plane is when we want to determine the period of the van der Pol equation, (5.8.1), in the limit

Fig. 5.80

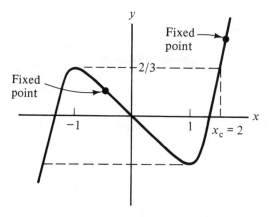

of large k (Fig. 5.80). Again we have $\Phi(x) = -1 + x^2$, so $F(x) = -x + (x^3/3)$, and we want to calculate the period around the indicated orbit. The period in the Liénard plane is the same as in $(x, \dot{x})$, and is given by

$$T' = \oint dt' = \int \frac{dy}{\dot{y}}. \tag{5.8.14}$$

The time required for the 'fast switches' goes to zero as $\varepsilon \to 0$, so this leaves only two equal intervals along $y = F(x)$. Thus

$$T' \simeq 2 \int \frac{dF(x)}{-x} = 2 \int_2^1 \frac{(1-x)^2}{x} \, dx = 2(\ln(x) - \tfrac{1}{2}x^2)|_2^1$$

$$= (3 - \ln 4) \simeq 1.614. \tag{5.8.15}$$

This, of course, is the time in terms of $t' = t/k$, so the period of (5.8.1) is $T \sim (3 - \ln 4)k$ (as $k \to \infty$). This is the leading term (in k) in the asymptotic expression for the period T. The corrections to this leading term involves considerable analysis. In 1947 Dorodnitsyn gave an asymptotic calculation for the van der Pol period which extends the result (5.8.15). As later corrected by Urabe (1963), it reads (for (5.8.11))

$$T = (3 - \ln 4) - 3\alpha\varepsilon^{2/3} - \tfrac{1}{6}\varepsilon \ln \varepsilon - 1.3246\varepsilon + O(\varepsilon^{7/6}) \tag{5.8.16}$$

where $\alpha \simeq -2.338$ is the largest zero of Airy's function, Ai(z). The corresponding expression for (5.8.1) is (as $k \to \infty$)

$$T = (3 - \ln 4)k - 3\alpha k^{-1/3} + \frac{1}{3k} \ln k - 1.3246/k + O(k^{-4/3}). \tag{5.8.17}$$

Exercise 5.33. Use the method of the path integral, (5.8.14), to determine the periods of:

(A) the oscillator $m\ddot{x} + kx = O(|x| < L)$, between two walls (at $x = \pm L$). At the walls it experiences an elastic collision ($\dot{x} \to -\dot{x}$). Hence the 'fast switches' occur at

$x = \pm L$, and are vertical in the usual $(x, \dot{x})$ phase plane. Obtain the period as a function of the energy, $T(E)$, for all E.

(B) the relativistic 'Hookian' oscillator, $d(m\gamma v)/dt = -kx$, $v = dx/dt$, and $\gamma = 1 - (v/c)^2)^{-1/2}$. The energy integral is $E = mc^2\gamma + 1/2kx^2$. Obtain the period, $T(\varepsilon)$, as a function of $\varepsilon = E/mc^2$, for large ε. The general result is involved, so first plot $E(x, \dot{x}) = $ constant for large γ, and then use ideas from part (A) to obtain the leading term for $T(\varepsilon)$.

The way that a series (5.8.17) arises can be illustrated by a model due to Stoker and Haag. This asymptotic singular perturbation analysis is discussed in Appendix I. They used a simplification of the above functions $F(x)$, $\Phi(x)$, which is based on a method widely used by many Russian investigators in the 1930s. It is the method of using *piecewise-linear systems* to approximate nonlinear systems. In more recent times this method has also been used by Levinson (1949) and Levi (1981) to simplify the studies by Cartwright and Littlewood (1945) of periodically forced self-exciting oscillators. We will return to this in Section 14.

The idea is to replace a nonlinear function, such as $\Phi(x) = x^2 - 1$ and $F(x) = -x + \frac{1}{3}x^3$, associated with the van der Pol equation, (5.8.1) and (5.8.10), by *piecewise-linear* functions. An example of this is illustrated in Fig. 5.81. That is,

$$\Phi(x) = +1, \quad \text{if} \quad |x| > 1; \quad \Phi(x) = -1, \quad \text{if} \quad |x| < 1. \tag{5.8.18}$$

or

$$F(x) = 2 + x \, (x \leqslant -1), \qquad F(x) = -x \, (-1 \leqslant x \leqslant 1),$$
$$F(x) = x - 2 \, (x \geqslant 1). \tag{5.8.19}$$

Fig. 5.81

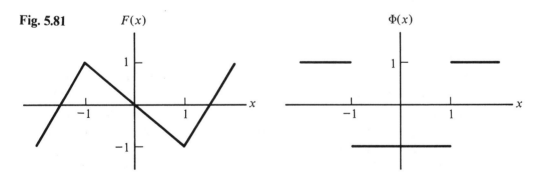

The advantage in this choice is that the nonlinear equations, (5.8.11), are then piecewise linear in space and easily integrated in each region. The difficulty, of course, is that the matching conditions at the discontinuities can only be performed in practice over a limited range of time. Therefore the apparent advantage is illusory unless it is coupled with a clever analysis, as was done by Levinson. Exercise 5.33 illustrates less profound, but nonetheless useful applications of piecewise-linear approximations. For a more significant application see Appendix I.

Exercise 5.34. To illustrate the difficulties with the matching conditions for the piecewise-linear systems, obtain the solution of (5.8.11) for $\varepsilon = 1$, using (5.8.19), and the initial conditions $(x = -2, y = 0)$. Obtain the solution up to the time when first $x = 0$. Can you determine this time?

5.9 Global bifurcations (homoclinics galore!)

The topological character of the phase portrait of a system can change when a control parameter is varied, so that a bifurcation occurs, without changing the type of any of the fixed points of the system. In this case the change in the phase portrait is not noticeable in the neighborhood of any fixed point, but can only be discerned on a global scale. Such *global bifurcations* are more difficult to detect and do not, for example, fall within the framework of elementary catastrophe theory or any generalization which considers only the local degeneracy of fixed points. As we will see, an intelligent (judicious) use of computer solutions of the equations of motion, together with analysis (Ulam's process of 'synergetics'), is an invaluable method for discovering such global bifurcations.

To illustrate a global bifurcation, consider the example (Arnold, 1972),

$$\dot{x}_2 = x_1; \quad \dot{x}_1 = c_1 + c_2 x_2 + x_2^2 + x_1 x_2. \tag{5.9.1}$$

The fixed points of this system are at

$$x_1 = 0, \quad x_2 = -\tfrac{1}{2}c_2 \pm \tfrac{1}{2}[c_2^2 - 4c_1]^{1/2}, \tag{5.9.2}$$

provided that $c_2^2 > 4c_1$. Clearly these fixed points are degenerate along the curve $c_2^2 = 4c_1$ in the control space. If the c values cross this curve, there is a localized change in the phase portrait, namely two fixed points come together and vanish. We know, from the consideration of the Poincaré index, that these simple singularities must be a saddle point (index $= -1$) and a node or focus (index $= +1$), since they must add up to zero (no singularity). This birth of a saddle point and node is referred to as a *saddle-node bifurcation*.

In addition to the bifurcations determined by this local degeneracy of fixed points, it turns out that, in the region $c_2^2 > 4c_1$, there is both a curve associated with a Andronov–Hopf bifurcation (involving a change of the stability of a fixed point and the birth/death of a limit cycle), and a curve associated with a global bifurcation. These curves are indicated in Fig. 5.82. The bifurcation lines are labelled (SN^-, SN^+, AH, SC), which divide the control space into regions (A, B, C, D). The local *saddle-node* and *Andronov–*

Fig. 5.82

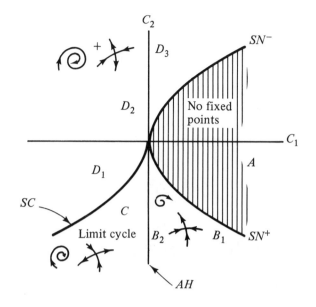

Hopf bifurcations occurs across the lines $SN^\pm$ and AH respectively, whereas the *global bifurcation*, called a *saddle-connection*, occurs across the curve SC. The phase portraits corresponding to the above regions and curves in the control space are illustrated in Fig. 5.83 (after Arnold (1972)). Starting with $4C_1 > C_2^2$, in region A, and moving clockwise around the origin, the first fixed points (at $x_2 > 0$) occur upon crossing SN^+. This gives birth to a saddle point and an unstable node. Progressing from B_1 to B_2, the spiral on the node 'tightens', and the Andronov–Hopf bifurcation takes place on AH. This yields a limit cycle in region C, in addition to a stable node and saddle point. The saddle-connection occurs along SC. On this control curve there is an orbit, $x(t)$, which tends to the same fixed point as $t \to -\infty$ or $t \to +\infty$; that is,

$$\lim_{t \to +\infty} x(t) = x^0 = \lim_{t \to -\infty} x(t).$$

Hence x^0 is a *homoclinic point*, and the orbit is a homoclinic orbit (Section 5.1).

Crossing into the region D, there is no longer a limit cycle, and one branch of the saddle point now flows into the second fixed point (compare the outflow connection in B_2 with the inflow connection in D_1). Crossing the curve SC has changed neither the saddle point nor the stable vortex. Progressing from D_1 to D_2, brings the saddle point and node closer together, until they fuse ($x_2 < 0$) on the curve SN^-.

Exercise 5.35. You might enjoy numerically exploring the saddle-connection of (5.9.1). Note that initial states which are slightly to the left of the saddle point ($x_1 \lesssim 0$) have drastically different behavior for large t as (C_1, C_2) crosses SC. To see this, compare the figures SC and D_1. Thus, if we take an initial state with x_2 given by the proper root of

Fig. 5.83

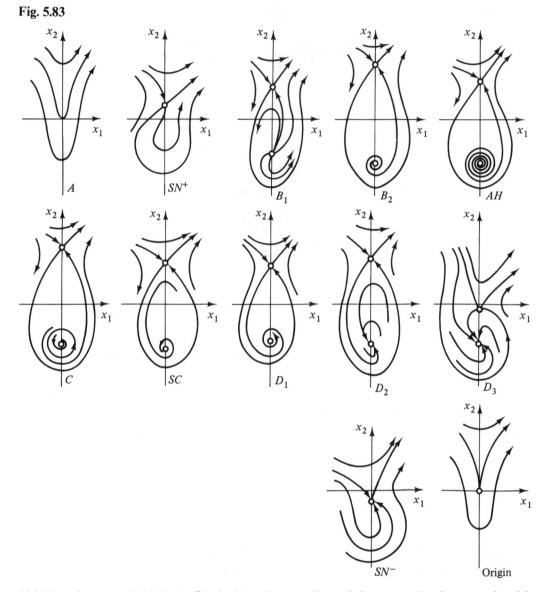

(5.9.2) and $x_1 = -0.01$ (say), fix C_1 (say $C_1 = -1$), and decrease C_2, then we should notice a dramatic change at some value of C_2. What approximate value do you find? Clearly this method can be improved and extended. How? A fairly detailed analysis of this global bifurcation (saddle-connection) can be found in Chow and Hale (1982; Chapter 13, Section 2) and Arnold (1983).

Exercise 5.36. Consider the system

$$\dot{x}_1 = x_2 + h_1(x_1, x_2, c)$$
$$\dot{x}_2 = x_1 - x_1^2 + h_2(x_1, x_2, c)$$

where $h_i(x_1, x_2, 0) = 0$ $(i = 1, 2)$. Show that this system has a homoclinic orbit (saddle-connection) at $c = 0$, by obtaining an algebraic expression for the homoclinic orbit. Obtain the time independent integral of this system when $c = 0$. Determine the behavior of the flow near the fixed point $(x_1 = 1, x_2 = 0)$ when $c = 0$.

As noted before, the discovery and clarification of global bifurcations frequently comes about from the intelligent use of computers (synergetics). While it is interesting to read about the discoveries of others, as just done, nothing is as much fun as uncovering new results yourself! As an example of this process, we now will extend the study begun in Exercise 5.29. (If you skipped that exercise, return to it!)

The system we are considering is a hybrid van der Pol–Duffing oscillator

$$\ddot{x} + (c_2 + x^2)\dot{x} + c_1 x + x^3 = 0. \tag{5.9.3}$$

As discussed in the comments concerning Exercise 5.29, when $c_1 < 0$ the origin is a saddle point, and there are two other fixed points, at $(x, \dot{x}) = (\pm(-c)^{1/2}, 0)$. The linear analysis in that exercise shows that these fixed points have the character illustrated in the schematic figure (Fig. 5.84).

Fig. 5.84

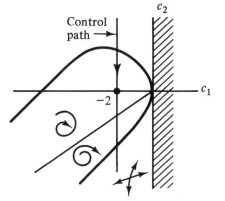

We are interested in the character of the global flow when there are three fixed points $(c_1 < 0)$, and choose $c_1 = -2$ for purposes of illustration. In this case there are two stable focii if $3 > c_2 > -2$. Moreover it is clear from (5.9.3) that, if $c_2 > 0$, the system (globally) damps to one or the others of the fixed points at $x = \pm(-c)^{1/2}$. The interesting behavior occurs when $c_2 < 0$, but $c_2 > c_1$, so that these fixed points remain stable. If, for example, we take $c_2 = -1.7$ we find that the two initial states $(x, \dot{x}) = (0.1, 0)$ and $(0.5, 0)$ have very different asymptotic limits. The first tends to a limit cycle, the second to the fixed point $(2^{1/2}, 0)$, as illustrated in Fig. 5.85. Since these solutions diverge there must be some initial state between them which neither goes 'left' or 'right' – in other words, there exists a periodic solution. This reasoning is illustrated in the schematic figure (Fig. 5.86). This periodic solution is not easy to find precisely, since it is an *unstable limit cycle* – not easy that is, unless you think of running time

Fig. 5.85

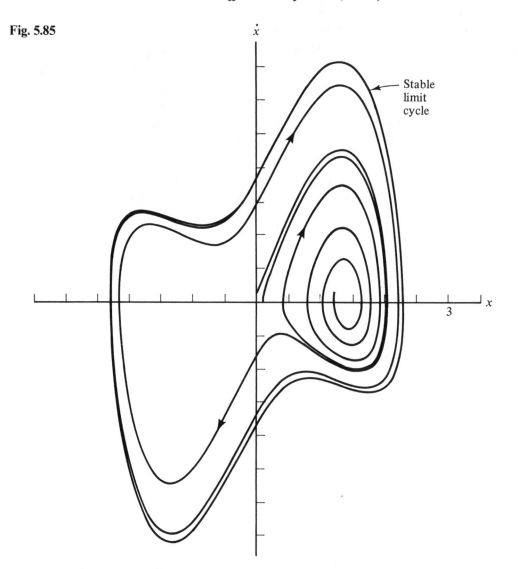

backwards (an advantage computations have over nature!). This is done by taking the time step to be negative in the Runge–Kutta iteration. The result of this, beginning at $(1, 0)$ for clarity, is illustrated in Fig. 5.87. Clearly there is indeed an unstable limit cycle.

Having found an unstable limit cycle, two bifurcation facts are clear. The first is that the loss of stability of the fixed points $(\pm (2)^{1/2}, 0)$ as $c_2 \to c_1$ $(c_2 \geqslant c_1)$ is a *subcritical Hopf bifurcation*, where the radius of the unstable limit cycle goes to zero as $c_2 \to c_1$ $(c_2 \geqslant c_1)$. That type of bifurcation is, of course, local in character, not global.

We know, however, that there must also be a global bifurcation, because something 'gave birth' to the stable limit cycle and to two unstable limit cycles (about $x = \pm (2)^{1/2}$)

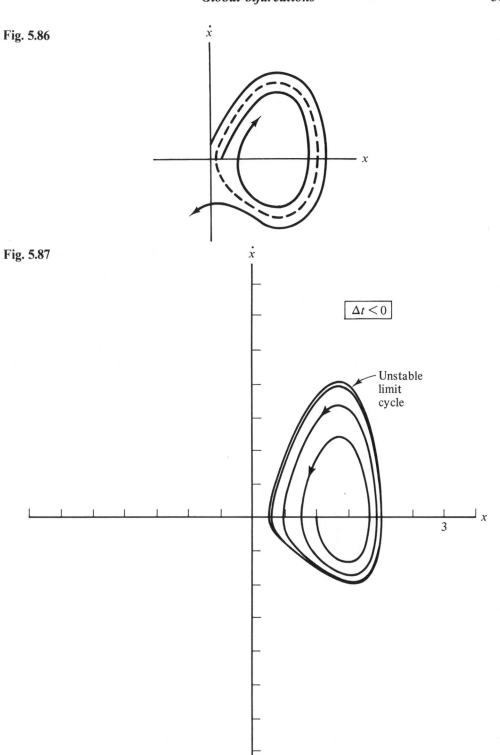

Fig. 5.86

Fig. 5.87

$\Delta t < 0$

Unstable
limit
cycle

3

Fig. 5.88

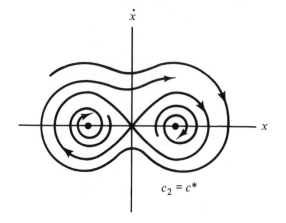

somewhere in the range $0 > c_2 > -2$). What is the character of that bifurcation? It is not difficult to see that, if c_2 is larger than -1.7, the unstable limit cycles must have a larger radius, and must intersect the fixed point at the origin if we further increase c_2. We might also guess that the stable limit cycle shrinks as c_2 is increased. Then it is clear that what happens is that two homoclinic orbits occur for some $c_2 = c^*$ (Fig. 5.88), and for $c_2 < c^*$ these two orbits bifurcate into two unstable limit cycles plus one enclosing stable limit cycle. (A strange birth; two identical twins plus a larger paternal twin!) This is schematically illustrated in Fig. 5.89. The exact value of c^* is not important, but a little experimentation indicates that it lies in the range $-1.65 > c^* > -1.66$.

Fig. 5.89

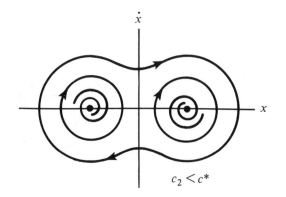

Exercise 5.37. The figure for the case $c_2 < c^*$ is incomplete. Sketch the orbits of the two incoming and two outgoing separatrices of the origin.

Note that these homoclinic orbits are not limit cycles, because the ω-limit set of all exterior points at $c = c^*$ is the origin (which is also the α-limit set of all interior points.

The two orbits, beginning at $(0.1, 0)$ and $(0.5, 0)$, are shown for $c_2 = -1.7$ (Fig. 5.90). The unstable limit cycles are already well separated from the origin, shrinking towards the fixed points $(\pm 2^{1/2}, 0)$.

Fig. 5.90

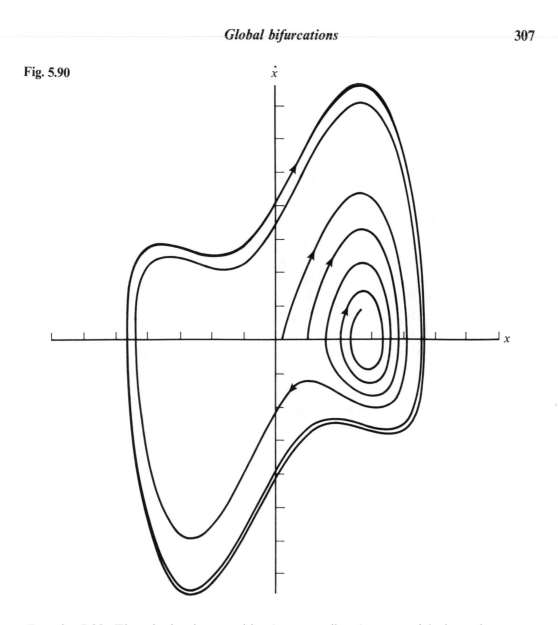

Exercise 5.38. There is clearly something 'unnatural', or 'nongeneric', about the system (5.9.3), because the potential is symmetric. That symmetry gives rise to the degeneracy of the two homoclinic orbits at $c_2 = c^*$. In such a situation it is nearly mandatory to understand the 'generic' nonsymmetric case

$$\ddot{x} + (c_2 + x^2)\dot{x} + c_1 x + c_3 x^2 + x^3 = 0. \tag{5.9.4}$$

We can take $c_3 > 0$, since $c_3 < 0$ amounts to replacing x by $-x$. To be specific, take $c_3 = 1$, $c_1 = -2$ and again determine the bifurcations as c_2 is decreased. You should identify the first bifurcation $(c_2 > -1)$ with one described above. The second bifurcation $c_2 = -1$, is hopefully familiar. The third $c_2 \simeq -3.5$ is a new type which may take some thought, whereas $c_2 = -4$ is an 'oldie'. This exercise verges on the edge of real, if not profound research, and should be used for that type of experience. If you did not find the previous results interesting and stimulating you might as well omit the present 'research'.

5.10 Periodically forced passive oscillators: a cusp catastrophe

We next consider a nonautonomous passive oscillator. The most important case in this category is when the applied force is periodic. We begin with a heuristic examination of this situation, and then follow this discussion with a more systematic approach (averaging method). It should be emphasized that there are again (as in the autonomous systems) two cases: (a) perturbative situations and (b) nonperturbative responses to the periodic force. The division between these is not altogether clear and simple, but one needs to be on guard against simplistic treatments which exclude, *ab initio*, effects which arise in (*b*). The following heuristic examination of a periodically forced oscillator is basically a simple extension of a perturbative approach, taking into account some nonperturbative effects in a simple fashion. More dramatic nonperturbative effects will be discussed in Section 14.

Consider first the case of a forced linear oscillator

$$\ddot{x} + 2\mu\dot{x} + \omega_0^2 x = A \cos \Omega t$$

where we have included a damping term. Its solution is

$$x = a_0 \exp\left(-\mu t\right) \cos\left(\left[\omega_0^2 - \mu^2\right]^{1/2} t + \theta_0\right) + \frac{A \cos\left(\Omega t + \delta\right)}{\left[(\omega_0^2 - \Omega^2)^2 + 4\mu^2\Omega^2\right]^{1/2}}$$

and $\tan \delta = -2\mu\Omega/(\omega_0^2 - \Omega^2)$. If $\mu t \gg 1$ this reduces to

$$x \simeq \frac{A \cos\left(\Omega t + \delta\right)}{\left[(\omega_0^2 - \Omega^2)^2 + 4\mu^2\Omega^2\right]^{1/2}}.$$

Fig. 5.91

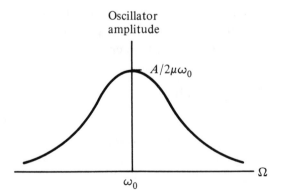

Oscillator
amplitude

$A/2\mu\omega_0$

ω_0

Ω

Thus the frequency and the amplitude of the oscillator depend on Ω and A. The amplitude behaves in the typical linear resonance fashion, as illustrated in Fig. 5.91.

Now consider the periodically forced nonlinear (Duffing) oscillator

$$\ddot{x} + 2\mu\dot{x} + \omega_0^2 x + \varepsilon x^3 = A\cos\Omega t.$$

After long times ($\mu t \gg 1$) we again expect this oscillator to oscillate at the frequency Ω, but with an amplitude which now depends on the nonlinear term. To estimate this effect, we note that to, $O(\varepsilon)$, the nonlinear oscillator can be represented by a linear oscillator

$$\ddot{x} + 2\mu\dot{x} + \omega^2 x = A\cos\Omega t, \tag{5.10.1}$$

where the *effective frequency*, ω, is given by (see (5.4.17), ff)

$$\omega = \omega_0 + \frac{3a^2}{8\omega_0}\varepsilon, \tag{5.10.2}$$

and a is the amplitude of the oscillation. This replacement of the original nonlinear equation by the linear equation is valid provided that $a^2 = O(1)$. The solution of (5.10.1) for $\mu t \gg 1$ is, as before,

$$x = \frac{A\cos(\Omega t + \delta)}{[(\omega^2 - \Omega^2)^2 + 4\mu^2\Omega^2]^{1/2}} \equiv a\cos(\Omega t + \delta)$$

that is, the amplitude satisfies

$$a^2 = \frac{A^2}{(\omega(a)^2 - \Omega^2)^2 + 4\mu^2\Omega^2}$$

so a depends on ω, but ω also depends on a, because of (5.10.2).

The most interesting case is near resonance

$$\Omega = \omega_0 + \varepsilon\Delta. \tag{5.10.3}$$

If we assume that the damping is weak, $\mu = O(\epsilon)$, then to lowest order in ϵ, $4\mu^2\Omega^2 \simeq 4\mu^2\omega_0^2$. Moreover, using (5.10.2) and (5.10.3),

$$(\omega^2 - \Omega^2)^2 = [(\omega - \Omega)(\omega + \Omega)]^2 \simeq \left(\frac{3a^2}{8\omega_0}\varepsilon - \Delta\varepsilon\right)^2 (2\omega_0)^2,$$

so the last equation for a^2 becomes

$$a^2 \simeq \frac{(A^2/4\omega_0^2\varepsilon^2)}{\left(\dfrac{3a^2}{8\omega_0} - \Delta\right)^2 + (\mu/\varepsilon)^2}.$$

Clearly, if $a^2 = O(1)$, we must require that $A = O(\varepsilon)$. When this is the case, the forcing term is referred to as *soft excitation*.

Let

$$\alpha = \frac{3a^2}{8\omega_0} \quad F = 3A^2/32\omega_0^3\varepsilon^2 \tag{5.10.4}$$

then we finally obtain the basic relationship

$$\alpha[(\alpha - \Delta)^2 + (\mu/\varepsilon)^2] = F. \tag{5.10.5}$$

This is a cubic equation for α, which is essentially the amplitude squared, as a function of F and Δ, the square of the applied force and the frequency of resonance respectively. It yields a very interesting effect. As the applied force is increased (larger F), the resonance shape of the oscillators amplitude, α, becomes distorted and ultimately becomes triple-valued, as illustrated in Fig. 5.92.

Fig. 5.92

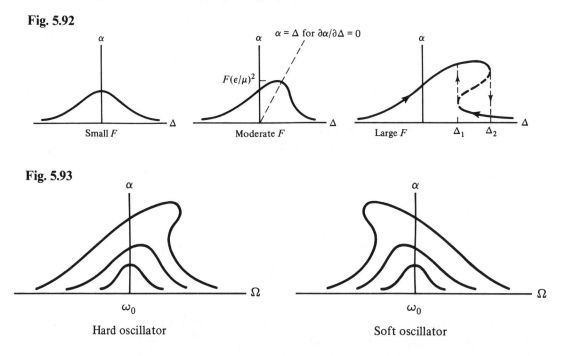

Fig. 5.93

Hard oscillator   Soft oscillator

This figure does not depend on the sign of ε, but the applied frequency, $\Omega = \omega_0 + \varepsilon\Delta$, does depend on this sign. Thus, if we draw α vs. Ω we obtain Fig. 5.93.

Exercise 5.39. The following analysis to determine the minimum critical force, F_c, which is required to produce three values of α, is incorrect. Determine the error in this analysis (the correct result is given below): α, in (5.10.5), has three real roots only if $d\alpha/d\Delta = -\infty$ for some value of Δ, (5.10.3). We have

$$\frac{d\alpha}{d\Delta}[(\alpha - \Delta)^2 + (\mu/\varepsilon)^2] + 2\alpha(\alpha - \Delta)\left(\frac{d\alpha}{d\Delta} - 1\right) = 0$$

so $d\alpha/d\Delta$ can be infinite only if

$$(\alpha - \Delta)^2 + (\mu/\varepsilon)^2 + 2\alpha(\alpha - \Delta) = 0 \qquad (5.10.6)$$

or $\Delta^2 - 4\alpha\Delta + 3\alpha^2 + [\mu/\varepsilon]^2 = 0$. This gives the two roots

$$\Delta_1 = 2\alpha - [\alpha^2 - (\mu/\varepsilon)^2]^{1/2}, \quad \Delta_2 = 2\alpha + [\alpha^2 - (\mu/\varepsilon)^2]^{1/2}. \qquad (5.10.7)$$

From the above figures, we see that the critical value of the force F is determined by the case when the inflection point (Δ_1, Δ_2) are equal $(\Delta_1 = \Delta_2)$. From (5.10.7) these are equal when $\alpha = \mu/\varepsilon$, and then $\Delta = 2\alpha$. The value of F for which this holds is, by (5.10.5)

$$F_c = \alpha[(\alpha - \Delta)^2 + (\mu/\varepsilon)^2] = (\mu/\varepsilon)\cdot 2(\mu/\varepsilon)^2 = 2(\mu/\varepsilon)^3$$

where F_c represents the critical value. The corresponding critical value of the forcing amplitude A is $A_c^2 = 64\omega_0^3\mu^3/3\varepsilon$. Again, these answers are wrong. Why?

Although the above values are incorrect, the basic idea illustrated in the figure is correct. In particular this system can oscillate with two different amplitudes for a range of frequencies (between Δ_1 and Δ_2). The system, therefore, has a *dynamic* form of *bistability*. This makes it possible for this system to exhibit a very interesting *hysteresis effect*. Namely, if F is greater than some critical value, then as Ω is slowly increased, the amplitude will increase until $\Delta = \Delta_2$, at which point the amplitude will 'drop' to a much smaller value. However, if Ω is then slowly decreased the amplitude will slowly increase, then 'jump' to a large amplitude when $\Delta = \Delta_1$, which is less than Δ_2. This is schematically illustrated in Fig. 5.94. In other words the system does not behave the same when Ω is increased as it does when Ω is decreased. There is a dynamic *hysteresis effect* – its oscillatory behavior at a given $\Omega_1 < \Omega < \Omega_2$ depends on its past history.

The results obtained from computations are also illustrated. The effect proves to be

Fig. 5.94

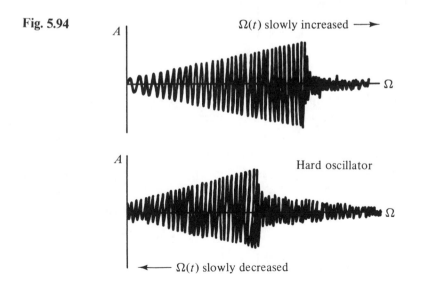

A $\Omega(t)$ slowly increased $\longrightarrow$ Ω

A Hard oscillator Ω

$\longleftarrow$ $\Omega(t)$ slowly decreased

Fig. 5.95

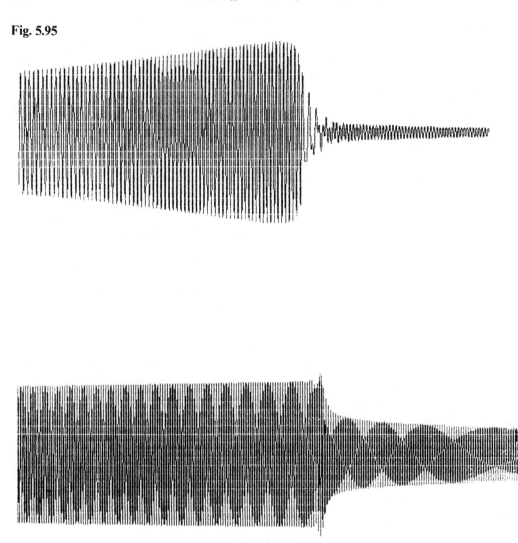

more difficult to produce, because the required 'adiabatic' variation of $\Omega(t)$ may be very time consuming. In the following figures $\Omega^{-2}\,d\Omega/dt \simeq 5 \times 10^{-4}$ was used.

A fuller appreciation of the structure of this *asymptotic amplitude*, α, is obtained by considering the entire *surface*, $\alpha = \alpha(F, \Delta)$, defined by (5.10.5). Clearly, if we put the third axis (F) on Fig. 5.95, the resulting surface $\alpha = \alpha(F, \Delta)$ in this control-phase space has a fold-over 'pleat' characteristic of a cusp catastrophe. This looks roughly as shown in Fig. 5.96.

The catastrophe set K in (F, Δ) plane is obtained from the simultaneous solution of (5.10.5) and

$$\frac{\partial}{\partial \alpha}[\alpha[(\alpha - \Delta)^2 + (\mu/\varepsilon)^2]] = 0$$

Fig. 5.96

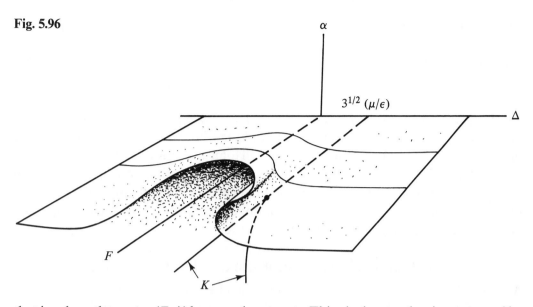

that is, where the roots $\alpha(F, \Delta)$ become degenerate. This algebra can be circumvented by using the fact that a cubic equation, such as (5.10.5), has two identical roots if the standard condition of the form $q^3 + r^2 = 0$ holds, which in the present case is

$$[\tfrac{1}{3}(\mu/\varepsilon)^2 - \tfrac{1}{9}\Delta^2]^3 + [\tfrac{1}{2}F - \tfrac{1}{27}\Delta^3 - \tfrac{1}{3}\Delta(\mu/\varepsilon)^2]^2 = 0,$$

or

$$F = \tfrac{2}{27}\Delta^3 + \tfrac{2}{3}\Delta(\mu/\varepsilon)^2 \pm \tfrac{2}{27}[(\Delta^2 - 3(\mu/\varepsilon)^2)^3]^{1/2}. \qquad (5.10.8)$$

This gives the two curves of the cusp catastrophe set. Also note that the location of the

Fig. 5.97

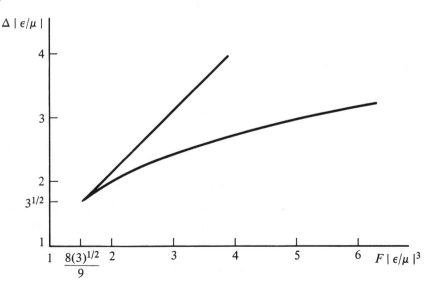

cusp $(\Delta_c, F_c) = (3^{1/2}(\mu/\varepsilon), (8/9) 3^{1/2}(\mu/\varepsilon)^3)$ differs from the values $(2(\mu/\varepsilon), 2(\mu/\varepsilon)^3)$ in Exercise 5.39.

If we use the scaled variables

$$\alpha' = \alpha\varepsilon/\mu, \quad \Delta' = \Delta\varepsilon/\mu, \quad F' = F\varepsilon/\mu^3,$$

which one can do if $\mu \neq 0$, there is one surface $\alpha'(F', \Delta')$ and hence one catastrophe set for all values of $(\mu/\varepsilon \neq 0)$. This invariant set is plotted in Fig. 5.97. A number of other figures, and more discussion can be found in Holmes and Rand (1976).

This example is another example of the fact that the classic catastrophe sets, which are the only ones which can stably occur for gradient systems, may also arise in a variety of other systems, such as the present nonautonomous ordinary differential equation system, or even partial differential equation systems.

5.11 Harmonic excitations: extended phase space

Another effect of the nonlinearity is that there can be resonances even when Ω is quite different from ω_0. The literature on this subject is very extensive (see references below), and will not be dealt with in any detail here. We will only note some basic aspects of this motion, and then consider methods of representing the motion which will be useful later.

To illustrate this phenomena, consider the ultraharmonic case when $\Omega \simeq \omega_0/3$. In this case we are interested in exciting a harmonic of the applied frequency.

The solution to

$$\ddot{x} + 2\mu\dot{x} + \omega_0^2 x + \varepsilon x^3 = A \cos \Omega t \tag{5.11.1}$$

in lowest order is

$$x^{(0)} = \frac{A \cos(\Omega t + \delta)}{[(\omega^2 - \Omega^2)^2 + 4\mu^2\Omega^2]^{1/2}}$$

if $\mu t \gg 1$. Here $\omega = \omega_0 + O(\varepsilon)$, and we consider A to be of order ε^0 (i.e., *hard excitation*). In the next order we have

$$\ddot{x}^{(1)} + 2\mu\dot{x}^{(1)} + \omega_0^2 x^{(1)} = -\varepsilon \frac{[A \cos(\Omega t + \delta)]^3}{[(\omega^2 - \Omega^2)^2 + 4\mu^2\Omega^2]^{3/2}}$$

$$= -\frac{\varepsilon A^3}{[(\omega^2 - \Omega^2)^2 + 4\mu^2\Omega^2]^{3/2}} \{\tfrac{1}{4} \cos 3(\Omega t + \delta) + \tfrac{3}{4} \cos(\Omega t + \delta)\}.$$

Since $3\Omega \simeq \omega$, the forcing term $\cos 3(\Omega t + \delta)$ is near resonance, and since it is proportional to ε, it is a *soft excitation*. Thus this is very similar to the resonance which we just considered. In fact it is not difficult to see that in general there are (or, at least, may be) resonances any time that

$$m\omega + n\Omega \simeq 0 \quad (m, n = \pm 1, \pm 2, \ldots). \tag{5.11.2}$$

If the driving frequency, Ω, bears this relationship to the frequency of the oscillator, ω, then the following descriptions are used (for integers n, m):

If $\omega = m\Omega$ the motion is *ultraharmonic*.
If $\omega = \Omega/n$ the motion is *subharmonic*.
If $\omega = m\Omega/n$ the motion is *ultrasubharmonic* $(m \neq 1, n \neq 1)$.

Fig. 5.98 contains schematic illustrations of possible relationships between the applied force (top figure) and $x(t)$ in three cases, for which $n = 2$, $m = 3$.

Fig. 5.98

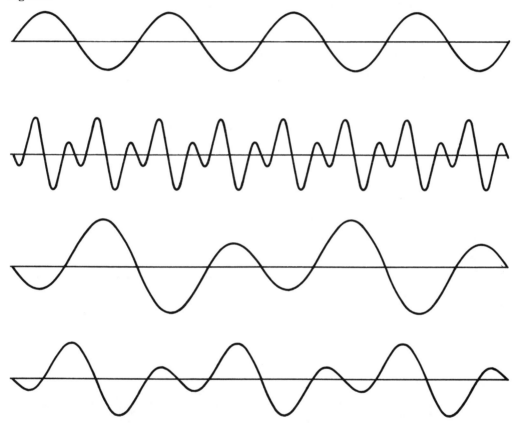

A simple mechanical experiment in the generation of *subharmonics* of the applied frequency is given in C.A. Ludeke, '*Resonance*', *J. Appl. Phys.*, **13**, 418–23 (1942). In this experiment a cantilever spring is kept in motion by an unbalanced motor mounted on its free end as shown in Fig. 5.99. The restoring force produced by the spring as a function of $z = R\theta$ was measured, as shown in Fig. 5.99, and found to be well represented by

$$F = \alpha z + K z^3$$

Fig. 5.99

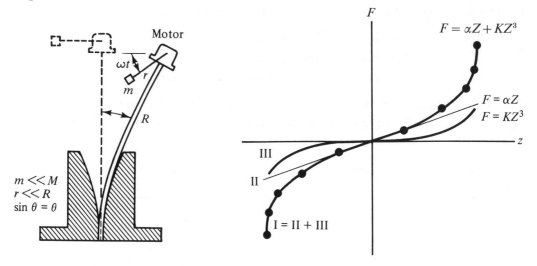

where $\alpha = 33\,300$ dynes/cm and $K = 1700$ dynes/cm^3. The equation of motion is therefore

$$(M + m)\ddot{z} + [\alpha - (M + m)g/R]z + Kz^3 = -mr\omega^2 \cos \omega t$$

which has the harmonic frequency $\omega_0^2 = \alpha/(M + m) - g/R$. The left group in Fig. 5.100 shows pictures taken when $\omega = 3\omega_0$ (pictures 1–4, 4–8, and 8–12 being approximately $2\pi/\omega_0$ apart in time). The right group shows a higher order subharmonic resonance $\omega = 4\omega_0$. Now 1–3, 3–6, 6–8, and 8–10 corresponds to $2\pi/\omega_0$. For many other physical examples, and details concerning the analysis of subharmonic oscillations, see Minorsky (1962), Hayashi (1985), Pippard (1978), Schmidt and Tondl (1986), and Jordan and Smith (1986). The discovery of subharmonic electrical oscillations by van

Fig. 5.100

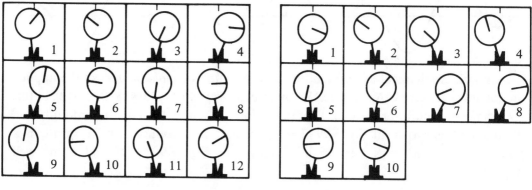

(a)   (b)

der Pol and van der Mark (1927) was instrumental in stimulating a series of theoretical research activities, which uncovered *bistable subharmonic* and 'chaotic families' of solutions. We will discuss these features in Section 5.14.

The representation of a nonautonomous periodic orbit in phase space looks that shown in Fig. 5.101. Here dots have been added to the trajectory to indicate a

Fig. 5.101

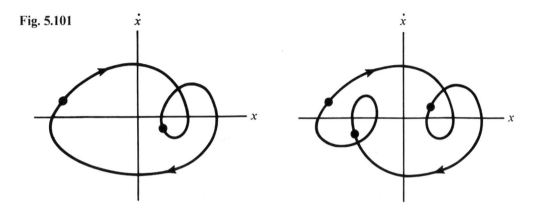

particular phase of the applied periodic force. Thus from one dot to the next is one period of the applied force, $T \equiv 2\pi/\Omega$. The period of the system $T_s \equiv 2\pi/\omega$, has the subharmonic relationship $\omega = \Omega/2$ and $\omega = \Omega/3$ in these two cases.

To make this relationship clearer it is useful to represent the dynamics in the *extended phase space*, at which one variable (axis) is the time, $x_3 = t$. In this space the equations are autonomous and trajectories do not intersect, so the subharmonic motion on the left (which is less violent than the above examples) becomes drawn out into a nonintersecting trajectory, as illustrated in Fig. 5.102. If, moreover, we consider the intersection of this trajectory with any two transverse planes, $(x, \dot{x})$, we obtain a *Poincaré map* in this space. The most useful Poincaré map for studying subharmonic, or more generally ultrasubharmonic motion ($\omega = m\Omega/n$), is obtained by separating the transverse planes by a 'distance' T, the forcing period, along the time axis. The

Fig. 5.102

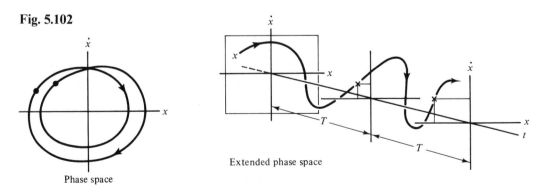

Phase space

Extended phase space

usefulness of this, as can be seen in Exercise 5.40, comes from a uniqueness property, because the applied force repeats at each plane. The disadvantage, is that it takes a vast number of planes to represent much dynamics in the extended phase space. Alternatively, we can picture this as dynamics occuring in the dark, and we turn on the light briefly every T seconds, yielding *a stroboscopic record*, which we then record on a single plane.

Another representation of this dynamics is to take a bounded region of the $(x, \dot{x})$ space (*a disk D*), and join the two transverse planes together, thereby forming a *solid torus region* $(D \times S^1)$, whose central axis is the time (Fig. 5.103). The great advantage is

Fig. 5.103

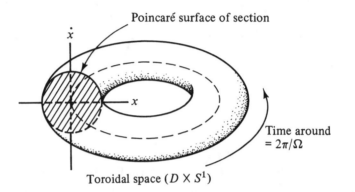

Poincaré surface of section

Toroidal space $(D \times S^1)$

Time around $= 2\pi/\Omega$

that now there is only one transverse plane which trajectories will repeatedly intersect. This toroidal region is really not a phase space, as the following exercise makes clear, but it is nonetheless very useful. Moreover the terminology of phase space is usually applied it, calling it a *Poincaré surface of section* as discussed in Chapter 2.

In this toroidal space the *ultraharmonic* $(\omega = m\Omega)$, *subharmonic* $(\omega = \Omega/n)$, and *ultrasubharmonic* motions $(\omega = m\Omega/n, n \neq 1, m \neq 1)$ are closed curves of the form shown in Fig. 5.104.

Fig. 5.104

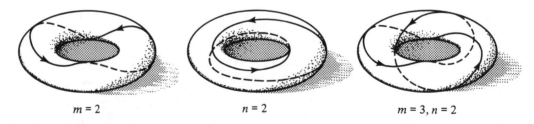

$m = 2$ $n = 2$ $m = 3, n = 2$

Exercise 5.40. Show that in the region $(D \times S^1)$: (a) if a trajectory intersects itself it must be a periodic solution of $\dot{x} = F(x, t)$; (b) if a trajectory is a periodic solution it need not be a closed curve in this space (when is it closed?).

As we consider increasing complicated dynamics, the extended phase space representation, together with its associated Poincaré map on the Poincaré surface of section, will prove to be of considerable value. We will return to this in Section 14, and particularly in Chapter 6.

5.12 Averaging method for nonautonomous system (KBM)

Now let us try to put these heuristic considerations into a more systematic method. We consider the case of a soft force ($O(\varepsilon)$), and introduce this explicitly in the equation of motion, so that

$$\ddot{x} + x = \varepsilon f(x, \dot{x}) + \varepsilon A \cos \Omega t \tag{5.12.1}$$

where now $A = O(1)$, and we have picked the units of time such that $\omega_0 = 1$. We proceed by again looking for solutions of the form

$$x = a \cos \psi + \varepsilon u_1 + \varepsilon^2 u_2 + \cdots \tag{5.12.2}$$

where $\dot{a} = O(\varepsilon)$ and $\psi = t + \theta$, with $\dot{\theta} = O(\varepsilon)$. Then we have

$$\dot{x} = -a(1 + \dot{\theta}) \sin \psi + \dot{a} \cos \psi + \varepsilon \dot{u}_1 + O(\varepsilon^2)$$

$$\ddot{x} = -2\dot{a} \sin \psi - a(1 + 2\dot{\theta}) \cos \psi + \varepsilon \ddot{u}_1 + O(\varepsilon^2).$$

Up to $O(\varepsilon^2)$, (5.12.1) becomes

$$\varepsilon(\ddot{u}_1 + u_1) - 2\dot{a} \sin \psi - 2a\dot{\theta} \cos \psi = \varepsilon f(a \cos \psi, -a \sin \psi) + \varepsilon A \cos \Omega t. \tag{5.12.3}$$

The basic problem is how to treat the last term, in which time appears explicitly. In the autonomous case we had in this order

$$\dot{a} = \varepsilon A_1(a) \quad \dot{\psi} = 1 + \varepsilon B_1(a), \quad \ddot{u}_1(a, \psi) \equiv u_{1,\psi\psi}$$

so that (5.12.3) (with $A = 0$) became

$$-2A_1(a) \sin \psi - 2aB_1(a) \cos \psi + (u_{1,\psi\psi} + u_1) = f(a \cos \psi, -a \sin \psi),$$

and we could treat a and ψ as independent variables. Now, however, the time appears explicitly. We might try to get rid of the term $A \cos \Omega t$ in (5.12.3) by simply setting

$$u_1 = \frac{A \cos \Omega t}{1 - \Omega^2} + \tilde{u}_1(a, \psi).$$

Then the resulting equation for $\tilde{u}_1(a, \psi)$ is the same as in the autonomous case. This in fact can be done provided that $A/(1 - \Omega^2) = O(\varepsilon^0)$, or in other words $1 - \Omega^2 = O(1)$. This is the *nonresonant case*. The real difficulty arises in the *resonant case*, when $1 - \Omega^2 = O(\varepsilon)$. Therefore let us consider this case.

We begin with the equation (5.12.1). The idea is to write the factor $\cos \Omega t$ in terms of ψ

and a slowly varying quantity. To do this we write

$$\cos \Omega t = \cos \{\psi + (\Omega t - \psi)\} = \cos \psi \cos (\Omega t - \psi) - \sin \psi \sin (\Omega t - \psi).$$

We note that the quantity $\Phi = \Omega t - \psi$ is slowly varying in time, because

$$\frac{d\Phi}{dt} = \Omega - 1 - \dot\theta = O(\varepsilon).$$

Now we will assume (and later verify) that we can take

$$\ddot{u}_1 = u_{1,\psi\psi} + O(\varepsilon).$$

Then (5.12.1) becomes

$$\varepsilon(u_{1,\psi\psi} + u_1) - 2\dot{a}\sin\psi - 2a\dot\theta\cos\psi = \varepsilon f(a\cos\psi, -a\sin\psi)$$
$$+ \varepsilon A\{\cos\psi\cos\Phi - \sin\psi\sin\Phi\}.$$

Since Φ is slowly varying in time (compared with ψ), we will treat Φ as being independent of ψ. In that case u_1 will be free from terms proportional to ψ (i.e., secular terms), only if the inhomogeneous terms do not contain $\cos\psi$ or $\sin\psi$. This immediately yields the conditions that

$$\dot{a} = -\frac{\varepsilon}{2\pi}\int_0^{2\pi} f(a\cos\psi, -a\sin\psi)\sin\psi\,d\psi + \frac{\varepsilon}{2}A\sin\Phi \qquad (5.12.4)$$

and

$$\dot\theta = -\frac{\varepsilon}{2\pi a}\int_0^{2\pi} f(a\cos\psi, -a\sin\psi)\cos\psi\,d\psi - \frac{\varepsilon}{2}\frac{A}{a}\cos\Phi$$

or

$$\dot\Phi = (\Omega - 1) + \frac{\varepsilon}{2\pi a}\int_0^{2\pi} f(a\cos\psi, -a\sin\psi)\cos\psi\,d\psi + \frac{\varepsilon}{2}\frac{A}{a}\cos\Phi \quad (5.12.5)$$

Note that $\dot\Phi$ is now a function of Φ as well as of a (in contrast with the autonomous case where $\dot\Phi = F(a)$).

Let us apply this method to the forced Duffing equation

$$\ddot{x} + x = -\varepsilon(x^3 + 2\mu\dot{x}) + \varepsilon A\cos\Omega t$$

so $f(x, \dot{x}) = -x^3 - 2\mu\dot{x}$, and

$$f(a\cos\psi, -a\sin\psi) = -a^3(\tfrac{1}{4}\cos 3\psi + \tfrac{3}{4}\cos\psi) + 2\mu a\sin\psi.$$

Then, from (5.12.4) and (5.12.5), the averaged equations are

$$\dot{a} = -\varepsilon\mu a + \tfrac{1}{2}\varepsilon A\sin\Phi$$
$$\dot\Phi = (\Omega - 1) - \frac{3\varepsilon a^2}{8} + \frac{1}{2}\frac{\varepsilon A}{a}\cos\Phi \qquad (5.12.6)$$

and

$$u_{1,\psi\psi} + u_1 = -\tfrac{1}{4}a^3 \cos 3\psi.$$

The fixed point solutions, corresponding to $\dot{a} = 0$, $\dot{\Phi} = 0$ $(\dot{\psi} = \Omega)$, are given by

$$\tfrac{1}{2}\varepsilon A \sin \Phi = \varepsilon\mu a$$

$$\tfrac{1}{2}\varepsilon A \cos \Phi = a(1 - \Omega) + \frac{3\varepsilon a^3}{8}$$

or

$$(\tfrac{1}{2}\varepsilon A)^2 = (\varepsilon\mu a)^2 + \left[a(1 - \Omega) + \frac{3\varepsilon a^3}{8} \right]^2$$

$$A^2 = (2\mu a)^2 + \left[\frac{2a}{\varepsilon}(1 - \Omega) + \frac{3a^3}{4} \right]^2.$$

Note that these are fixed points in the (a, Φ) plane, not fixed points in the $(x, \dot{x})$ phase space. If we set $\alpha = 3a^2/8$, $F = 3A^2/32$, $\Omega = 1 + \varepsilon\Delta$

$$F = \alpha\mu^2 + \alpha(\alpha - \Delta)^2 \tag{5.12.7}$$

this is the same equation as (5.10.5) (except that the present (μ, A) equals the previous (μ/ε) and (A/ε)).

We can obtain more information than this from the above averaged equations, because they can be explicitly integrated (if $\mu = 0$). Let

$$P \equiv a \sin \Phi, \quad Q \equiv -a \cos \Phi.$$

Differentiating and using (5.12.6) yields

$$\dot{P} = \tfrac{1}{2}\varepsilon A + \left[\Omega - 1 - \frac{3\varepsilon}{8}(P^2 + Q^2) \right]Q$$

and

$$\dot{Q} = (1 - \Omega)P + \frac{3\varepsilon}{8}(P^2 + Q^2)P$$

'Eliminating dt' between $\dot{P}$ and $\dot{Q}$ gives

$$\left[1 - \Omega + \frac{3\varepsilon}{8}(P^2 + Q^2) \right]P \, dP + \left\{ -\tfrac{1}{2}\varepsilon A + \left[-\Omega + 1 + \frac{3\varepsilon}{8}(P^2 + Q^2) \right]Q \right\} dQ = 0.$$

This is an exact differential

$$dH = \frac{\partial H}{\partial P} \, dP + \frac{\partial H}{\partial Q} \, dQ = 0,$$

so we have the first integral

$$H(P, Q) \equiv (1 - \Omega)(P^2 + Q^2) - \varepsilon A Q + (3\varepsilon/16)(P^2 + Q^2)^2 = \text{constant}.$$

Since $P^2 + Q^2 = a^2$, this gives $\cos \Phi$ in terms of a,

$$\varepsilon A a \cos \Phi = (1 - \Omega)a^2 + (3\varepsilon/16)a^4 - H.$$

Moreover, from $\dot{P}$ and $\dot{Q}$ we readily obtain

$$\dot{a} = \tfrac{1}{2}\varepsilon A \sin \Phi = \pm \tfrac{1}{2}\varepsilon A [1 - \cos^2 \Phi]^{1/2}$$

or

$$\frac{da^2}{dt} = \pm \varepsilon A \left[a^2 - \frac{1}{\varepsilon A} \left\{ a^2(1 - \Omega)a^2 + \frac{3\varepsilon}{16}a^4 - H \right\}^2 \right]^{1/2} \tag{5.12.8}$$

This shows that a^2 can be expressed in terms of elliptic functions (see Struble, 1962, for some details).

This analysis does not give the stability of the asymptotic state obtained above, (5.12.7), because $\mu = 0$. One should be able to determine this stability by using the usual linear analysis of $\dot{a} = G(a, \Phi)$, $\dot{\Phi} = H(a, \Phi)$, namely by determining the characteristic exponents associated with the matrix $\partial(G, H)/\partial(a, \Phi)$, evaluated at the fixed point.

5.13 Forced van der Pol equations – frequency entrainment

We will now consider the case when a periodic force is applied to a *self-exciting system*, as illustrated by a van der Pol oscillator. The difference from the situation of the forced Duffing oscillator is that now the oscillator would oscillate around the limit cycle even if no forcing term is present. A phenomena of great interest concerns the conditions under which such an oscillator will 'give up' its independent mode of oscillation and acquire the frequency of the applied oscillating force – a phenomenon known as *entrainment*. A simple model of this phenomena was obtained by using the circle map in Chapter 4. Phenomena of this type have been known at least since the time of Huygens, who observed (according to van der Pol, 1927) that two clocks on the same wall (which provides the coupling mechanism) tend to keep synchronous time, provided their independent frequencies are not too far apart. Rayleigh observed a similar 'locking' of two organ pipes or tuning forks with nearly the same independent frequencies. Another very pretty mechanical example, involving piano strings, will be discussed at the end of this section. Also, of course, there is the well-known phenomena of 'two hearts that beat as one' which, however, may be another phenomena!

On a more serious level, there are many important and interesting occurrences of entrainment, both applied and in nature. One important application is the control of any erratic behavior of a heart by a periodic electric pacemaker (Fig. 5.105). In this case the heart, which may have tendencies to exhibit intermittent or erratic oscillations, can be entrained to follow the nearby period of the oscillator, provided that the electric signal is sufficiently strong. There may also be a number of important circadean

Fig. 5.105

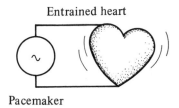

Entrained heart

Pacemaker

rhythms in living systems (i.e., 'endogenous' physiological cycles of approximately 24 hour periods) which are also frequently referred to as 'pacemakers'. These pacemakers interact with each other, and may be entrained by environmental (exogenous) cycles, such as night–day light or temperature cycles. These environmental cycles are referred to as 'zeitgebers' (time-givers), and many studies have shown the influence, and sometimes entrainment of a pacemaker by a zeitgeber – possibly through another pacemaker (e.g., Moore-Ede, Sulzman, and Fuller, 1982).

Fig. 5.106

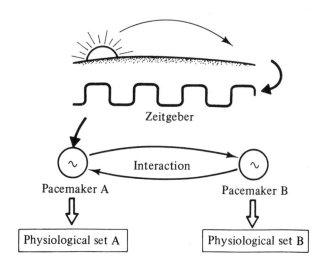

Zeitgeber

Interaction

Pacemaker A   Pacemaker B

Physiological set A   Physiological set B

The following analysis is considerably simpler than many of these situations, because the forcing term is given. Hence there is no reaction-force acting back on the driving system, and the situation is therefore not symmetric, as in the case of coupled clocks. Nonetheless the following results will give the basic idea of entrainment, if not many realistic details.

The basic equation is now

$$\ddot{x} - \varepsilon(1 - x^2)\dot{x} + \omega_0^2 x = A \cos \Omega t \quad (\varepsilon > 0). \tag{5.13.1}$$

We will just consider the case of a *soft excitation*, $A = O(\varepsilon)$, but near resonance $\Omega \simeq \omega_0$, specifically

$$\Omega = \omega_0 + \varepsilon\Delta \quad (\Delta = O(1)).$$

Recall that, to $O(\varepsilon)$, the frequency of the van der Pol oscillator is ω_0. That is, there is no first order frequency shift. We now introduce the *van der Pol variables*, $(a(t), b(t))$, through the definition

$$x = a(t) \cos \Omega t + b(t) \sin \Omega t \quad (\dot{a} = O(\varepsilon) = \dot{b}), \tag{5.13.2}$$

thereby looking for the possibility of a solution near the driving frequency. Hence, if (a, b) asymptotically approach constants (i.e., have stable singular points), then (5.13.2) represents the phenomena of *entrainment*. That is, the x-oscillator, is 'entrained' by the driving force, because it oscillates at the driving frequency.

Now, using (5.13.2)

$$\dot{x} = -(\Omega a - \dot{b}) \sin \Omega t + (\Omega b + \dot{a}) \cos \Omega t$$

$$\ddot{x} = -(2\Omega \dot{a} + \Omega^2 b) \sin \Omega t + (2\Omega \dot{b} - \Omega^2 a) \cos \Omega t + O(\varepsilon^2).$$

Moreover, we need $\frac{1}{3} d(x^3)/dt$, which appears in (5.13.1). If we write $x = r \cos (\Omega t + \phi)$ then $x^3 = \frac{3}{4} r^3 \cos (\Omega t + \phi)$, plus terms going as $\cos 3(\Omega t + \phi)$ or $\sin 3(\Omega t + \phi)$, which we neglect (these are important for harmonic excitation). Thus, to $O(\varepsilon)$,

$$\frac{1}{3} \frac{d}{dt} (x^3) = \frac{1}{4} \frac{d}{dt} [(a^2 + b^2)(a \cos \Omega t + b \sin \Omega t)]$$

$$= \tfrac{1}{4}(a^2 + b^2)(-\Omega a \sin \Omega t + \Omega b \cos \Omega t) + O(\varepsilon)$$

If we substitute these results into the van der Pol equation, the coefficients of $\cos (\Omega t)$ and $\sin (\Omega t)$ yield respectively the following two equations

$$2\Omega \dot{b} - \Omega^2 a - \varepsilon \Omega b + \frac{\varepsilon}{4}(a^2 + b^2)\Omega b + \omega_0^2 a = A$$

$$-2\Omega \dot{a} - \Omega^2 b + \varepsilon \Omega a - \frac{\varepsilon}{4}(a^2 + b^2)\Omega a + \omega_0^2 b = 0$$

Using

$$\omega_0^2 - \Omega^2 = (\omega_0 - \Omega)(\omega_0 + \Omega) \simeq -\varepsilon \Delta (2\Omega) + O(\varepsilon^2)$$

these reduce to

$$\dot{a} = -\varepsilon \Delta b + \frac{\varepsilon}{8} a(4 - a^2 - b^2)$$

$$\dot{b} = \varepsilon \Delta a + \frac{\varepsilon}{8} b(4 - a^2 - b^2) + \frac{A}{2\Omega} \tag{5.13.3}$$

The singular points of (5.13.3) are given by the solutions of

$$-\varepsilon \Delta a_0 - \frac{\varepsilon}{8} b_0(4 - a_0^2 - b_0^2) = \frac{A}{2\Omega}$$

$$- \varepsilon \Delta b_0 + \frac{\varepsilon}{8} a_0 (4 - a_0^2 - b_0^2) = 0. \tag{5.13.4}$$

If $r^2 = a^2 + b^2$, then by squaring and adding these equations we obtain

$$\Delta^2 r_0^2 + \tfrac{1}{64} r_0^2 (4 - r_0^2)^2 = (A/2\varepsilon\Omega)^2.$$

Introducing

$$\sigma = (2\Delta), \quad \rho = r_0^2/4, \quad \text{and} \quad F^2 = (A/2\varepsilon\Omega)^2 \tag{5.13.5}$$

This becomes

$$\sigma^2 \rho + \rho(1 - \rho)^2 = F^2 \tag{5.13.6}$$

which was obtained by van der Pol.

To determine the number of positive real roots, ρ, consider the derivative of the left side of (5.13.6)

$$\sigma^2 + (1 - \rho)^2 - 2\rho(1 - \rho) = 0$$

or

$$3\rho^2 - 4\rho + 1 + \sigma^2 = 0.$$

Hence, the extrema of the left side of (5.13.6) are given by

$$\rho = \tfrac{2}{3} \pm \tfrac{1}{6}[16 - 12(1 + \sigma^2)]^{1/2}$$
$$= \tfrac{2}{3} \pm \tfrac{1}{3}[1 - 3\sigma^2]^{1/2}$$

and there are three singular points (i.e., solutions of (5.13.6), for some F) provided that $\sigma^2 < \tfrac{1}{3}$. If $\sigma^2 = \tfrac{1}{3}$ then the inflection point occurs for $\rho = \tfrac{2}{3}$, in which case

$$F^2 = \tfrac{1}{3}\tfrac{2}{3} + \tfrac{2}{3}(\tfrac{1}{3})^2 = \tfrac{8}{27}.$$

In fact, if $F^2 > \tfrac{8}{27}$ there is only one root of (5.13.6) for all σ (Fig. 5.107).

Fig. 5.107

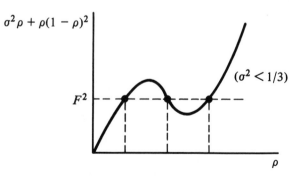

Exercise 5.41. Prove this statement.

To determine the stability of the singular points, set

$$a = a_0 + \alpha, \quad b = b_0 + \beta$$

and linearize equation (5.13.3) to obtain

$$\dot{\alpha} = -\varepsilon\Delta\beta + \tfrac{1}{2}\varepsilon\alpha(1-\rho) + \tfrac{1}{8}\varepsilon a_0(-2a_0\alpha - 2b_0\beta)$$
$$\dot{\beta} = \varepsilon\Delta\alpha + \tfrac{1}{2}\varepsilon\beta(1-\rho) + \tfrac{1}{8}\varepsilon b_0(-2a_0\alpha - 2b_0\beta)$$

The matrix of the right hand coefficients is

$$\begin{pmatrix} \tfrac{1}{2}\varepsilon(1-\rho) - \tfrac{1}{4}\varepsilon a_0^2 & -\varepsilon\Delta - \tfrac{1}{4}\varepsilon a_0 b_0 \\ \varepsilon\Delta - \tfrac{1}{4}\varepsilon a_0 b_0 & \tfrac{1}{2}\varepsilon(1-\rho) - \tfrac{1}{4}\varepsilon b_0^2 \end{pmatrix}$$

and according to the discussion at the beginning of this chapter, the stability properties are determined by the trace (p) and determinant (q) of this matrix. That is, the characteristic exponents ($\alpha, \beta \sim \exp(st)$) are given by

$$s = \tfrac{1}{2}p \pm \tfrac{1}{2}(p^2 - 4q)^{1/2}, \tag{5.13.7}$$

where now, using (5.13.5),

$$p = \varepsilon(1-\rho) - \tfrac{1}{4}\varepsilon(a_0^2 + b_0^2) = \varepsilon(1 - 2\rho)$$

$$q = \frac{\varepsilon^2}{16}[2(1-\rho) - a_0^2][2(1-\rho) - b_0^2] + \frac{\varepsilon^2}{16}(4\Delta + a_0 b_0)(4\Delta - a_0 b_0).$$

So we obtain

$$p = \varepsilon(1 - 2\rho) \tag{5.13.8a}$$
$$q = \tfrac{1}{4}\varepsilon^2\sigma^2 + \tfrac{1}{4}\varepsilon^2(1 - 4\rho + 3\rho^2). \tag{5.13.8b}$$

The values of ρ to be used in (5.13.8) are those given by (5.13.6), and the character of each of the singular points for given (F, σ) is thereby determined.

The standard procedure is to represent the characteristics of the singular points in the (ρ, σ) plane, rather than in the original (p, q) plane. This amounts to the mapping

$$(p, q) \rightarrow (\rho, \sigma)$$

given by (5.13.8). Thus the saddle point region has the map

$$q < 0 \rightarrow \sigma^2 + 1 - 4\rho + 3\rho^2 < 0,$$

whereas the unstable nodes and focus are in the region

$$p > 0 \rightarrow 1 - 2\rho > 0 \quad (\text{if } \varepsilon > 0),$$

and the stable region is the portion of (ρ, σ) not contained in the above regions. These are shown in Fig. 5.108.

In this figure several lines of constant F have been shown. These are given by the equation for the allowed singular points (5.13.6). Note that, for a given σ (i.e., Ω) there is one or possibly three values of ρ depending on whether $F^2 > \tfrac{8}{27}$ or $F^2 < \tfrac{8}{27}$ respectively.

Fig. 5.108

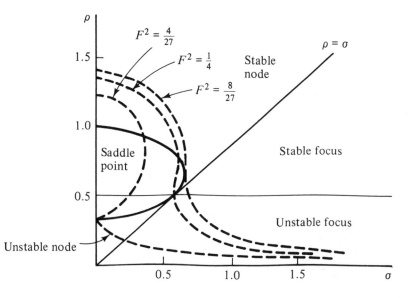

The physically interesting variables are the control variables (F, σ), rather than (ρ, σ) shown in the last figure. The following figure shows the types of singular points in the (F, σ) plane, labelled as follows:

SN: Stable Node	SF: Stable Focus
UN: Unstable Node	UF: Unstable Focus
SP: Saddle Point	

An important point to note is that, for large (a, b), the equations (5.13.3) reduce to

$$\dot{a} \simeq -\frac{\varepsilon}{8}(a^2 + b^2)a, \quad \dot{b} \simeq -\frac{\varepsilon}{8}(a^2 + b^2)b$$

so all integral curves are directed toward the origin. Thus, if there is only one unstable singular point (i.e., $F^2 > \frac{8}{27}, \rho < \frac{1}{2}$) then, by the *Poincaré–Bendixson theorem*, there must exist a limit cycle in the van der Pol plane. This means that the motion $x(t)$, given by (5.13.12), is not simply periodic, but involves several frequencies (almost periodic). In lowest order this would involve only the beat frequencies $\omega_0 \pm \Omega$, and this motion is called *heterodyning*.

On the other hand if $F^2 > \frac{8}{27}$, but $\rho > \frac{1}{2}$, there is only one singular point and it is stable. Note that this physically corresponds to a large force and not too large a difference between the applied and natural frequency

$$\Omega - \omega_0 = \varepsilon\Delta.$$

Since the singular point is stable in the van der Pol plane, this means that $x(t)$ will asymptotically go to the applied frequency (rather than ω_0, or some combination frequencies). Thus this is the case of *entrainment*, which is indicated in Fig. 5.109.

Fig. 5.109

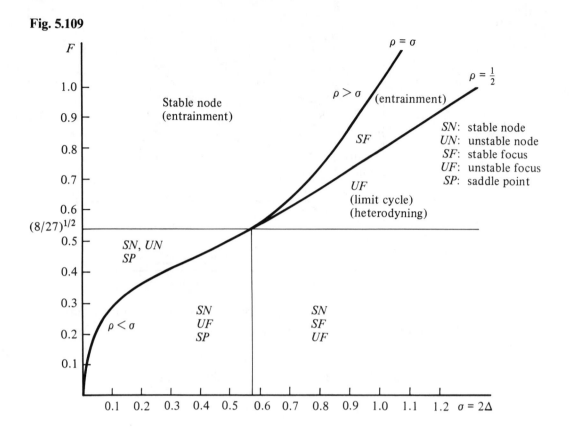

Fig. 5.110

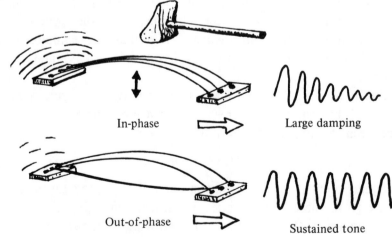

In-phase   ⇒   Large damping

Out-of-phase   ⇒   Sustained tone

A very interesting application of entrainment is used in the tuning of pianos (Weinreich, 1979). The objective is to produce a sustained aftertone, when the three piano strings are struck by the hammer (Fig. 5.110). The problem is that, if the strings oscillate in phase, then they exert a coherent (additive) force on the bridge at one end of the strings. This in turn causes a large transmission of energy to the bridge, which rapidly dampens the oscillations of the strings. The net result is that the tone rapidly decays.

To avoid this, it would be useful for two strings to be out of phase, so that their forces would tend to cancel at the bridge. On the other hand , we want them to be at the same frequency. Moreover, they are excited in phase, by the hammer blow. How can they become out of phase, and yet oscillate at a common frequency? Apparently this can be accomplished by tuning the strings at slightly different frequencies, so that after they are struck they become out of phase. However, the nonlinear coupling of the bridge causes them to become entrained at a common frequency (after they become out of phase!). The net result is a sustained aftertone. A pretty idea, even if difficult to prove!

5.14 Nonperturbative forced oscillators

When nonlinear oscillators are driven by 'large' periodic forces, a variety of phenomena can occur, depending both on the type of oscillator and the frequency and amplitude of the driving force. The primary classes of oscillators which have been studied are:

(1) *Self-exciting (autocatalytic) oscillators*;

$$\ddot{x} + k(x^2 - 1)\dot{x} + x = A\cos(\Omega t) \tag{5.14.1}$$

$$\varepsilon\ddot{x} + \Phi(x)\dot{x} + \varepsilon x = A\cos(\Omega t) \tag{5.14.2}$$

$$\dot{x} = a + x^2 y - (1 + b)x + A\cos(\Omega t)$$
$$\dot{y} = bx - yx^2 \tag{5.14.3}$$

(2) *Damped hard-polynomial oscillators*;

$$\ddot{x} + \mu\dot{x} + x + x^3 = A\cos(\Omega t) \tag{5.14.4}$$

$$\ddot{x} + \mu\dot{x} + x^3 = A\cos(\Omega t) \tag{5.14.5}$$

(3) *Damped 'inverted' hard-polynomial oscillators*;

$$\ddot{x} + \mu\dot{x} - x + x^3 = A\cos(\Omega t) \tag{5.14.6}$$

(4) *The damped pendulum and soft-polynomial oscillators*;

$$\ddot{\theta} + \mu\dot{\theta} + \sin\theta = A\cos(\Omega t) \tag{5.14.7}$$

$$\ddot{x} + \mu\dot{x} + x - x^3 = A\cos(\Omega t) \tag{5.14.8}$$

(5) *Area-preserving (undamped) hard-polynomial oscillators*;

$$\ddot{x} + x + x^3 = A\cos(\Omega t). \tag{5.14.9}$$

A systematic understanding of the response of these various oscillators to strong periodic forces does not presently exist. We can, however, begin by noting that the response of these oscillators fall into several general categories:

(A) *Period solutions*; involving harmonics, subharmonics, or ultraharmonics of the applid frequency;

(B) *Almost periodic solutions*; which are represented by a Fourier series involving incommensurable frequencies;

(C) *Several coexisting (multi-stable) periodic solutions*; *Nonperiodic and highly unstable solutions*; the latter solutions belong to '*chaotic' families of solutions*. They may, or may not be attractors. [Note carefully: individual solutions, $x(t)$, are deterministic and smooth (at least (C^1); it is a family of solutions which can be characterized as 'chaotic', as we will see].

The historic study of forced oscillators dealt exclusively with the dynamics in categories (A) and (B), some of which have been discussed in previous sections. Most of these topics are exhaustively discussed in many fine books (see references), and will not be further pursued extensively here. Our interests will largely focus on the transitions from the solutions of type (A) and (B) to those of type (C). However, it should be made clear that much remains to be learned in this area – which, of course, makes it all the more interesting!

The original impetus which led to the discovery of solutions of type (C), was an experimental study by van der Pol and van der Mark (1927). This study led to a remarkable chain of subsequent research activities. The electric circuit which they used is shown in the figure below. Here Ne represents a neon glow lamp (i.e., a discharge tube), E a battery ($\simeq 200$ volts), R a resistor (several megaohms), and an applied EMF ($E_0 \simeq 10$ volts). In the absence of the EMF the period of the system increases with increasing capacitance, C. They discovered three effects, but only one was of primary interest to them, namely the production of forty or more subharmonics (Ω/n) of the applied frequency, Ω. Moreover, these subharmonics were found to be *entrained*, in the sense that the frequency of the system remained fixed while the capacitance (and hence the natural frequency) was varied over a limited range. As C was further varied, the frequency changed discontinuously to another subharmonic, as illustrated in their figure (Fig. 5.111).

Equally remarkable was their observation of bands of 'noise' in the regions of many transitions of the frequency, which they regarded as 'a subsidiary phenomena'. Also

Fig. 5.111

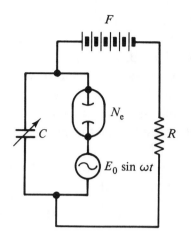

Fig. 5.112

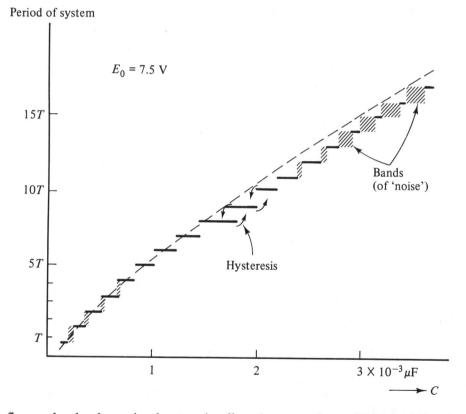

their figure clearly showed a hysteresis effect, hence a dynamical bistability in the system, about which they made no comment. (Fig. 5.112).

While much of the recent theoretical research concerning strongly forced nonlinear oscillators is based on computer studies, the experiment of van der Pol and van der Mark apparently sparked the early analytical analysis of Cartwright and Littlewood

(1945), and of Levinson (1949). The details of these studies are quite complicated, and do not make for easy reading. However their research was the first to show that the abstract chaotic dynamics, which Birkhoff had described much earlier (1922), could indeed occur in simple physical systems (the forced van der Pol oscillator). Therefore their research is of great physical importance and warrants the required effort. A few more details of their research are given in Chapter 6 and Appendix J. Here an abbreviated discussion of their results, and the more recent results of Levi (1981), will be described.

The studies of Cartwright and Littlewood were based on the classic van der Pol equation, (5.14.1), which they wrote in the form

$$\ddot{x} + k(x^2 - 1)\dot{x} + x = b\Omega k \cos(\Omega t) \qquad (5.14.10)$$

where k is large. They casually noted that, if $b > \frac{2}{3}$ and $k > k_0(b, \Omega)$, (15.14.10) has a stable periodic solution of period $2\pi/\Omega$, to which all trajectories converge as $t \to +\infty$ (a globally stable limit cycle). The first published proof of this was apparently some twenty-five years later (N.G. Lloyd, 'On the Non-autonomous van der Pol Equation with Large Parameter', *Proc. Camb. Phil. Soc.*, **72**, 213–27, 1972).

However, the more interesting parameter region is when $b < \frac{2}{3}$, and k is large. Briefly stated, what they first established is that, in this range of the parameter b, there are two sets of alternating intervals, $\{A_k\}$ and $\{B_k\}$, separated by small gaps, as illustrated in Fig. 5.113. When b is in any interval A_k, the equations have a stable and an unstable

Fig. 5.113

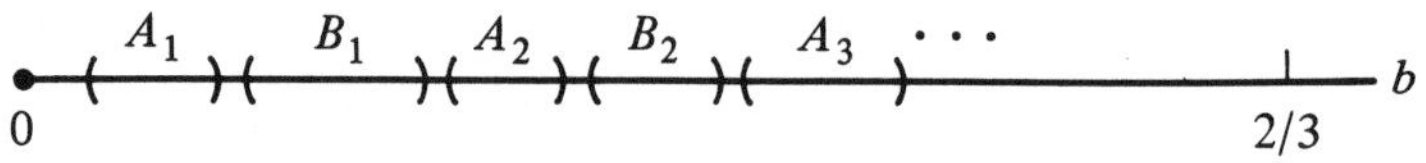

subharmonic solution, $\omega = \Omega/(2n(k) + 1)$, where $n(k)$ is an integer of order k. Moreover these solutions are the limit cycles of most solutions (as $t \to +\infty, -\infty$, respectively).

The more interesting case, however, is when $b \in B_k$, in which case there are two stable subharmonic solutions, for each b, with periods $(2n(k) + 1)T$ and $(2n(k) - 1)T$, where $T = 2\pi/\Omega$ is the period of the applied force, (5.14.10). We will call these two solutions P_1 and P_2. Almost all solutions tend to either P_1 or P_2 as $t \to +\infty$. There are also two unstable periodic solutions with these periods. The stable periodic solutions turn out not to be the most 'exciting' solutions in the parameter intervals $\{B_k\}$, but they are obviously important as $t \to +\infty$. So let us first consider these solutions, to better appreciate the more spectacular properties of the dynamics of other solutions in these parameter intervals.

An extensive discussion of many of these regular solutions can be found in Hayashi's classic book (reissued, 1985). More recently Flaherty and Hoppensteadt (1978) made a numerical study of the subharmonic solutions of the system

$$\varepsilon\dot{x} = x - x^3/3 + y, \quad \dot{y} = -\varepsilon x + b\cos(t). \qquad (5.14.11)$$

This system is related to (5.14.10) by setting $\Omega = 1$ and $k = 1/\varepsilon$.

They obtain the *rotation number* $\rho \equiv \omega/\Omega$ $(\Omega = 1)$ of the subharmonics, as a function of ε and b. Recall (Section 8) that, for large k, the period of such a free relaxation oscillator (5.14.10) (that is, when $b = 0$) is given by (5.8.17)

$$T = (3 - \ln 4)k + 7.014k^{-1/3} + \frac{1}{3k}\ln k - 1.3246/k + 0(k^{-4/3}) \quad (k \to \infty).$$

Since the forcing period in (5.14.11) is 2π, the particular subharmonic oscillations, with periods $T = 2\pi n$ at $b = 0$, are expected to approximately occur at (using $\varepsilon = 1/k$)

n	2	3	4	5	6	7	8	9	10	11	12
ε	0.184	0.103	0.073	0.056	0.046	0.039	0.034	0.030	0.027	0.024	0.022

For $b \neq 0$ the regions of *subharmonic entrainment* are indicated in the figure. The rotation number $\rho = \omega/\Omega = 2\pi/T$ of some of the single stable subharmonics are indicated by the numbers in the clear regions (Fig. 5.114). These regions correspond to (some of) the parameter regions $\{A_k\}$ in Fig. 5.113.

Fig. 5.114

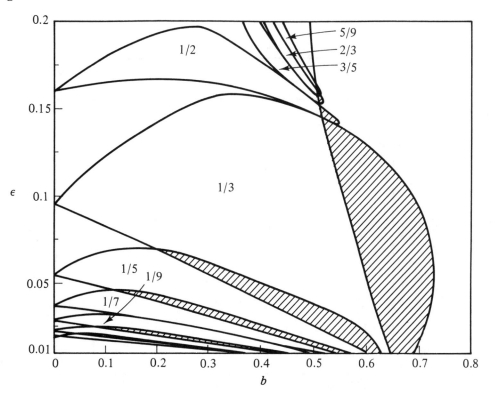

In the case of larger values of ε (which does not correspond to strong relaxation oscillations), we see that there are also ultrasubharmonic solutions, $t = 2\pi n/m$. These do not enter into the Cartwright–Littlewood or Levinson analyses.

The shaded regions in the above figures indicate where two stable subharmonics

Fig. 5.115

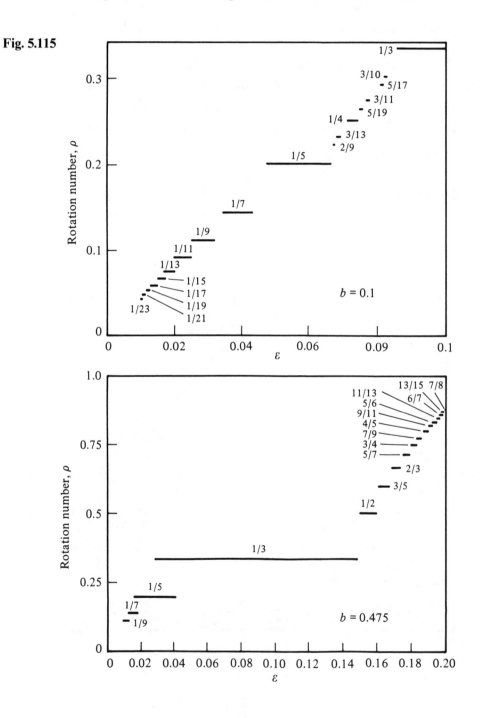

exist. These correspond to the previous parameter regions $\{B_k\}$. The very small gaps which exist between the intervals $\{A_k\}$ and $\{B_k\}$ are not resolved in this figure.

Flaherty and Hoppensteadt also obtained the following figures (Fig. 5.115) illustrating some of the rotation numbers which exist, for fixed values of b, as a function of ε.

Note the overlap of the entrained subharmonics (e.g., $\frac{1}{3}$ and $\frac{1}{5}$, or $\frac{1}{5}$ and $\frac{1}{7}$), corresponding to the shaded regions above. Such subharmonic *bistability*, of course, produces hysteresis effects if ε (or b) is increased and then decreased.

Of course similar 'steps' and hysteresis effects also occur, for fixed ε, as a function of b. Such a hysteresis effect is also apparent in van der Pol and van der Mark's experiment, as noted above.

Levi (1981) pointed out that the hysteresis, caused by varying the parameter b, can be easily understood in terms of the Cartwright–Littlewood and Levinson analyses. Recall that, when $b \in B_k$, there are two stable periodic solutions (P_1 and P_2), whereas there is one is the intervals A_k. Thus, as b is increased from A_k to B_k, the periodic motion has period $(2n+1)2\pi$. If b is further increased to A_{k+1}, the only stable periodic solution has period $(2n-1)2\pi$, because $n(k+1) = n(k) - 1$. Thus, if b is subsequently decreased from A_{k+1} back to B_k, it will have period $(2n-1)2\pi$, thereby exhibiting hysteresis (Fig. 5.116).

Fig. 5.116

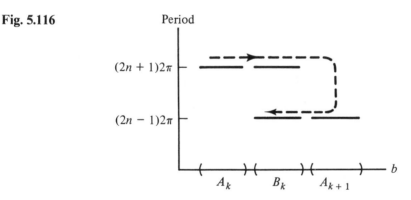

How does subharmonic motion occur in relaxation oscillators? This is illustrated (Fig. 5.117) for the case $\rho = \frac{1}{3}$, where there are three oscillations of the driving force for one oscillation of the system. The reason this occurs is that the force is strong enough to cause the system to reverse its motion and go through a 'loop' along the slow phase of its motion. Recall that this is in the *Liénard plane*, so $y(\neq \dot{x})$ does not change sign when the x-motion reverses direction. Many accurate figures can be found in Flaherty and Hoppensteadt.

The extended phase space, discussed in Section 5.11, is useful for representing these periodic solutions. Thus Fig. 5.118(a) shows the ultrasubharmonic motion, $\rho = \frac{2}{3}$, in the Liénard plane. Now there is only one small 'loop' (which occurs between 1 and 2)

Fig. 5.117

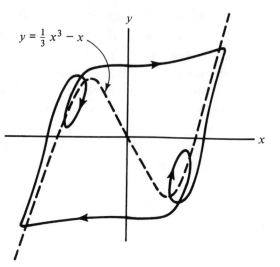

$$y = \tfrac{1}{3} x^3 - x$$

Fig. 5.118

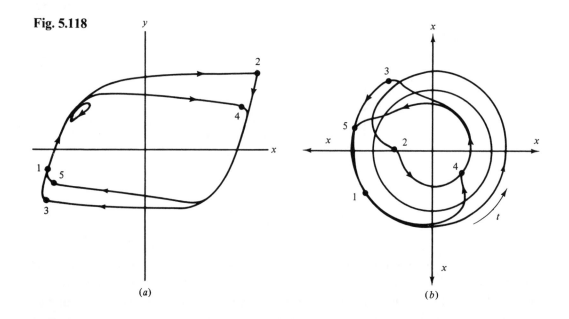

(a)

(b)

within the two larger loops, that are requested for the periodicity. Figure 5.118(*b*) is the same motion when viewed looking down on the extended toroid region ($D \times S^1$). Here *x* is radially outward (everywhere) and the *y*-axis points out of the figure. The toris, which is schematically illustrated in Fig. 5.119, now has a squared-off cross section, characteristic of the relaxation oscillator.

Having considered the periodic solutions which occur, we now turn to the other solutions which exist in the invervals $\{B_k\}$, where two stable subharmonic solutions coexist (P_1 and P_2). Cartwright and Littlewood and also Levinson established that there also exists two unstable periodic solutions with the the same period. To 'observe' them, we must run time backwards (on the computer, of course!). More unexpected, and interestingly, is that there exists a family of solutions, called F by Levinson, whose members can be uniquely associated with any doubly infinite binary sequence

$$\{\ldots, +1, -1, +1, +1, -1, +1, -1, -1, -1, \ldots\}. \tag{5.14.12}$$

In other words, the family F, has members which can be associated with any infinite sequence of coin tosses (a *Bernoulli sequence*). As was discussed in some detail

Fig. 5.119

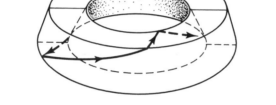

in Chapter 4, Section 5, such an association means that this *family of solutions* warrants being called '*chaotic*'. In the present case of continuous flows, as contrasted with the maps of Chapter 4, the solutions in the family F can all be found in a small neighborhood of the phase space. To repeat:

> *There is a family of solutions with nearly the same initial conditions, which can produce behaviors that are as dissimilar as any infinite sequence of coin tosses.*

Thus the behavior of the members of F is 'unpredictable' in the sense that a small uncertainty in the initial conditions can yield an entirely different behavior of $x(t)$.

A member of this chaotic family looks something like the (very) schematic illustrations in Fig. 5.120, one for x vs. t (a), and the other in phase space (b). The connection between such solutions and the Bernoulli sequence (5.14.12) was established by Levinson, using the piecewise-linear form (5.14.2), where Φ is illustrated (recall the discussion in Section 8). He considered the times, $\{t_k\}$, when a solution crosses $x = +1$ (at t_i) and next crosses $x = -1$ (at t_{i+1}), where the points $x = \pm 1$ arise

Fig. 5.120

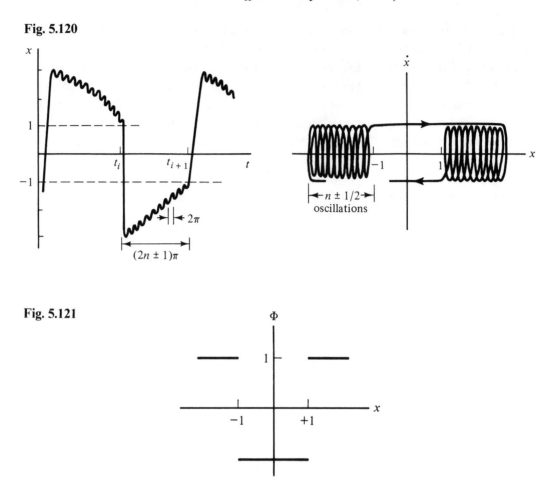

Fig. 5.121

from the definition of $\Phi(x)$, Fig. 5.121. Two such crossings are illustrated in Fig. 5.120. He showed that one can find a solution such that $(t_{i+1} - t_i) = (2n + \sigma_i)\pi + O(\varepsilon)$ (note, $\pi = T/2$), for all i and where σ_i can, for each i, arbitrarily be assigned the value $+1$ or -1. That arbitrary assignment of ± 1 means that these solutions, which are members of F, can be associated with an arbitrary Bernoulli sequence (5.14.12).

The existence of such a highly unstable (chaotic) family of solutions would be remarkable, but largely of only academic interest, if it were not for the fact that all solutions of the equations (5.14.2) are attracted toward the region of phase space (as $t \to \infty$), which contains this chaotic family. While most solutions ultimately tend to the periodic solutions P_1 and P_2 (described above), they do so in a very irratic fashion, since they must 'pass through' the region containing the chaotic F-family. It is difficult, at best, to try to picture such dynamics, and it is wise to recall the fact that computer-generated 'solutions' can likewise only exhibit such features to a limited degree (depending on their roundoff errors). We therefore again see the need to consider the relationship between mathematics, computations, and empirical sciences, discussed in Chapter 4, Section 11.

The representation, or attempted representation, of the chaotic family of solutions is best done with Poincaré maps on various transverse sections of the toroidal space. These representations will be discussed in the next chapter, which concerns second order difference equations (maps in R^2).

However, another characterization of this chaotic family is made particularly clear by Levi's (1980) analysis, and has also been noted and illustrated by Abraham and Shaw (1983). The source of chaos, as in one-dimensional maps (see Chapter 4, Section 5), is due to a '*folding-and-stretching*', or 'kneading' action between nearby states as time progresses. This repeated *kneading action* in phase space is the same action used on bakery dough or a taffy-pull, in order to thoroughly mix the material (read here: thoroughly mix the states).

This *kneading action* of a periodically forced relaxation oscillator is illustrated in

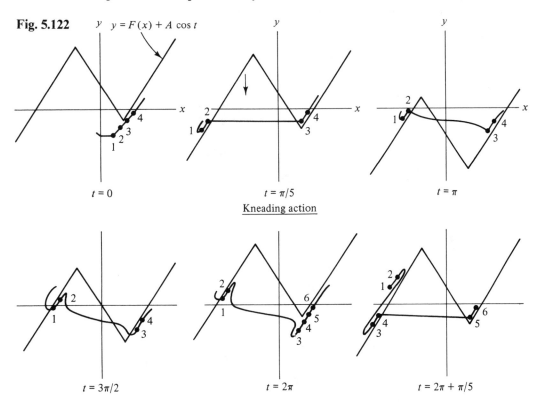

Fig. 5.122

$y = F(x) + A\cos t$

Kneading action

$t = 0$ $t = \pi/5$ $t = \pi$

$t = 3\pi/2$ $t = 2\pi$ $t = 2\pi + \pi/5$

Fig. 5.122, where we use the representation

$$\dot{y} = -\varepsilon x, \quad \varepsilon\dot{x} = y - F(x) - A\cos(t).$$

We consider a 'string' (continuum) of initial states ($t = 0$), a few of which are numbered

$(1,2,\ldots)$ for reference. When ε is small, the states tend to stay very close to the curve $y = F(x) + A\cos(t)$, unless they 'fall off' the sharp corners, whence they make a rapid excursion to the other side (e.g., states 1 and 2 in Fig. 5.123). Because the curve $y = F(x) + A\cos(t)$ varies up and down more rapidly in the Liénard space than the slow dynamics along the vertical portions of the curve, the oscillation produces the kneading action which is shown (after Levi, 1981)

Figure 5.122 does not do justice to the effectiveness of such kneading action, because many more folds can be produced on the 'slow slopes' of $y = F(x) + A\cos(t)$ for the higher order subharmonics. Moreover, of course, a folded surface itself becomes folded onto another folded surface at later times. Other more detailed representations of this action can be found in Abraham and Shaw (Part Two, 1983) and in Levi (1981).

It is rather amusing, even if not too informative, to attempt to represent two periodic solutions P_1 and P_2 with periods $2\pi/3$ and $2\pi/5$ which are 'clothed' in a chaotic family of trajectories. The result looks somewhat like a (Cantor!) fuzzy lady's hat, (see Fig. 5.123) from around 1910. Another (more profound) perspective of this behavior will be given in Chapter 6.

Fig. 5.123

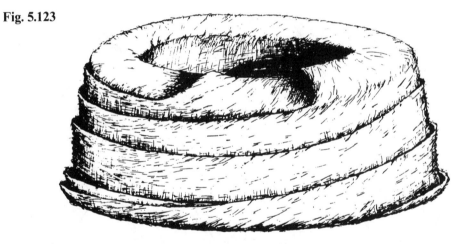

While this torus structure does exist, it is apparently very difficult to observe in numerical or analog solutions of the equations of motion. However, R. Shaw (1981) discovered a variant of the forced van der Pol equation which exhibits this folding action very vividly. Shaw's equations are

$$\dot{x} = -y + F\cos(\omega t)$$
$$\dot{y} = kx + u(a - x^2)y. \tag{5.14.13a}$$

By scaling $(y \to (ka)^{1/2}y,\ x \to a^{1/2}x,\ t \to k^{-1/2}t)$ these may also be put in the three-parameter form

$$\dot{x} = -y + A\cos(\Omega t)$$
$$\dot{y} = x + \mu(1 - x^2)y \qquad\qquad (5.14.13b)$$

where $A = F(ka)^{-1/2}$, $\mu = au\, k^{-1/2}$, and $\Omega = \omega k^{-1/2}$. The effect of using (5.14.13), rather than the usual forced form

$$\dot{x} = -y, \quad \dot{y} = x + \mu(1 - x^2)y + A\cos(\Omega t)$$

was discovered by Shaw when he moved a connection on an analog computer. Note that the 'force' in (5.14.13) acts to change the 'position', x, rather than the 'velocity', y (in the usual mechanical interpretation of these equations). A simple physical example of such an action remains to be discovered. Shaw (1981) reported a number of 'strange attractors' (a concept to be discussed below), as well as indicating the importance of the above folding action, which can be clearly seen by using (5.14.13). Many other results have been imaginatively illustrated by his brother, C. Shaw, in Abraham and Shaw (1983).

Rather than the tight folding structure, illustrated above by Levi (for the family of chaotic solutions), now most solutions tend to a 'surface' (fractal) on which the folds are 'macroscopic', and can be relatively easily viewed. However, this structure is most clearly seen with the help of Poincaré maps, which will be discussed more fully in the next chapter. If the trajectory of a solution of (5.14.13) is simply plotted in the (x, y) phase space, it might look as illustrated in Fig. 5.124. ($A = .932$, $\mu = 1.18$, $\Omega = 1.86$). It

Fig. 5.124

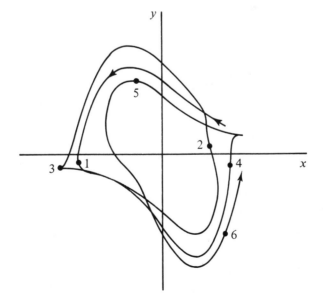

can be seen that this trajectory forms 'tip loops' in the first and third gradient, and that these disappear to yield a smooth curve, only to return again in subsequent cycles. Because of the chaotic crossings in the phase space, the structure of Shaw's strange attractor is best appreciated in the extended phase space, using various Poincaré sections; that is, in plane sections perpendicular to the time axis (these are Poincaré maps – see Chapter 6). This clearly shows that Shaw's 'torus-attractor' has more macroscopic (observable) structure than the Levinson-Levi torus. This process is schematically illustrated in Fig. 5.125 (also see Abraham and Shaw, 1983). The actual

Fig. 5.125

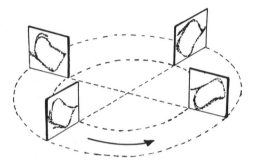

Fig. 5.126

form of these maps on various Poincaré sections will be shown and discussed in the next chapter. These intersections are joined in a schematic manner in Fig. 5.126. This represents a fractal-torus which contains regions where folds are formed (cusp-catastrophe structures), then stretched, and finally compressed back onto the 'torus'. This stretch-fold-squeeze operation of the flow in the extended phase space is also the basic way that autonomous flows in R^3 produce strange attractors, as we will see in Chapter 7.

Returning now to the list of equations at the beginning of this chapter, the equations (5.14.3) are known as the *Brusselator equations* (Prigogine and Lefever, 1968), after the city of their birth. These equations also have a limit cycle behavior when $A = 0$, and had their physical origin in the study of *chemical oscillations*, but they imply the presence of trimolecular reactions (the terms yx^2), which is uncommon. They have been numerically studied in the forced context ($A \neq 0$) by Kai and Tomita (1979) using $a = 0.4$, $b = 1.2$, and varying (A, Ω). The dynamics of this system is very rich, and contains subharmonic, ultrasubharmonic, and 'chaotic' regions of (A, Ω). This dynamics is also most readily analyzed by using mapping methods discussed in Chapter 6. The relationship between this autocatalytic system and (5.14.1) is not clear at present. We will consider somewhat more realistic models of chemical oscillations in Chapter 7.

Having considered some aspects of forced self-exciting oscillators, we now consider some of the responses of more passive (non-autocatalytic) oscillators, (5.14.4)–(5.14.9). The most classic system is perhaps the Duffing equation

$$\ddot{x} + \mu\dot{x} + x + \varepsilon x^3 = A \cos(\Omega t) \qquad (5.14.14)$$

where $\mu > 0$, and with either hard ($\varepsilon > 0$) or soft ($\varepsilon < 0$) nonlinearities. If $A = 0$ (autonomous case) these two potentials, and their associated global phase portraits (Fig. 5.127) are quite different:

Fig. 5.127

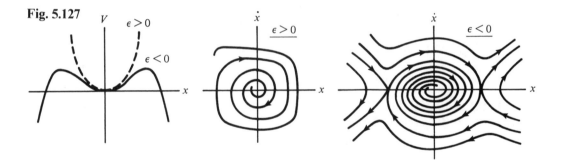

A variation of this is the 'nonharmonic' Duffing equation

$$\ddot{x} + \mu\dot{x} + x^3 = A \cos(\Omega t) \qquad (5.14.15)$$

which is missing the linear restoring force. For large amplitude motion (e.g., large A), (5.14.15) is essentially the same as (5.14.14), with $\varepsilon = 1$, because the harmonic force presumably has little influence. The degree to which that is true for finite A is not obvious, however, as we will see (see Exercise 5.42). The potential and autonomous phase portrait ($A = 0$) now look like those shown in Fig. 5.128. Such nonharmonic oscillators have been briefly considered in Exercise 5.32.

Most of the systems (5.14.4)–(5.14.9) are special cases of

$$\ddot{x} + \mu\dot{x} + \alpha x + \varepsilon x^3 = A \cos(\Omega t), \qquad (5.14.16)$$

Fig. 5.128

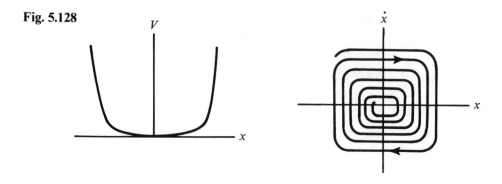

where $(\alpha = 1, 0, -1)$ respectively. Indeed we note that, by the scaling of x and the time, it is always possible to set both

$$\alpha = 0, \text{ or } +1, \text{ or } -1 \quad \text{and} \quad \varepsilon = 0, \text{ or } +1, \text{ or } -1 \qquad (5.14.17)$$

in (5.14.16). This means that, after the selection (5.14.17), this system still has three control parameters (e.g., μ, A, Ω). It is not surprising therefore that the dynamic possibilities are very rich and not well understood at present.

Writing (5.14.16) as the pair of equations

$$\dot{x} = y, \quad \dot{y} = -\mu y - \alpha x - \varepsilon x^3 + A \cos(\Omega t) \qquad (5.14.18)$$

we note they are invariant to the transformation

$$x' = -x, \quad y' = -y, \quad t' = t + (\pi/\Omega). \qquad (5.14.19)$$

Hence, if $(x_1(t), y_1(t))$ is a solution of (5.14.18), then so is $x_2(t) \equiv -x_1(t + \pi/\Omega)$, $y_2(t) = -y_1(t + \pi/\Omega)$. Of course it may be that $x_1(t) \equiv x_2(t)$, in which case the orbit has a reflective symmetry in phase space, but this is not necessarily the case (see below).

Certainly the least complicated autonomous phase portrait is for the 'nonharmonic' Duffing equation (5.14.15). This has been extensively studied by Ueda in the more restricted form

$$\ddot{x} + k\dot{x} + x^3 = B \cos(t). \qquad (5.14.20)$$

In terms of the parameters in (5.14.15) this implies an interrelation $\mu = k\Omega$, $A = B\Omega^3$ (with $t \to \Omega t$, $x \to x/\Omega$). This unfortunately differs from many studies where A and μ are fixed, while Ω is continuously varied (Section 12), which makes it difficult to relate his results to other studies. However Ueda (1980) has presented many interesting phase space pictures of steady state trajectories, and has generalized the above right side to $B_0 + B_1 \cos(t)$, which no longer is invariant under (5.14.19).

Of particular interest from his 'picture book' are examples of *multistable* (not just bistable) periodic solutions shown in Fig. 5.129. The mark on the trajectories corresponds to $t = 2\pi n$, which fixes the phase of the force. These five trajectories, all at

Fig. 5.129

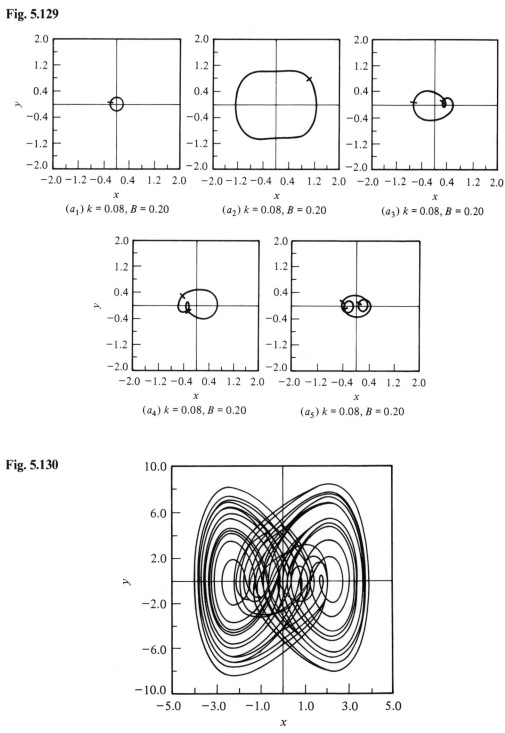

(a_1) $k = 0.08$, $B = 0.20$ (a_2) $k = 0.08$, $B = 0.20$ (a_3) $k = 0.08$, $B = 0.20$

(a_4) $k = 0.08$, $B = 0.20$ (a_5) $k = 0.08$, $B = 0.20$

Fig. 5.130

(a_1) $k = 0.10$, $B = 12.0$

$k = 0.08$, $B = 0.2$, show two cases with $\omega = \Omega(=1)$, but with different amplitudes. Also shown are two cases, $(a_3), (a_4)$, at $\omega = \Omega/2$ (subharmonic of order 2) which are related by the invariance property, and one subharmonic $\omega = \Omega/2$, (a_5). Clearly the situation is much richer than a simply hysteresis region, represented by a cusp catastrophe surface (Section 12).

For $15 \gtrsim B \gtrsim 5$ Ueda also found regions of 'chaotic' motion, illustrated in Fig. 130.

Exercise 5.42. As noted before, it is much more fun and instructive if you explore some of the dynamics yourself:

(a) Use a simple Runge–Kutta program to integrate (5.14.20), and recover Ueda's results in the above figures. You can obtain the approximate initial conditions from the marked locations on the figures (the damping will attract you to the nearest limit cycle). How would you determine the *basins of attraction* of these various limit cycles?

(b) Now do some (unpublished) exploration, by (slightly!) increasing the linear coefficient (α in (5.14.16)) from zero, thereby making (5.14.20) more 'generic'. You will find a dramatic influence on the multistable character of the solutions. Many features of the standard Duffing oscillator, (5.14.14), have been computed by Parlitz and Lauterborn (1985).

In contrast with the forced van der Pol oscillator, there exists no mathematical analysis which establishes the precise character of the 'chaotic' solutions of (5.14.20), nor of many examples which follow. However the evidence indicates that the limiting trajectories, towards which most solutions tend as $t \to +\infty$ (that is, their ω-limit set (s); Appendix *A*), are not periodic orbits. In contrast to the strong autocatalytic action of the van der Pol oscillator, which drives essentially all solutions to stable limit cycles (one, or two), the damping action of (5.14.20) apparently only insures that most solutions are attracted to a limited region of phase space. It apparently does not always force them toward periodic behavior, at least not with periods of detectable length.

To be slightly more precise, we will mean by the *attractor A* of a flow, f^t, an invariant ($f^t A = A$) compact set of points in the phase (or extended phase) space, which is the ω-limit set of almost every point in some neighborhood of *A*. In other words, there is some neighborhood, *N*, of *A* such that $\bigcap_{t \geq 0} f^t N = A$. The larger set of points, $B \supset N$, which all have *A* as their ω-limit set (i.e., $B = \{x \mid f^t x \to A \text{ as } t \to +\infty\}$) is called the *basin of attraction* of *A*. The simplest attractors are fixed points, or limit cycles. Not all points in R^2 need to be in some basin of attraction, as illustrated in Fig. 5.131. Also, more refined concepts of attraction are possible (Ruelle, 1981), but will not be pursued here.

The attractors we are now encountering (in the extended phase space, where solutions are unique) are not as simple as the illustrated examples. Indeed, it is not at all

Fig. 5.131

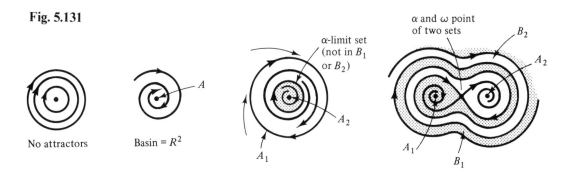

No attractors Basin = R^2

clear what is their 'precise', much less observable, characteristics. The terminology *strange attractor* is presently used by many people in a rather loose and casual fashion to describe attractors, A, with one or more of the following properties:

(SA.I)

The set A is a fractal; that is, it has a fractional dimension. This clearly eliminates all of the above simple attractors.

(SA.II)

The solutions in some attracting neighborhood, N, of A (described above) have positive Lyapunov exponents. In other words, these solutions are 'sensitive to their initial conditions'. The solutions do not tend to uniformly converge in time to the set A.

(SA.I) is a *geometric characterization* of a strange attractor, whereas (SA.II) is a *dynamic characterization*. There are other possible variations on these characterizations, but these give the flavor of some rather commonly accepted meanings of *strange attractors*. It is frequently a matter of taste or circumstance, what meaning is given to the term 'strange attractor'. One should always approach it with some caution, and not read too much into this term. This is particularly true since most properties of these attracting sets are only determined by very limited computations of unknown accuracy.

We note that the van der Pol attractor does not have the properties (SA.I) or (SA.II) when $b \in \{B_k\}$, but it does possess a strange attracting region. We might, therefore, find it useful to focus on the character of the basins of attraction (two or more), rather than the attractor. This leads to the possibly useful distinction of *strange basins*:

(SB) Two or more basins of attraction are strange if some region of their boundary is a fractal.

It appears that the van der Pol attractors (limit cycles P_1 and P_2) have such strange basins, as suggested by Levi's representation of the kneading action in the dynamics, illustrated above. A further discussion of this will be given in Chapter 6, where other (more abstract) explicit examples of (SB) will be illustrated. We simply note here that, if a system has strange basins, then it may be impossible to control initial conditions so as to insure that the asymptotic state is a preselected attractor. In this sense we are back to a single 'flip-of-a-coin-dynamics'.

As we proceed down the list of oscillator systems to (5.14.6)–(5.14.9), we encounter increasingly 'delicate' systems, which are more readily influenced by the applied periodic force. The first of these is the 'inverted' Duffing system:

$$\ddot{x} + \mu\dot{x} - x + x^3 = A\cos(\Omega t), \qquad (5.14.21)$$

which has a negative linear coefficient (similar to an inverted pendulum), but remains bounded because of the cubic restoring force (in contrast to the pendulum). The potential and autonomous phase portrait ($A = 0$) now look like those in Fig. 5.132. It is

Fig. 5.132

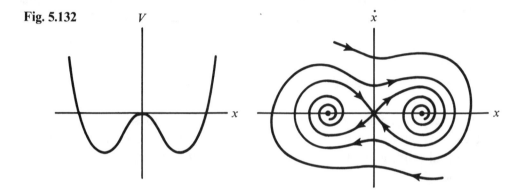

fairly clear from these figures, that this system could easily become chaotic when $A \neq 0$ in (5.14.21), because it can be 'flipped' from one basin of attraction to another. The 'inverted Duffing' oscillator has been studied by Holmes (1979), and also by Moon and Holmes (1979), in a nice experimental study, using a vibrating steel strap between two fixed magnets (see Fig. 5.133). The central position is an unstable equilibrium as in the

Fig. 5.133

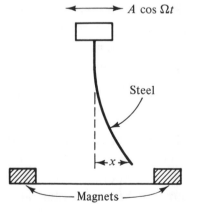

previous figure. As expected, this system is quite capable of exhibiting 'chaotic' motion. A somewhat more complicated oscillator has also been investigated experimentally by Moon (1980). Much of the analysis of this system has been done with the use of Poincaré maps, and this topic is taken up in the next chapter.

Even more delicate systems are the soft-polynomial oscillator

$$\ddot{x} + \mu\dot{x} + x - x^3 + A\cos{(\Omega t)}, \qquad (5.14.22)$$

and the damped pendulum

$$\ddot{\theta} + \mu\dot{\theta} + \sin{\theta} = A\cos{(\Omega t)}. \qquad (5.14.23)$$

The reason for this delicacy is a presence of the separatricies and their associated saddle points. Put more physically, when a pendulum is upside down, it doesn't take much disturbance to drastically change its future course of motion. Thus, once these systems are driven to the 'top of the potential energy hill', a small change can produce a major effect – hence, possible chaos (Fig. 5.134). These systems are in fact so delicate that

Fig. 5.134

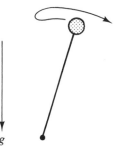

there is not general agreement on many aspects of their chaotic motion.

The soft nonlinear oscillators are much more difficult to study than the hard oscillators, because the solutions can 'escape' from the primary separatrix region ($|x| < 1$ for (5.14.23) and $|\theta| < \pi$ for (5.14.23)) if they are driven too hard. On the other hand, their most interesting dynamics occurs only when they move near the saddlepoint region in the phase space.

An early numerical study of the soft nonlinear oscillator, (5.14.22), was done by Huberman and Crutchfield (1979) on an analog computer, using $A = 0.23$, $\mu = 0.4$, and varying Ω. Their results are schematically illustrated in Fig. 5.135.

As the driving frequency is lowered below the linear resonance frequency ($\omega_0 = 1$), the amplitude rises as predicted by the simple perturbation theory (Section 12), shown in the lower right figure. However with the stronger driving force, as Ω is further decreased, the symmetric mode becomes an asymmetric mode (symmetry-breaking 'bifurcation'), which satisfies (5.14.19), but still maintains the period $T = 2\pi/\Omega$. For smaller Ω, this bifurcates into a period $2T$ mode. They found several period-two

Fig. 5.135

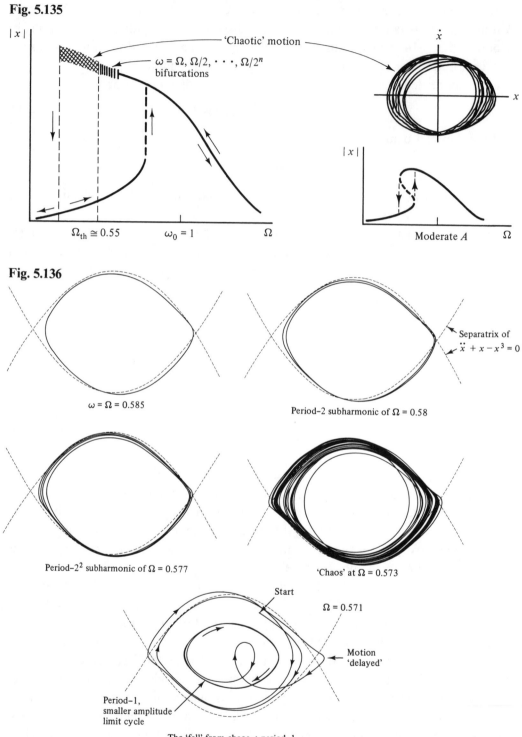

'Chaotic' motion

$\omega = \Omega, \Omega/2, \cdots, \Omega/2^n$
bifurcations

$\Omega_{th} \cong 0.55$ $\omega_0 = 1$ Ω

Moderate A Ω

Fig. 5.136

$\omega = \Omega = 0.585$

Period-2 subharmonic of $\Omega = 0.58$

Separatrix of
$\ddot{x} + x - x^3 = 0$

Period-2^2 subharmonic of $\Omega = 0.577$

'Chaos' at $\Omega = 0.573$

Start

$\Omega = 0.571$

Motion
'delayed'

Period-1,
smaller amplitude
limit cycle

The 'fall' from chaos → period-1

bifurcations, followed by chaotic motion when $\Omega \lesssim 0.55$ (also see Steeb, Erig, and Kunick, 1983; Elgin, Forster, and Sarkar, 1983).

This sequence of events is illustrated in greater detail in the Fig. 5.136, obtained for the case $\mu = 0.5$, and $A = 0.3$. In order to give a meaningful scale to the figures, the separatrix of the undamped, unforced system

$$\ddot{x} + x - x^3 = 0$$

has been superimposed. The first figure ($\Omega = 0.585$) shows a broken symmetry mode, which passes near a saddle point, but retains the driving period $2\pi/\Omega$. When Ω is lowered to $\Omega = 0.580$, this motion has bifurcated to period-2. Lowering Ω to 0.577 appears to generate period-2^2 motion. When $\Omega = 0.573$, the motion appears quite 'chaotic', as illustrated in Fig. 5.136. It is quite time consuming on a personal computer to search for higher order bifurcations.

The behavior of the forced damped pendulum, (5.14.23), might be expected to have many of the dynamic features of (5.14.22). It should be noted, however, that although we found a limited relationship between these systems in Section 4, when $\mu = 0$ and $A = 0$, no such simple connection now holds. In particular, we note that the force goes to zero for a pendulum when $\theta = \pi$, which is roughly three times larger than the corresponding value, $x = 1$, for (5.14.22). Thus, for the same value of μ, we expect that the comparable value of A in (5.14.23) should be roughly three times the value of A in (5.14.22). The bifurcation frequencies would, of course, only be expected to be roughly the same. By way of illustration, Fig. 5.137 shows the period-two subharmonic motion for $\mu = 0.5$ and $A = 1$, now at $\Omega = 0.63087$. Once again a separatrix ($\frac{1}{2}\dot{\theta}^2 - \cos\theta = 1$) is

Fig. 5.137

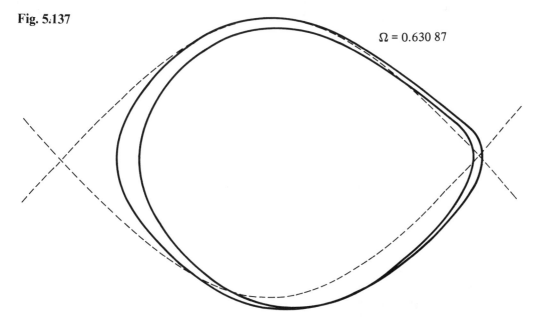

$\Omega = 0.63087$

Fig. 5.138

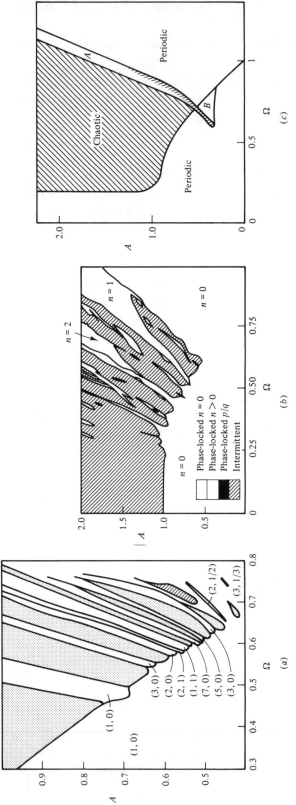

superimposed, for reference. The 'split' in the orbit is more pronounced than in the case of (5.14.22), but there is no essential difference. Lowering Ω to 0.625 is found to produce 'chaotic' features.

The complexities involved in numerically establishing the entrainment (phase-locking) and 'chaotic' features of a forced pendulum, (15.14.23), are illustrated by Fig. 5.138 on this subject by different investigators.

Figure 5.138(a) is based on Pedersen and Davidson's (1981) findings for the case $\mu = 0.2$. A number of subharmonic responses, Ω/n, are illustrated by shaded 'stripes', together with a 'chaotic' region, which is hatched (a number of their details have been omitted here, for simplicity).

Figure 5.138(b) is based on the analog computer results of D'Humieres, Beasley, Huberman and Libchaber (1982), now for $\mu = 0.25$. The phase-locked (entrained) regions are now the clear regions, with 'intermittent' regions being hatched.

Figure 5.138(c) is based on the numerical results by Huberman, Crutchfield and Packard (1980), with a $\mu = 0.5/(6.4)^{1/2} = 0.1976$. In the region B the pendulum motion remains in one potential well, and contains the period-doubling bifurcation discussed above.

The similarities and differences in these results, all for roughly the same value of μ, are clear. They illustrate the delicacy required in numerical calculations of such soft systems.

A careful numerical study of (5.14.22), with $\mu = 0.4$ and $A = 0.23$, has been made by Räty, Isomäki and von Boehm (1984). They noted that, for any solution $x(t)$, there either exists another distinct solution $\tilde{x}(t) = -x(t + \frac{1}{2}T)$, where $T = 2\pi/\Omega$, or else $x(t) = -x(t + \frac{1}{2}T)$. The latter case they call inversion-symmetric (IS), whereas the former case implies that $x(t)$ is asymmetric in the phase plane. In the asymmetric case (AS) there are two distinct solutions for the same applied frequency, Ω. In the case $\mu = 0.4$ and $A = 0.23$, they find that there is a symmetry-breaking bifurcation (IS → AS) as Ω is decreased through the value $\Omega \simeq 0.535$. Such a symmetry-breaking bifurcation is also found in the autonomous Lorenz system, where it will be shown that this generates linked stable limit cycles in R^3 (Section 7.9). The analogous dynamics here, of course, is the extended phase space, $R^2 \times T^1$. They found that, for twenty bifurcations (period doublings, windows with their bifurcations and chaotic states) in the range $0.53 > \Omega > 0.5268$, all of the states are asymmetric, and hence double-states. Rather remarkably they found that, as Ω is decreased through 0.5268, the two asymmetric chaotic states merge into one inversion-symmetric chaotic state. For six subsequent bifurcations below this frequency ($\Omega > 0.5252$), they found a return to only inversion-symmetric states. They also found that the periodic windows occur in the sequence 6, 5, 3, 8, 7, and 5, which follows the universal ordering of one-dimensional maps (Metropolis, Stein and Stein, 1973).

Following Räty, Isomäki, and von Boehm, if we change the x-scale by a factor of 2, so that (5.14.22) becomes

$$\ddot{x} + \mu\dot{x} + x - 4x^3 = f\cos(\Omega t). \tag{5.14.25}$$

They computed the solid curve in Fig. 5.139 from the dynamics, whereas (5.14.25) is

Fig. 5.139

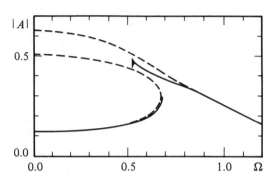

represented by the dashed curve. The twenty-eight bifurcations that they found on the upper curve lie between $0.53005 > \Omega > 0.525256$, which is the small diamond-shaped region in their figure. This width represents either the asymmetric broadening of $x(t)$, or its range in symmetric cases. Thus Fig. 5.140(a) is one of the two AS chaotic attractor at

Fig. 5.140

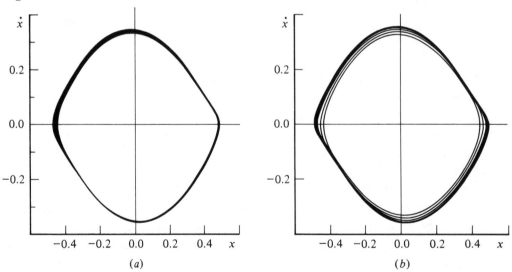

(a)   (b)

$\Omega = 0.527$, whereas Fig. 5.140(b) shows the IS period-5 attractor at $\Omega = 0.526$. In the present x-scale, the hyperbolic points are at $x = \pm 0.5$.

The last group of forced oscillators we will briefly consider are the conservative, area-preserving systems. Because such oscillator systems are neither damped nor autocatalytic, they do not produce attracting orbits when forced, and hence do not give rise to strange attractors. They can, however, have very chaotic dynamics of a different form. Indeed, conservative systems respond more strongly to applied forces than do the damped or autocatalytic systems. Conservative systems are quite 'delicate', but they have sets of 'tough orbits', whose periodic character is not changed by sufficiently weak forces. These resilient sets of orbits are called KAM *sets*, after Kolmogorov, Arnold, and Moser, who established their existence under suitable conditions. These sets will be encountered frequently in the following chapters.

The trajectory of a conservative oscillator can be represented on the surface of a torus, T^2, in the extended phase space associated with any period $2\pi/\Omega$. All the trajectories on a given torus differ in phase but have the same period, $2\pi/\omega$, since they do not intersect. Since the frequency, $\omega(E)$, generally depends on the energy, different tori usually have trajectories with different periods. The trajectories on a torus are closed only if $\omega(E) = m\Omega/n$, for some integers (m, n). Otherwise the trajectory ergodically covers the torus. When a periodic force, Ω, is applied to this system, these different solutions may respond very differently, depending on their phase and frequency relationship to the applied force. Generally we might expect different solutions, all originally on one torus (one ω) to 'go their separate ways', yielding a 'tangled-torus', schematically illustrated in Fig. 5.141. That, indeed, is frequently the

Fig. 5.141

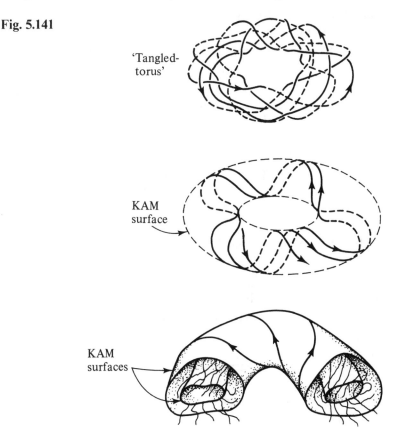

'Tangled-torus'

KAM surface

KAM surfaces

case. However, among the ergodically covered tori, there are special 'resilient sets' of KAM *orbits* whose tori do not shatter in this fashion, but instead only become distorted in shape. These sets are called KAM *surfaces* or KAM *tori*. When the force is weak, the KAM-tori and tangled-tori occupy alternating regions in this space, which, however, is difficult to depict. The description 'tangled-torus' is, of course, only picturesque, for this set of orbits does not lie on any smooth manifold. Moreover these sets occur densely in the extended phase space, and not with large gaps between KAM surfaces, as depicted in the lower figure. A much more accurate account of this structure in the phase space can be given with the help of Poincaré maps, which we will consider in Chapter 6.

5.15 Experimental Poincaré (stroboscopic) maps of forced passive oscillators

The real world does not, of course, necessarily behave like the standard simple models enjoyed by theoreticians. One objective of experimentalists is to determine the common, and dissimilar features between physical systems and these various dynamic models, with the hope of discovering what universal dynamic aspects may exist in apparently diverse physical systems. In this section we will briefly consider a few mechanical and electrical systems which have been studied.

One system, which has been studied by several experimentalists, is the simple RLC circuit shown in Fig. 5.142, with the nonlinear element being a varactor diode. Under

Fig. 5.142

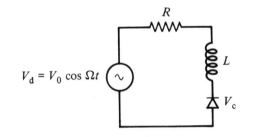

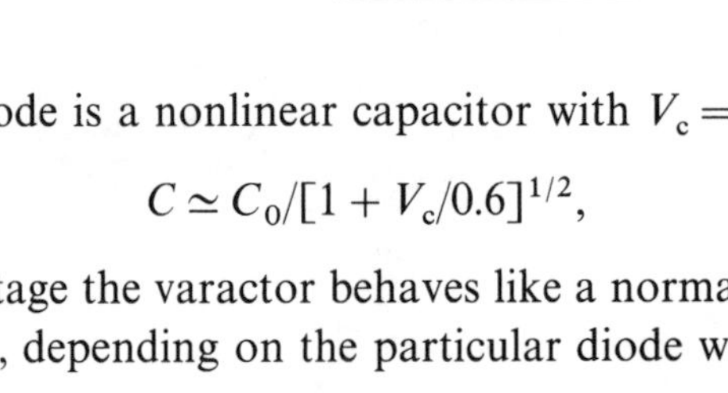

reverse voltage, this diode is a nonlinear capacitor with $V_c = q/C$, where

$$C \simeq C_0/[1 + V_c/0.6]^{1/2},$$

and under forward voltage the varactor behaves like a normal conducting diode (see Fig. 5.143). In addition, depending on the particular diode which is employed, there

Fig. 5.143

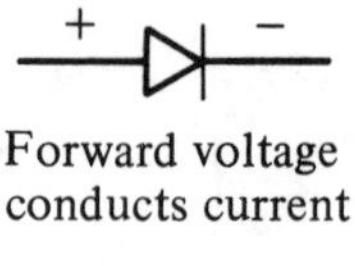

$$+ \vartriangleright\!\!\!|\ -$$

Forward voltage
conducts current

Fig. 5.144

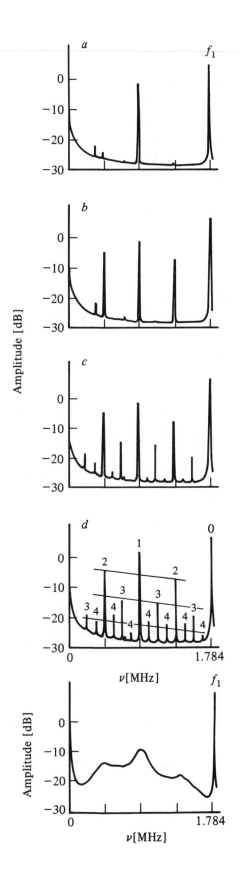

may be other physical effects which play important roles in the dynamics of such systems. Thus such systems are not necessarily simple analog computers of known equations. It might be remarked that many nonlinear 'analog computers' may likewise not be what they are intended to be.

Linsay (1981) found that, at a low drive voltage, V_0, the RLC circuit behaved linearly, with a resonance at a frequency $f_1 = 1.78$ MHz. As the voltage V_0 was increased at this frequency, the usual harmonic production was first observed. However, when $V_0 \simeq 1.9$ volts, the first subharmonic, $f_1/2_D$ was observed. Further increasing V_0, he found a sequence of period doubling bifurcations, $f_n = f_1/2^n$ ($n = 1, 2, 3, 4$) leading to a 'chaotic' dynamics. This is illustrated in the voltage spectra in Fig. 5.144. Using the bifurcation

Fig. 5.145

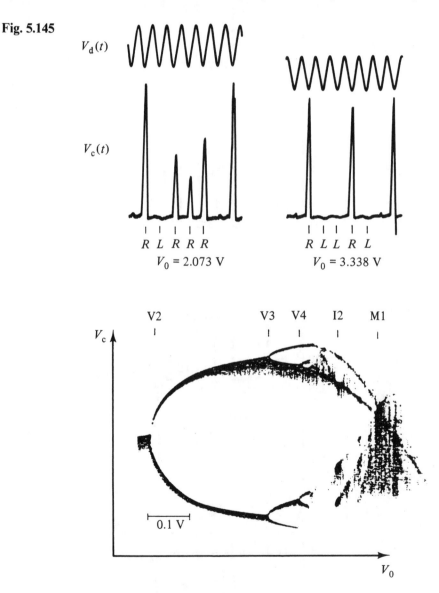

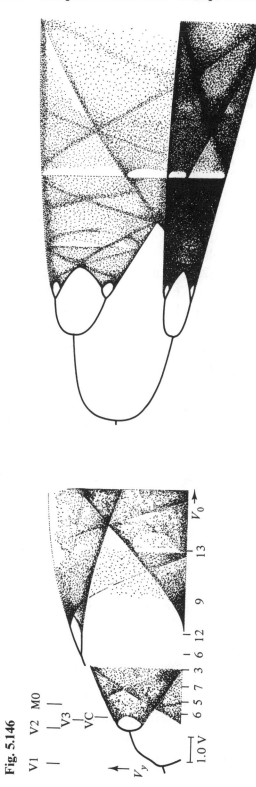

Fig. 5.146

values, V_{on}, he found that $(V_{on+1} - V_{on})/(V_{on+2} - V_{on+1}) \equiv \delta_n$ gave $\delta_1 = 4.4 \pm 0.1$, $\delta_2 = 4.5 \pm 0.6$ which may be compared with Feigenbaum's value $\delta = 4.669\ldots$ (see (4.3.10)). The 'fully developed chaotic spectrum' in the figure is not white noise, but is enhanced at $f_1/4$, $f_1/2$, and $3f_1/4$. Linsay also found other (e.g., period tripling) bifurcations for larger V_0, and other chaotic regions. These features indicate that more subtle physical processes may be important in such systems.

Testa, Pérez and Jeffries (1982) used essentially the same circuit as Linsay. They likewise recorded a series of bifurcations, but also recorded both the voltage patterns across the varactor diode. The top figure shows two period-six windows observed at different driving voltages. V_0. The observed on/off (or R/L) responses in fact agree with the first two period-six patterns given by Metropolis, Stein and Stein (1973) for rather general maps in R^1 (see Chapter 4). Fig. 5.145 indicates the magnitudes of V_c peaks, as a function of the driving voltage $V_d = V_0 \cos \Omega t$. For a limited range of V_0 they found a series of period doublings (2, 4, 8) and 'windows'. Over a wider range of V_0, they observed bifurcations with periods 6, 5, 7, 3, 6, 12, 9 and 13, as indicated in the lower left figure. While the enlarged portion looks very much like the results obtained from the one-dimensional logistic map, it is clear from the figure that the dynamics cannot be represented by such a simple one-dimensional map. Even more beautiful experimental examples of such bifurcation features have been more recently obtained, as illustrated on the right (Jefferies, 1985) (Fig. 5.146).

Questions have been raised by Rollins and Hunt (1982) concerning the physical origin of some bifurcation mechanisms. More information has been shed on this question by the experiments of Brorson, Dewey and Linsay (1983). Using a similar electrical circuit, they measured the current at each cycle of the driving period, and studied its bifurcation pattern as a function of the applied voltage V_0 (1–10 volts). One of their results is shown in Fig. 5.147(a). A series of cascading bifurcations, beginning with successive subharmonics, each with its own chaotic regions, *window structures*, and *crises*, and indicates a behavior which is significantly more sophisticated than the logistic map. They in fact obtained results indicating a higher dimensional attractor.

They proposed a physical model whose numerical integration yields a bifurcation pattern, shown in Fig. 5.147(b), which is rather similar to the experimental result. The simplest model which they found would produce similar results is

$$Ld^2q/dt^2 + Rdq/dt + V(q) = V_0 \sin(\Omega t) \qquad (5.15.1)$$

where $V(q)$ is shown in Fig. 5.148, and whose physical origin is discussed. The macroscopic 'self-replicating' character of the bifurcation sequences are quite spectacular, and presently unanalyzed.

Mechanical experiments involve their own subtleties which rarely yield dynamics as simple as their 'model' equations. There have been several experiments related to an 'upside-down' pendulum, or 'inverted' Duffing equation. One nice series of experiments

Fig. 5.147

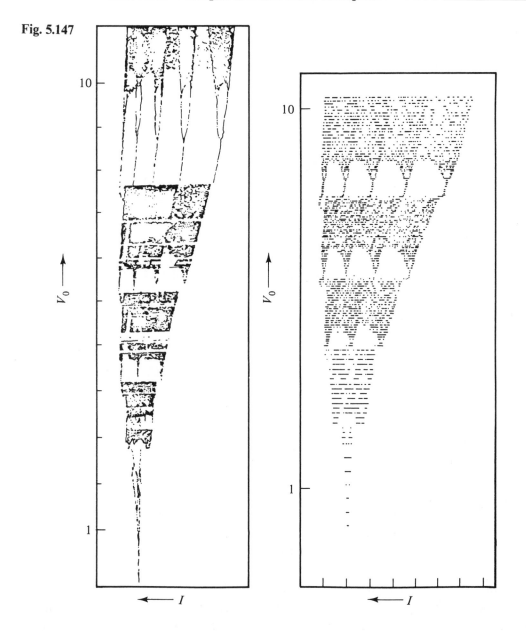

(Moon and Holmes, 1979; Moon 1980) involved a steel strap vibrated between two fixed magnets, as indicated in Fig. 5.149. The central point is an unstable equilibrium, so it has the desired 'inverted' property. The clearest experimental results come from using stroboscopic methods, (Poincaré maps), which are discussed in Chapter 6.

Another 'upside-down' pendulum involved a wheel driven by a motor and torsional spring, as illustrated in Fig. 5.150 (Beckert, Schock, Schulz, Weidlich and Kaiser, 1985). The equilibrium positions were at $\pm 60°$, and the motion was damped by an eddy

Fig. 5.148

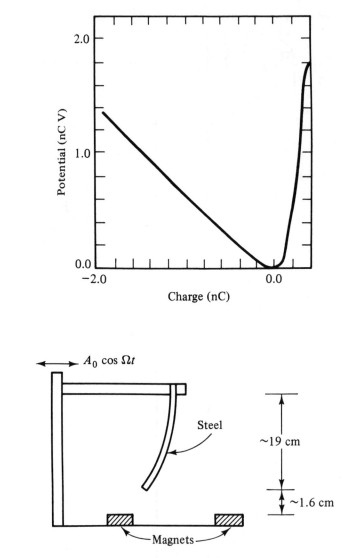

Fig. 5.149

$A_0 \cos \Omega t$

Steel

~19 cm

~1.6 cm

Magnets

current brake (using an electromagnet, controlled by a current I). The richness in the dynamics of this system is illustrated on the right in the bifurcation diagram of frequency vs. maximum ϕ as f is decreased toward $0.32\,\text{Hz}$, one sees the period 2^n bifurcations ($n = 0, 1, 2, 3$), followed by a nonperiodic region (NP). They even indicate a period three window, before the system begins to exhibit its own unique set of bifurcations.

The usual ('right-side-up') pendulum could be studied with similar equipment, or the following torsional pendulum arrangement. An aluminum disk, with an off-center brass disk is suspended by a piano wire (Fig. 5.151). The oscillations are again produced by a stepping motor and eddy current damping can be used.

Fig. 5.150

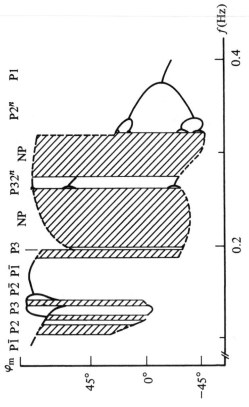

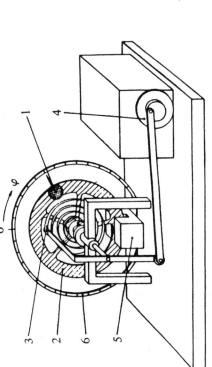

Experimental arrangement: 1. additional mass;
2. turning wheel; 3. torsional spring; 4. stepper
motor with excenter; 5. eddy current brake;
6. dial

Fig. 5.151

Fig. 5.152

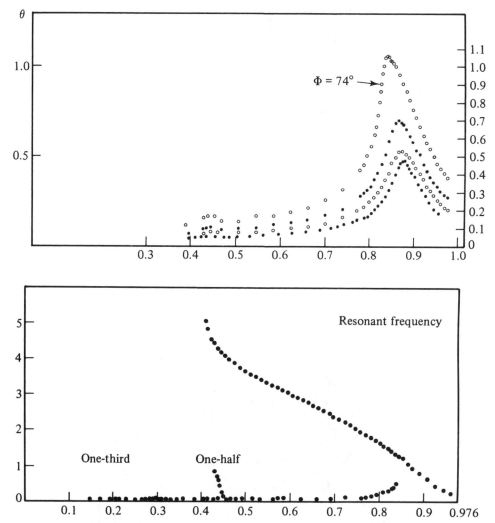

While the hysteresis effect is readily obtained (see the following experimental data, due to M. Zimmer, Fig. 5.152), which even indicates the subharmonic resonances, it is very difficult to control bifurcations, or sustain the chaotic motion. Small vibrations of the mounting and lateral oscillations of the piano wire strongly influence these dynamic features.

5.16 Epilogue

Fig. 5.153

An Oscillator Overview

Passive Oscillators
(e.g., Duffing Oscillator)

Self-Exciting Oscillators
(e.g., van der Pol Oscillator)
↳ Limit Cycles ←

Autonomous

Pertubative Analysis
Averaging Methods
Amplitude dependent
Frequencies (soft/hard)
Nonperturbative Analysis
Phase Plane Analysis
Phase integrals: $T = \oint dx/v$
Separatrices; Homoclinic
Heteroclinic orbits
Inverted Duffing Oscillator
Symmetry Breaking
Global bifurcations
Poincaré Index

Autonomous

Perturbative Analysis
Averaging Methods
Limit cycle dynamics
Nonperturbative Analysis
Liénard phase plane
Relaxation Oscillations-
Rapid release of energy
Fast and Slow Manifolds
Dynamic Catastrophes
Singular Perturbations
Poincaré-Bendixon
Andronov-Hopf bifurcation
Center Manifolds

Nonautonomous (Ω)

Perturbative
Averaging; Hysteresis
Dynamic Bistability
Dynamic Cusp Catastrophe
Harmonic, Subharmonic, $\omega = \frac{m\Omega}{n}$
ultrasubharmonic dynamics
Nonperturbative
Dynamic Multistability
Extended Phase space
Period doubling bifurcations
Conservative oscillators
→ KAM surfaces

Nonautonomous (Ω)

Perturbative
Entrainment
Phase locking
Zeitgebers
Nonperturbative
Entrained subharmonics
Subharmonic hysteresis
van der Pol-van der Mark
Cartwright-Littlewood
Levinson-Chaotic
Family of solutions
⟺ Bernoulli
sequence
Strange Attractors
Strange Basins

Comments on exercises

(5.1) If you set $s_i = s + (-1)^i \varepsilon$, $s = p/2$, and let $\varepsilon \to 0$ you obtain

$$x = x_0 \exp(st) + [\tfrac{1}{2}(a - d)x_0 + by_0]t \exp(st)$$
$$y = y_0 \exp(st) + [\tfrac{1}{2}(d - a)y_0 + cx_0]t \exp(st)$$

Thus the solution is not simply exponential in time.

(5.2) $d/dt(x + y) = (x + y)^2$; Divide by $(x + y^2)$; integrate; $(x + y) = (x_0 + y_0)/(1 - (x_0 + y_0)t)$. Solution only for $t < (x_0 + y_0)^{-1}$.

(5.3)   One possibility is $\dot{x} = px - v$, $\dot{v} = qx$; where $v \equiv d\,x - b\,y$.

(5.4)   We have $\begin{pmatrix} \dot{x}' \\ \dot{y}' \end{pmatrix} = AJ \begin{pmatrix} x \\ y \end{pmatrix} = AJA^{-1} \begin{pmatrix} x' \\ y' \end{pmatrix}$ so $p' = \mathrm{Tr}\ |AJA^{-1}|$ and $q' = \det|AJA^{-1}|$. A straightforward computation verifies that $p' = p$, and $q' = q$.

(5.5)   The artistry is left to you, but two cases should be clear, namely a center and a *special* saddle point (Fig. 5.154). Liouville's theorem requires $\nabla \cdot F = a + d = 0$.

Fig. 5.154

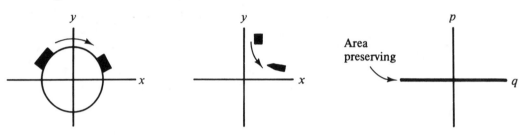

Thus $p = s_1 + s_2 = 0$ (contraction rate equals expansion rate). See Hamiltonian systems at the end of Section 1.

(5.6) $\frac{1}{2}(d/dt)(x^2 + y^2) = x^4 + y^4$, so unstable in all senses. In the second case, $\frac{1}{2}(d/dt)(x^2 + y^2) = 2xy - (x^4 + y^4)$ is stable in the sense of Laplace, because $(x^2 + y^2)$ cannot become arbitrarily large.

(5.7)   If $c > 0$, there is no fixed point and all solutions go to $x = +\infty$ in a finite time. If $c < 0$, there are fixed points at $(x = \pm c^{1/2}, y = 0)$, and these are a saddle and a stable node. At $c = 0$, the fixed point is called a *saddle-node* (see Ex. 5.14).

(5.8) $\dot{x} = -y(a + bx)$, $\dot{y} = cx + \frac{1}{2}by^2$ $(a, b, c > 0)$. Then $K = \frac{1}{2}bxy^2 + \frac{1}{2}cx^2 + \frac{1}{2}ay^2$ is a constant of the motion. One fixed point, $(x, y) = (0,0)$: linearize, $\dot{x} = -ay$, $\dot{y} = cx$, so $p = 0, q = ac > 0$; hence $(0,0)$ is a center. Two other fixed points, $x = -a/b$, $y = \pm(2ac)^{1/2}/b$. Set $x = -(a/b) + u, y = \pm(2ac)^{1/2}b + v$, then $\dot{u} = \mp(2ac)^{1/2}u$ any $\dot{v} = cu \pm (2ac)^{1/2}\,v$ (Fig. 5.155). Hence $p = 0$, $q = -2ac < 0$, so these are saddle points. The heteroclinic orbits go between the two saddle points, so $K = ca^2/2b^2$. One orbit is $x = -(a/b)$, $y^2 < (2ac/b^2)$; the other is given by $-(a/b) \leqslant x \leqslant (a/b)$ any $y = \pm[(c/b)((a/b) - x)]^{1/2}$.

(5.9) $\mu x[1 + y^2]^{1/2}\,dx + \mu y[1 + x^2]^{1/2}\,dy = d([1 + x^2]^{1/2} + [1 + y^2]^{1/2})$, if $\mu = [1 + x^2]^{-1/2}$ $[1 + y^2]^{-1/2}$. The period is $T = \int dt = 2\int_{x_{min}}^{x_{max}} dx/\dot{x} = 2\int_{x_{min}}^{x_{max}} dx[y(x)[1 + x^2]^{1/2}]^{-1}$, where $y(x) = [(K - (1 + x^2)^{1/2})^2 - 1]^{1/2}$, and $y(x_{max}) = y(x_{min}) = 0$.

(5.10) $\mu = x$ and $K = x^4 + 2x^2y^2$. Although this is non-negative, K is not a Lyapunov

Fig. 5.155

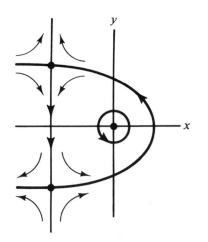

function, since $K = 0$ when $x = 0$, $y \neq 0$. The origin is unstable. This can be seen easily from the solution $x \equiv 0$, $y > 0$.

(5.11) The equations are homogeneous, with $P = x^3 - y^3$, $Q = -x^2 y$, so an integrating factor is $\mu = (y^4 - 2yx^3)^{-1}$; since $\partial K/\partial x = \mu Q$, we find that $K = G(y) - \frac{1}{2}\ln(2yx^3 - y^4)$; using $\partial K/\partial y = -\mu P$ yields $G = -\frac{1}{3}\ln(y)$, so $\exp(-6K) = y^3(2x^3 - y^3)$. A simpler integrating factor is $\mu = y^2$, which yields the integral $K'(x, y) = y^3(2x^3 - y^3)/6$, which is simply a function of $K(x, y)$.

(5.12) $\mu = [cx^2 + (d - a)xy - by^2]^{-1}$; $\mathrm{d}K = \mu(cx + \mathrm{d}y)\mathrm{d}x - \mu(ax + by)\mathrm{d}y$ let $u = x/y$; $\mathrm{d}u = (\mathrm{d}x/y) - (x/y^2)\mathrm{d}y = (\mathrm{d}x/y) - u(\mathrm{d}y/y)$. Then

$$\frac{cu + d}{cu^2 + (d - a)u - b}\frac{\mathrm{d}y}{y} = \frac{au + b}{cu^2 + (d - a)u - b}\frac{\mathrm{d}y}{y}; \text{ eliminate } (\mathrm{d}x/y),$$

$$\frac{cu + d}{cu^2(d - z)u - b}\mathrm{d}u = -\frac{\mathrm{d}y}{y}; \text{ so}$$

$$K = \ln y + \int^u \frac{cz + d}{cz^2 + (d - a)z - b}\mathrm{d}z$$

(5.13) This is left for a solid exercise!

(5.14) (A) $I = 0$, (B) $I = 2$. The Poincaré index does not depend on the orientation of the flow, so it is the same for stable and unstable flows.

(5.15) If $(\partial P/\partial x) + (\partial Q/\partial y) > 0$ the integral does not vanish. However, if we assume that γ is an orbit, then $P\mathrm{d}y - Q\mathrm{d}x = 0$ on γ, so we have a contradiction. Hence γ cannot be an orbit.

(5.16) The index is n. Only the case $n = 1$ is a linear singularity.

(5.17) (a–c) Simply follow from $I_\gamma = 1$ for a periodic γ. Since $I_s = \pm 1$, $I_\gamma = 1$ only for an odd number of singularities. If there are three singularities, one must be a saddle point. If the other two are unstable, we can have the enclosing stable limit cycle

Fig. 5.156

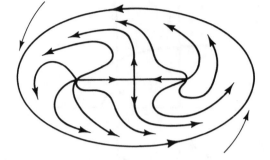

shown in Fig. 5.156. Now make one node stable, and show that it is still possible to have an enclosing limit cycle!

(5.18) (a) A node has index $+1$, so a saddle $(I_s = -1)$ must also vanish, since I_γ does not change for an enclosing curve. (See Ex. 5.7.) This is a *saddle-node bifurcation*.

(b) The index of the fixed point changed from $+1$ to -1, so its neighborhood must have picked up additional singularities with total index $+2$. An example is illustrated in Fig. 5.157.

Fig. 5.157 $q < 0$

$q > 0$

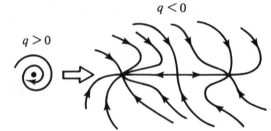

(5.19) The analysis is similar to Exercise (5.15).

(5.20) One possibility is

$$\dot{x} = x + y\,(x^2 + y^2 - 1); \qquad \dot{y} = y - x\,(x^2 + y^2 - 1).$$

So, if $x = r\cos\theta$, $y = r\sin\theta$, then $\dot{r} = r$ and $\dot{\theta} = (1 - r^2)$.

(5.21) (A) The position (x, y) is constrained to satisfy $x = l\sin\theta$, $y = l(1 - \cos\theta) + A\sin(\omega t)$. Obtain the kinetic and potential energy, $K = 1/2\,m\,(\dot{x}^2 + \dot{y}^2)$, $V = mgy$, and thereby the Lagrangian, $L = K - V$. The equation of motion is

$$\frac{\mathrm{d}t}{\mathrm{d}t}(\partial L/\partial\dot{\theta}) - \partial L/\partial\theta = 0,$$

or

$$\ddot{\theta} + (g/l)(1 - (A\omega^2/g)\sin(\omega t))\sin\theta = 0.$$

This type of excitation, involving a periodic coefficient, is called a *parametric excitation* (McLachlan, 1964).

(B) Similarly, using $x = l \sin \theta + A \cos(\omega t)$ and $y = l(1 - \cos \theta) + A \sin(\omega t)$, one obtains

$$\ddot{\theta} + (g/l) \sin \theta - (A\omega^2/l) \cos(\theta - \omega t) = 0.$$

(5.22) The fixed points at $x = \pm \varepsilon^{1/2}$ are centers, whereas $x = 0$ is a saddle point. The flow looks like the 'Daruma' face in the last section (but now on $\mathbf{R}^2$).

(5.23) (a) The 'energy' is $E = \frac{1}{2}\dot{x}^2 + \frac{1}{2}x^2 + \frac{1}{3}x^3$ ($dE/dt = 0$). The saddle point is at $\dot{x} = 0$, $x = -1$, so any trajectory having this as a limit point has $E = \frac{1}{2} - \frac{1}{3} = \frac{1}{6}$. Hence, at $x = 0$, we must take $\dot{x} = 3^{-1/2}$.

(b) On the homoclinic orbit $\dot{x} = (\frac{1}{3} - x^2 - \frac{2}{3}x^3)^{1/2}$, so

$$t - t_0 = 3^{1/2} \int_0^x (1 - 32z^2 - 2z^3)^{-1/2} \, dz.$$

(5.24) $A^{-1} V(r = 1 + x) = -1 + 36x^2 - 252x^3 + 1,113x^4 + 0(x^5)$.

(5.25) (a) Set $t = \omega_0\tau$, $y = \omega_0^2 x/A$, then $\varepsilon = B\omega_0^2/A^2$.

(b) Fixed points are $x = 0$ (center) and possibly $x = -(\frac{1}{2}\varepsilon) \pm (\frac{1}{2}\varepsilon)(1 - 4\varepsilon)^{1/2}$ so: (i) $\varepsilon > \frac{1}{4}$, (ii) $\varepsilon < \frac{1}{4}$, and (iii) none.

Fig. 5.158

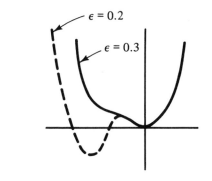

(c) There are two homoclinic orbits, if $\varepsilon < \frac{1}{4}$. The saddle point is at $x_s = -(\frac{1}{2}\varepsilon) + (\frac{1}{2}\varepsilon)(1 - 4\varepsilon)^{1/2}$, and $\dot{x} = 0$. Both homoclinic orbits have the same energy, $\frac{1}{2}x_s^2 + \frac{1}{3}x_s^3 + (\varepsilon/4)x_s^4$ (Fig. 5.158).

(5.26) Using $\Delta t = 0.05$,

$$E = 0.02, \ 0.02, \ 0.05, \ 0.10, \ 0.15, \ 0.20,$$
$$T(E) = 12.7, \ 12.8, \ 13.2, \ 14, \ 15.4, \ 45.1,$$
$$E = 0.25, \ 0.30, \ 0.40, \ 0.50, \ 1.0$$
$$T(E) = 29.8, \ 26.8, \ 23.9, \ 22.2, \ 18.5$$

Fig. 5.159

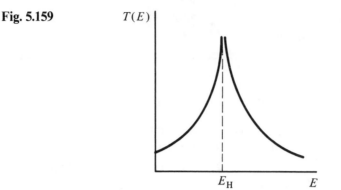

Using $\Delta t = 0.01$ at $E = 0.20$ yields $T(E) = 44.6$ $T(E) \to \infty$, as $E \to E_{\rm H}$, the homoclinic orbit; the saddle point is at $\dot{x} = 0$, $x = -1.270$, so $E_{\rm H} = \frac{1}{2}x^2 + \frac{1}{3}x^3 + \frac{0.1}{4}x^4 = 0.1887$ (Fig. 5.159).

(5.27) One finds that $A_2 = 0$ and $B_2 = (-1/8) + (a^2/8) - (7a^4/256)$, so $a(t)$ is the same as in the first order. Since $\dot{\psi} = 1 - \varepsilon^2 B_2(t)$, $\psi(t)$ can be determined. One finds, as

$$t \to \infty, \quad \psi(t) \sim \left(1 - \frac{\varepsilon^2}{16}\right)t + \left(\frac{\varepsilon}{8}\right)\ln\left(\frac{a_0}{2}\right) + \left(\frac{7\varepsilon}{16}\right)\left(1 - \left(\frac{a_0}{2}\right)^2\right).$$

Thus this predicts a period $T = 2\pi/(1 - (\varepsilon^2/16))$. If $\varepsilon = \frac{1}{2}$, $T = 6.38$ (compared with actual 6.39), and if $\varepsilon = 1$, $T = 6.70$ (actual 6.65), so the accuracy is fairly good.

(5.28) See Fig. 5.160.

(5.29) (a) $(x = 0, \dot{x} = 0)$ is always a fixed point. $(\pm(-c_1)^{1/2}, 0)$ are fixed points if $c_1 < 0$.

(b) Linearize the equations about each fixed point. Near $(0,0)$ the characteristic equation is

$$S = -\tfrac{1}{2}c_2 \pm \tfrac{1}{2}(c_2^2 - 4c_1)^{1/2}$$

whereas, near $(\pm(-c_1)^{1/2}, 0)$

$$s = \tfrac{1}{2}(c_1 - c_2) \pm \tfrac{1}{2}[(c_1 - c_2)^2 + 8c_1]^{1/2}.$$

The character of the flows are all illustrated in Fig. 5.161.

(c) If $c_1 > 0$, $c_2 > 0$, it is a damped nonlinear oscillator.

If $c_1 > 0, c_2 < 0$, it is a nonlinear form of a van der Pol oscillator, with a limit cycle. If $c_1 < 0$, $c_2 > 0$, it has damped motion to the fixed points $(\pm(-c_1)^{1/2}, 0)$ as illustrated in Fig. 5.167. If $c_1 < 0$ and $c_2 < 0$, then there are new and interesting bifurcation features to be discovered. You should try to make such discoveries, with the help of a computer. Specifically, try to understand the bifurcation which occurs when $c_1 = -2$ and you decrease c_2. The above results show that the origin is always a saddle point, but $(\pm(-c_1)^{1/2}, 0)$ changes from stable to unstable at $c_2 = c_1$. The question is what is the character of the global flow, when $c_2 \gtrsim c_1$? In particular, set $c_1 = -2$, $c_2 = -1.7$ and obtain the solutions for the two ini-

Fig. 5.160

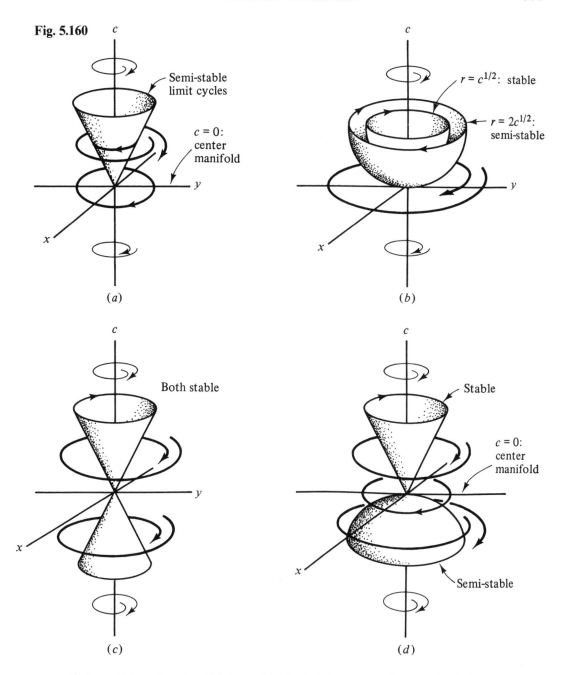

(a)   Semi-stable limit cycles $c = 0$: center manifold

(b) $r = c^{1/2}$: stable $r = 2c^{1/2}$: semi-stable

(c)   Both stable

(d)   Stable $c = 0$: center manifold   Semi-stable

tial conditions $(x, \dot{x}) = (0.1, 0)$ and $(0.5, 0)$. What must the flow look like between these solutions? You have the background to answer this – do you have the imagination? If so, then go to the next step and determine how this structure arises when c_2 is varied, and whether any other structure arises concurrently. If you get stuck, see Section 9 – but only after you have tried!

Fig. 5.161

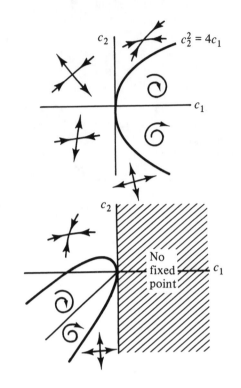

$c_2^2 = 4c_1$

Fig. 5.162

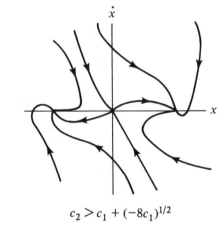

$$c_2 > c_1 + (-8c_1)^{1/2}$$

(5.30) Published figures should always be checked – it is educational! (5.7.4) yields $\dot{M}_1 = 0$ at $M_2 = 1$, which does not agree with the figure. The integral of (5.7.5) yields an implicit relationship, $\alpha_1(\log M_2 - M_2) + \alpha_2(\log M_1 - M_2) = \beta$, which does not help integrate (5.7.4).

(5.31) We obtain $a = \pm a_1 - 2a_3x_0 \pm a_4y_0 = -a_3x_0 - (a_0 + a_2y_0 + a_5y_0^2)/x_0$, so $a < 0$ and $d = \pm b_1 - 2b_3y_0 \pm b_4x_0 = -b_3y_0 - (b_0 + b_2x_0 + b_5x_0^2)/y_0 < 0$. The second equalities are obtained from (5.7.8).

Fig. 5.163

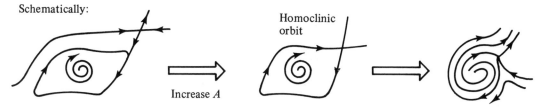

Schematically:

Increase A

Homoclinic orbit

(5.32) For many more interesting details, see Abraham and Simó (1985).

(5.33) (A) The energy is $E = \frac{1}{2}m\dot{x}^2 + \frac{1}{2}kx^2$. If $E < \frac{1}{2}kL^2$, then $x = A \sin(\omega t + \phi)$, with $\omega = (k/m)^{1/2}$, so $T(E) = 2\pi(m/k)^{1/2}$. If $E > \frac{1}{2}kL^2$

Fig. 5.164

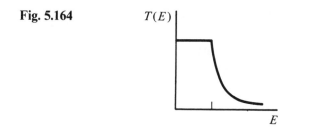

$T(E)$

E

$$T(E) = 2 \int_{-L}^{L} \left(\frac{2E}{m} - \frac{kx^2}{m} \right)^{-1/2} \mathrm{d}x = 4(m/k)^{1/2} \sin^{-1}(L(k/2E)^{1/2}).$$

(B) The turning points of the oscillator (where $\gamma = 1$) are $x_{\pm} = \pm[(\varepsilon - 1) \cdot 2mc^2/k]^{1/2}$, so we could use $T(E) = 2\int_{x_-}^{x_+} v^{-1}\,\mathrm{d}x = (4/c) \int_0^x (\varepsilon - \alpha x^2)$ $[(\varepsilon - \alpha x^2) - 1]^{-1/2} [(\varepsilon - \alpha x^2) + 1]^{-1/2}\,\mathrm{d}x$ where $\alpha = (k/2mc^2)$, but this is complicated. This becomes simpler in terms of γ, using $v/c \equiv \beta = (\gamma^2 - 1)^{1/2}/\gamma$ and $x = ((\varepsilon - \gamma)/\alpha)^{1/2}$, we obtain $T(E) = (2/c) \quad \alpha^{1/2} \int_1^\varepsilon \gamma[(\gamma^2 - 1)(\varepsilon - \gamma)]^{-1/2}\,\mathrm{d}\gamma$. This is still rather involved because of the end singularities. An asymptotic result ($\varepsilon \to \infty$) can be easily obtained by recognizing that the energy surface in the phase plane becomes a rectangle as $\varepsilon \to \infty$, so this is like part (A) except the length $L = x_+$ (above) depends

Fig. 5.165

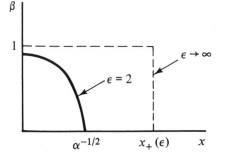

β

1

$\varepsilon \to \infty$

$\varepsilon = 2$

$\alpha^{-1/2}$ $x_+(\varepsilon)$ x

Fig. 5.166

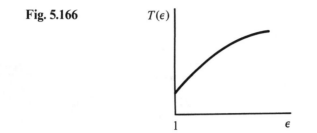

on the energy (Fig. 5.165). Since the particle's velocity is $v \simeq c$, its period is $T \simeq 4x_+/c \sim 4(2m\varepsilon/k)^{1/2} = (2/c)\alpha^{1/2}[2\varepsilon^{1/2}]$. The last bracket is the asymptotic value of the previous integral (Fig. 5.161).

(5.34) When $x \leqslant -1$, the solution is $y = 2 + e^{-t/2}[-2\cos(\omega t) + \omega^{-1}\sin(\omega t)]$, and $x = -\exp(-t/2)[2\cos(\omega t) + \omega^{-1}\sin(\omega t)]$ where $\omega = 3^{1/2}/2$. Similarly, for $|x| \leqslant 1$, $x = \exp(t/2)[\alpha\cos(\omega t) + \beta\sin(\omega t)]$, where (α, β) are determined by the continuity of x and $\dot{x}$, when $x = -1$.

Finally, $x = 0$ when $\omega t = \tan^{-1}(-\alpha/\beta)$. Now we only need α and β!

(5.35) $c_2 \simeq -2.325$. For high accuracy, take the initial point displaced from the saddle point along the outgoing separatrix, obtained by a linear analysis. More interestingly, we can obtain the appropriate curve SC by varying c_1, and obtain $c_2(c_1)$.

(5.36) At $c = 0$, $K = x_1^2 - \frac{2}{3}x_1^3 - x_2^2$ is a constant of the motion. The homoclinic orbit connects with the saddle point at the origin ($K = 0$, so $x_2^2 = x_1^2 - \frac{2}{3}x_1^3 (x_1 \geqslant 0)$. Inside the homoclinic loop each integral curve ($K > 0$) forms a closed curve, so all solutions are periodic; that is, this region is a center manifold (Fig. 5.167).

Fig. 5.167

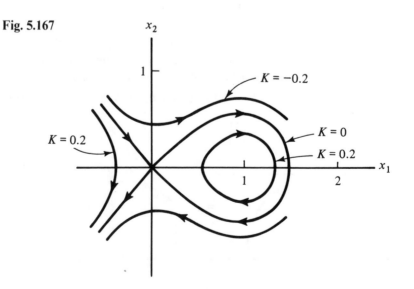

(5.37) The outgoing separatrices both tend toward the outer stable limit cycle, whereas each incoming separatrix spins off the nearest unstable limit cycle.

(5.38) The fixed points are at the origin (a saddle point, if $c_1 < 0$) and at $x_0 = -(c_{3/2}) \pm (\frac{1}{2}) [c_3^3 - 4c_1]^{1/2}$, or $x_0 = 1$, -2 in our example. The characteristic exponents are

$$S = \tfrac{1}{2}(c_1 + c_3 x_0 - c_2) \pm \tfrac{1}{2}[(c_1 + c_3 x_0 - c_2)^2 + 8c_1 + 4c_3 x_0]^{1/2}$$

so they are centers along the line $c_2 = c_1 + c_3 x_0 = -2 + x_0$. We know that Hopf bifurcations occur here. As c_2 decreases the first Hopf bifurcation is at $x_0 = 1$, $c_2 = -1$, so there must be an unstable limit cycle for $c_2 > -1$. This limit cycle is 'born' from a homoclinic orbit at some $c_2 > -1$ (find it!). If $c_2 < -1$, $x_0 = 1$ is unstable whereas $x_0 = -2$ is stable (draw the phase portrait). $x_0 = -2$ there is a Hopf bifurcation at $c_2 = -4$, so again there must be an unstable limit cycle born around $x_0 = -2$ for some $c_2 > -4$. Now, however, there must also be born (concurrently) a stable limit cycle around both $x_0 = 1$ and $x_0 = -2$! How does that happen? Discover it for yourself, or write me a letter!

(5.39) We are interested in the two roots (5.10.6) at the same value of F, where F is critical. If the $\alpha(\Delta)$ roots from (5.10.6) are substituted into (5.10.5), one obtains (5.10.8).

(5.40) (a) If a trajectory intersects itself, then $x(t_1) = x(t_2)$ where $t_2 = t_1 + nT$ (n:integer). Also $F(x(t_1), t_1) = F(x(t_2), t_1) = F(x(t_2), t_1 + nT) = F(x(t_2), t_2)$, because $F(x, t + nT) = F(x, t)$. Since the initial conditions and F are the same at t_1 and t_2, we must obtain $x(t_2 + nT) = x(t_2)$, and so on for any $t_1 + KnT$ ($K = 1, 2, \ldots$).

(b) It will be closed only if the period is a rational multiple, (m/n), of the period T.

(5.41) No comment.

(5.42) (a) Approximate initial conditions are: $(a_2)x = 1.0467$, $\dot{x} = 0.7902$; $(a_3)x = -0.65$, $\dot{x} = 0.02$; $(a_5)x = -0.3634$, $\dot{x} = -0.0764$. Basins of attraction can be roughly obtained by using a lattice of initial conditions, and seeing what attractor they approach (stopping the integration when the attractor is felt to be identified). A second approach is to begin at different points near an attractor, and integrate backwards to see the region of its origin. The former methods appears best suited for obtaining accurate boundaries of basins, but it can be time consuming. See the basins of the Lorentz system (Chapter 7), determined in this fashion. Clearly this method will not yield much information about strange basins, discussed below.

(b) It appears difficult to find any multiple-stable orbits when $\alpha > 0.1$, and $k = 0.08$, $A = 0.2$ (Ueda's values).

Appendix A

A brief glossary of mathematical terms and notation

This elementary glossary attempts to briefly describe a number of concepts which are more frequently encountered in dynamical theories. They have been arranged sequentially so that a term is defined only by concepts which proceeds it. For an extensive listing, see *Encyclopedic Dictionary of Mathematics* (Mathematical Society of Japan, S. Iyanaga and Y. Kawada (eds.), MIT Press, 1980).

Sets, X, Y, A, B,....: a set is a collection of elements (or objects) of one's choosing, defined by some property, such as a collection of points in phase space; $\emptyset$ is the *empty* or *null*, set.

$\{x | P(x)\}$: The set of points (elements), $(x_1, x_2, \ldots)$, each which has the property $P(x)$; e.g., $\{x | 1 > x > 0\}$, $\{x | x = 1/n, n = 1, 2, \ldots\}$, etc.

$x \in B, A \subset B$: x is an element of B; A is a *subset* of B (all elements of A are also elements of B).

$A \cup B$: The *union* of the sets A and B, which contains all elements which are in either A or B (or both); $A \cup B = \{x | x \in A$ or $x \in B\}$.

$A \cap B$: The *intersection* of the sets A and B; contains all elements which are common to A and B; $A \cap B = \{x | x \in A$ and $x \in B\}$.

$A \backslash B = A - B$: The *complement* of B realtive to A is the set of points in A which are not in B. The notation $A - B$ is called the *difference*.

$f : X \to Y = A$ *map*, or transformation, which associates with each element $x \in X$ one element $y \in Y$. This is denoted in various ways; $f : X \to Y$, or $x \rightsquigarrow F(x)$, $y = f(x)$. If each element of Y is associated with some element of X, then the mapping is said to be *onto*, $f(X) = Y$, or *surjective* otherwise it is *into*, $F(X) \subset Y$. If two elements of Y

are not associated with the same element of X then the map is said to be *one-to-one*, or *injective*.

$f^{-1}: Y \to X$: The *inverse* of the map $f: X \to Y$ is the collection of all of the points in X such that $y = f(x)$. $f^{-1}(y)$ is a map only if $f(x)$ is both one-to-one and onto.

bijection: A one-to-one and onto map, but not necessarily continuous.

homeomorphism: A bijection for which both f and f^{-1} are continuous. If $X \equiv Y$ this map is sometimes called an automorphism.

C^m: m-continuously differentiable functions (e.g., $f(x) = |x|$ is C^0; $f(x) = \text{sign}(x)x^2$ is C^1).

diffeomorphism: A homeomorphism such that f and f^{-1} are at least C^1 (usually C^∞).

Examples. Let X and Y be the entire real axis, R, with the mapping $y = f(x)$. Then:

(a) $f(x) = x^3$ is a homeomorphism but not a diffeomorphism.
(b) $f(x) = x^2$ is C^∞, but not one-to-one, nor onto.
(c) $f(x) = x^2 + x^3$ is onto, but not one-to-one.
(d) $f(x) = \exp(x)$ is one-to-one but not onto (hence it is into).

However, if

$$Y = \{y | y > 0\}, \quad \text{then } f(x) \text{ is a diffeomorphism.}$$

An example of a continuous map, f, such that f^{-1} is not everywhere continuous, is the map of $X = \{x | 0 \leqslant x < 1\}$, note $x \neq 1$, onto the unit circle, $y_1 = \cos(2\pi x)$, $y_2 = \sin(2\pi x)$ which is one-to-one. Hence f is a continuous bijection. However, $f^{-1}(y, y_2) = x$, where (y_1, y_2) are in the above space Y, is not continuous at $y_1 = 1$, $y_2 = 0$, because f^{-1} goes to 0 or to 1 as $(y_1, y_2) \to (1, 0)$ in two directions (Fig. A.1).

Fig. A.1

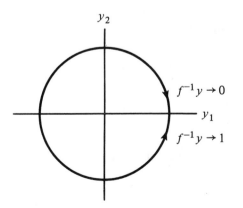

$f^{-1}y \to 0$

y_1

$f^{-1}y \to 1$

R^n: The set of all ordered n-tuples of *real* numbers $x = (x_1, x_2, \ldots, x_n)$. This is the n-dimensional *Euclidean* space, also designated by E_n or E^n. The Euclidean *vector space*, with elements x, has the usual definition of vector addition and scalar multiplication.

Metric: A metric ('distance') between two points, x and y, in R^n is any function $d(x, y) \geqslant 0$, such that $d(x, y) = d(y, x)$ satisfies the triangle inequality $d(x, y) + d(y, z) \geqslant d(x, z)$, and $d(x, y) = 0$ if and only if $x = y$. In E^n the usual metric is $d(x, y) = [(x_1 - y_1)^2 + \cdots + (x_n - y_n)^2]^{1/2}$ or sometimes $d(x, y) = |x_1 - y_1| + \cdots + |x_n - y_n|$ (the 'taxicab distance', since it counts blocks). The distance is also denoted by $|x - y|$.

Measure: In the phase space R^n a measure $\mu(A)$ of a set $A \subset R^n$ is a real number associated with the set A, with the properties: (*a*) for any bounded $A \subset R^n$, $0 \leqslant \mu(A) < \infty$, (*b*) $\mu((\emptyset)) = 0$, where $\emptyset$ is a null (empty) set, (*c*) for all $A, B \subset R^n$, $\mu(A \cup B) = \mu(A) + \mu(B)$ if $A \cap B = \emptyset$. The measure we will usually use is simply $\mu(A) = \int_A \rho(x) \, dx_1 \cdots dx_n$, where $\rho(x) \geqslant 0$, and is integrable over any bounded region A (so that $\mu(A) < \infty$). If the measure of all the phase space of interest is bounded (e.g., a bounded subspace, R^{n-1}, on which a similar measure is defined), then it is commonly normalized to unity. For a more exotic (Hausdorff) measure, see below.

Open set: A set $x \subset R^n$ is open if every $x \in X$ has some ε-neighborhood (i.e., all y such that $|x - y| < \varepsilon$) inside X. An open set does not contain its boundary points.

Dense subset of Y: The set $x \subset Y$ is a dense subset of Y if, for every $y \in Y$ and every $\delta > 0$, there is some $x \in X$ satisfying $|x - y| < \delta$. X is a dense subset of Y if, for any open subset S of Y, the intersection $S \cap X$ is nonempty ($\neq \emptyset$).

Limit point: A point $x \in X$ is a limit point of X if every open set containing x also contains another point of X.

Neighborhood of x in a set S: A set of points of S which contains an open set that contains x.

Set dense in itself: A set $X \subset R^n$ is dense in itself if, in every neighborhood of any point $x \in X$, there is another point of X (in other words, if every point of X is a limit point).

Derived set: The set of limit points of X is called the derived set of X, denoted as X'.

Perfect set: A set X is a perfect set if it is identical to its derived set, $X = X'$ (each point of X is a limit point).

Closure: The closure of X is the union of X and X', $\bar{X} = X \cup X'$. If the closure of X is Y, then X is dense in Y.

Nowhere dense: A subset X of Y is nowhere dense in Y if for each open set O of Y, $O \cap X$ is not dense in 0.

Compact set: A set X which either contains only a finite number of points, or is such that every infinite set of points of X has at least one limit point in X.

Cantor set: A nowhere dense, perfect, compact set in R.

Covering of a set: A covering, σ, of a set $A \subset X$ is a family of sets $A_k \subset X$, $\sigma = \{A_k\}$, such that $A \subset U_k A_k$.

Hausdorff dimension: Let $A \, R^n$ and σ be a covering of A by a countable number of balls of radius $d_1, d_2, \ldots$, all of which satisfy $0 < d_n < \varepsilon$. Let $K > 0$, and consider the sum $\sum d_n^K$ for various coverings σ. By $\inf_\sigma \sum d_n^K$ is meant the smallest value assumed by the sum for all possible coverings satisfying the above conditions (the greatest lower bound of the sum). In the limit $\varepsilon \to 0$ this will diverge if $K < H$, and go to zero if $K > H$, for some H (not necessarily an integer). H is called the *Hausdorff dimension* of the set A. If the limit exists (for $K = H$)

$$\lim_{\varepsilon \to 0} \mathrm{Inf}_\sigma \sum d_n^H = \mu_\mathrm{H}(A) \quad (0 < \mu_\mathrm{H} < \infty)$$

then $\mu_\mathrm{H}(A)$ is called the *Hausdorff measure* of the set A.

Topological space: First of all, briefly, a topological space is a set of points, together with a collection of subsets, which allows one to define 'neighboring points' and 'continuous functions', without introducing a metric. More specifically:

Let S be a set, and $\{s_i\}$ a collection of subsets, which are called *open sets*, and which satisfy the axioms:

A1: both S and the empty set, $\varnothing$, are open sets;
A2: the union of any number of open sets is an open set;
A3: the intersection of a finite number of open sets is an open set.

The particular collection of subsets, $\{s_i\}$, which is used to define a topology (topologizes the set S), and the set S together with the topology defines a *topological space*. See Appendix B for more discussion.

A function $f: X \to Y$: is *continuous* provided that, if p is a limit point of a subset A of X, $f(p)$ is a limit point of $f(A)$. In other words limit points are preserved by f.

A Hausdorff space: is a topological space such that any two distinct points possess disjoint neighborhoods.

Topological dimension

D1: an empty set, and only an empty set, has dimension -1;

D2: A Hausdorff space X has a dimension $\leqslant n$ at a point p, if p has arbitrarily small neighborhoods whose boundaries have dimension $\leqslant (n-1)$:

D3: X has dimension $\leqslant n$, dim $X \leqslant n$, if the dimension of $X \leqslant n$ at each point p of X;

D4: dim $X = n$ if dim $X \leqslant n$ is true and dim $X \leqslant (n-1)$ is false.

Manifold: An *n-dimensional topological manifold*, M, is a Hausdorff space such that every point has an open neighborhood N_i homeomorphic to an open set of E^n (metric Euclidean space), $h: N_i \to U_i$ .... It is a *differentiable manifold* of class C^r if the (single) homeomorphism, h, is C^r.

A generic property: of a topological space, X, is any property which is possessed by points which form a subset that is dense in X.

Boundary point: p is a boundary point of a subset X of S if every open set of S which contains p also contains points of X as well as points of $S - X$.

Dense open set, X: if every point in the compliment of X can be approximated arbitrarily close by points in X (because X is dense in R^n), but no point in X can be approximated arbitrarily close by points in the compliment of X (because X is open).

Example: consider some curve on the plane R^2 (e.g., $x = y^2$). The compliment of this curve in R^2 is a dense open set, X. If $x_0 \neq y_0^2$ there are $(x, y) \in X$ such that, if $|x - x_0|$ and $|y - y_0|$ are sufficiently small, then $x \neq y^2$ (so X is open). Also, if $(x_0, y_0) \in R^2$ one can find an $(x, y) \in R^2$ as close to (x_0, y_0) as desired and such that $x \neq y^2$ (proving that X is dense).

Baire set: is a countable number of open dense sets of a normed space X.

Generic properties: of a diffeomorphism is a property that is true for diffeomorphisms belonging to some Baire set of diff(X).

Structural stability: (Andronov and Pontriagin): A dynamical system $\dot{\mathbf{x}} = \mathbf{F}(\mathbf{x})$ is structurally stable if its phase portrait is topologically equivalent to the portrait of $\dot{\mathbf{x}} = \mathbf{F}(\mathbf{x}) + \delta\mathbf{F}$, where $\delta\mathbf{F}$ is an arbitrary smooth (C^r) function which is sufficiently 'small' (e.g., $\max \|\delta\mathbf{F}(\mathbf{x})\| \leqslant \varepsilon$ for all $\mathbf{x} \in R^n$) (for an excellent introductory survey see M.M. Peixoto, 'Generic properties of ordinary differential equations', p. 52, in Studies in Ordinary Differential Equations (J. Hale, editor), *Studies in Math.*, **14** (Math. Assoc. Amer. 1977).

Bifurcation: A *bifurcation* of a dynamical system $\dot{\mathbf{x}} = \mathbf{F}(\mathbf{x}, \mathbf{c})$, which has *unique* solutions $(\mathbf{x}(t, \mathbf{x}^0, \mathbf{c}))$, occurs when a change in the total control parameters, $\mathbf{c}$, produces a change in the topology of the (unique) phase portrait. The partitioning of the control space, $\mathbf{c} \in R^k$, into regions where the phase portraits are topologically equivalent is called the *bifurcation diagram* of the system $\dot{\mathbf{x}} = \mathbf{F}(\mathbf{x}, \mathbf{c})$ (Arnold, 1972, p. 90). The boundary of these regions (in the control space) is called the *catastrophe set* (Thom, 1975). Thus the system is *not* structurally stable on the catastrophe set.

Bifurcation – of the solutions of a system $\dot{\mathbf{x}} = \mathbf{F}(\mathbf{x}, \mathbf{c})$, subject to conditions $\mathbf{K}(\mathbf{x}, \mathbf{c}') = 0$ which do *not* generally produce *unique* solutions, occurs when a change in the parameters $\mathbf{c}$ or $\mathbf{c}'$ changes the *number* of solutions satisfying the conditions $\mathbf{K} = 0$. [e.g., $\dot{x}_1 = -x_2, \dot{x}_2 = cx_1$ subject to $\mathbf{x}(t + c') - \mathbf{x}(t) = 0$]

p is a *recurrent point* if there is an infinite set $\{n_i\}$, such that $f^{n_i}(p_0) \to p$ as $|n_i| \to \infty$.

p *wanders* under f if it has a neighborhood, U, such that $f^{-n}U\ U = 0$ for all $n \neq 0$.

p is *non-wandering* under f if for any neighborhood U of p there exists an $n \neq 0$ such that $f^{-n}U, U \neq \varnothing$.

The set of all non-wandering points of f is called the non-wandering set $\Omega(f)$. It is a closed invariant set of f, and is conjugacy invariant also

$$\Omega(h \cdot f \cdot h^{-1}) = h(\Omega(f))$$

Two diffeomorphisms f, g are Ω-*conjugate* ($f \tilde{\Omega} g$) if there is a homeomorphism h between their non-wandering sets $h : \Omega(f) \to \simeq (g)$ $\{f \tilde{\Omega} g$ means that the sets of fixed points, periodic points and recurrent points and recurrent points for f are taken by h onto the corresponding sets for $g\}$.

The endomorphism T is *mixing* if, for any two measureable sets (A_0, A_1),

$$\lim_{n \to \infty} \mu(T^{-k_n^0} A_0 T^{-k_n^1} A_1) = \mu(A_0)\mu(A_1)$$

for any nonegative integer sequences $(k_1^0, k_2^0, \ldots)$ $(k_1', k_2, \ldots)$ satisfying $\lim_{n \to \infty} |k_n^0 - k_n^1| = \infty$. Mixing implies *ergodicity*.

An endomorphism T is called *ergodic* if every measurable set A which is invariant under T (so $T^{-1}A = A$) has either measure zero or 1.

Hyperbolic point: (fixed point or periodic point) – If $\mathbf{x}_0$ is a fixed point of $\dot{\mathbf{x}} = \mathbf{F}(\mathbf{x})$ (so $\mathbf{F}(\mathbf{x}_0) = 0$) and A is the matrix $A_{ij} = (\partial F_i/\partial x_j)\mathbf{x}_0$, then $\mathbf{x}_0$ is a hyperbolic point if A has no purely imaginary eigenvalues.

Appendix B

Notes on topology, dimensions, measures, embeddings and homotopy

Topology had its origins in what some people would now call mathematical recreations. An early example was the problem of the seven bridges of Königsberg, solved by Euler in 1736, which involved showing that it is impossible to transverse each of seven bridges connecting two islands in a river to the banks exactly once. A much more profound problem, apparently first posed by Francis Guthrie to his teacher Augustus de Morgan (1852), is to prove that four colors suffice to distinguish the countries on any possible map (see, e.g., K.O. May, 'The origin of the four-color conjecture', Isis. **56**, 346–8 (1965). It might be noted that the controversial concept of a 'computer assisted proof' of this conjecture has been employed by Appel and Haken (1977, 1986), using over 1000 hours of computer time). The term 'topology' was first introduced by J.B. Listing in 1847, and was roughly associated with the analysis of placements, and hence was frequently referred to as '*analysis situs*' by Poincaré and others. Poincaré however is generally credited with being responsible for raising this method of analysis to the level of a branch of mathematics. Topology is now a diverse field of research, with a number of specialized branches. Our needs in topology are very modest, and the present discussion is largely limited to these needs. Fortunately there exists a number of good introductions to topology, some of which are listed in the references, which the reader should consult to augment this meager introduction.

Modern topology is a branch of general set theory, which was introduced by Georg Cantor around 1880. As Cantor defined it, a set is a collection of well-defined objects (or elements) of our perception or thought. However the sets we will usually be interested in are simply the collection of points in some phase space. Onto such a set, S, is then placed a 'structure' or *topology*, which allows us to define *limit points* and thereby *continuity*. In general, the structure, or topology, is introduced into *any* set S by specifying a collection of subsets, $\{s_i\}$, called (i.e., by definition) *open sets*, which satisfy certain axioms:

A1: both S and the empty set are open,
A2: the union of any collection of open sets is an open set,
A3: the intersection of a finite number of open sets is an open set.

The set S, together with the topology, defined by the subsets s_i satisfying these axioms, defines a *topological space*.

This process can be made reasonably transparent, if we keep in mind the important special case of a Euclidean space, with the usual Euclidean metric (distance), defined in Appendix A. We will denote this metric Euclidean space by R^n, and take the set S to be all the points in some open n-cell (any set homeomorphic to the open unit sphere, $\sum^n x_i^2 < 1$). In this case, the subsets s_i might be open spheroids about points, x_i. That is, the points x which are within a distance Δ_i of x_i, $d(x, x_i) < \Delta_i$. In addition, the collection $\{s_i\}$ will contain the other open sets needed to satisfy the axioms A1–A3. Note that, in the case of R^n, we have in addition introduced a metric, which is not necessary to define a topological space. However, in most of our applications in phase spaces, such a metric is quite natural to use.

Example 1: Let S be the points $0 < x < 1$ in R^1. The 'usual topology' consists of S, $\varnothing$, and all open intervals $s_i = \{x \,|\, a_i < x < b_i\}$, where $0 < a_i < b_i < 1$. This satisfies all of the axioms A1–A3. Note that, if A3 had allowed an *infinite* number of intersections, then we could obtain any *point*, say x_0, by using the sets $s_n = \{x \,|\, x_0 - 1/n < x < x_0 + 1/n;\ n = N, N+1, \ldots\}$, where N is large enough to keep s_n in S, However, in this so-called discrete topology every point is an open set; by A2, every set is then open, and the concept becomes redundant.

Example 2: An abstract (and not very profound) example is the set $S = \{a, b, c, d\}$. This can be topologized using the sets $s_1 = S$, $s_2 = \varnothing$, $s_3 = \{a, b\}$, and $s_4 = \{b\}$. No metric is involved in this case. Note that if we had taken $s_4 = \{c\}$, this would not be a topology, because $s_3 \cup s_4$ would not then be an open set (i.e., one of the s_i).

A topological space which also satisfies

A4: For any two distinct elements (x, y) of the space there are two disjoint open sets (s_k, s_l) containing x and y respectively, is called a *Hausdorff space*. Example 1 is Hausdorff, but Example 2 is not.

The importance of the above structure is that it allows us to define *limit points* in the set, which are of central importance because of their relationship to *continuity*, and continuity is at the heart of topology. We next turn to these definitions:

> *limit point*: A 'point' is any element of S. A limit point, p, of a subset X of S is a point such that every open set containing p also contains a point of X distinct from p.

Thus the limit points in example 2 are a, c, and d. Using this general definition we can introduce the concept of *continuity* of a transformation, or *map*:

> *map (or transformation)*: If X and Y are two sets, a map $f : X \rightarrow Y$ associates with

each element of X one element of Y, denoted by $y = f(x)$. If two or more elements of X are not associated with the same element of Y, the map is said to be one-to-one. If each element of Y is associated with some element of X, then the map is said to be *onto*, otherwise it is *into*. The map $f:S \to T$ is said to be *continuous* ('mapping') if, when p is a limit point of any subset X of S, then $f(p)$ is a limit point of $f(X)$. Finally, by $f^{-1}(y)$ is meant the set of all points x in X such that $y = f(x)$. Note that $f^{-1}(y)$ is a map only if $f(x)$ is both one-to-one and onto.

This brings us to the concept of *topological equivalence*, which is usually defined through a special mapping, namely a *homeomorphism*. A one-to-one mapping of X onto Y, $f:X \to Y$, is a homeomorphism if both f and f^{-1} are continuous. If such a map exists between X and Y, these sets are said to be homeomorphic. A property which is common to each of a collection of mutually homeomorphic sets is called a *topological property*, or *topological invariant*. Thus we see that topological properties are directly related to map which retain the property of continuity in the set X when mapped into Y, as well as the uniqueness (one-to-oneness).

One of the fundamental historical problems was to introduce the concept of dimension, in such a way that it would be a 'topologically invariant concept'. This involved both obtaining a topological definition of dimension and discovering that it required the use of a homeomorphism to retain its invariance properties (and hence the definition of the last paragraph). Before the introduction of set theory, 'dimension' was used in a rather vague sense. Whatever the ultimate definition might be, however, it is clearly desired to have a line be one dimensional, and a plane be two dimensional. This suggested that a continuum be said to be n-dimensional if the least number of parameters 'needed to describe its points', in some unspecified way, equals n. The difficulty with this approach, and the reason it is necessary to consider homeomorphisms, is illustrated by the following two celebrated discoveries.

(I) Peano showed that it is possible to construct a *continuous* mapping of the points on an interval, say $(0, 1)$, *onto* the whole unit square. The construction of such maps is described in many places, and so we simply sketch the beginning construction process in Fig. B.1 (one possibility). The consequence of this discovery is that the dimension of a continuum (say the points in a unit square, with dimension two) is not equal to the least number of continuous real parameters which can be used to label the points. It should be emphasized, however, that the Peano map cannot be made one-to-one (two or more points in the interval *must* refer to the same point in the square).

(II) Cantor, on the other hand, showed that it is possible to obtain a *one-to-one* correspondence between the points of the interval $(0, 1)$ and the points in a unit square. Let the (x, y) point in the unit square be expressed in terms of nonterminating decimals. For example

$$x = 0.400\,361\,040\,07\ldots.$$
$$y = 0.010\,230\,064\,21\ldots.$$

Fig. B.1

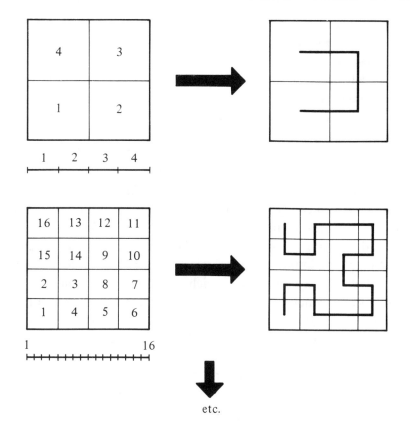

etc.

Break these into 'complexes', consisting of the groups of numbers up to, and including the next nonzero number. In the above example the complexes are

$$x = 0.4 \quad 003 \quad 6 \quad 1 \quad 04 \quad 007 \ldots .$$
$$y = 0.01 \quad 02 \quad 3 \quad 006 \quad 4 \quad 1 \quad 1 \ldots .$$

From these two collections of complexes we construct

$$z = 0.014\,020\,033\,600\,61 \ldots ,$$

which is a unique point in the interval $0 < z < 1$. Thus we have a one-to-one map between the interval and the square, but of course this map is *not continuous*. The lesson from Cantor's map is that the dimension can be changed by a one-to-one map. Put another way, the dimension of a set does not determine whether the set has more or less points than another set of different dimension.

Since the desired concept of dimension is not invariant under either a one-to-one map nor a continuous map, a more restrictive map is required to preserve the dimension. A homeomorphism, which is both one-to-one and continuous, is a possibility, since it removes both of the above difficulties. However it was not until 1911 that Brouwer proved that the Euclidean spaces R^n and R^m are not homeomorphic unless $n = m$ ('Beweis der Invarianz der Dimensionenzahl', *Math. Ann.* **70**, 161–4, 1911). This clearly is a necessary result if a homeomorphism is to be the condition for dimensional invariance.

The topological definition of dimension was first considered by Poincaré, and he wrote on the subject rather extensively. In the briefest possible excerpt, we quote here from his *Mathematics and Science: Last Essays* (translated by J.W. Bolduc, Dover Pub., 1963):

'We now know that a continuum of n dimensions is. A continuum has n dimensions when it is possible to divide it into many regions by means of one or more cuts which are themselves continua of $n - 1$ dimensions. The continuum of n dimensions is thus defined by a continuum of $n - 1$ dimensions. This is a definition by recurrence.... a continuum which can thus be cut up by limiting ourselves to a finite number of elements will have one dimension.' Thus he implicitly took elements, or points, to be of dimension zero. More importantly he introduced the *definition of dimension by recurrence.* This was a major advance, which is retained in the modern definition of *topological dimension* (which is clearly always an integer). Note that a continuum of dimension n may require more than one 'cut' of dimension $n - 1$ to divide the continuum into two regions, as illustrated by a torus (Fig. B.2).

Fig. B.2

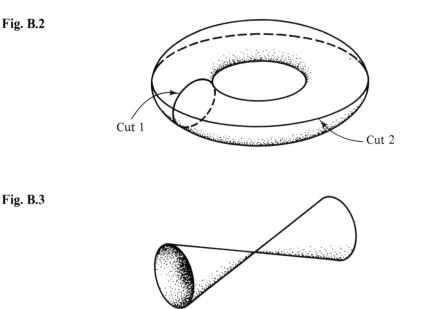

Cut 1

Cut 2

Fig. B.3

Brouwer pointed out that Poincaré's definition fails to give the correct dimension for the two-dimensional continuum consisting of the surface of a double cone (Fig. B.3). Here the removal of the vertex point (dimension zero) suffices to divide the continuum into two parts, but the cone surface does not have dimension one, as given by Poincare's definition. This difficulty can be removed, however, if the definition involves regions ('neighborhood') of the points in the continuum. Indeed Lesbeque conjectured that a space, if covered by sufficiently small 'cubes', must have at least $(n + 1)$ of the cubes with a point in common, where n is the dimension of the space. The modern definition of topological dimension (Brouwer; Menger; Urysohn) is based on these ideas. In full detail it now reads:

Topological dimension (in R^m)

(1) The empty set, and only the empty set, has dimension -1.
(2) A space X has dimension $\leqslant n$ $(n \geqslant 0)$ at a point p if p has arbitrarily small neighborhoods whose boundaries have dimension $\leqslant n - 1$.
(3) X has dimension $\leqslant n$, dim $X \leqslant n$, if X has dimension $\leqslant n$ at each of its points.
(4) X has dimension n at a point p if it is true that X has dimension $\leqslant n$ at p, and false that X has dimension $\leqslant n - 1$ at p.
(5) X has dimension n, if dim $X \leqslant$ is true and dim $X \leqslant n - 1$ is false.

Thus consider the above space X, which consists of Brower's double cone surface. If we take a sufficiently small neighborhood of any point of X, including the troublesome vertex, the boundary points (in X) always consist of closed curves. Since these have dimension one (proved by a separate induction), the space X has dimension two, as desired.

The nontopological dimensions, discussed in Chapter 2, and the Hausdorff dimension defined in Appendix A, are all based in a crucial way on the metric which is introduced in R^n. As a consequence these dimensions, as well as measures of the sets, are *not* generally invariant under a homeomorphic map. Thus these concepts are not topologically invariant, based on homeomorphic comparisons, but they are invariant concepts when the maps are restricted to *diffeomorphisms* – that is, with the additional condition that they have continuous derivatives, C^m $(m > 0)$. The nontopological dimensions can, of course, have noninteger values, and Mandelbrot has suggested that sets with dimension $d > d_{\text{topological}}$ be referred to as *fractals*.

To make any association between dimensions and measures of sets, we need a more precise definition of some measures of a set. The measure of a set X in E^n can be related to the metric, $d(x, x')$ (Appendix A) in the following general way. Let $X = A_1 \cup A_2 \cup \cdots$ be any decomposition of X into a countable number of subsets. Let the 'diameter' of a set A be defined as

$$\delta(A) = \sup_{x, x' \varepsilon A} d(x, x'),$$

where sup stands for the supremum (least upper bound). We require that the decomposition subsets satisfy $\delta(A_i) < \varepsilon$, for a given $\varepsilon > 0$. Next we define

$$m_p^\varepsilon(X) = \inf_{\{A_i\}} \sum_{i=1}^{\infty} [\delta(A_i)]^p,$$

where $\infty > p \geqslant 0$, and inf stands for the infimum (greatest lower bound). Finally we define the p-dimensional measure of X to be

$$m_p(X) = \sup_{\varepsilon > 0} m_p^\varepsilon(X).$$

Hopefully it is clear that, for a given set X, if p is taken too large, $m_p(X) = 0$ (think, for example, of the three dimensional measure of a plane segment).

With this definition we can now state a theorem which relates the topological dimension of a set with a measure associated with the set (see Hurewicz and Wallman):

Theorem: A necessary and sufficient condition that a space X has a topological dimension $\leqslant n$ (n:integer) is that X be homeomorphic to a subset A of the region $\{x \mid |x_i| \leqslant 1, \ i = 1, \ldots, 2n + 1\}$ of R^{2n+1}, whose $(n + 1)$-dimensional measure is zero, $m_{n+1}(A) = 0$.

This p-dimensional measure can also be used to define the Hausdorff dimension of a set X in R^n now for general p). The Hausdorff dimension, d_{H}, is the supremum (least upper bound) of all p such that $m_p > 0$. One therefore has Hausdorff dim $X \geqslant$ topological dim X.

It should be made clear, if it is not already obvious, that it is one thing to state the above definitions and quite another thing to actually apply them to specific sets X. For that reason we confine our applications of dimensions to the 'simpler', but sometimes unsatisfactory, dimensions d_c and d_I, discussed in Chapter 2.

It will be noted that, in the last theorem, the set X was related to a set in E^m by a homeomorphism. In general a homeomorphic map of a space X into a space Y is called an homeomorphic *embedding* of X and Y. The above theorem implicitly uses the fact of other famous theorems (Menger, 1926; Nöbeling, 1930), which state that:

Embedding (homeomorphic) theorem.

> If X has dim $X \leqslant n$, then X can be embedded in Y if dim $Y = 2n + 1$. Furthermore, the homeomorphisms are dense in the sense that every mapping of X into Y can be made a homeomorphism by an arbitrarily small modification.

Later this embedding result was significantly generalized by Whitney (1936, 1944) to address the case of *diffeomorphisms* (for which now *all* dimensions are invariant). The term 'embedding' is presently frequently used to refer to diffeomorphisms. More specifically:

A differentiable map $f: X \to Y$ of a smooth manifold X into Y is called an *embedding* of X in Y if $f(X) \subset Y$ is a differentiable submanifold of Y and $f: X \to f(X)$ is a diffeomorphism.

Then we paraphrase a theorem due to Whitney (1944):

Theorem.

A m-dimensional differentiable manifold can be embedded in R^n if $n \geqslant 2m$.

The condition that $n \geqslant 2m$ can be seen to be necessary (i.e., n cannot generally be reduced below $2m$), by the simple examples of a circle, or a Klein bottle (Fig. B.4). On

Fig. B.4

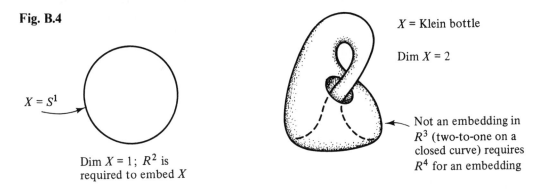

$X = S^1$

Dim $X = 1$; R^2 is required to embed X

X = Klein bottle

Dim $X = 2$

Not an embedding in R^3 (two-to-one on a closed curve) requires R^4 for an embedding

the other hand, of course, a set with dimension n *may* possibly be embedded in R^S where $2n \geqslant S \geqslant n$ (e.g., a torus, T^2, embeds in R^3).

In dynamics, the points and trajectories are always in a phase space, which can be taken to be R^n (i.e., a metric is readily defined). Therefore, from the outset, a trajectory is always embedded in some R^n. Moreover a trajectory is one-dimensional, and it is necessarily differentiable (i.e., along the trajectory), because it comes from a solution of differential equations (the velocity vector is defined and continuous). A *family* of trajectories, defined by some time independent constant of the motion,

$$K(x) = K_0$$

defines a surface in R^n which may, or may not be a differentiable at all points (Fig. B.5).

Fig. B.5

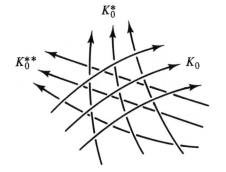

K_0^*

K_0^{**}

K_0

For example, consider

$$\dot{x} = -y - x, \quad \dot{y} = y - x, \quad \dot{z} = -(x^2 + y^2)^{1/2}$$

so $\dot{r} = -r$ $(r \equiv (x^2 + y^2)^{1/2})$ and therefore $r = r_0 \exp(-t)$, $z = z_0 + r_0(\exp(-t) - 1)$, where (r_0, z_0) are the initial conditions $(t = 0)$. One constant of the motion is $K(x) = (x^2 + y^2)^{1/2} - z = K_0 (= r_0 - z_0)$, which defines a family of cones, as K_0 is varied. These surfaces do not have a derivative at $x = y = 0$, so they are not differentiable manifolds. Indeed, if $\dot{z} = -z$, then the integral surfaces are all of the form of Brouwer's double cone. This, however, presents no difficulty in embedding the manifold in R^e. Similarly with other surfaces which are not everywhere differentiable (as illustrated in Fig. B.6).

Fig. B.6

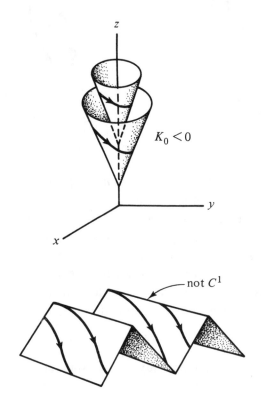

Since all trajectories of an nth order system are in R^n, the same is true of any family, and hence also any 'integral manifold' (not implying differentiability everywhere). We may be able to conclude much more about these manifolds than is given by Whitney's theorem. Namely any integral manifold of an nth order system must be embedded in R^n. If the manifold is of dimension m it *may* be possible to embed it in some R^S for which $n \geqslant s \geqslant m$. This upper bound on s may be much less than Whitney's value of $2m$. As an example, for the Hamiltonian system

$$H = \tfrac{1}{2}(p_1^2 + p_2^2) + \omega_1^2 q_1^2 + \omega_2^2 q_2^2,$$

the integral manifolds

$$H(p_1, p_2, q_1, q_2) = K_0$$

are three dimensional, but can only be embedded in R^4 (which, however, is significantly lower than R^6, guaranteed by Whitney's theorem). Thus the 'embedding' in R^3 of the Hénon–Heiles energy manifold, illustrated in Chapter 6, is improper, for it is double-valued in P_1.

Another aspect of embedding may not be obvious. If we draw some set in R^3 (say) it may be embedded in R^2 even if it seems to 'require' R^3. An example is the case of a knot. If we construct a knot out of string it requires three dimensions (Fig. B.7). However, it is

Fig. B.7

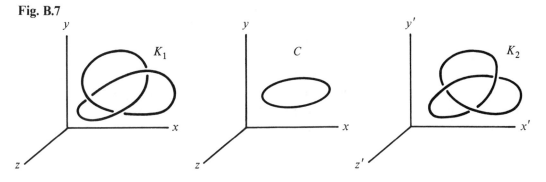

a one dimensional set, and hence can be embedded in R^2. In other words, there is a diffeomorphism connecting a knot with a set in R^2, namely a circle. Moreover, there are diffeomorphisms connecting 'different' knots (in a sense that will be described shortly). Thus the knots K_1 and K_2 can be mapped into each other

$$x' = x_0 - x, \quad y' = y, \quad z' = z$$

for a suitable x_0. Therefore all of the sèts K_1, C, and K_2 are equivalent in this sense. However, they would hardly be considered to be physically equivalent. Why? Simply because we cannot take the string knot K_1 and continuously deform it into either C or K_2 (the latter is perhaps not obvious, but it is true). The process of making a continuous deformation of a set, which is most frequently associated with topology, is related to an area of topology called homotopy theory.

The process of continuously deforming a set is not related to a single map, but rather involves a *family* of maps, which moreover can be continuously related to each other by some parameter (Fig. B.8). That is, a continuous deformation corresponds to a

Fig. B.8

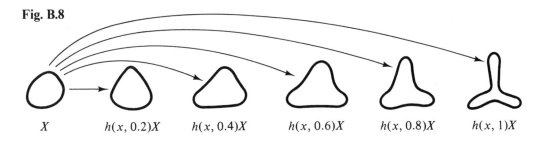

X $h(x, 0.2)X$ $h(x, 0.4)X$ $h(x, 0.6)X$ $h(x, 0.8)X$ $h(x, 1)X$

parameterized family of maps, $h(x, t)$, continuous in both x and $0 \leqslant t \leqslant 1$. This is illustrated in the figure, and leads to the concept of homotopy:

Let X and Y be two topological spaces, and f and g two continuous maps

$$f : X \rightarrow Y, \quad g : X \rightarrow Y.$$

Then f is said to be *homotopic* to g if there exists a map $h(x, t)$, continuous in (x, t), such that $h(x, 0) = f$ and $h(x, 1) = g$. The concept of homotopy is to maps, essentially what homeomorphism is to sets, with $h(x, t)$ being the homotopy between f and g. Two spaces X and Y are called homotopically equivalent if there exists maps $f : X \rightarrow Y$ and $g : Y \rightarrow X$ such that the composite maps $fg : Y \rightarrow Y$ and $gf : X \rightarrow X$ are homotopic to their respective identity maps $I : Y \rightarrow Y$ and $I : X \rightarrow X$.

In the case of dynamic systems we want to distinguish between sets which cannot be continuously deformed into each other, and we see that this requires more than a diffeomorphism between particular (embedded) sets. Indeed a more general condition was imposed by our requirement of *topological orbital equivalence (TOE)*, discussed in Chapter 2.

The essential point is that, in dynamic systems, we are not just concerned with a few trajectories, but rather the whole phase portrait. We are therefore concerned with maps defined on the *entire phase space*, and we require that these maps be *orientation preserving*. A diffeomorphism $f : R^n \rightarrow R^n$ is *orientation preserving* if the Jacobian

$$\det |\partial f_i / \partial x_j| > 0$$

at every point. By the use of these orientation preserving diffeomorphisms one obtains an equivalence class of knots, or other trajectories, which have the desired homotopic property, and similarly we obtain the physically desired equivalence classes of dynamical systems. In other words, if we take two dynamic systems to be topologically and orbitally equivalent only if their phase portraits can be related by an orientation preserving diffeomorphism, then we have insured that TOE systems only have trajectories which are continuously deformable into one another.

Appendix C

Integral invariants

We begin with the integral discussed in Section 4 of Chapter 2,

$$I(t) = \int \cdots \int_{\Omega(t)} \rho(x, t)\, dx_1, \dots, dx_n. \tag{C.1}$$

where $\rho(x, t) > 0$ is any function, and $\Omega(t)$ is a domain in the phase space whose points move according to the equations of motion

$$\dot{x}_k = F_k(x_1, \dots, x_n, t) \quad (k = 1, \dots, n) \tag{C.2}$$

starting from some arbitrary initial region, Ω_0. The function $I(t)$ is an integral invariant, by definition, if

$$dI(t)/dt = 0. \tag{C.3}$$

As discussed in (2.4.3), we can obtain dI/dt in the form

$$\frac{dI}{dt} = \int_{\Omega_0} \cdots \int \frac{d}{dt}\left\{ \rho(x, t)\left|\frac{\partial(x_1, \dots, x_n)}{\partial(x_1^0, \dots, x_n^0)}\right|\right\} dx_1^0 \cdots dx_n^0 \tag{C.4}$$

and, if (C.3) is to hold for arbitrary regions, Ω_0, then the integrand must vanish identically. On the other hand, the integrand can be differentiated, using (C.2), to give

$$\frac{\partial \rho}{\partial t} + \sum_k \dot{x}_k \frac{\partial \rho}{\partial x_k}\left|\frac{\partial(x_1, \dots, x_n)}{\partial(x_1^0, \dots, x_n^0)}\right| + \rho(x)\sum_k \left|\frac{\partial(x_1, \dots, \dot{x}_k, \dots, x_n)}{\partial(x_1^0, \dots, x_n^0)}\right|$$

$$= (F \cdot \nabla \rho)\left|\frac{\partial(x_1, \dots, x_n)}{\partial(x_1^0, \dots, x_n^0)}\right| + \rho \sum_k \left|\frac{\partial(x_1, \dots, F_k, \dots, x_n)}{\partial(x_1^0, \dots, x_n^0)}\right|. \tag{C.5}$$

The kth row of the kth determinant in the last sum contains the following terms:

$$\sum_l \frac{\partial F_k}{\partial x_l}\frac{\partial x_l}{\partial x_1^0}, \quad \sum_l \frac{\partial F_k}{\partial x_l}\frac{\partial x_l}{\partial x_2^0}, \dots, \sum_l \frac{\partial F_k}{\partial x_l}\frac{\partial x_l}{\partial x_n^0}.$$

A determinant with such a row can be written as the sum of determinants, each with the

kth row containing one of the above l values, from which can then be factored $\partial F_k/\partial x_l$. Hence the sum in (C.5) equals

$$\sum_{k,l} \frac{\partial F_k}{\partial x_l} \frac{\partial(x_1,\ldots,x_{k-1},x_{k+1},\ldots,x_n)}{\partial(x_1^0,\ldots,x_n^0)}.$$

Now, unless $l=k$ in this sum, the determinant will vanish, because two rows will be identical (namely the kth row and the lth row). Hence only the term $l=k$ survives, and so the last expression equals

$$\sum_k \frac{\partial F_k}{\partial x_l} \frac{\partial(x_1,\ldots,x_n)}{\partial(x_1^0,\ldots,x_n^0)} \equiv \nabla\cdot F \frac{\partial(x_1,\ldots,x_n)}{\partial(x_1^0,\ldots,x_n^0)}.$$

Substituting this back into the sum in (C.5), we see that the integrand of (C.4) becomes simply

$$\partial\rho/\partial t + \nabla\cdot(\rho(x,t)F(x,t)),$$

so that (C.3) is generally satisfied only if this vanishes. This proves Liouville's theorem.

It should be emphasized that, if $\rho(x,t)$ is a conserved quantity, such as the probability density of an ensemble of systems, we know *ab initio* that $dI/dt = 0$. Then we can use Liouville's theorem to conclude that $\rho(x,t)$ must obey the equation

$$\partial\rho/\partial t + \nabla\cdot(\rho(x,t)F(x,t)) = 0 \qquad (C.6)$$

This is the general form of the so-called Liouville equation, which is valid even in the case of forced, and dissipative systems. This greatly extends the applicability of this equation, which is frequently derived only for Hamiltonian systems.

There are many other possible integral invariants, as was noted by Poincaré in the case of Hamiltonian systems. To date, very little use has been made of such invariants, but they may prove to be useful in the future. We therefore will consider briefly some other integrals.

Consider an integral over some region $\omega(t)$ of the (x_1, x_2) plane which again moves according to the equations (C.2) – note that the phase space is still n-dimensional.

$$I_{12}(t) \equiv \iint_{\omega(t)} dx_1 dx_2. \qquad (C.7)$$

To evaluate the time derivative of $I_{12}(t)$, it would be helpful to integrate over the fixed domain of initial conditions. However, $\omega(t)$ is not just determined by the initial conditions (x_1^0, x_2^0), but also depends on all the rest, $(x_3^0,\ldots,x_n^0)$. We can simplify things at first, by considering the case where all initial conditions lie on a two-dimensional manifold in the phase space, so all $x_k^0(s_1,s_2)$ can be taken to be functions of only two variables, (s_1,s_2). The initial two-dimensional region in the phase space is specified by some region σ in the space (s_1, s_2). In this case (C.7) can be written

$$I_{12}(t) = \int_\sigma \int ds_1 ds_2 \frac{\partial(x_1, x_2)}{\partial(s_1, s_2)}. \tag{C.8}$$

Following Poincaré, in the case of a Hamiltonian system, consider

$$J(t) \equiv \sum_i \int \int_{\omega_i(t)} dp_i dq_i = \int \int_\sigma \sum_i \frac{\partial(p_i, q_i)}{\partial(s_1, s_2)} ds_1 ds_2. \tag{C.9}$$

Now, as above, to investigate dJ/dt, we consider

$$\frac{d}{dt} \sum_i \frac{\partial(p_i, q_i)}{\partial(s_1, s_2)} = \sum_i \left\{ \frac{\partial(-\partial H/\partial q_i, q_i)}{\partial(s_1, s_2)} + \frac{\partial(p_i, \partial H/\partial p_i)}{\partial(s_1, s_2)} \right\}$$

$$= \sum_{i,k} \left\{ \frac{\partial}{\partial p_k}\left(-\frac{\partial H}{\partial q_i}\right)\frac{\partial(p_k, q_i)}{\partial(s_1, s_2)} + \frac{\partial}{\partial q_k}\left(-\frac{\partial H}{\partial q_i}\right)\frac{\partial(q_k, q_i)}{\partial(s_1, s_2)} \right.$$

$$\left. + \frac{\partial}{\partial p_k}\left(\frac{\partial H}{\partial p_i}\right)\frac{\partial(p_i, p_k)}{\partial(s_1, s_2)} + \frac{\partial}{\partial q_k}\left(\frac{\partial H}{\partial p_1}\right)\frac{\partial(p_i, q_k)}{\partial(s_1, s_2)} \right\}.$$

The first and last sums cancel, and the second and third sums vanish individually because they change sign when i and k are interchanged. Thus we have another Poincaré integral invariant

$$dJ/dt = 0.$$

This invariant only holds for Hamiltonian systems, but other integrals, for example over four-dimensional domains

$$\int dx_1 dx_2 dx_3 dx_4,$$

can be invariant for more general systems (as well as Hamiltonian, for $dp_1 dq_1 dp_2 dq_2$). For a further discussion, see Whittaker (1944).

Appendix D

The Schwarzian derivative

(The concept is due to H.A. Schwarz, 1869; see E. Hille, *Lectures on Ordinary Differential Equations*; Appendix D; for many properties and applications to conformal mappings, univalent functions, etc.).

The Schwarzian derivative of $f(x)$ is defined as

$$\{f, x\} \equiv \frac{f'''}{f'} - \frac{3}{2}\left(\frac{f''}{f'}\right)^2.$$

Among its properties, it is invariant under linear fractional transformations (projective group)

P*I*) $$\left\{\frac{af + b}{cf + d}, x\right\} = \{f, x\}$$

which leads into the areas of automorphic functions and conformal mappings.

A property of greater present interest is connected with the maps $f(x)$ and its iterates $f^N(x)$:

P*II*) Let $f(x)$: $[0, 1] \to [0, 1]$ satisfy $\{f, x\} < 0$ for all x. Then, for any integer $N \geqslant 1$, $\{f^N, x\} < 0$, where $f^{k+1}(x) \equiv f(f^k(x))$, $f^0(x) = x$ (to show this, prove that $\{f(g(x)), x\} = \{g, x\} + \{f, g\}(g')^2$).

The importance of the Schwarzian to the bifurcation of maps was noted by D. Singer (1978).

For the 'standard' bifurcation of $f(x)$ at its fixed point, p, the slope of $f^2(x)$ at p is a maximum (Fig. D.1). So the condition is

$$\left.\frac{d^2 f^2}{dx^2}\right|_p = 0, \quad \text{and} \quad \left.\frac{d^2}{dx^2}\left(\frac{df^2}{dx}\right)\right|_p < 0, \quad \text{when} \quad \left.\frac{df}{dx}\right|_p = -1.$$

Now

$$\frac{d}{dx} f^2 = f'(x)f'(f(x))$$

Fig. D.1

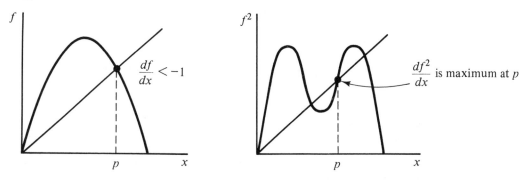

and

$$\frac{d^2}{dx^2} f^2 = f''(x)f'(f(x)) + (f'(x))^2 f''(f(x));$$

and finally,

$$(d^3/dx^3)f^2(x) = f'''(x)f'(f) + 3f'(x)f''(x)f''(f) + (f'(x))^3 f'''(f).$$

Evaluating this at p, when $f'(p) = -1$, yields $-2f'''(p) - 3(f''(p))^2 \equiv 2\{f,p\}$. Hence the standard bifurcation occurs only if $\{f,p\} < 0$.

However it is possible to have an $f(x)$ with a single critical point which bifurcates in the fashion shown in Fig. D.2. An explicit example is Singer's function,

$$f = cF(x); \quad f(x) = 7.86x - 23.31x^2 + 28.75x^3 - 13.3x^4,$$

Fig. D.2

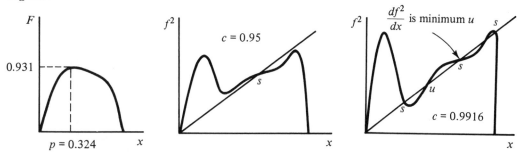

where $c < 1.0646$ for bounded solutions. This has a stable period-1 solution for $c \leqslant 1.035$. However, below this value there also exist many other periodic solutions (e.g., the period-2 solution for $c = 0.9916$, shown above, which bifurcates into a stable period-4 solution near $c = 1.012$, while still maintaining the stable period-1 solution. A period-8 solution appears near 1.02. In other words, the bifurcations are very 'nonstandard'.

The complete statement of Singer's theorem is:

Let $F(x)$ be a smooth (C^3) automorphism of $[0, 1]$ satisfying

(a) $F(0) = F(1) = 0$
(b) $F(x)$ has only one critical point x_0 in $[0, 1]$, $F'(x_0) = 0$
(c) $\{F, x\} < 0$ everywhere in $[0, 1]$.

Then there is at most one stable periodic orbit of $F(x)$ in $(0, 1)$ (open set); if the stable periodic orbit exists it is the set of limit points of x_0.

As another rather different example (due to A. Bondeson), we note that the Schwarzian depends on the first three derivatives of $f(x)$, and hence can be influenced by a short wavelength perturbation, such as

$$f(x) = Cx(1 - x)(1 + A \sin Bx).$$

For large B and small A this can locally change the sign of the Schwarzian without influencing the global properties of $f(x)$. It is important to note that such a

Fig. D.3

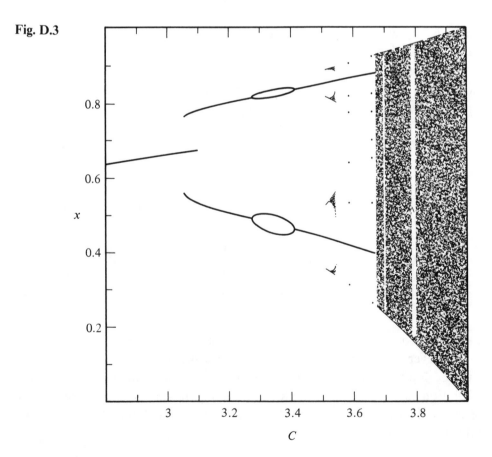

modification of a map might be very difficult to detect by empirical measurements, and yet have a major influence on the bifurcation sequence. Thus, if $A = -0.015$, $B = 31.6$, the bifurcation properties are shown in Fig. D.3. The 'unusual' period-2 and period-4 bifurcations occur at tangent bifurcations, due to a positive Schwarzian of $f(x)^2$, as well as of $f(x)$. Also, for $3.27 < C < 3.4$ there is a period-4 solution which appears and then disappears by a pitchfork bifurcation with period-2 solutions. Other anomalies are also evident in the figure.

Appendix E

The digraph method

To illustrate the digraph method (Straffin, 1978), consider a map which has a period-5, and assume that the iterates of p have the following values which define intervals I_k, as shown in Fig. E.1.

Fig. E.1

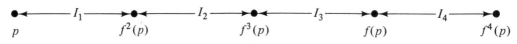

All of the points in I_3 can be obtained by the map of some points in I_1, that is $f(I_1) \supset I_3$. This follows from the fact that the end points of I_3 are $f(p)$ and $f^3(p)$, obtained from the map of p and $f^2(p)$, the end points of I_1. We can then use the Intermediate Value Theorem (e.g., G.H. Hardy, *A Course in Pure Mathematics* (Cambridge, 1952) p. 190) which states that, if $f(x)$ is continuous and $f(x_1) = p_1 < f(x_2) = p_2$ then, for any $p_3, p_1 \leqslant p_3 \leqslant p_2$, there is an x_3 between x_1 and x_2 such that $f(x_3) = p_3$. Therefore, $f(I_1)$ must contain all values between $f(p)$ and $f^3(p)$, so $f(I_1) \supset I_3$. Similarly

$$f(I_2) \supset I_3 \cup I_4 \text{(the union of } I_3 \text{ and } I_4\text{)}$$
$$f(I_3) \supset I_2 \cup I_3 \cup I_4; \quad f(I_4) \supset I_1$$

The *digraph representation* of this is shown in Fig. E.2. The desired theorem is

Fig. E.2

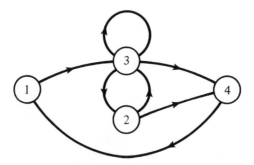

Theorem. *If there is a non-repetitive closed path in this diagram which has l 'steps', then there is a point in these sets which has period l.*

For example in the above figure there is a closed path, 133 323 241, so there is a point with period-8, and a path 1341, so there is a period-3 point. The non-repetitive condition simply means that only a proper periodic (with no subperiods) is allowed. For example, 134 134 is not a legitimate period-6. The proof of this is quite straightforward and based on the two lemmas:

Lemma 1

> *Suppose that I and J are closed intervals, f is continuous, and $f(I) \supset J$. Then there is closed $Q \subset I$ such that $f(Q) = J$.*

Lemma 2

> *Suppose I is a closed interval, f continuous, and $f(I) \supset I$. Then f has a fixed point in I.*

Now consider some sequence (non-repetitive) through the above graph which is closed, and write it as $I^0, I^1, \ldots, I^l = I^0$ (*l* steps). It may not be obvious that such a sequence can be accomplished, much less that there is a fixed point in this sequence $(x = f^l(x))$. To obtain this one has to work backwards. Since $f(I^{l-1}) \supset I^l$, then there is a $Q^{l-1} \subset I^{l-1}$ such that $f(Q^{l-1}) = I^l$ (Lemma 1). But then there is a $Q^{l-2} \subset I^{l-2}$ such that $f^2(Q^{l-2}) = I^l$. We can continue to 'back up', obtaining regions $Q^{l-k} \subset I^{l-k}$ which map in *k* steps back onto I^l. So finally we take $k = l$ and get $f^k(Q^0) = I^l$ where $Q^0 \subset I^0$.

Fig. E.3

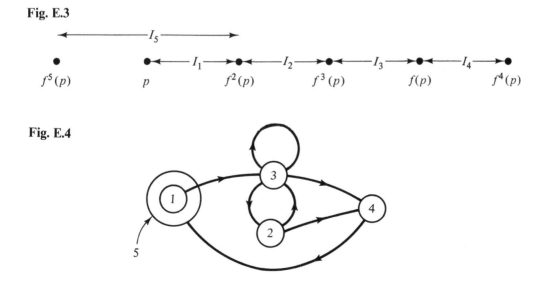

Fig. E.4

Fig. E.5

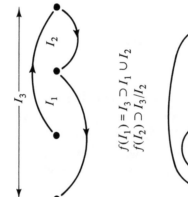

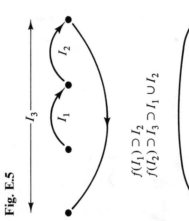

$f(I_1) \supset I_2$
$f(I_2) \supset I_3 \supset I_1 \cup I_2$

L_i–Yorke case

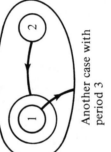

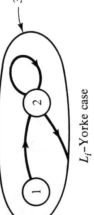

$f(I_1) = I_3 \supset I_1 \cup I_2$
$f(I_2) \supset I_3/I_2$

Another case with
period 3

But $I^l = I^0$, so $f^k(Q^0) = I^0 \supset Q^0$. Therefore, by Lemma 2, the set Q^0 must have a fixed point to the lth iterate map, or a period-l point to $f(x)$. This proves the theorem.

Actually this method can be usefully generalized to include nonperiodic sequences of the form shown in Fig. E.3. (now $f^5(p) \leqslant p$), which can be represented by a generalized digraph (Fig. E.4). Using the previous reasoning, there is a set $Q_1 I_1$ such that $f^k(Q_1) = I_5 Q_1$ (i.e., one does not require identity of I_5 and I_1). This again proves periodicity l points exist (in particular, $l = 5$). Thus the generalized digraph for the Li–Yorke case (Fig. E.5). Once one has period-3, then the digraph is which generates all periods. For a number of applications of this method, see Frauenthal (1979).

Appendix F

Elliptic integrals and elliptic functions

(e.g., see P.F. Byrd and M.D. Friedman, *Handbook of Elliptic Integrals for Engineers and Scientists*, Springer-Verlag, 1971).

First we will consider the simple equation

$$\left(\frac{dy}{dx}\right)^2 = a + by + cy^2 \tag{F.1}$$

where we are interested in only *real solutions*. Let $\Delta = b^2 - 4ac$, then we can write this equation in particularly transparent forms depending on whether $\Delta > 0$ or $\Delta < 0$. Let

$$\binom{\alpha}{\beta} = -\frac{b}{2c} \pm \frac{\Delta^{1/2}}{2c},$$

so that $dy/dx = 0$ if $y = \alpha$ or $y = \beta$ (which requires that α and β be real). Then, *for $\Delta > 0$*, we write equation (F.1) in the form

$$\left(\frac{dy}{dx}\right)^2 = (-c)(\alpha - y)(y - \beta) \tag{F.2}$$

and clearly, if $c < 0$, then $\alpha \geqslant y \geqslant \beta$ whereas, if $c > 0$, $y \geqslant \alpha$ or $\beta \geqslant y$. On the other hand, if $\Delta < 0$.

$$\left(\frac{dy}{dx}\right)^2 = c(y - \alpha)(y - \alpha^*) = c|y - \alpha|^2$$

and there are *only real solutions if $c > 0$.*

The solutions in the case (F.1) are then found to be either circular or hyperbolic functions. Specifically:

$$\left.\begin{array}{ll} y = -\dfrac{\Delta^{1/2}}{2c}\sin((-c)^{1/2}(x + x_0)) - \dfrac{b}{2c} & \Delta > 0, \quad c < 0 \\[2ex] y = \alpha + (\alpha - \beta)\sinh^2(\tfrac{1}{2}c^{1/2}(x + x_0)) & \Delta > 0, \quad c > 0, \quad y \geqslant \alpha \\[1ex] y = \beta - (\alpha - \beta)\sinh^2(\tfrac{1}{2}c^{1/2}(x + x_0)) & \Delta > 0, \quad c > 0, \quad \beta \geqslant y \\[2ex] y = \dfrac{(-\Delta)^{1/2}}{2c}\sinh(c^{1/2}(x + x_0)) - \dfrac{b}{2c} & \Delta < 0, \quad c > 0 \end{array}\right\}. \tag{F.3}$$

It might be noted that the first three are the *unique* solutions of (F.2) only if one requires that d^2y/dx^2 be continuous – because (F.2) does not satisfy the Lipschitz condition at the points $y = \alpha$ or $y = \beta$. On the other hand, the fourth solution is unique.

A natural generalization of (F.1) is

$$\left(\frac{dy}{dx}\right)^2 = a + by + cy^2 + dy^3 + ey^4$$

or, in terms of the roots $(\alpha, \beta, \gamma, \delta)$ of the right side

$$\left(\frac{dy}{dx}\right)^2 = \frac{e(y-\alpha)(y-\beta)(y-\gamma)(y-\delta) \quad (e \neq 0)}{d(y-\alpha)(y-\beta)(y-\delta) \quad\quad (e = 0)} \tag{F.4}$$

Of the great variety of such equations, two canonical forms have historically been selected, namely

$$\left(\frac{dy}{dx}\right)^2 = (1 - y^2)(1 - k^2 y^2) \quad (k^2 < 1) \tag{F.5}$$

$$\left(\frac{dy}{dx}\right)^2 = (1 - y^2)/(1 - k^2 y^2) \quad (k^2 < 1) \tag{F.6}$$

The relationship between (F.4) and (F.5) will be described shortly.

In both (F.5) and (F.6), either $y^2 < 1$ or $y^2 > 1/k^2$, because of the positive left side. Our primary interest will be with the solutions of equation (F.5), since these are related to nonlinear oscillators. We mention (F.6), briefly, only for some modest generality. The solutions of (F.5) and (F.6) are related to the *elliptic integrals* of the *first and second kind* respectively. The elliptic integral of the *first kind* is

$$F(x, k) = \int_0^x \frac{dy}{[(1 - y^2)(1 - k^2 y^2)]^{1/2}} \quad \text{or} \quad \int_0^\phi \frac{d\theta}{[1 - k^2 \sin^2 \theta]^{1/2}} = \bar{F}(\phi, k) \quad (k^2 < 1) \tag{F.7}$$

where the transformation $x = \sin \phi$ connects the two variables. The elliptic integral of the *second kind* is

$$E(x, k) = \int_0^x \left(\frac{1 - k^2 y^2}{1 - y^2}\right)^{1/2} dy \quad \text{or} \quad \bar{E}(\phi, k) = \int_0^\phi [1 - k^2 \sin^2 \theta]^{1/2} d\theta \quad (k^2 < 1) \tag{F.8}$$

and we will not discuss the properties of this function. The quantity k is called the *modulus* of the elliptic integral and k', given by $k'^2 = 1 - k^2$, the complementary modulus. Frequently α (not to be confused with the root in (F.4)) is introduced by the definitions

$$k = \sin \alpha \quad k' = \cos \alpha. \tag{F.9}$$

Finally the *complete elliptic integrals* (complimentary integrals) are defined by

$$K = F(1, k) = \bar{F}(\tfrac{1}{2}\pi, k) \quad (\text{or } K' = F(1, k') = \bar{F}(\tfrac{1}{2}\pi, k')) \tag{F.10}$$

The relationship between equation (F.4) and the canonical form (F.5) can be established by means of various transformations. Thus let

$$z^2 = \left(\frac{\beta - \delta}{\alpha - \delta}\right)\left(\frac{y - \alpha}{y - \beta}\right), \quad k^2 = \left(\frac{\beta - \gamma}{\alpha - \gamma}\right)\left(\frac{\alpha - \delta}{\beta - \delta}\right) \tag{F.11}$$

then, if y satisfies the first equation of (F.4), z satisfies the equation

$$\left(\frac{dz}{dx}\right)^2 = \frac{e}{4}(\beta - \delta)(\alpha - \gamma)(1 - z^2)(1 - k^2 z^2)$$

which, with a scaling of the variable x, is equation (4.5). Obviously (4.11) imposes a number of conditions on y and the roots $(\alpha, \beta, \gamma, \delta)$. Under different conditions, the following relationship may be possible

$$z^2 = \frac{\alpha - \gamma}{y - \gamma}, \quad k^2 = \frac{\beta - \gamma}{\alpha - \gamma}. \tag{F.12}$$

Then if y satisfies the second equation of (F.4), z satisfies the equation

$$\left(\frac{dz}{dx}\right)^2 = \frac{d}{4}(\alpha - \gamma)(1 - z^2)(1 - k^2 z^2)$$

which is again effectively (F.5). This illustrates some possible connections between (F.4) and (F.5).

The elliptic functions are the inverse of elliptic integrals. Thus

$$sn^{-1}(x, k) = \int_0^x \frac{dy}{[(1 - y^2)(1 - k^2 y^2)]^{1/2}} = F(x, k) \qquad (0 \leqslant x \leqslant 1)$$

$$cn^{-1}(x, k) = \int_x^1 \frac{dy}{[(1 - y^2)(k'^2 + k^2 y^2)]^{1/2}} = F([1 - x^2]^{1/2}, k) \qquad (0 \leqslant x \leqslant 1)$$

$$dn^{-1}(x, k) = \int_x^1 \frac{dy}{[(1 - y^2)(k'^2 + k^2 y^2)]^{1/2}} = F([(1 - x^2)]^{1/2}/k^2, k) \qquad (k' \leqslant x \leqslant 1) \quad (F.13)$$

$$tn^{-1}(x, k) = \int_0^x \frac{dy}{[(1 - y^2)(1 + k'^2 y^2)]^{1/2}} = F([x^2/(1 + x^2)]^{1/2}), k) \qquad (0 \leqslant x \leqslant \infty)$$

$$am^{-1}(\phi, k) = \int_0^\phi \frac{d\theta}{[1 - k^2 \sin^2 \theta]^{1/2}} = \bar{F}(\phi, k) = F(\sin \phi, k)$$

and if

$$F(x, k) = u$$

$$sn(u, k) = x \equiv \sin \phi; \quad cn(u, k) = \cos \phi; \quad dn(u, k) = [1 - k^2 \sin^2 \phi]^{1/2}$$
$$tn(u, k) = \tan \phi; \quad am(u, k) = \phi \, (\text{amplitude}). \tag{F.14}$$

For $k \ll 1$ and $k \simeq 1$ ($k' \ll 1$) we have the approximate relationships

$$\left.\begin{aligned}
sn(u, k) &\simeq \sin(u) - 1/4 \, k^2 \cos(u)(u - \sin(u)\cos(u)) \\
cn(u, k) &\simeq \cos(u) + 1/4 \, k^2 \sin(u)(u - \sin(u)\cos(u)) \quad k \ll 1 \\
dn(u, k) &\simeq 1 - 1/2 \, k^2 \sin^2(u)
\end{aligned}\right\} \tag{F.15}$$

and

$$sn(u, k) \simeq \tanh(u) + 1/4k'^2 \operatorname{sech}^2 u(\sinh(u)\cosh(u) - u)$$
$$cn(u, k) \simeq \operatorname{sech}(u) - 1/4k'^2 \tanh(u)\operatorname{sech}(u)(\sinh(u)\cosh(u) - u; \quad k' \ll 1$$
$$dn(u, k) \simeq \operatorname{sech}(u) + 1/4k'^2 \tanh(u)\operatorname{sech}(u)(\sinh(u)\cosh(u) + u)$$

The complete elliptic integral of the first kind has the expansion

$$K(k) = \frac{\pi}{2}\left[1 + (\tfrac{1}{2})^2 k^2 + \left(\frac{1 \cdot 3}{2 \cdot 4}\right)^2 k^4 + \left(\frac{1 \cdot 3 \cdot 5}{2 \cdot 4 \cdot 6}\right)^2 k^6 + \cdots \right] \tag{F.16}$$

It is also referred to as the *quarter period* (corresponding to $\pi/2$ for the circular functions), because

$$sn(u + 2K, k) = -sn(u, k); \quad cn(u + 2K, k) = -cn(u, k) \tag{F.17}$$

For small values of k $sn(u) \simeq \sin(u)$, $cn(u) \simeq \cos(u)$, $dn(u) \simeq 1$ whereas for larger values

Fig. F.1

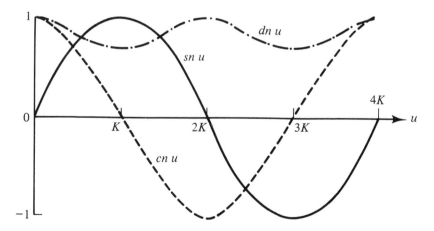

of k they approach $\tanh(u)$, $\operatorname{sech}(u)$, and $\operatorname{sech}(u)$ respectively (Fig. F.1). Recall that $k = \sin\alpha$ defines α (or k). One has the values

$$
\begin{array}{cccccc}
\alpha = & 0° & 10° & 20° & 30° & 40° \\
K = 1.571(=\pi/2) & 1.5828 & 1.6200 & 1.6857 & 1.7868 & \\
K/\pi/2 = \quad 1 & & 1.008 & 1.03 & 1.07 & 1.135
\end{array}
\tag{F.18}
$$

$$
\begin{array}{cccccc}
a = & 50° & 60° & 70° & 80° & 89° \\
K = 1.9356 & & 2.1565 & 2.5045 & 3.1534 & 5.435 \\
K/\pi/2 = 1.23 & & 1.37 & 1.65 & 2.00 & 3.46
\end{array}
\tag{F.19}
$$

Appendix G

The Poincaré–Bendixson theorem and Birkhoff's α and ω limit sets

As discussed in Chapter 5, Section 7, the Poincaré–Bendixson theorem concerns a general system in the plane

$$\dot{x} = P(x, y), \quad \dot{y} = Q(x, y) \tag{G.1}$$

which is assumed to have unique solutions in some region D of the plane (e.g., $P(x, y)$ and $Q(x, y)$ satisfy a Lipschitz condition in D). Moreover it is assumed that these solutions exist for all $0 \leqslant t < \infty$. Let K be a compact subset of D, and assume that there is some solution, γ, of (A) which remains in this region, K, as time goes to infinity

$$\lim_{t \to \infty} (x(t), y(t)) \in K; \quad \gamma = \{x(t), y(t); 0 \leqslant t < \infty\}.$$

The theorem states that the region K must contain at least one fixed point of (G.1) or else a periodic orbit, Γ (there may be any number of each, but there must be one or the another in K). Moreover the solution γ must tend either to the fixed point or to the periodic solution as $t \to +\infty$, in which case it is the limit cycle of the solution.

In this appendix, we will discuss the Poincaré–Bendixson theorem in more detail, and get some practice in the use of the limit sets and the point groups of Birkhoff (1922). As usual, the discussion is expository rather than rigorous mathematics, but hopefully some of the beauty will shine through.

Recall that the orbit $\gamma(x_0)$ of $\dot{x} = F(x)$ associated with the initial value $(t = 0)$ point $x_0 \in R^n$ is the set $\gamma(x_0) = \{x = f^t x_0: -\infty < t < +\infty\}$, which we assume to exist and to be unique. The *positive semiorbit* of x_0, $\gamma^+(x_0)$, is the set of points generated by the solution in the future, $\gamma^+(x_0) = \{x = f^t x_0: 0 \leqslant t < \infty\}$, and similarly for the *negative semiorbit*, $\gamma^-(x_0) = \{x = f^t x_0: -\infty < t \leqslant 0\}$.

Associated with any finite segment of an orbit (not containing a singular point) is an open *path cylinder* which is the cross-product of an $(n-1)$ dimensional solid sphere transverse to γ and an interval along γ (see Fig. G.1, where the 2-sphere is a disk). Because of the uniqueness of the solutions, and their differentiability with respect to initial conditions, a finite cylinder exists in which all orbits have the same orientation

Fig. G.1

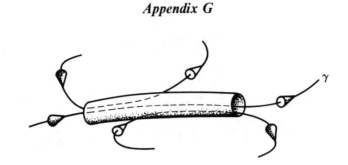

as γ. The important point here is that this cylinder has a finite 'cross-section' in which all the orbits progress along the interval in the same direction as the oriented γ.

Finally we introduce the concept of the ω-*limit* (and α-*limit*) *set of an orbit* γ, where we now drop the indication of the point x_0 if it is not important. These sets are indicated as $\omega(\gamma)$ and $\alpha(\gamma)$ respectively. A point q is contained in the set $\omega(\gamma)$ (or $\alpha(\gamma)$) if there exists a sequence of times $\{t_k\}$, where $t_k \to \infty$ as $k \to \infty$ (or else $t_k \to -\infty$ as $k \to \infty$, respectively) and such that q is the limit of the orbit $\gamma(x_0)$ in the future (or in the past, respectively):

$$
\left.
\begin{aligned}
q \in \omega(\gamma) &\quad \text{if} \quad \lim_{k \to \infty} x(t_k, x_0) = q \quad (t_k \to +\infty) \\[2mm]
q \in \alpha(\gamma) &\quad \text{if} \quad \lim_{k \to \infty} x(t_k, x_0) = q \quad (t_k \to -\infty)
\end{aligned}
\right\}. \tag{G.2}
$$

In other words, the distance between $x(t_k, x_0)$ and q goes to zero as $k \to \infty$. This concept may appear rather abstract and unnecessarily obtuse at first, but it turns out to be both useful and necessarily precise for the description of future dynamics.

These sets were defined and discussed first by Birkhoff (1922), in Chapter 5 of a lengthy (119 page) paper. In this remarkable chapter he discusses the possible behavior of an infinite set of points,

$$
\ldots, T^{-2}(P), \quad T^{-1}(P), \quad P, \quad T(P), \quad T^2(P), \ldots,
$$

generated by a one-to-one analytic map, T, of any point P in a closed space, S (e.g., a bounded region of R^n). The surprising fact is that this intellectual exercise uncovered exotic forms of dynamics which later were found to actually occur in fairly simple physical systems (see the studies of Cartwright and Littlewood, and of Levinson).

It is perhaps useful to pause here to give some illustrations, before discussing the Poincaré–Bendixson theorem further. In the case of the illustrated saddle point and the indicated orbits we have the limit sets: $\omega(\gamma_1) = \alpha(\gamma_2) = (0,0)$, whereas, if γ_1 and γ_2 are unbounded, $\alpha(\gamma_1) = \omega(\gamma_2) = \varnothing$ (the empty set). Similarly, if γ_3 is unbounded, $\omega(\gamma_3) = \varnothing = \alpha(\gamma_3)$. In the case of the two nodes at $(0,0)$ and $(1,0)$ $\omega(\gamma_4) = \omega(\gamma_1) = \omega(\gamma_2) = (0,0)$ and $\alpha(\gamma_1) = \alpha(\gamma_2) = \alpha(\gamma_3) = (1,0)$. Finally $\alpha(\gamma_4) = \varnothing = \omega(\gamma_3)$, if there γ are unbounded. If γ is a periodic orbit, then every point belongs to both limit sets, $\alpha(\gamma) = \omega(\gamma) = \gamma$.

A set of points $\{P_k\}$ is called a *minimal set* if it is nonempty, closed (contains its limit

Fig. G.2

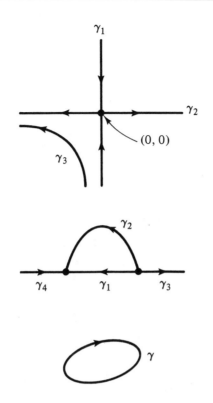

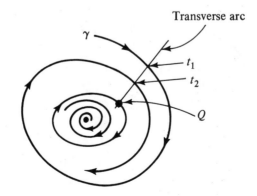

points), invariant and has no subset possessing these properties. The α and ω sets of bounded orbits are both minimal. If a minimal set is bounded then it is either a fixed point or a periodic orbit. The semistable limit cycle in Fig. G.2 is the ω-limit set for any exterior orbit. This is because, to any point, Q, on the cycle, there is clearly a sequence of times when γ crosses a transverse arc (see Fig. G.3) and these are such that $x(t_k)$ approaches Q as $t_k \to +\infty$. Since this is true for any Q on the limit cycle, it is the ω-limit set for γ. Likewise the limit cycle is the α-limit set for any interior orbit, whereas the origin is the ω-limit set for these orbits. Again these sets are minimal.

Fig. G.3

Fig. G.4

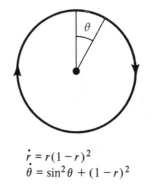

$$\dot{r} = r(1-r)^2$$
$$\dot{\theta} = \sin^2\theta + (1-r)^2$$

Exercise. Consider the system indicated in Fig. G.4. Determine the α and ω-limit sets for the orbits with the following 'initial' conditions: (a) $r_0 > 1$, (b) $r_0 = 0$, (c) $1 > r_0 > 0$, (d) $r_0 = 1$, $\theta_0 = \pi/2$. Are any of these limit sets not minimal? (Comments on this are in Chapter 5.)

We now return to the Poincaré–Bendixson theorem, applying these point group concepts of Birkhoff. Let γ^+ be a positive semiorbit in a bounded region K which contains no singular points. Let the set of times $\{t_n\}$ satisfy $t_n \to \infty$ as $n \to \infty$. If $\gamma^+ = \{x(t): t > 0\}$ then the infinite set of bounded points $\{x(t_n)\}$ must have at least one subset with a limit point (the Bolzano–Weierstrass theorem). Call it p. The fact that this limit exists means that the ω-limit set, $\omega(\gamma^+)$, is not empty. Now either p is a point of γ^+ or else it is not. We show first that p is not in γ^+. If p were in γ^+ then there must be a sequence $\{t_n\}$ such that $\lim_{t_n \to \infty} x(t_n) = p$; in other words, γ^+ must revisit p (hence itself) in a limiting fashion. Consider a transverse arc (Fig. G.5) through γ^+ at p. In order to approach p in a limiting fashion, γ^+ must cross some transverse arc in both directions arbitrarily close to p. Since γ^+ does not have a singular point, it has a path cylinder, and hence this limiting behavior is impossible.

Fig. G.5

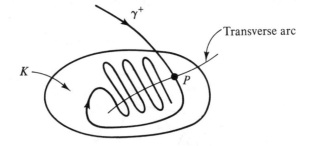

Transverse arc

γ^+

K

P

Therefore, associated with p is another orbit, Γ (Fig. G.6) distinct from γ^+, $\Gamma = \{f^t p: t > 0\}$. But if p is a limit point of some set $\{x(t_n)\} \in \gamma^+$ then the set $\{x(t_n + t)\} \in \gamma^+$ must have as its limit point $f^t p$. This follows from the uniqueness and continuity of the

Fig. G.6

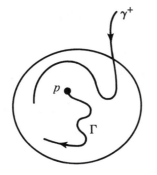

solutions of $\dot{x} = F(x)$. Thus all the points of Γ are limit points of some set $\{x(t_n)\}$ from γ^+. Put another way, Γ must be part of the ω-limit set of γ^+, $\omega(\gamma^+)$. Since γ^+ is bounded, Γ must also be bounded (inside K). The claim now is that Γ must be a closed orbit. This is required in order for γ^+ to revisit each point of Γ in a limiting fashion, as you can convince yourself without much difficulty. Formally, the set Γ is a minimal set, since it is invariant ($f^t\Gamma = \Gamma$) and has no invariant subset. Bounded minimal sets are either singular points or periodic orbits (see e.g., Hale (1980)), and Γ is not a singular point. Finally for a bounded γ^+, the set $\omega(\gamma^+)$ must be both compact and connected (not in several disconnected parts), so Γ is all of the $\omega(\gamma^+)$ set. This completes a 'plausibility proof' of the Poincaré–Bendixson theorem.

Appendix H

A modified fourth-order Runge–Kutta integration method

A simple, and usually very adequate, numerical integration method is the following fourth-order Runge–Kutta method (for a listing of many integration programs, see Press *et al.* (1986)). The method applies to a system of ODE

$$\dot{x} = F(x, t) \quad (x \in R^m), \tag{H.1}$$

and uses a fixed time step, $t = n\Delta$ ($n = 0, 1, 2, \ldots$), with $x_n \equiv x(n\Delta)$. The accuracy of the method at each time step is $C\Delta^5$, where C depends on the function in (H.1).

The values of x_{n+1} are obtained from the previous values, x_n, by the following prescription:

$$x_{n+1} = x_n + \tfrac{1}{6}(K_1 + 2K_2 + 2K_3 + K_4) + O(\Delta^5) \tag{H.2}$$

where

$$\left. \begin{array}{l} K_1 \equiv \Delta \cdot F(x_n, n\Delta) \\ K_2 \equiv \Delta \cdot F(x_n + \tfrac{1}{2}K_1, (n + \tfrac{1}{2})\Delta) \\ K_3 \equiv \Delta \cdot F(x_n + \tfrac{1}{2}K_2, (n + \tfrac{1}{2})\Delta) \\ K_4 \equiv \Delta \cdot F(x_n + K_3, (n + 1)\Delta). \end{array} \right\} \tag{H.3}$$

Note that the x_n, K_i, and F are all in R^m. To handle these vector components, we can define the functions $K(J, I) \equiv \Delta \cdot F_I(y(J), T(J))$ ($I = 1, \ldots, m$: the number of first order ODE) in a subroutine. Then by sequentially defining the floating variables (y, T), so as to match the arguments of F in (H.3), namely $y(1) = x_n$, $T(1) = n\Delta$; $y(2) = x_n(I) + \tfrac{1}{2}K(1, I)$, $T(2) = (n + \tfrac{1}{2})\Delta$, etc., the variables

$$x_{n+1}(I) = x_n(I) + \tfrac{1}{6}(K(1, I) + 2K(2, I) + 2K(3, I) + K(4, I)) \tag{H.4}$$

can be computed for all $I = 1, \ldots, m$.

When the dynamics involves *fast* and *slow manifolds*, it is usually necessary to use a variable size time step. This permits the iteration to take larger time steps on the slow manifolds, where the action progresses very slowly and the accuracy is not diminished by the use of larger time steps. There are many such methods, but they need to be used with care, since they can 'run away' (become unstable) in some cases.

One such variable time step method is the Merson modification of the above Runge–Kutta scheme. It involves the introduction of one more iteration, plus a check for accuracy. In this method (H.3) is replaced by

$$
\left.
\begin{aligned}
K_1 &= \Delta \cdot F(x_n, t_n) \\
K_2 &= \Delta \cdot F(x_n + \tfrac{1}{3}K_1, t_n + \tfrac{1}{3}\Delta) \\
K_3 &= \Delta \cdot F(x_n + \tfrac{1}{6}(K_1 + K_2), t_n + \tfrac{1}{3}\Delta) \\
K_4 &= \Delta \cdot F(x_n + \tfrac{1}{8}(K_1 + 3K_3), t_n + \tfrac{1}{2}\Delta) \\
K_5 &= \Delta \cdot F(y_5, t_n + \Delta)
\end{aligned}
\right\}
\tag{H.5}
$$

where

$$
y_5 \equiv x_n + (1/2)K_1 - (3/2)K_3 + 2K_4.
$$

Finally, we obtain

$$
y_6 \equiv x_n + \tfrac{1}{6}(K_1 + 4K_4 + K_5),
\tag{H.6}
$$

and obtain the following estimate of the error of x_{n+1},

$$
E = \tfrac{1}{5}(y_5 - y_6).
\tag{H.7}
$$

If ε is the maximum acceptable error, then:

$$
\begin{aligned}
&(1)\ \text{If } |E| < \varepsilon, \quad \text{set } x_{n+1} = y_6, \quad \text{and 'increase } \Delta\text{';} \\
&(2)\ \text{If } |E| > \varepsilon, \quad \text{set}
\end{aligned}
\tag{H.8}
$$

$$
\Delta_{\text{new}} = 0.8(\varepsilon/|E|)^{0.2}\Delta_{\text{old}},
\tag{H.9}
$$

which reduces the time step. Then repeat (H.5) with the new Δ. It is sometimes suggested that (H.9) can be used to 'increase Δ' in (H.8) when $|E| < \varepsilon$. However, they may lead to a large increase, and instability. One possibility is to simply set $\Delta_{\text{new}} = 1.001\,\Delta_{\text{old}}$ (or some such fixed multiple). For more sophisticated approaches, see Press Flannery, Teukolsky and Vetterling (1986).

Appendix I

The Stoker–Haag model of relaxation oscillations

Singular perturbation calculations are generally quite laborious. The following model, however, illustrates the essential aspects of such problems, while requiring a limited amount of algebra. We consider the equation

$$\ddot{x} + F(\varepsilon \dot{x}) + x = 0 \qquad (I.1)$$

where, setting $v = \varepsilon \dot{x}$, the function $F(v)$ is given by

$$F(v) = -v \quad |v| \leqslant 1$$

$$F(v) = -(2-v) \quad \text{if} \quad v \geqslant 1; \quad \text{and} \quad F(v) = -(2+v) \quad \text{if} \quad v \leqslant -1.$$

This piecewise linear function, which was discussed in Chapter 5, is illustrated in Fig. I.1.

Fig. I

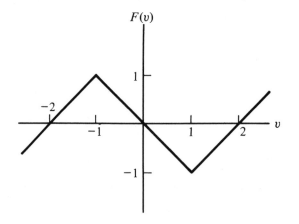

We are interested in obtaining the period, $T(\varepsilon)$, of the periodic (limit cycle) solution of (I.1) in the limit $\varepsilon \to \infty$. As discussed in Chapter 5, we expect that, in this limit, $x = -F(v)$ for $|v| \geqslant 1$ (the 'slow manifolds'), and $x = +1$ or -1 if $|v| \leqslant 1$ (the 'fast manifolds'). We thereby easily obtain the period in the limit $\varepsilon \to \infty$,

$$T(\varepsilon) \simeq 2\varepsilon \int_{-1}^{1} \frac{dx}{v} = 2\varepsilon \int_{-1}^{1} \frac{dx}{2-x} = -2\varepsilon \ln(2-x)|_{-1}^{1} = 2\varepsilon \ln(3).$$

To obtain the period for finite ε, we need to obtain a periodic solution by connecting solutions in the regions 1 and 2 in (Fig. I.1). We take as an initial condition

$$v(0) = +1 \quad x(0) = 1 + \Delta(\varepsilon) \quad (\Delta(\varepsilon) \text{ to be determined})$$

and let $x = x_1(t)$, $x = x_2(t)$ be the solutions of

$$\ddot{x}_1 - (2 - \varepsilon \dot{x}_1) + x_1 = 0, \quad \ddot{x}_2 - \varepsilon \dot{x}_2 + x_2 = 0 \tag{I.2}$$

in regions 1 and 2 respectively (Fig. I.2). The solutions will be periodic provided that

$$x_1(t_1) = -x_2(t_2),$$

where (t_1, t_2) are given by $v_1(t_1) = +1 \ (x_1 < 0); \ v_2(t_2) = -1$.

Fig.I.2

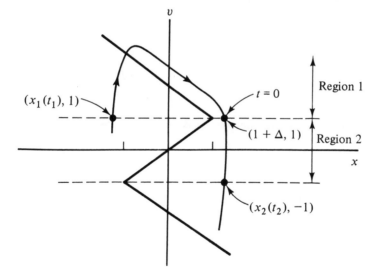

$(x_1(t_1), 1)$

$t = 0$

Region 1

$(1 + \Delta, 1)$ Region 2

x

$(x_2(t_2), -1)$

In general the solutions of (I.2) are

$$\begin{aligned}
x_1(t) &= 2 + c_3 \exp(-\lambda_1 t) + c_4 \exp(-\lambda_2 t), \\
x_2(t) &= c_1 \exp(\lambda_1 t) + c_2 \exp(\lambda_2 t),
\end{aligned} \tag{I.3}$$

where

$$\begin{pmatrix} \lambda_1 \\ \lambda_2 \end{pmatrix} = \tfrac{1}{2}\varepsilon \pm \tfrac{1}{2}(\varepsilon^2 - 4)^{1/2}. \tag{I.4}$$

Equating x_1 and x_2 at $t = 0$, yields

$$c_1 + c_2 = 2 + c_3 + c_4 = 1 + \Delta \equiv x(0), \tag{I.5}$$

whereas continuous derivatives require,

$$\varepsilon(\lambda_1 c_1 + \lambda_2 c_2) = \varepsilon(-\lambda_1 c_3 - \lambda_2 c_4) = 1 \equiv v(0). \tag{I.6}$$

To have a periodic solution we need the times (t_1, t_2), such that $v_1(t_1) = +1$,

$$\varepsilon(-\lambda_1 c_3 \exp(-\lambda_1 t_1) - \lambda_2 c_4 \exp(-\lambda_2 t_1)) = +1 \quad (t_1 < 0), \tag{I.7}$$

and $v_1(t_2) = -1$,

$$\varepsilon(\lambda_1 c_1 \exp(\lambda_1 t_2) + \lambda_2 c_2 \exp(\lambda_2 t_2)) = -1 \quad (t_2 > 0) \tag{I.8}$$

and satisfying $x_1(t_1) = -x_2(t_2)$,

$$2 + c_3 \exp(-\lambda_1 t_1) + c_4 \exp(\lambda_2 t_1) = -(c_1 \exp(\lambda_1 t_2) + c_2 \exp(\lambda_2 t_2)). \tag{I.9}$$

We now have the seven conditions needed to determine the unknowns $(c_1, c_2, c_3, c_4, \Delta, t_1, t_2)$, and the quantity of interest is of course the period, $T = 2(t_2 - t_1)$. For large ε, we obtain from (I.4)

$$\lambda_1 = \varepsilon\left(1 - \frac{1}{\varepsilon^2} - \frac{1}{\varepsilon^4} - \frac{2}{\varepsilon^6} - \frac{5}{\varepsilon^8} + \cdots\right)$$

$$\lambda_2 = \frac{1}{\varepsilon}\left(1 + \frac{1}{\varepsilon^2} + \frac{2}{\varepsilon^4} + \frac{5}{\varepsilon^6} + \frac{14}{\varepsilon^8} + \cdots\right)$$

and $\lambda_1 \lambda_2 = 1$. Moreover, since the period is $O(\varepsilon)$ and t_1 is the interval along the slow manifold, we expect $t_1 = O(\varepsilon)$, so

$$-\lambda_1 t_1 = O(+\varepsilon^2); \quad -\lambda_2 t_1 = O(1).$$

From (I.7) it follows that $\varepsilon \lambda_1 c_3 = O(\exp(-\varepsilon^2))$, and hence (I.6) yields to exponential order

$$c_4 = -\frac{1}{\varepsilon \lambda_2}. \tag{I.10}$$

Moreover, from (I.5) and (I.10), to exponential order

$$\Delta \sim 1 - \frac{1}{\varepsilon \lambda_2} \equiv 1 - \frac{\lambda_1}{\varepsilon} \tag{I.11}$$

Next, using (I.5) and (I.6)

$$\varepsilon(\lambda_1 - \lambda_2)c_1 = 1 - \varepsilon \lambda_2(1 + \Delta) = 2(1 - \varepsilon \lambda_2)$$

$$= -\frac{2}{\varepsilon^2}\left(1 + \frac{2}{\varepsilon^2} + \frac{5}{\varepsilon^6} + \frac{14}{\varepsilon^8} + \cdots\right)$$

or

$$c_1 \varepsilon^2 \left(1 - \frac{2}{\varepsilon^2} - \frac{2}{\varepsilon^4} - \frac{4}{\varepsilon^6} - \frac{10}{\varepsilon^8} - \cdots \right) = -\frac{2}{\varepsilon^2} \left(1 + \frac{2}{\varepsilon^2} + \frac{5}{\varepsilon^6} + \frac{14}{\varepsilon^8} + \cdots \right)$$

so

$$c_1 \sim -\frac{2}{\varepsilon^4} \left(1 + \frac{4}{\varepsilon^2} + \frac{15}{\varepsilon^4} + \cdots \right) \qquad (I.12)$$

Similarly from (I.5) and (I.6),

$$\varepsilon(\lambda_2 - \lambda_1)c_2 = 1 - \varepsilon\lambda_2(1 + \Delta)$$

or

$$c_2 \sim 1 + \frac{1}{\varepsilon^2} + \frac{3}{\varepsilon^4} + \frac{10}{\varepsilon^6} + \frac{35}{\varepsilon^8} + \cdots. \qquad (I.13)$$

The development of t_2 can now be obtained from (I.8), using the fact that

$$\lim_{\varepsilon \to \infty} t_2 = 0, \quad \text{so} \quad \lim_{\varepsilon \to \infty} \exp(\lambda_2 t_2) = 1$$

for the purpose of iteration. Thus (I.8) gives

$$\exp(\lambda_1 t_2) = -\frac{1}{\varepsilon\lambda_1 c_1} - \frac{\lambda_2 c_2}{\lambda_1 c_1} \exp(\lambda_2 t_2) \sim$$

$$-\frac{(\lambda_1 - \lambda_2)}{2\lambda_1(1 - \varepsilon\lambda_2)} + \frac{\lambda_2 + \lambda_1 - 2\varepsilon}{2(1 - \varepsilon\lambda_2)} \exp(\lambda_2 t_2)$$

Now, using (I.12) and (I.13), iteration yields

$$t_2 = \frac{1}{\varepsilon} \left[2\ln(\varepsilon) + \frac{3\ln(\varepsilon)}{\varepsilon^2} - \frac{2}{\varepsilon^2} + \frac{1}{2}\frac{\ln^2(\varepsilon)}{\varepsilon^4} + O\left(\frac{\ln(\varepsilon)}{\varepsilon^4}\right) \right]. \qquad (I.14)$$

To obtain t_1 we cannot proceed in the same way from (I.7), because of the troublesome term $c_3 \exp(-\lambda_1 t_1)$ which may tend to $O(1)$ as $\varepsilon \to \infty$, but whose value is not known. It is here that the periodicity condition, (I.9), must be introduced to eliminate $c_3 \exp(-\lambda_1 t_1)$. Note that everything which has been obtained so far follows from the single assumption that $\lim_{\varepsilon \to \infty} t_1 = O(\varepsilon)$, and not from any periodicity requirement. Making this elimination between (I.9) and (I.7) yields

$$\varepsilon c_4(\lambda_1 - \lambda_2) \exp(-\lambda_2 t_2) = 1 - \lambda_1\varepsilon(2 + c_1 \exp(\lambda_1\lambda_2) + c_2 \exp(\lambda_2 t_2))$$

$$= 1 - \varepsilon\lambda_1 \left(2 - \frac{1}{\varepsilon\lambda_1} + c_2 \left(1 - \frac{\lambda_2}{\lambda_1} \right) \exp(\lambda_2 t_2) \right)$$

so

$$c_4(\lambda_1 - \lambda_2) \exp(-\lambda_2 t_1) = 2 - 2\varepsilon\lambda_1 - \varepsilon c_2(\lambda_1 - \lambda_2) \exp(\lambda_2 t_2)$$

$$-\frac{1}{\varepsilon\lambda_2}(\lambda_1 - \lambda_2) \exp(\lambda_2 t_1) = 2 - 2\varepsilon\lambda_1 + \left[1 - \varepsilon\lambda_1 \left(2 - \frac{1}{\varepsilon\lambda_2} \right) \right] \exp(\lambda_2 t_2).$$

This gives

$$t_1 = -\varepsilon\left[\ln(3) + \frac{2}{3}\frac{\ln(\varepsilon)}{\varepsilon^2} + \left(\frac{4}{3} - \ln(3)\right)\frac{1}{\varepsilon^2} + \frac{4}{9}\frac{\ln^2(\varepsilon)}{\varepsilon^4} + O\left(\frac{\ln(\varepsilon)}{\varepsilon^4}\right)\right]$$

so, finally

$$T(\varepsilon) = 2(t_2 - t_1) \sim 2\left[\varepsilon\ln(3) + \frac{8}{3}\frac{\ln(\varepsilon)}{\varepsilon}\right.$$

$$\left. + \left(\frac{4}{3} - \ln(3)\right)\frac{1}{\varepsilon} + \frac{4}{9}\frac{\ln^2(\varepsilon)}{\varepsilon^4} + O\left(\frac{\ln(\varepsilon)}{\varepsilon^3}\right)\right] \qquad (I.15)$$

A comparison of (I.15) can be made with the result given in Section 8 of Chapter 5, for a smooth function $F(v)$ in (I.1), rather than the present piecewise linear function.

Another detailed example, for a simplified form of Prigogine's biochemical reaction model (the *Brusselator model*), namely

$$\dot{x} = A - (B+1)x + x^2 y; \quad \dot{y} = Bx - x^2 y \quad (A, B > 0)$$

can be found in J.A. Boa, Asymptotic calculation of a limit cycle, *J. Math. Anal. Appl.* **54**, 115–37 (1976).

Bibliography

Ablowitz, M.J. (1981). Remarks on nonlinear evolution equations and inverse scattering transform. In *Nonlinear Phenomena in Physics and Biology*, ed. R. Enns *et al.*

Ablowitz, M.J. & Fokas, A.S. (1983). Comments on the inverse scattering transform and related nonlinear evolution equations. In *Nonlinear Phenomena*. Lecture Notes in Physics 189, ed. Wolf, K.B. pp. 4–23. Springer-Verlag.

Ablowitz, M.J. & Nachman, A.I. (1986). Multidimensional nonlinear evolution equations and inverse scattering, *Physica* **18D**, 223–41.

Ablowitz, M.J., Kaup, D.J., Newell, A.C. & Segur, H. (1973). Method for solving the Sine–Gordon Equations, *Phys. Lett.* **30**, 1262–4.

Ablowitz, M.J., Kaup, D.J. & Newell, A.C. (1974). Coherent pulse propagation, a dispersive irreversible phenomena. *J. Math. Phys.* **15**, 1852–8.

Ablowitz, M.J., Kaup, D.K. Newell, A.C. & Segur, H. (1974). The inverse scattering transform-fourier analysis for nonlinear problems. *Studies Appl. Math.* **53**, 249–315.

Ablowitz, M.J. & Satsuma, J. (1978). Solitons and rational solutions of nonlinear evolution equations. *J. Math. Phys.* **19**, 2180–6.

Ablowitz, M.J., Ramani, A. & Segur, H. (1978). Nonlinear evolution equations and ordinary differential equations. *Lett. Nuovo. Cimento* **23**, 333–8.

Ablowitz, M.J., Ramani, A. & Segur, H. (1980). A connection between nonlinear evolution equations and ordinary differential equations of p-type. I. *J. Math. Phys.* **21**, 715–21.

Ablowitz, M.J. & Satsuma, J. (1978). Solitons and rational solutions of nonlinear evolution equations. *J. Math. Phys.* **19**, 2180–6.

Ablowitz, M.J. & Segur, H. (1977a). Exact linearization of a Painlevé transcendent. *Phys. Rev. Letters* **38**, 1103–6.

Ablowitz, M.J. & Segur, H. (1977b). Asymptotic solutions of the Korteweg–deVries equation. *Studies Appl. Math.* **57**, 13–44.

Ablowitz, M.J. & Segur, H. (1981). *Solitons and the Inverse Scattering Transform*. SIAM.

Abraham, E. & Smith, S.D. (1982). Optical bistability and related devices. *Rep. Prog. Phys.* **45**, 815–85.

Abraham, R. & Marsden, J.E. (1978). *Foundations of Mechanics*, 2nd edn. Benjamin/Cummings. (Contains: A.N. Kolmogorov's 1954 Address to the International Congress of Mathematicians: 'The General Theory of Dynamical Systems and Classical Mechanics'.)

Abraham, R. & Shaw, C. *Dynamics: The Geometry of Behavior*; Part one: *Periodic Behavior* (1982); Part two: *Chaotic Behavior* (1983); Part three: *Global Behavior* (1983); Part four: *Bifurcation behavior* (1984). Aerial Press: Santa Cruz, CA.

Abraham, R. & Simó, C. (1985). Bifurcations and chaos in forced van der Pol systems. In *Singularities and Dynamical Systems*, ed. S.N. Pnevmatikos, pp. 313–44. Elsevier Sci. Pub.

Abraham, N.B., Gollub, J.P. & Swinney, H.L. (1984). Testing nonlinear dynamics, *Physica* **11D**, 252–64.

Adey, W.R. (1975). Evidence for cooperative mechanisms in the susceptibility of cerebral tissue to environmental and intrinsic electric fields. In *Functional Linkage in Biomolecular Systems*, ed. F.O. Schmit, D.M. Schneider & D.M. Crothers, Raven Press: New York.

Adler, R.L., Konheim, A.G. & McAndrew, M.H. (1965). Topological Entropy, *Trans. Amer. Math. Soc.* **114**, 309–19.

Ahlers, G. & Behringer, R. (1978a). Evolution of turbulence from the Rayleigh–Benard Instability. *Phys. Rev. Lett.* **40**, 712–16.

Ahlers, G. & Behringer, R.P. (1978b). The Rayleigh–Benard, instability and the evolution of turbulence. *Prog. Theor. Phys. Suppl.* **64**, 186–201.

Aikawa, T. (1987). The Pomeau–Manneville intermittent transition to chaos in hydrodynamic pulsation models, *Astrophys. Space Sci.* **139**, 281–93.

Aizawa, Y. & Nishikawa, I. (1986). Toward the classification of the patterns generated by one-dimensional cell automata. In *Dynamical Systems and Nonlinear Oscillations*, ed. G. Ikegami, pp. 210–22, World Scientific.

Albrecht, F., Gatzke, H., Haddad, A. & Wax, N. (1974). The dynamics of two interacting populations, *J. Math. Anal. and Appl.* **46**, 658–70.

Alekseev, V.M. & Yakobson, M.V. (1981). Symbolic dynamics and hyperbolic dynamic systems. *Phys. Rep.* **75**, 287–325.

Alfaro, V. de & Rasetti, M. (1978). Structural stability theory and phase transition models. *Fortschritte der Physik* **26**, 143–73.

Alfsen, K.H. & Frøyland, J. (1984). *Systematics of the Lorenz Model at $\sigma = 10$*. Univ. of Oslo, Institute of Physics: Norway.

Amann, H., Bazley, N. & Kirchgassner, K. (Eds.) (1981). *Applications of Nonlinear Analysis in the Physical Sciences*. Pitman: Bath.

Amari, S. & Arbib, M.A. (Eds.) (1982). *Competition and Cooperation in Neutral Nets*, Lecture Notes in Biomathematics **45**. Springer-Verlag.

Ames, W.F. (1965, 1972). *Nonlinear Partial Differential Equations in Engineering*, Vol. I, Vol. II, Academic Press.

Ames, W.F. (1978). Nonlinear superposition for operator equations. In *Nonlinear Equations in Abstract Spaces* ed. V. Lakshmikantham, pp. 43–66. Academic Press.

Anderson, D.L.T. (1971). Stability of time-dependent particle-like solutions in nonlinear field theories II. *J. Math. Phys.* **12**, 945–52.

Anderson, P.W., Arrow, K.J. & Pines, D. (Eds.) (1988). *The Economy as an Evolving Complex System*, Santa Fe Institute Studies in the Sciences of Complexity, Addison–Wesley.

Anderson, R.L., Harnad, J. & Winternitz, P. (1982). Systems of ordinary differential equations with nonlinear superposition principles. *Physica* **4D**, 164–82.

Anderson, R.L. & Ibragimov, N.H. (1979). *Lie–Backlund Transformations in Applications*, SIAM. (Contains historical references.)

Andrews, J.G. & McLone, R.R. (Eds.) (1976). *Mathematical Modeling*. Butterworths.

Andronov, A.A., Leontovich, E.A., Gordon, I.I. & Maier, A.G. (1971). *Theory of Bifurcations of Dynamic Systems on a Plane*. US Dept. Commerce; Israel Program for Scientific Translations.

Andronov, A.A.,Khaikin, S.E., Vitt, A.A. (1966). *Theory of Oscillators*, Pergamon.

Anosov, D.V. (1985). Smooth dynamical systems. Introductory article, *Amer. Math. Soc. Transl. (2)* **125**, 1–20.

Appel, K. & Haken, W. (1977). Every planar map is four colorable. I. Discharging. II. With J. Koch. Reducibility *Ill. J. Math.* **21**, 429–90; 491–567.

Appel, K. & Haken, W. (1986). The four color proof suffices. *Math. Intelligencer* **8**, 10–20.

Aris, R. (1978). *Mathematical Modelling Techniques*. Pitman: Bath.

Arnold, V.I. (1961). The stability of the equilibrium position of a Hamiltonian system of ordinary differential equations in the general elliptic case. *Sov. Math. Dokl.* **2**, 247–9.

Arnold, V.I. (1963). Small denominators and problems of stability of motion in classical and celestial mechanics. *Russian Math. Surveys* **18**, No. 6, 85–189.

Arnold, V.I. (1964). *Dokl. Akad. Nauk. SSSR* **156**, 9.

Arnold, V.I. (1965). Small denominators. I. Mappings of the circumference onto itself. *Trans. Amer. Math. Soc. 2nd Series* **46**, 213–84.

Arnold, V.I. (1972). Lectures on bifurcations in versal families. *Russian Math. Surveys* **27**, 54–123.

Arnold, V.I. (1978a). *Mathematical Methods of Classical Mechanics*. Springer-Verlag.

Arnold, V.I. (1978b). *Ordinary Differential Equations*. M.I.T. Press.

Arnold, V.I. (1983). *Geometrical Methods in the Theory of Ordinary Differential Equations*. Springer-Verlag.

Arnold, V.I. (1984). *Catastrophic Theory*. Springer-Verlag.

Arnold, V.I. & Avez, A. (1968). *Ergodic Problems of Classical Mechanics*. W.A. Benjamin.

Arnold, V.I., Kozlov, V.V. & Neishtadt, A.I. (1988). Mathematical aspects of classical and celestial mechanics. In *Dynamical Systems* **III**, ed. V.I. Arnold, Springer-Verlag.

Aronson, D.G., Chory, M.A., Hall, G.R. & McGehee, R.P. (1982). Bifurcations from an invariant circle for two-parameter families of maps of the plane. *Commun. Math. Phys.* **83**, 303–54.

Aubert, K.E. (1984). Spurious mathematical modelling. *Math. Intelligencer* **6**, 54–60.

Aubin, J.P., Saari, D. & Sigmund, K. (Eds.) (1985). *Dynamics of Macrosystems*. Lecture Notes in Economic and Mathematical Systems **257**. Springer-Verlag.

Aubry, S. (1983). The twist map, the extended Frenkel–Kontrova model and the devil's staircase. *Physica* **7D**, 240–58.

Babloyantz, A. (1986). *Molecules, Dynamics and Life*, Wiley.

Bai-line, H. (1984). *Chaos*. World Scientific: Singapore.

Balazs, N.L. & Voros, A. (1986). Chaos on the pseudosphere, *Phys. Reports* **143**, No. 3, 109–240.

Balian, R. & Peube, J.L. (Eds.) (1977). *Fluid Dynamics*. Gordon and Breach.

Barber, M.N. & Ninham, B.W. (1970). *Random and Restricted Walks*. Gordon and Breach.

Bard, J.B.L. (1981). A model for generating aspects of zebra and other mammalian coat patterns. *J. Theor. Biol.* **93**, 363–85.

Bardos, C. & Bessis, D. (Eds.) (1980). *Bifurcation Phenomena in Mathematical Physics and Related Topics*. D. Reidel.

Barenblatt, G.I., Iooss, G. & Joseph, D.D. (Eds.) (1983). *Nonlinear Dynamics and Turbulence*. Pitman.

Barnsley, M. (1988). *Fractals Everywhere*, Academic Press.

Barnsley, M.F. (1986). Making chaotic dynamical systems to order. In *Chaotic Dynamics*, ed. M.F. Barnsley & S.G. Demko, pp. 53–68. Academic Press.

Barnsley, M.F. & Demko, S. (1985). Iterated function system and the global construction of fractals. *Proc. R. Soc. Lond.* **A399**, 243–75.

Barnsley, M.F. & Demko, S.G. (Eds.) (1986). *Chaotic Dynamics and Fractals*. Academic Pres.

Barone, A., Exposito, F., Magee, C.J. & Scott, A.C. (1971). Theory and applications of the Sine–

Gordon equation. *Nuovo Cimento Series 2,* **1,** 227–67. (Contains extensive references to all applications.)

Barrow, J.D. (1982). Chaotic behavior in general relativity. *Phys. Rep.* **85,** 1.

Bartlett, J.H. (1975). *Classical and Modern Mechanics.* Univ. of Alabama Press.

Basar, E., Durusan, R., Gorder, A.P. Ungan, P. (1979). Combined dynamics of E.E.G. and evoked potential. *Biol. Cybernetics* **34,** 21.

Beau, W., Metzler, W. & Ueberla, A. (1987). *The Route to Chaos of Two Coupled Logistic Maps.* Univ. of Kassel.

Bedford, T. & Swift, J. (Eds.) (1988). *New Directions in Dynamical Systems.* Cambridge University Press.

Bednar, J.B. et al. (Eds.) (1983). *Conference on Inverse Scattering: Theory and Applications.* SIAM.

Behringer, R.P. & Ahlers, G. (1987). Heat transport and critical slowing down near the Rayleigh–Benard instability in cylindrical containers. *Phys. Letters* **62A,** 329–31.

Bélair, J. & Glass, L. (1985). Universality and self-similarity in the bifurcations of the circle maps. *Physica* **16D,** 2, 143–54.

Belousov, B.P. (1959). *Sb. Ref. Radiats. Med.,* 1958. Medgiz: Moscow, 145.

Benjamin, T.B. (1978). Bifurcation phenomena in steady flows of a viscous fluid I. Theory, II. Experiment. *Proc. Roy. Soc. Lond.* **A359,** 1–26, 27–43.

Bennett, C.H. (1986). On the nature and origin of complexity in discrete, homogeneous, locally-interacting systems. *Foundation Physics* **16,** 585–92.

Benettin, G.A., Casartelli, G.M., Galgani, L., Giorgilli, A. & Strelcyn, J.M. (1978). On the reliability of numerical studies of stochasticity. *Nuovo Cimento* **44B,** 183–5.

Bennettin, G., Galgani, L. & Strelcyn, J.M. (1976). Kolmogorov entropy and numerical experiments. *Phys. Rev.* **A14,** 2338–45.

Bennettin, G., Galgani, L. & Giorgilla, A. (1985). A proof of Nekhoroshev's theorem for the stability times in nearly-integrable Hamiltonian systems, *Celest. Mech.* **37,** 1–25.

Bennettin, G., Galgani, L. & Giorgilli, A. (1987). Realization of holonomic constraints and freezing of high frequency degrees of freedom in light of classical perturbation theory, *Commun. Math. Phys.* **113,** 87–103.

Bennetin, G., Giorgilli, A. & Strelcyn, J.-M. (1980). Lyapunov characteristic exponents for smooth dynamical systems and for Hamiltonian systems: a method for computing all of them. *Meccanica* **15,** 9.

Benzinger, H.E., Burns, S.A. & Palmore, J.I. (1987). Chaotic complex dynamics and Newton's method. *Phys. Letters* **119A,** 441–46.

Bergé, P. (1979). *Experiments on Hydrodynamic Instabilities and the Transition to Turbulence.* Lecture Notes in Physics 104, 289–308. Springer-Verlag.

Bergé, P. & Dubois, M. (1984). Rayleigh–Bénard convection. *Contemp. Phys.* **25,** 535–82.

Bergé, P., Pomeau, Y. & Vidal, C. (1984). *Order Within Chaos: Towards a Deterministic Approach to Turbulence.* Wiley.

Berlekamp, E.R., Conway, J.H. & Guy, R.K. (1982). *Winning Ways, for your Mathematical Plays,* Vol. 2, Chap. 25. Academic Press.

Berry, M.V. (1976). Waves and Thom's theorem. *Adv. in Phys.* **25,** 1–26.

Berry, M.V. (1977). Semi-classical mechanics in phase space: a study of Winger's function. *Phil. Trans. R. Soc. Lond.,* **287,** 237–71.

Berry, M.V. (1978). Regular and irregular motion, pp. 16–120. In *Topics in Nonlinear Dynamics,* ed. S. Jorna. AIP Conference Proceedings, 46. American Institute of Physics.

Berry, M.V. (1979a). Diffractals. *J. Phys.* **A12**, 781.

Berry, M.V. (1979b). Catastrophe and stochasticity in semiclassical quantum mechanics; Catastrophe and fractal regimes in random waves. In *Structural Stability in Physics*. Springer-Verlag.

Berry, M. (1981). Singularities in waves and rays, pp. 456–541. In *Les Houches*. Session 35: *Physics of Defects*, ed. R. Balian, M. Kléman & J.-P. Poirier. North-Holland.

Billingsley, P. (1965). *Ergodic Theory and Information*. Wiley.

Birkhoff, G.D. (1913). Proof of Poincaré's geometric theorem. *Trans. Amar. Math. Soc.* **14**, 14–22.

Birkhoff, G.D. (1917). Dynamical systems with two degrees of freedom. *Trans. Amer. Math. Soc.* **18**, 199–300.

Birkhoff, G.D. (1922). Surface transformations and their dynamical application. *Acta Math.* **43**, 1–119.

Birkhoff, G.D. (1927). *Dynamical Systems*. Amer. Math. Soc.

Birkhoff, G. (1955). *Hydrodynamics*. Dover.

Birman, J.S. & Williams, R.F. (1983). Knotted periodic orbits in dynamical systems. I: Lorenz's equations. *Topology* **22**, 47–82.

Bishop, A., Campbell, D. and Nicolaenko, B. (Eds.) (1981). *Nonlinear problems: Present and Future*. Math Studies **61**. North-Holland.

Bishop, A.R. & Schneider, T. (Eds.) (1978). *Solitons and Condensed Matter Physics*. Springer-Verlag.

Bishop, G. (1985). Pattern selection and low-dimensional chaos in systems of coupled nonlinear oscillation. In *Dynamical Problems in Soliton Systems*, ed. S. Takeno, pp. 250–7. Springer-Verlag.

Block, L., Guckenneimer, J., Misiurewicz, M. & Young, L.S. (1980). Periodic points and topological entropy of one dimensional maps. In *Global Theory of Dynamical Systems*, eds. A. Dold & B. Eckmann, *Lecture Notes in Mathematics*, **819**, Springer-Verlag.

Bluman, G.W. & Cole, J.D. (1974). *Similarity Methods for Differential Equations*. Springer-Verlag.

Blumenfeld, L.A. (1981). *Problems of Biological Physics*. Springer-Verlag.

Bocchieri, P., Scotti, A., Bearzi, B. & Loinger, A. (1970). Anharmonic chain with Lennard–Jones interaction. *Phys. Rev.* **A2**, 2013–9.

Bogoliubov, N. & Mitropolsky, Y.A. (1961). *Asymptotic Methods in the Theory of Nonlinear Oscillations*. Gordon and Breach.

Bogoliubov, N.N., Mitropoliskii, Ju. A. & Samoilenko, A.M. (1976). *Methods of Accelerated Convergence in Nonlinear Mechanics*. Springer-Verlag.

Bohm, D. (1984). Causality and Chance in Modern Physics, Univ. of Pennsylvania Press.

Bohm, D. & Peat, F.D. (1987). *Science, Order, and Creativity*, Bantam.

Bohr, T., Bak, R. & Jensen, M.H. (1984). Transition to chaos by interaction of resonances in dissipative systems. *Phys. Rev.* **A30**, 1970–81.

Boon, J.P. & Noullez, A. (1986). Development, Growth, and Form in Living Systems. In *On Growth and Form*, eds. H.F. Stanley & N. Ostrowsky, pp. 174–83, Martinus Nijhoff Pub.

Borrelli, R.L. & Coleman, C.S. (1987). *Differential Equations: A Modeling Approach*. Prentice-Hall.

Bountis, T.C. (1984). A singularity analysis of integrability of chaos in dynamical systems. In *Singularities and Dynamical Systems*, ed. S. Pneumatikos. North-Holland.

Bountis, T. & Segur, H. (1982). Logarithmic singularities and chaotic behavior in Hamiltonian systems. *AIP Conference Proceedings*, 88, ed. M. Tabor & Y.M. Treve, pp. 279–92. American Institute of Physics. Mathematical methods in hydrodynamics and integrability in dynamical systems.

Bountis, T., Segur, H. & Vivaldi, F. (1982). Integrable Hamiltonian systems and the Painlevé property. *Phys. Rev.* **A25**, 1257–64.

Bowen, R. (1975). *Methods of Symbolic Dynamics*. Lecture Notes in Mathematics, **470**. Springer Berlin.

Bowen, R. (1978). On axiom a diffeomorphisms. *CBMS Regional Conference Series in Math.* 35.

Brandstäter, A., Swift, J., Swinney, H.L., Wolf, A., Farmer, J.D., Jen, E. & Crutchfield, P.J. (1983). Low-dimensional chaos in a hydrodynamic system. *Phys. Rev. Letters* **51**, 1442–5.

Brillouin, L. (1963). *Science and Information Theory*. Academic Press.

Brillouin, L. (1964). *Scientific Uncertainty and Information*. Academic Press.

Bröcker, Th. & Jänich, K. (1982). *Introduction to Differential Topology*. Cambridge Univ. Press.

Broomhead, D.S. and King, G.P. (1986). Extracting qualitative dynamics from experimental data. *Physica* **20D**, 217–36.

Broomhead, D.S. & King, G.P. (1986). On the qualitative analysis of experimental dynamical systems, pp. 113–44. In *Nonlinear Phenomena and Chaos*, ed. S. Sarkar, Adam Hilger Ltd.

Brown, M. & Neumann, W.D. (1977). Proof of the Poincaré–Birkhoff fixed point theorem. *Mich. Math. J.* **24**, 21–31.

Buck, R.C. (1981). The solutions to a smooth PDE can be dense in C(1). *J. Diff. Eq.* **41**, 239–44.

Budinsky, N. & Bountis, T. (1983). Stability of nonlinear modes and chaotic properties of 1D Fermi-Pasta-Ulam lattices. *Physica* **8D**, 445–54.

Bullard, E. (1978). The disk dynamo. In *Topics in Nonlinear Dynamics*. AIP Conference Proceedings, ed. S. Jorna, pp. 373–89. American Institute of Physics.

Bullough, R.K. & Caudrey, P.J. (Eds.) (1980). Solitons Topics in Current Physics **17**. Springer-Verlag.

Bullough, R.K. & Dodd, R.K. (1979). Solitons in physics: basic concepts; Solitons in mathematics. In *Structural Stability in Physics*, ed. W. Guttinger & H. Eikemier, pp. 219–53. Springer-Verlag.

Burgers, J.M. (1948). A mathematical model illustrating the theory of turbulence: In *Advances in Applied Mechanics* 1, ed. R. von Mises & T. von Kármán, pp. 171–99. Academic Press.

Burks, A.W. (Ed.) (1970). *Essays on Cellular Automata*. Univ. of Illinois Pres: Urbana.

Busse, F.H. (1978). Magnetohydrodynamics of the earth's dynamo. *Ann. Rev. Fluid Mech.* **10**, 435–62.

Busse, F.H. (1980). *Science* **208**, 173–5, Convection in a rotating layer.

Busse, F.H. (1980). Convections in a rotating layer: a simple case of turbulence. *Science* **208**, 173–5.

Calogero, F. (1983). Integrable dynamical systems and related mathematical results. In *Nonlinear Phenomena*. Lecture Notes in Physics **189**; ed. K.B. Wolf, pp. 49–95. Springer-Verlag.

Campbell, J.E. (1903). *Introductory Treatise on Lie's Theory of finite Continuous Transformation Groups*. Oxford: Clarendon Press. Reprinted by Chelsea, 1966.

Campbell, D. & Rose, H. (Eds.) (1983). Order in chaos. *Physica* **7D**, 1–362.

Campbell, D.K., Newell, A.C., Schrieffer, R.J. & Segur, H. (Eds.) (1986). Solitons and coherent structures. *Physica* **18D**, 1–480.

Capasso, V. (1981). Periodic solutions for a system of nonlinear differential equation modeling the evolution of oro-faecal diseases. In *Nonlinear Differential Equations*, eds. P. deMottoni & L. Salvadori. Academic Press.

Carrigan, C.R. & Gubbins, D. (1979). The source of the earth's magnetic field. *Sci. Amer.* **240**, (2), 118.

Cartwright, M.L. & Littlewood, J.E. (1945). On non-linear differential equations of the second order. I. The equation $\ddot{y}+k(1-y^2)\dot{y}+y=b\lambda k \cos(\lambda t + a), k$ large. *J. Lond. Math. Soc.* **20**, 180–9.

Casati, G. & Ford, J. (1975). Stochastic transition in the unequal-mass Toda lattice, *Phys. Rev.* **A12**, 1702–9.

Casati, G., Chirikov, B.V. & Ford, J. (1980). Marginal local instability of quasi-periodic motion, *Phys. Lett.* **77A**, 91–4.

Cesari, L. (1963). *Asymptotic Behavior and Stability Problems in Ordinary Differential Equations*. Academic Press.

Cesari, L. (1976). *Functional Analysis, Nonlinear Differential Equations, and the Alternative Method.* Lecture Notes in Pure and Applied Mathematics **19**, 1–197. Dekker.

Chandrasekhar, S. (1961). *Hydrodynamic and Hydromagnetic Stability.* Oxford Univ. Press.

Chang, S.J. & McCown, J. (1985). Universality behaviors and fractal dimensions associated with M-furcations. *Phys. Rev.* **31A**, 3791–801.

Chang, S.J. & Wright, J. (1981). Transitions and distribution functions for chaotic systems. *Phys. Rev.* **A22**, 1419–33.

Chang, S.J., Wortis, M. & Wright, J.A. (1981). Iterative properties of a one-dimensional quartic map: critical lines and tricritical behavior. *Phys. Rev.* **A24**, 2669–84.

Chang, Y.F., Tabor, M. & Weiss, J. (1982). Analytic structure of the Hénon-Heiles Hamiltonian in integrable and nonintegrable regimes. *J. Math. Phys.* **23**, 531–8.

Channell, P.J. (1978). An illustrative example of the Chirikov criterion for stochastic instability. In *Topics in Nonlinear Dynamics.* AIP Conference Proceedings **46**, ed. S. Jorna, pp. 248–59. American Institute of Physics.

Chapman, S. & Cowling, T.R. (1970). *The Mathematical Theory of Nonuniform Gases.* Cambridge Univ. Press.

Charney, J.G., & DeVore, J.G. (1979). Multiple flow equilibrium in the atmosphere and blocking. *J. Atmos. Sci.* **36**, 1205–16.

Chernikov, A.A., Sagdeev, R.Z., Usikov, D.A., Zakharov, M.Yu. & Zaslavsky, G.M. (1987). Minimal chaos and stochastic webs. *Nature* **326**(9), 559–63.

Chevalley, C. (1946). *Theory of Lie Groups.* Princeton Univ. Press.

Chillingworth, D.R.J. (1976). Structural stability of mathematical models: the role of the catastrophe method. In *Mathematical Modeling*, ed. J.G. Andrews & R.R. McLone, pp. 231–58. Butterworths.

Chinn, W.G. & Steenrod, N.E. (1966). *First Concepts of Topology.* M.A.A.

Chirikov, B.V. (1979). A universal instability of many-dimensional oscillator systems. *Phys. Rep.* **52**, 265–376.

Chirikov, B.V., Keil, E., Seisler, A.M. (1971). Stochasticity in many-dimensional nonlinear oscillator systems. *J. Stat. Phys.* **3**, 307–21.

Chirikov, B.V., Ford, J. & Vivaldi, F. (1980). Some numerical studies of Arnold diffusion in a simple model. In *Nonlinear Dynamics and the Beam–Beam Interaction*, pp. 323–40, Amer. Inst. Phys.

Chirikov, B.V., Izrainev, F.M., & Tryursky, V.A. (1973). Numerical experiments on the statistical behavior of dynamical systems with a few degrees of freedom. *Comp. Phys. Comm.* **5**, 11–6.

Choquet-Bruhat, Y., Dewitt-Morette, C. & Dillard-Bleick, M. (1977). *Analysis, Manifolds and Physics.* North-Holland.

Chow, S.N. & Hale, J.K. (1982). *Methods of Bifurcation Theory.* Springer-Verlag.

Ciliberto, S. & Gollub, J.P. (1984). Pattern competition leads to chaos. *Phys. Rev. Letters*, 922–5.

Clark, J.W., Rafelski, J. & Winston, J.V. (1985). Brain without mind: computer simulation of neutral networks with modifiable neuronal interactions. *Phys. Rep.* **123**(4), 215–73.

Clark, J.W., Winston, J.V. & Rafelski, J. (1984). Self-organization of neural networks. *Phys. Letters* **10A**, 207–11.

Coodington, E.A. & Levinson, N. (1955). *Theory of Ordinary Differential Equations.* McGraw-Hill.

Coggeshall, S.V. & Axford, R.A. (1986). Lie group invariance properties of radiation hydrodynamic equations and their associated similarity solutions. *Phys. Fluids* **28**, 2398–420.

Cohen, A. (1911). *Introduction to the Lie Theory of One-Parameter Groups.* D.C. Heath.

Cole, J.D. (1968). *Perturbation methods in applied mathematics.* Blaisdell.

Coles, D. (1965). Transition in circular couette flow. *J. Fluid Mech.* **21**, 385–425.

Collet, P. & Eckmann, J.-P. (1980). *Iterated Maps on the Interval as Dynamic Systems.* Birkhauser.

Collet, P., Eckmann, J.-P. & Koch, H. (1981). On universality for area-preserving maps of the plane. *Physica* **3D**, 457–67.

Colvin, J.T. & Stapleton, H.J. (1985). Fractal and spectral dimensions of biopolymer chains: solvent studies of electron spin relaxation rates in myoglobin azide, *J. Chem. Phys.* **82**, 4699–706.

Contopoulous, G. (1963). A classification of the integral of motion. *Ap. J.* **138**, 1297–1305.

Cook, A.E. & Roberts, P.H. (1970). The Rikitake two-disc dynamo system. *Proc. Camb. Phil. Soc.* **68**, 547–69.

Cooley, J.W. & Dodge, F.A. (1966). Digital computer solutions for excitation and propagation of the nerve impulse. *Biophys. J.* **6**, 583–99.

Cooper, L.N. (1973). A possible organization of animal memory and learning. In *Proceedings of the Nobel Symposium on Collective Properties of Physical Systems* **24**, 252–64, ed. B. Lundquist & L. Lundquist, Academic Press.

Coven, E.M., Kan, I. & Yorke, J.A. (1986). Pseudo-orbit shadowing in the family of tent maps.

Cox, A. (1982). Magnetostratigraphic time scale. In *Geologic Time Scale*, ed. W.B. Harland *et al.*, p. 63: Cambridge Univ. Press.

Crawford, J.D. & Omohundro, S. (1984). On the global structure of period doubling flows. *Physica* **13D**, 161–80.

Creveling, H.F., dePaz, J.F., Baladi, J.Y. & Schoenhals, R.J. (1975). Stability characteristics of a single-phase free convection loop. *J. Fluid Mech.* **65**, 67.

Cronin, J. (1981). *Mathematics of Cell Electrophysiology*. Lecture Notes in Applied Mathematics 63. Dekker: New York.

Crutchfield, J.P., Farmer, J.D., Packard, N.H. & Shaw, R.S. (1986). Chaos. *Sci. Amer.* **255**(6), 46–57.

Crutchfield, J., Farmer, D., Packard, N., Shaw, R., Jones, G. & Donnelly, J. (1980). Power spectral analysis of a dynamical system. *Phys. Letters* **76A**, 1–4.

Cunningham, W.J. (1964). *Introduction to Nonlinear Analysis*. McGraw-Hill.

Curry, J.H. (1978). A generalized Lorenz system. *Commun. Math. Phys.* **60**, 193–204.

Curry, J.H. (1979). On the Hénon transformation. *Commun. Math. Phys.* **68**, 129–40.

Cutland, N. (1980). *Computability* Cambridge Univ. Press.

Cvitanović, P. (Ed.) (1984). *Universality in Chaos*. Adam Hilger. Bristol.

Davidson, M. (1983). *Uncommon Sense*. J.P. Tarcher: Los Angeles.

Davis, H.T. (1962). *Introduction to Nonlinear Differential and Integral Equations*. Dover.

Davis, R. (1988). *The Cosmic Blueprint*, Simon and Schuster.

Davydov, A.S. (1985). *Solitons in Molecular Systems*. D. Reidel.

Demongeot, J., Golés, E. & Tchuente, M. (Eds.) (1985). *Dynamical Systems and Cellular Automata*. Academic Press.

Denman, H.H. (1965). Invariance and conservation laws in classical mechanics, *J. Math. Phys.* **6**, 1611–16.

De Santo, J.A., Saenz, A.W. & Zachary, W.W. (Eds.) (1980). *Mathematical Methods and Applications of Scattering Theory*. Lecture Notes in Physics 130. Springer-Verlag.

Deutsch, D. (1985). Quantum theory, the Church–Turing principle and universal quantum computer. *Proc. R. Soc. Lond.* **A400**, 97–117.

De Vogelaere, R. (1958). On the structure of symmetric period solutions of conservative systems, with applications. In *Contributions to the Theory of Nonlinear Oscillations*, **4**, eds. S. Lefshetz, pp. 53–84.

Devaney, R.L. (1986). *An Introduction to Chaotic Dynamical Systems*. W.H. Benjamin.

Dickson, L.E. (1924). Differential equations from the group standpoint. *Ann. Math.* **25**, 287.

Dinaburg, E.I. (1971). On the relations among various entropy characteristics of dynamical systems. *Math. USSR. Izvestija* **5**, 337–78.

Diner, S., Fargue, D. & Lochak, G. (Eds.) (1986). *Dynamical Systems*. World Scientific.

Dodd, R.K., Eilbeck, J.C., Gibbon, J.D. & Morris, H.C. (1982). *Solitons and Nonlinear Wave Equations*. Academic Press.

Dorizzi, B., Grammatieos, B. & Ramani, A. (1983). A new class of integrable systems. *J. Math. Phys.* **24**, 2282–8.

Dorizzi, B., Grammaticos, B. & Ramani, A. (1984). Explicit integrability for Hamiltonian systems and the Painlevé conjecture. *J. Math. Phys.* **25**, 481–5.

Dorodnitsyn, A.A. (1947). Asymptotic solution of the van der Pol equation. *Inst. Mech. Acad. Sci. USSR*, **11**, (Russian).

Dragt, A.J. & Finn, J.M. (1976). Insolubility of trapped particle motion in a magnetic dipole field. *J. Geophys. Res.* **81**, 2327–40.

Drazin, P.G. & Reid, W.H. (1981). *Hydrodynamic Stability*. Cambridge Univ. Press.

Dresden, M. (1962). A study of models in non-equilibrium statistical mechanics. In *Studies in Statistical Mechanics*, ed. J. de Boer & G.E. Uhlenbeck, pp. 303–43. North-Holland.

Duffin, R.J. (1981). Rubel's universal differential equation. *Proc. Nat. Acad. Sci.* **78**, 4661–2.

Dym, C.L. & Ivey, E.S. (1980). *Principles of Mathematical Modeling*. Academic Press.

Dyson, F.J. (1982). A model for the origin of life. *J. Mol. Evol.* **18**, 344–50.

Dyson, F.J. (1985). *Origins of Life*. Cambridge Univ. Press.

Dziewonski, A.M. (Ed.) (1980). *Physics of the Earth's Interior*. Proc. International School of Physics Enrico Fermi. Articles by F.H. Busse, Motions in the earth's core and the origin of geomagnetism, pp. 493–507 & J.A. Jacobs. The evolution of the earth's core and geodynamics, pp. 508–30. North-Holland.

Ebeling, W. & Perschel, M. (Eds.) (1985). *Lotka–Volterra: Approach to Cooperation and Competition in Dynamic Systems*. Academic–Verlag: Berlin.

Ebert, K.H., Deuflhard, P. & Jäger, W. (1981). *Modelling of Chemical Reaction Systems*. Springer-Verlag.

Eckhaus, W. (1973). *Matched Asymptotic Expansions and Singular Perturbations*. North-Holland.

Eckhaus, W. (1977). Formal approximations and singular perturbations. *SIAM Review* **19**.

Eckhaus, W. & Van Harten, A. (1981). *The Inverse Scattering Transformation and the Theory of Solitons*. North-Holland.

Eckmann, J.-P. (1981). Roads to turbulence in dissipative dynamical systems. *Rev. Mod. Phys.* **53**, 643.

Eckmann, J.-P. & Ruelle, D. (1985). Ergodic theory of chaos and strange attractors. *Rev. Mod. Phys.* **57**, 617–56.

Eglund, J.C., Snapp, R.R. & Schieve, W.E. (1984). Fluctuations, instabilities, and chaos in the laser-driven nonlinear ring cavity. *Prog. in Optics* **21**, ed. E. Wolf, pp. 357–441. North-Holland.

Eigen, M. (1971). Self-organization of matter and the evolution of biological macro-molecules. *Naturwissenschaften* **33a**, 465–522.

Eigen, M. (1970). *Mathematical Models in Cell Biology and Cancer Chemotherapy*. Lecture Notes in Biomathematics 3. Springer-Verlag.

Eigen, M. (1984). The origin and evolution of life at the molecular level. In *Aspects of Chemical Evolution*, ed. G. Nicolis, pp. 119–37. Wiley.

Eigen, M. (1985). Macromolecular evolution. In *Emerging Synthesis in Science*, ed. D. Pines, pp. 25–69. Santa Fe Institute.

Eigen, M. (1986). The physics of molecular evolution. *Chemica Scripta* **26B**, 13–26.

Eigen, M., Gardiner, W.C. & Schuster, P. (1980). Hypercycles and compartments. *J. Theor. Biol.* **85**, 407–11.

Eigen, M. & Schuster, P. (1974). *The Hypercycle: A Principle of Natural Self-organization.* Springer-Verlag.

Eigen, M. & Schuster, P. (1979). *The Hypercycle: A Principle of Natural Self-organization.* Springer-Verlag.

Eilenberger, G. (1981). *Solitons.* Springer-Verlag.

Eisenhart, L.P. (1933). *Continuous Ground of Transformations.* Princeton Univ. Press.

Eminhizer, C.H., Helleman, R.H.G. & Montroll, E.W. (1976). On a convergent nonlinear perturbation theory without small denominators or secular terms. *J. Math. Phys.* **17**, 121–.

Enns, R.H., Jones, B.L., Miura, R.M. & Rangnekar, S.S. (Eds.) (1981). *Nonlinear Phenomena in Physics and Biology.* Plenum.

Epstein, I.R. (1984). New chemical oscillators. In *Non-Equilibrium Dynamics in Chemical Systems*, ed. C. Vidal & A. Pacault, pp. 24–34. Springer-Verlag.

Erdélyi, A. (1956). *Asymptotic Expansions.* Dover.

Ermentrout, E.B. & Cowan, J.D. (1979a). A mathematical theory of visual hallucination patterns. *Biol. Cybernetics* **34**, 137–50.

Ermentrout, E.B. & Cowan, J.D. (1979b). Temporal oscillations in neuronal nets. *J. Math. Biol.* **7**, 265–80.

Ermentrout, E.B. & Cowan, J.D. (1980). Large scale spatially organized activity in neural nets. *SIAM J. Appl. Math.* **38**, 1–21.

Falconer, K.J. (1985). *The Geometry of Fractal Sets*, Cambridge Univ. Press.

Farmer, D., Crutchfield, J., Froehling, H., Packard, N. & Shaw, R. (1980). Power spectra and mixing properties of strange attractors. *Ann. N.Y. Acad. Sci.* **357**, 453–72.

Farmer, D., Lapedes, A., Packard, N. & Wendroff, B. (Eds.) (1986). Evolution, games and learning; models for adaptation in machines and nature, *Physica* **22D** (Nos. 1–3).

Farmer, D., Toffoli, T. & Wolfram, S. (Eds.) (1984). Cellular automata. *Physica* **10D**, 1–246.

Farmer, J.D., Ott, E. & Yorke, J.A. (1983). The dimension of chaotic attractors. *Physica* **7D**, 153–80.

Fenstermacher, P.R., Swinney, H.L., Benson, S.V. & Gollub, J.P. (1979). Bifurcations to periodic, quasiperiodic and chaotic regimes in rotating and convecting fluids. *Ann. N.Y. Acad. Sci.* **316**, 652–66.

Fenstermacher, P.R., Swinney, H.L. & Gollub, J.P. (1979). Dynamic instabilities and the transition to chaotic Taylor vortex flow. *J. Fluid Mech.* **94**, 103–28.

Feigenbaum, M.J. (1978). Quantitative universality for a class of nonlinear transformations. *J. Stat. Phys.* **19**, 25–52.

Feigenbaum, M.J. (1979a). The universal metric properties of nonlinear transformations. *J. Stat. Phys.* **22**, 186–223.

Feigenbaum, M.J. (1979b). The onset spectrum of turbulence. *Phys. Letters* **74A**, 375–8.

Fein, A.P., Heutmaker, M.S. & Gollub, J.P. (1985). Scaling and transition from quasiperiodicity to chaos in a hydrodynamic system. *Physica Scripta* **T9**, 79–84.

Feit, S.D. (1978). Characteristic exponents and strange attractors. *Commun. Math. Phys.* **61**, 249–60.

Fermi, E. (1923). Beweis dass ein mechanisches Normalsystem im Allgemeinen quasi-ergodisch ist, *Phys. Zeits.* **24**, 261–5.

Fermi, E., Pasta, J.R. & Ulam, S.M. (1965). Studies of nonlinear problems. I. Los Alamos report LA-1940, May 1955. In *Collected Works* E. Fermi, Vol. 2, 978–88. Univ. of Chicago Press.

Fesser, K., McLaughlin, D.W., Bishop, A.R. & Holian, B.L. (1985). Chaos and nonlinear modes in perturbed Toda chain. *Phys. Rev.* **A31**, 2728–31.

Feynman, R.P. (1967). *The Character of Physical Law*, pp. 57–8. M.I.T. Press: Cambridge Mass.

Feynman, R.P. (1982). Simulating physics with computers. *Int. J. Theor. Phys.* **21**, 467–88.

Feynman, R. P. (1986). Quantum mechanical computers, *Foundation Phys.* **16**, 506–31.

Field, R.J. & Noyes, R.M. (1974). Oscillations in chemical systems. IV. Limit cycle behavior in a model of a real chemical reaction. *J. Chem. Phys.* **60**, 1877–84.

Fitzhugh, R. (1961). Impulses and physiological states in theoretical models of nerve membrane. *Biophys. J.* **1**, 445–64.

Fitzhugh, R. (1969). Mathematical models of excitation and propagation in nerves. In *Biological Engineering*, ed. H.P. Schwan. McGraw-Hill.

Flaherty, J.E. & Hoppenstedt, F.C. (1978). Frequency entrainment or a forced Van der Pol oscillator. *Studies Appl. Math.* **58**, 5–15.

Flaschka, H. (1975). Discrete and periodic illustrations of some aspects of the inverse method. In *Dynamic Systems Theory and Applications*, ed. J. Moser, pp. 441–66. Springer-Verlag.

Flaschka, H. (1974a). The Toda lattice. I. Existence of integrals. *Phys. Rev.* **B9**, 1924–5.

Flaschka, H. (1974b). On the Toda lattice. *Prog. Theor. Phys.* **51**, 703–16.

Flaschka, H. & McLaughlin, D.W. (Eds.) (1978). Conference on the theory and application of solitons. *Rocky Mountain J. Math.* **8**, 1–428.

Flaschka, H. & Newell, A.C. (1975). Integrable systems of nonlinear evolution equations. In *Dynamic Systems, Theory and Applications*, ed. J. Moser, pp. 355–440. Springer-Verlag.

Fokas, A.S. & Ablowitz, M.J. (1983). The inverse scattering transform for multi-dimensional $(2 + 1)$ problems. In *Nonlinear Phenomena*. Lecture Notes in Physics **189**, ed. K.B. Wolf, pp. 139–54. Springer-Verlag.

Ford, J. (1961). Equipartition of energy for nonlinear systems. *J. Math. Phys.* **2**, 387–93.

Ford, J. (1973). The transition from analytic dynamics to statistical mechanics. *Adv. Chem. Phys.* **24**, 155–85.

Ford, J. (1975). The statistical mechanics of classical analytic dynamics. In *Fundamental Problems in Statistical Mechanics*, ed. E. Cohen, pp. 215–55. North-Holland.

Ford, J. (1977). Relevance of computer experiments to the pure mathematics of integrable and ergodic dynamical systems. In *Dynamical Systems*. I. Warsaw, pp. 75–92. Société Mathematique de France.

Ford, J. (1983). How random is a coin toss?, *Physics Today*, April, pp. 40–47.

Ford, J. (1988). What is chaos, that we should be mindful of it? In *The New Physics*, ed. P.C. Davis, Cambridge Univ. Press.

Ford, J., Stoddard, S.D. & Turner, J.S. (1973). On the integrability of the Toda lattice. *Prog. Theor. Phys.* **50**, 1547–60.

Fowler, A.C. & McGuinness, M.J. (1982a). A description of the Lorenz attractor at high Prandtl number. *Physica* **5D**, 149–82.

Fowler, A.C. & McGuinness, M.J. (1982b). Hysteresis in the Lorenz equations. *Phys. Letters* **92A**, 103–6.

Fowler, D.H. (1972). The Riemann–Hugoniot catastrophe and Van der Waal's equation. *Towards a Theoretical Biology* **4**, 1–7, ed. C.H. Waddintgon.

Fox, R.F. (1988). *Energy and the Evolution of Life*. W.H. Freeman.

Fox, S. (1988). *The Emergence of Life*. Basic Books.

Francis, G.K. (1987). *A Topological Picturebook*, Springer-Verlag.

Fraser, S. & Kapral, R. (1982). Analysis of flow hysteresis by a one-dimensional map. *Phys. Rev.* **A25**, 3223–33.

Frauenfelder, H., Petsko, G.A. & Tsenoglu, D. (1979). Temperature-dependent X-ray diffraction as a probe of protein structural dynamics. *Nature* **280**, 558–63.

Frauenthal, J.C. (1979). *Introduction to Population Modeling*. (EDC/umap/55 Chapel St., Newton Mass.

Fredkin, E. (1982). Digital Information Mechanics (talk given January, 1982, Moskito Island, BVI); Also see: *The Atlantic Monthly*, April 1988, p. 29; *Three scientists and their Gods*, by R. Wright, Times Books, 1988.

Fredkin, E. & Toffoli, T. (1982). Conservative logic. *Int. J. Theor. Phys.* **21**, 219–53.

Freedman, H.I. (1980). *Deterministic Mathematical Models in Population Ecology*. Dekker.

Frisch, U. (1986). Fully developed turbulence: where do we stand? In *Dynamical Systems: A Renewal of Mechanism*, ed. S. Diner, D. Fargue & G. Lochak, pp. 13–28. World Scientific.

Frisch, U., Hasslacher, B. & Pomeau, Y. (1986). Lattice-gas automata for the Navier-Stokes equation. *Phys. Rev. Letters* **56**, 1505–8.

Froehling, H., Crutchfield, J.P., Farmer, D., Packard, N.N. & Shaw, R. (1981). On determining the dimension of chaotic flows. *Physica* **3D**, 605–17.

Froeschilé, C. (1978). Connectance of dynamical systems with increasing number of degrees of freedom. *Phys. Rev.* **A18**, 277–81.

Frøyland, J. & Alfsen, K.H. (1984). Lyapunov-exponent spectra for the Lorenz model. *Phys. Rev.* **A29**, 2928–31.

Furstenberg, H. (1983). Poincaré recurrence and number theory. In *The Mathematical Heritage of Henri Poincaré*. Part 2, pp. 193–216. Amer. Math. Soc.

Gantmacher, R.R. (1959). *The Theory of Matrices*. Chelsea.

Gardner, C.S., Greene, J.M., Krushal, M.D. & Miura, R.M. (1967, 1974). Method for solving the Korteweg-deVries equation. *Phys. Rev. Letters* **19**, 1095–7. Korteweg-deVries equation and generalization. VI. Methods for exact solution. *Commun. Pure Appl. Math.* **27**, 97–133.

Gardner, M. (1970, 1971). *Sci. Amer.* **223**(10), 120; **224**(2), 112; **224**(4), 117.

Gardner, M. (1976). *Knotted Doughnuts and Other Mathematical Entertainments*. W.H. Freeman.

Gardner, M. (1981). The Laffer curve and other laughs in current economics. *Sci. Amer.* **245**(6), 18.

Gardner, M. (1983). *Wheels, Life, and Other Mathematical Amusements*, Freeman.

Gardner, M. (1986). *Knotted Doughnuts and Other Mathematical Entertainments*, W.H. Freeman.

Garrido, L. (Ed.) (1983). *Dynamic Systems and Chaos*. Lecture Notes in Physics **179**. Springer-Verlag.

Gazis, D.C. (1967). Mathematical theory of automobile traffic. *Science* **157**, 273–81.

Gazis, D.C. (Ed.) (1974). *Traffic Science*, Wiley.

Greene, J.M. (1979). A method of determining a stochastic transition. *J. Math. Phys.* **20**, 1183–1201.

Gibbon, J.D., Freeman, N.C. & Johnson, R.S. (1978). Correspondence between the classical gd^4. Double and single Sine-Gordan equations for three-dimensional solitons. *Phys. Letters* **65A**, 380–4.

Gierer, A. & Meinhardt, H. (1972). A Theory of Biological Pattern Formation, *Kybernetik* **12**, 30–39.

Gilmore, R. (1981). *Catastrophe Theory for Scientists and Engineers*. Wiley.

Gilpin, M.E. (1973). Do hares eat lynx? *Amer. Nat.* **107**, 727–30.

Glansdorff, P. & Prigogine, I. (1971). *Thermodynamic Theory of Structure Stability and Fluctuations*. Wiley-Interscience.

Glass, L. & Mackey, M.C. (1988). From Clocks to Chaos; The Rythms of Life. Princeton University Press.

Glass, L. & Perez, R. (1982). Fine structure of phase locking. *Phys. Rev. Letters* **48**, 1772–5.

Goldanskii, V.I., Anikin, S.A., Avetisov, V.A. & Kuz'min, V.V. (1987). On the decisive role of chiral purity in the organization of the biosphere and its possible destruction. *Comments. Mol. Cell. Biphys.* **4**, 79–98.

Goel, N.S., & Thompson, R.L. (1988). *Computer Simulations of Self-Organization in Biological Systems*. Macmillan Pub.

Goel, N.S., Maitra, S.C. & Montroll, E.W. (1971). On the Volterra and other non-linear models of interacting populations. *Rev. Mod. Phys.* **43**, 231–76.

Gollub, J.P. & Benson, S.V. (1978). Chaotic response to period perturbation of convecting fluid. *Phys. Rev. Letters* **41**, 948.

Gollub, J.P. & Freilich, M.H. (1974). Optical heterodyne study of the Taylor instability in a rotating fluid. *Phys. Rev. Letters* **33**, 1465–8.

Gollub, J.P. & Meyer, C.W. (1983). Symmetry-breaking instabilities on a fluid surface, *Physica* **6D**, 337–46.

Gollub, J.P. & Swinney, H.L. (1975). Onset of turbulence in a rotating fluid. *Phys. Rev. Letters* **35**, 927–30.

Gollub, J.P. & Benson, S.V. (1978). Chaotic response to periodic perturbation of convecting fluid. *Phys. Rev. Letters* **41**, 948.

Gollub, J.P. & Meyer, C.W. (1983). Symmetry-breaking instabilities on a fluid surface. *Physica* **6D**, 337–46.

Golubev, V.V. (1953). *Lectures on Integration of the Equations of Motion of a Rigid Body about a Fixed Point*. State Publishing House; Moscow.

Goursat, E. (1964). *A Course in Mathematical Analysis*. Dover.

Grabert, H. (1982). *Projection Operator Techniques in Nonequilibrium Statistical Mechanics*. Springer Tracts in Modern Physics **95**. Springer-Verlag.

Grasman, J. & Veling, E.J.M. (1979). *Asymptotic Methods for the Volterra-Lotka Equation*. Lecture notes in Mathematics **11**, ed. F. Verhulst. Springer-Verlag.

Grassberger, P. & Procaccia, I. (1983). Measuring the strangeness of strange attractors. *Physica* **9D**, 189–202.

Grebogi, C., Ott, E. & Yorke, J.A. (1983). Fractal basin boundaries, long-lived chaotic and unstable–unstable pair bifurcations, *Phys. Rev. Letters* **50**, 935–8.

Greene, J.M. (1979). A method of determining a stochastic transition. *J. Math. Phys.* **20**, 1183–1201.

Greene, J.M. (1986). How a swing behaves. *Physica* **18D**, 427–47.

Greene, J.M., MacKay, R.S., Vivaldi, F. & Feigenbaum, M.J. (1981). Universal behavior in families of area-preserving maps. *Physica* **3D**, 468–86.

Greenspan, D. (1980). *Arithmetic Applied Mathematics*. Pergamon.

de Groot, S.R. (1961). *Thermodynamics of Irreversible Processes*. North-Holland.

Gu, Y., Tung, M., Yuan, J.-M., Feng, D.H. & Narducci, L.M. (1984). Crises and hysteresis in coupled logistic maps. *Phys. Rev. Letters* **52**, 701–4.

Gubbins, D. (1974). Theories of the geomagnetic and solar dynamos. *Rev. Geophys. Sp. Phys.* **12**, 137–54.

Guckenheimer, J. (1977). On the bifurcation of maps of the internal. *Inventiones Math.* **39**, 165–78.

Guckenheimer, J. (1979). The bifurcation of quadratic functions. In *Bifurcation Theory, Ann. N.Y. Acad. Sci.* **316**, pp. 78–85.

Guckenheimer, J. (1979). Sensitive dependence to initial conditions for one dimensional maps. *Commun. Math. Phys.* **70**, 133–60.

Guckenheimer, J. & Holmes, P. (1983). *Nonlinear Oscillations, Dynamical Systems, and Bifurcations of Vectors Fields*. Springer-Verlag.

Guckenheimer, J., Moser, J. & Newhouse, S.E. (1980). *Dyanmical Systems*. Birkhauser.

Guckenheimer, J. Oster, G. & Ipaktchi, A. (1977). The dynamics of density dependent population models. *J. Math. Biol.* **4**, 101–47.

Guevara, M.R. & Shrier, A. (1983). Bifurcation and chaos in periodically stimulated cardiac oscillator. *Physica* **7D**, 89–101.

Guillemin, V. & Pollack, A. (1974). *Differential Topology*. Prentice-Hall.

Gurel, G. & Rossler, O.E. (Eds.) (1979). Bifurcation theory and applications in scientific disciplines. *Ann. N.Y. Acad. Sci.* **316**.

Güttinger, W. (1986). Bifurcation geometry in physics. In *Frontiers of Nonequilibrium Statistical Physics*, G.T. Moore & M.O. Scully, pp. 57–82. Plenum.

Guttinger, W. & Eikemeier, H. (Eds.) (1979). *Structural Stability in Physics*. Springer-Verlag.

Hagihara, Y. (1974). *Celestial Mechanics*. Vol. I–Vol. V. *Japan Soc. Promotion of Science*, Tokyo Press.

Haken, H. (1977). *Synergetics: An Introduction*. Vol. 2. *A Workshop*. Springer-Verlag.

Haken, H. (Ed.) (1981). *Chaos and Order in Nature*. Springer-Verlag.

Haken, H. (1983). *Advanced Synergetics*. Springer-Verlag.

Hale, J. (1977). *Theory of Functional Differential Equations*. Springer-Verlag.

Hale, J.K. (1980). *Ordinary Differential Equations*. R.E. Krieger.

Hasley, T.C., Jensen, M.H., Kadanoff, L.P., Procassia, I. & Shraiman, B.I. (1986). Fractal measures and their singularities: the characterization of strange sets, *Phys. Rev.* **A33**, 1141–51.

Hammel, S.M., Yorke, J.A. & Grebogi, C. (1987). Do numerical orbits of chaotic dynamical processes represent true orbits? *J. Complexity* **3**, 136–45.

Hansen, A.G. (1964). *Similarity Analyses of Boundary Value Problems in Engineering*. Prentice-Hall.

Hardy, G.E. (1916). Weierstrass's non-differentiable function. *Trans. Amer. Math. Soc.* **17**, 301–25.

Harmon, L.D. (1961). Studies with artificial neurons, I Properties and functions of an artificial neuron, *Kybernetik* **1**, 89–101.

Hartman, P. (1964). *Ordinary Differential Equations*. Wiley. Reprinted by Hartman, 1973.

Hasegawa, A. (1985). Self-organization processes in continuous media. *Adv. in Phys.* **34**, 1–42.

Hastings, S.P. & Murray, I.D. (1975). The existence of oscillatory solutions in the Field-Noyes model of the Belousov-Zhabotinskii reactions. *J. Appl. Math.* **28**, 678–88.

Hawking, S.W. (1988). *A Brief History of Time*, Bantam.

Hayashi, C. (1985). *Nonlinear Oscillations in Physical Systems*. Princeton Univ. Press.

Heims, S.J. (1984). *John Von Neumann and Norbert Wiener*, MIT Press.

Helleman, R.H.G. (1977). On the iterative solution of a stochastic mapping. In *Statistical Methods, Theory and Applications*, ed. U. Landman, Plenum.

Helleman, R.H.G. (1980). Self-generated chaotic behavior in nonlinear mechanics. In *Fundamental Problems in Statistical Mechanics*, ed. E.G.D. Cohen, pp. 165–233. North-Holland.

Helleman, R.H.G. (Ed.) (1980). Nonlinear dynamics. *Ann. N.Y. Acad. Sci.* **257**.

Hénon, H. (1969). Numerical study of quadratic area-preserving mappings. *Quart. Appl. Math.* **27**, 291–312.

Hénon, H. (1974). Integrals of the Toda lattice. *Phys. Rev.* **B9**, 1921–3.

Hénon, M. (1976). A two dimensional mapping with a strange attractor. *Commun. Math. Phys.* **50**, 69–77.

Hénon, M. (1982). On the numerical computation of Poincaré maps. *Physica* **5D**, 412–44.

Hénon, M. & Heiles, C. (1964). The applicability of the third integral of motion: some numerical experiments. *Astrophys. J.* **69**, 73–9.

Herman, M.R. (1977). Measure DeLebesque et nombre de rotation, pp. 271–93. In *Geometry and Topology*, ed. J. Palis. Springer-Verlag.

Hibberd, F.H. (1979). The origin of the earth's magnetic field. *Proc. R. Soc. Lond.* **A369**, 31–45.

Hill, E.L. (1951). Hamilton's Principle and the conservation theorems of mathematical physics. *Rev. Mod. Phys.* **23**, 253–60.

Hill, J.M. (1982). *Solution of Differential Equations by Means of One-Parameter Groups*. Pitman.

Hirooka, H., Saito, N. & Ford, J. (1984). Chaos around hyperbolic fixed points. *J. Phys. Soc. Japan* **53**, 895–98.

Hirota, R. (1971). Exact solution of the Korteweg-de Vreies equation for multiple collisions of solitons. *Phys. Rev. Lett.* **27**, 1192–4.

Hirota, R. (1976). Direct methods of finding exact solutions of nonlinear evolution equations. In *Backlund Transformations*, ed. R.M. Miura. Lecture Notes in mathematics **515**. Springer-Verlag.

Hirota, R. (1985). Fundamental properties of the binary operators in soliton theory and their generalization. In *Dynamic Problems in Soliton Systems*, ed. S. Takeno. pp. 42–9. Springer-Verlag.

Hirota, R. (1980). Direct methods in soliton theory. In *Solitons. Topics of Modern Physics*, ed. R.K. Bullough & P.J. Candrey. Springer-Verlag.

Hirsch, M.W. (1984). The dynamical systems approach to differential equations. *Bull. Amer. Math. Soc.* **11**, 1–64.

Hirsch, M.W. & Smale, S. (1974). *Differential Equations, Dynamical Systems and Linear Algebra.* Academic Press.

Hodgkin, A.L. & Huxley, A.F. (1952). A quantitative description of membrane current and its application to conduction and excitation in nerves. *J. Physiol.* **117**, 500–44.

Hofbauer, J. (1981). On the occurance of limit cycles in the Volterra-Lotka equation. *Nonlinear Analysis T.M.A.* **5**, 1003–7.

Hogg, T. & Huberman, B.A. (1984). Generic behavior of coupled oscillators. *Phys. Rev.* **A29**, 275–81.

Holden, A.V. (1986). *Chaos.* Princeton Univ. Press.

Holmes, P.J. (1980). Averaging and chaotic motions. In Forced Oscillations. *SIAM J. Appl. Math.* **38**, 65–80. (Correction in *New Approaches to Nonlinear Problems in Dynamics*, ed. P.J. Holmes. SIAM.

Holmes, P. & Marsden, J.E. (1979). Qualitative techniques for bifurcation analysis of complex systems. *Ann. NY Acad. Sci.* **316**, 608–22.

Holmes, P.J. & Moon, F.C. (1983). Strange attractors and chaos in nonlinear mechanics. *J. Appl. Mech.* **50**, 1021–132.

Holmes, P.J. & Rand, D.A. (1976). The bifurcations of Duffing's Equation: an application of catastrophe theory. *J. Sound and Vib.* **44**, 237–53.

Holmes, P.J. & Whitley, D.C. (1984). Bifurcations of one- and two-dimensional maps. *Phil. Trans. R. Soc. Lond.* **A311**, 43–102; *Erratum, ibid.*, 601.

Holmes, P. & Williams, R.F. (1985). Knotted periodic orbits in suspensions of Smale's horseshoe: Torus knots and bifurcation sequences, *Arch. Ration. Mech. Anal.* **90**, 115–94.

Hoover, W.G. (1986). *Molecular Dynamics.* Lecture Notes in Physics **258**. Springer-Verlag.

Hopf, E. (1948). A mathematical example displaying features of turbulence. *Commun. Pure Appl. Math.* **1**, 303–22.

Hopf, F.A. & Hopf, F.W. (1986). Darwinian evolution in physics and biology: the dilemma of missing links. In *Frontiers of Nonequilibrium Statistical Physics*, ed. G.T. Moore & M.O. pp. 411–9. Scully, Plenum.

Hopfield, J.J. (1984). Neural networks and physical systems with emergent collective computational abilities. *Proc. Nat. Acad. Sci.* **79**, 2554–8.

Hopfield, J.J. (1984). Neurons with graded response have collective computational properties like those of two-state neurons. *Proc. Nat. Acad. Sci.* **81**, 3088–92.

Hoppensteadt, F.C. *Mathematical Theories of Populations: Demographics, Genetics, and Epidemics.* SIAM.

Hoppensteadt, F.C. (Ed.) (1979). *Nonlinear Oscillations in Biology.* Lectures in Applied Mathematics **17**, Amer. Math. Soc.

Hoppensteadt, F.C. (Ed.) (1981). *Mathematical Aspects of Physiology*. Lectures in Applied Mathematics **19**, Amer. Math. Soc.

Houtappel, R.M.F. Van Dam, H. & Wigner, E.P. (1965). The conceptual basis and use of the geometric invariance principles. *Rev. Mod. Phys.* **37**, 595–632.

Hsu, C.S. (1987). *Cell-To-Cell Mapping*, Springer-Verlag.

Hunt, E.R. (1982). Comment on a driven nonlinear oscillator. *Phys. Rev. Letters* **49**, 1054. Testa, Perez, and Jeffries respond. *Phys. Rev. Letters*.

Hurewicz, W. & Wallman, H. (1948). *Dimension Theory*. Princeton Univ. Press.

Ichikawa, Y.H. & Ino, K.H. (1985). Lax-pair operators for squared-sum and squared-difference Eigen-functions. *J. Math. Phys.* **26**, 1976–8.

Ichikawa, Y.H., Kamimura, T. & Karney, C.F.F. (1983). Stochastic motion of particles in tandom mirror devices. *Physica* **6D**, 233–40.

Ince, E.L. (1956). *Ordinary Differential Equations*. Dover.

Inglis, D.R. (1981). Dynamo theory of the earth's varying magnetic field. *Rev. Mod. Phys.* **53**, 481–96.

Inselberg, A. (1972). Superpositions for nonlinear operators. I. Strong super-positions and linearizability. *J. Math. Anal. Appl.* **40**, 494–508.

Iooss, G., Helleman, R.H.G. & Stora, R. (Eds.) (1983). *Chaotic behavior of Deterministic Systems*. North-Holland.

Iooss, G. & Joseph, D.D. (1980). *Elementary Stability and Bifurcation Theory*. Springer-Verlag.

Irwin, M.C. (1980). *Smooth Dynamical Systems*. Academic Press.

Iyanaga. S. & Kawada, Y. (Eds.) (1980). *Encyclopedic Dictionary of Mathematics*. Math. Soc. Japan. M.I.T. Press.

Jackson, E.A. (1963). Nonlinear coupled oscillators. I. Perturbation theory: ergodic problem. *J. Math. Phys.* **4**, 551–8; Nonlinear coupled oscillators. II. Comparison of theory with computer solutions. *J. Math. Phys.* **4**, 686–700.

Jackson, E.A. (1978). Nonlinearity and irreversibility in lattice dynamics. *Rocky Mountain J. Math.* **8**, 127–46.

Jackson, E.A. (1984). Radiation reaction dynamics in an electromagnetic wave and constant magnetic field, *J. Math. Phys.*, 1584–91.

Jackson, E.A. (1985). The lorenz system. I. The global structure of its stable manifolds. *Physica Scripta* **32**, 469–75; The lorenz system. II. The homoclinic convolution of the stable manifolds. *Physica Scripta* **32**, 476–81.

Jackson, E.A., Pasta, J.R. & Waters, J.F. (1968). Thermal conductivity of one-dimensional lattices, *J. Comput. Phys.* **2**, 207–27.

Jacobs, A.J. (1984). *Reversals of the Earth's Magnetic Field*. Heyden Philadelphia.

Jacobs, A.J. (1984). *Reversals of the Earth's Magnetic Field*. Heyden Philadelphia.

Jeffries, C.D. (1985). Chaotic dynamics of instabilities in solids. *Physica Scripta* **T9**, 11–26.

Jeffrey, A. & Taniuti, T. (1964). *Nonlinear Wave Propagation*. Academic Press.

Jen, E. (1986). Global properties of cellular automata. *J. Stat. Phys.* **43**, 219–42; Strings and pattern recognizing properties of one-dimensional cellular automata. *J. Stat. Phys.* **43**, 243–65.

Jensen, M.H., Kadanoff, L.P. Libchaber, A. Procaccia, I. & Stavans, J. (1985). Global universality at the onset of chaos: results of a forced Rayleigh-Bénard experiment. *Phys. Rev. Letters* **55**, 2798–801.

Jensen, R.V. & Oberman, C.R. (1982). Statistical properties of dynamical systems. *Physica* **4D**, 183–96.

Jorna, S. (Ed.) (1978). *Topics in nonlinear dynamics*. AIP Conference Proceedings, 46. American Institute of Physics.

Joseph, D. (1976). *Nonlinear Stability of Fluid Motions*. Springer-Verlag.

Jowett, J.M., Month, M. & Turner, S. (1985). *Nonlinear Dynamics Aspects of Particle Accelerators*. Lecture Notes in Physics **247**, Springer-Verlag.

Kac, M. (1959). *Probability and Related Topics in Physical Sciences*. Interscience Pub.

Kadomtsev, B.B. & Petviashvili, V.I. (1970). *Sov. Phys. Dokl.* **15**, 539–41.

Kai, T. & Tomita, K. (1979). Stroboscopic phase portrait of a forced nonlinear oscillator. *Prog. Theor. Phys.* **61**, 54–73.

Kako, F. & Yajima, N. (1980). Interaction of ion-acoustic solitons in two-dimensional space. *J. Phys. Soc. Japan.* **49**, 2063–71.

Kalia, R.K. & Vashista, P. (Eds.) (1982). *Melting, Localization, and Chaos*. North-Holland.

Kaneko, K. (1983). Transition from torus to chaos accompanied by frequency lockings with symmetry breaking. *Prog. Theor. Phys.* **69**, 1427–42.

Kaneko, K. (1984). Period-doubling of kink-antikink patterns. *Prog. Theor. Phys.* **72**, 480–6.

Kaplan, J.L. & Yorke, J.A. (1979). Preturbulence: a regime observed in fluid flow model of lorenz. *Commun. Math. Phys.* **67**, 93–108.

Kaplan, J.L. & Yorke, J.A. (1979b). Chaotic behavior of multidimensional difference equations, pp. 204–27. In *Functional Differential Equations*, ed. H.O. Peitgen & H.D. Waither. Lecture Notes in Mathematics **730**, Springer-Verlag.

Kaplan, J.L. & Yorke, J.A. (1979c). The onset of chaos in a fluid flow model of lorenz. *Ann. N.Y. Acad. Sci.* **316**, 400–7.

Kaplan, J.L., Mallet-Paret, J. & Yorke, J.A. (1984). The Lyapunov Dimension of a Nowhere Differentiable Attracting Torus, *Ergod. Th. and Dynam. Sys.* **4**, 261–81.

Kapral, R. (1985). Pattern formation in two-dimensional arrays of coupled, discrete time oscillators. *Phys. Rev.* **A31**, 3868–79.

Karney, C.F.F. (1983). Long-time correlations in the stochastic regime, *Physica* **8D**, 360–80.

Kaup, D.J. (1976). The three-wave interaction: a nondispersive phenomena. *Studies Appl. Math.* **55**, 9–44.

Kay, I. & Moses, H.E. (1982). *Inverse Scattering Papers 1955–1963*, **12**, *Lie Groups, History Frontiers and Applications*, Mathematical Science.

Keeler, J.D. & Farmer, J.D. (1986). Robust space-time intermittency and 1/f noise. *Physica* **23D**, 413–35.

Keller, J.B. & Antman, S. (Eds.) (1969). *Bifurcation Theory and Nonlinear Eigenvalue Problems*. W.A. Benjamin.

Khinchin, A.I. (1949). *Mathematical Foundations of Statistical Mechanics*, Dover.

Kruskal, M. (1982). The Korteweg-deVries Equation. In *Laser-Plasma Interaction*, eds. R. Balian & J.C. Adam, pp. 788–808, North-Holland.

Kolata, G.B. (1977). Catastrophe theory: the emperor has no clothes. *Science* **196**, 287.

Kolata, G. (1984). Order out of chaos in computers. *Science* **223**. 917–9.

Kolmogoroff, A. (1936). Sulla teoria di volterra della lotta per l'esistenza. *Giornale dell'Istituto Italiano degli Attuari* **7**, 74–80.

Kolmogorov, A.N. (1954). *The General Theory of Dynamical Systems and Classical Mechanics* (English translation in *Foundations of Mechanics* by R. Abraham & J.E. Marsden.)

Kolmogorov, A.N. (1958). A new invariant of transitive dynamical systems, *Dokl. Akad. Nauk. SSSR* **119**, 861–65.

Kolmogorov, A.N. (1984). *Automata and life*. In *Cybernetics Today*, ed. I.M. Makarov, pp. 20–41. MIR Pub. Moscow.

Kolmogorov, A.N. & Uspensky, V.A. (1987). Algorithms and randomness. In *Proceedings of the 1st*

World Congress of the Bernoulli Society, eds. Yu.A. Prohorov & V.V. Sazonov, pp. 3–53, VNU Science Press.

Korteweg, D.J. & deVries, G. (1895). On the change of form of long waves advancing in a rectangular canal, and a new type of long stationary waves. *Phil. Mag.* **39**, 422–43.

Koschmieder, E.L. (1981). Experimental aspects of hydrodynamic instabilities. In *Order and Fluctuations in Equilibrium and Nonequilibrium Statistical Mechanics*, ed. G. Nicolis, G. Dewel & J.W. Turner, pp. 159–88. Wiley.

Kozak, J.J. (1979). *Nonlinear Problems in the Theory of Phase Transitions*. Adv. Chem. Phys. **40**, 229–368. Interscience Publ.

Krinskii, V.I., Pertsov, A.M. & Reshetilov, A.N. (1972). Investigation of one mechanism of origin of the etopic focus of excitation in modified Hodgkin-Huxley equations. *Biofizika* **17**, 271–7.

Kruskal, M. (1962). Asymptotic theory of Hamiltonian systems. *J. Math. Phys.* **3**, 806–28.

Kruskal, M.D., Miura, R.M., Gardner, C.S. & Zabusky, N.J. (1970). Uniqueness & nonexistence of polynomial conservation laws. *J. Math. Phys.* **11**, 952–60.

Kruskal, M.D. & Zabusky, N.J. (1964). Stroboscopic perturbation procedure for treating a class of nonlinear wave equations. *J. Math. Phys.* **5**, 231–44.

Kryloff, N. & Bogliuboff, N. (1947). *Introduction to Non-Linear Mechanics*. Annals of Math. Studies. Princeton Univ. Press.

Kuramoto, Y. (1983). *Chemical Oscillations, Waves & Turbulence*. Springer-Verlag.

Kusnirenko, A.G., Kotok, A.B. & Alekseev, V.M. (1981). *Three Papers on Dynamical Systems*. Amer. Math. Soc.

Kuznetsov, S.P. (1986). Universality and scaling in the behavior of coupled feigenbaum systems. *Radiophys. Quantum Electronics* **28**, 681–95.

Lamb, G.L. Jr. (1971). Analytical descriptions to ultrashort optical pulse propagation in resonant media. *Rev. Mod. Phys.* **43**, 99–124.

Lamb, G.L. Jr. (1974). Bäcklund transformations for certain nonlinear evolution equations. *J. Math. Phys.* **15**, 2157.

Lamb, G.L. Jr. (1980). *Elements of Soliton Theory*. Wiley-Interscience.

Lamb, H. (1945). *Hydrodynamics*. Dover.

Lancaster, P. (1985). *Theory of Matrices*. Academic Press.

Landau, L.D. & Lifshitz, E.M. (1959). *Fluid Mechanics*. Addison-Wesley.

Lanford, O.E. (1977). In *Turbulence Seminar*. Lecture Notes in Mathematics **615**, ed. P. Bernard & T. Rativ, p. 114, Springer-Verlag.

Landford, O.E. (1987). Circle Mappings. In *Recent Developments in Mathematical Physics*, ed. H. Mitter, & L. Pittner. pp. 1–17, Springer-Verlag.

Langley, P., Simon, H.A., Bradshaw, G.L. & Zythow, J.M. (1987). *Scientific Discovery*, MIT Press.

Langton, C.G. (1984). Self-reproduction in cellular automata. *Physica* **10D**, 135–44.

Langton, C.G. (1986). Studying artificial life with cellular automata, *Physica* **22D**, 120–49.

LaSalle, J. (1949). Relaxation oscillations. *Quart. J. App. Math.* **7**, 1.

LaSalle, J.P. & Lefschetz, S. (Eds.) (1963). *Nonlinear Differential Equations & Nonlinear Mechanics*. Academic Press.

Laslow, E. (1987). *Evolution: The Grand Synthesis*. Shambhala Pub. Boston.

Lasota, A. (1977). Ergodic problems in biology. In *Dynamical Systems 2*, Warsaw, 239–250. Sociéte Mathématique de France.

Lauterborn, W. & Crames, E. (1981). Subharmonic route to chaos observed in acoustics. *Phys. Rev. Letters* **47**, 1445–48.

Lax, P.D. (1968). Integrals of nonlinear equations of evolution and solitary waves. *Commun. Pure Appl. Math.* **21**, 467–90.

Lax, P.D. (1986). Mathematics & computing. *J. Stat. Phys.* **43**, 749–56.

Lebowitz, N.R. & Schaar, R. (1975). Exchange of stabilities in autonomous systems. *Studies Appl. Math.* **54**, 229–60.

Lee, T.D. (1985). Discrete mechanics. In *How Far Are We From the Gauge Forces?*, ed. A. Zichichi, pp. 15–144, Plenum Press.

Lee, T.D. (1986). Solutions of discrete mechanics near the continuum limit, in *Rationale of Beings*, eds K. Ishikawa, Y. Kawazoe, Y. Matsuzaki, K. Takahashi, p. 324, World Scientific.

Ledrappier, F. & Young, L.-S. (1985). The metric entropy of diffeomorphisms Part I: Characterization of Measures Satisfying Pesin's entropy Formula, *Ann. Math.* **122**, 509–39; Part II: Relations between entropy, exponents, and dimension, *ibid.*, 540–74.

Lefever, R. (1968). Stabilité des structures dissipatives, *Bull. Classe Sci., Acad. Roy. Belg.* **54**, 712–19.

Lefschetz, S. (1959). *Differential Equations: Geometric Theory.* Interscience Pub.

Leggett, A.J. (1987). *The Problems of Physics*, Oxford Univ. Press.

Leisegang, S. (1953). Zum Astigmatismus von Electronenlinsen. *Optik* **10**, 5–14.

Levi, M. (1981). Qualitative analysis of the periodically forced relaxation oscillations. *Memoirs Amer. Math. Soc.* **32**, 244, 1–145.

Levinson, N. (1949). A second order differential equation with singular solutions. *Ann. Math.* **50**, 127–53.

Levinson, N. & Smith, O.K. (1942). A general equation for relaxation oscillations, *Duke Math. J.* **9**, 287–403.

Li, T.Y. & Yorke, J.A. (1975). Period three implies chaos. *Amer. Math. Monthly* **82**, 983–92.

Lichtenberg, A.J. & Lieberman, M.A. (1983). *Regular and Stochastic Motion.* Springer-Verlag.

Lie, S. (1893). *Vorlesungen ueber Continuierliche Gruppen mit Geometrischen und Anderen Anwendugen.* B.C. Teubner: Leipzig.

Linkens, D.A., Biological Systems (1980). In *Modelling of Dynamical Systems*, Vol. 1 ed. N. Nicholson, pp. 193–223. Peregrinus.

Linsay, P.S. (1981). Period doubling and chaotic behavior in a driven anharmonic oscillator. *Phys. Rev. Letters* **47**, 1349–52.

Little W.A. (1974). The existence of persistent states in the brain. *Math. Biosci.* **19**, 101–20.

Livi, R., Pettini, M., Ruffo, S., Sparpaglione, M. & Vulpiani, A. (1983). Relaxation to different stationary states in the Fermi-Pasta-Ulam model. *Phys. Rev.* **A28**, 3544–52.

Livi, R., Pettini, M., Ruffo, S., Sparpaglione, M. & Vulpiani, A. (1985). Equipartition threshold in nonlinear large Hamiltonian systems: The Fermi-Pasta-Ulam model. *Phys. Rev.* **A31**, 1039–45.

Longren, K & Scott, A. (Eds.) (1975). *Solitons in Action.* Academic Press.

Lorenz, E.N. (1963). Deterministic nonperiodic flow. *J. Atmos. Sci.* **20**, 130–41; The mechanics of vacillation. *J. Atmos. Sci.* **20**, 448–64.

Lorenz, E.N. (1964). The problem of deducing the climate from the governing equations. *Tellus* **16**, 1–11.

Lorenz, E.N. (1979). On the prevalence of aperiodicity in simple systems. In *Global Analysis*, ed. M. Grmela & J.E. Marsden. Lecture Notes in Mathematics **755**, pp. 53–75. Springer-Verlag.

Lorenz, E.N. (1980). Attractor sets and quasi-geostrophic equilibrium. *J. Atmos. Sci.* **37**, 1685–99.

Lorenz, E.N. (1984). The local structure of a chaotic attractor in four dimensions. *Physica* **13D**, 90–104.

Lotka–Volterra (1985). *Approach to Cooperation & Competition in Dynamic Systems.* Ebeling, W. & Peschel, M. (Eds.) Academic–Verlag: Berlin.

Lu, Y.C. (1976). *Singularity Theory & an Introduction of Catastrophy Theory.* Springer-Verlag.

Lugiato, L.A. (1984). *Theory of Optical Bistability.* Progress in Optics **21**, ed. Wolf. E. pp. 71–211. North-Holland.

Lundquist, S. (Ed.) (1985). The physics of chaos and related problems, *Physica Scripta* **T9**, 1–219.

Lunsford, G.H. & Ford, J. (1972). On the stability of periodic orbits for nonlinear oscillator systems in regions exhibiting stochastic behavior. *J. Math. Phys.* **13**, 700–5.

Macdonald, K.C. & Luyendyk, B.P. (1981). The crest of the East Pacific rise. *Sci. Amer.* **244**(5), 100–16.

MacDonald, R.A. & Tasi, D.H. (1978). Molecular dynamics calculations of energg transport in crystalline solids. *Phys. Rep.* **46**, 1–48.

MacKay, R.S. (1986). Transition to chaos in area-preserving maps. *Nonlinear Dynamics Aspects of Particle Accelerators*, ed. J.M. Jowett, M. Month. & S. Turner. pp. 390–454. Springer-Verlag.

MacKay, R.S. & Percival, I.C. (1985). Converse KAM: theory and practice. *Commun. Math. Phys.* **98**, 469–512.

Mackay, R.S. & Tresser, C. (1986). Transitions to topological chaos for circle maps. *Physica* **19D**, 206–37.

Makarov, I.M. (Ed.) (1984). *Cybernetics Today.* Mir Pub.

Makhankov. V.G. (1978). Dynamics of classical solitons (in non-integrable systems). *Phys. Rep.* **35**, 1. (See under Backlund transformations.)

Mammerville, P. & Pomeau, Y. (1979). Intermittency and the Lorenz model. *Phys. Letters* **75A**, 1–2.

Manakov, S.V. (1974). Complete integrability and stochasticity of discrete dynamical systems. *Sov. Phys. JETP* **40**, 269–74.

Manakov, S.V. (1974). Nonlinear fraunhofer diffraction. *Sov. Phys. JETP* **38**, 693–6.

Manakov, S.V. (1975). Complete integrability and stochastization of discrete dynamical systems. *Sov. Phys. JETP* **40**, 269–74.

Manakov, S.V. & Zakhorov, V.E. (Eds.) (1981). *Soliton Theory. Physica* **3D**, No. 1 & 2.

Mandelbrot, B.B. (1977). *Fractals: Form, Chance & Dimension.* W.H. Freeman.

Mandelbrov. B.B. (1983). *The Fractal Geometry of Nature.* W.H. Freeman.

Manneville, P. & Pomeau, Y. (1980). Different ways to turbulence in dissipative dynamic systems. *Physica* **1D**, 219–26.

Maradudin, A.A., Montroll, E.W., Weiss, G.H. & Ipatova, I.P. (1971). *Theory of Lattice Dynamics in the Harmonic Approximation*, 2nd edn. Academic Press.

Margolus, N. (1984). Physic-like models of computation, *Physica* **10D**, 81–95.

Margolus, N., Toffoli, T. & Vichniac, G. (1986). Cellular–automata supercomputers for fluid-dynamics modeling. *Phys. Rev. Letters* **56**, 1694–6.

Markley, N.G., Martin, J.C. & Perrizo, W. (1978). *The Structure of Attractors in Dynamical Systems.* Springer-Verlag.

Marotto, F.R. (1978). Snap-back repellers imply chaos in R^n. *J. Math. Anal. Appl.* **63**, 199–223.

Marsden, J.E. & McCracken, M. (1976). *The Hopf Bifurcation & its Applications.* Applied Math. Sci. **19**, Springer-Verlag.

Martin, P.C. (1975). The onset of turbulence: a review of recent developments in theory and experiment. In *Proceedings of the International Conference on Statistical Physics, Budapest 1975*, ed. L. Pal. & P. Szepfalusy. pp. 69–96. North-Holland.

Martin, O., Oldzko, A.M. & Wolfram, S. (1984). Algebraic properties of cellular automata. *Comm. Math. Phys.* **93**, 219–58.

Martinez, H.M. (1972). Morphogenesis & chemical dissipative structures. *J. Theor. Biol.* **36**, 479–501.

Martini, R. (Ed.) (1980). *Geometrical Approaches to Differential Equations.* Lecture Notes in Mathematics **810**. Springer-Verlag.

Maselko, J. & Swinney, H.L. (1985). A complex transition sequence in the Belousov–Zhabotinskii reaction, *Physica Scripta* **T9**, 35–9.

Mather, J.N. (1984). Non-existence of invariant circles. *Ergod. Th. & Dynam. Sys.* **4**, 301–9.

Mayer-Kress, G. & Haken, H. (1987). An explicit construction of a class of suspensions and autonomous differential equations for diffeomorphisms in the plane. *Commun. Math. Phys.* **111**, 63–74.

May, R.M. (1976). Simple mathematical models with very complicated dynamics. *Nature* **261**, 459–67. (Review article.)

Matsuno, K., Dose, K., Harada, K. & Rohlfing (Eds.) (1984). *Molecular Evolution and Protobiology*, Plenum Press.

May, R.M. (1973). *Stability and Complexity in Model Ecosystems*, Princeton University Press.

Mayr, E. (1988). *Toward A New Philosophy of Biology*, Belknap Press of Harvard Univ. Press.

Maynard Smith, J. (1982). *Evolutionary Game Theory*. Cambridge Univ. Press.

Maynard Smith, J. (1986). *The Problems of Biology*. Oxford Univ. Press.

McCauley, J.L. Jr. & Palmore, J.I. (1986). Computable chaotic orbits. *Phys. Letters A* **15**, 433.

McDonald, S.W., Grebogi, C., Ott, E. & Yorke, J.A. (1985). Fractal basin boundaries. *Physica* **7D**, 125–53.

McGuiness, M.J. (1985). A computation of the limit capacity of the Lorenz attractor. *Physica* **16D**, 265–75.

McLachlan, N.W. (1964). *Theory and Application of Mathieu Functions*. Dover.

McLaughlin, D.W. (1975). Four examples of the inverse method as a canonical transformation. *J. Math. Phys.* **16**, 96–9.

McLaughlin, J. (1976). Successive bifurcations leading to stochastic behavior. *J. Stat. Phys.* **15**, 307–26.

McLaughlin, J.B. & Martin, P.C. (1975). Transition to turbulence in a statistically stressed fluid system. *Phys. Rev.* **A12**, 186–203.

Meadows, D.L. *et al.* (1973). *Toward Global Equilibrium*. Wright–Allen Press: Cambridge, Mass.

Meadows, D.L. *et al.* (1974a). *Dynamics of Growth in a Finite World*. Wright–Allen Press: Cambridge, Mass.

Meadows D.L. *et al.* (1974b). *The Limits to Growth*. Universe Books: NY.

Meinhardt, H. (1982). *Models of Biological Pattern Formation*. Academic Press.

Melekhin, V.N. (1976). Phase dynamics of particles in a microtron & the problem of stochastic instability of nonlinear systems. *Sov. Phys. JETP* **41**, 803–8.

Melnikov, V.K. (1963). On the stability of the center for time periodic perturbations. *Trans. Moscow Math. Soc.* **12**, (1), 1–57.

Merrill, R.T. & McElminny, M.W. (1983). *The Earth's Magnetic Field: Its History, Origin & Planetary Perspective* International Geophysics Series, 32. Academic Press.

Metropolis, N., Stein, M. & Stein, P. (1973). On finite limit sets for transformations on the unit interval. *J. Combinatorial Theory* **15**, 25–44.

Meyer, R.E. (Ed.) (1981). *Transition and Turbulence*. Academic Press.

Meyer, R.E. & Parter, S.V. (Eds.) (1980). *Singular Perturbations & Asymptotics*. Academic Press.

Minorsky, N. (1947). *Introduction to Nonlinear Mechanics*. J.W. Edwards: Ann Arbor, Mich.

Minorsky, N. (1962). *Non-Linear Oscillations*. Van Nostrand.

Misiurewicz, M. (1976). Topological conditiond entropy, *Studia Mathematica* **55**, 175–200.

Misiurewicz, M. & Szlenk, W. (1980). Entropy of Piecewise Monotone Mappings, *Studia Mathematica* **57**, 45–63.

Mistriotis, M.D. (1986). From determinism to stochasticity, Thesis, Dept. Phys., Univ. Illinois, Urbana.

Mistriotis, A.D. & Jackson, E.A. (1987). Transition to stochasticity for periodically perturbed area preserving system. *Physica Scripta* **35**, 97–104.

Mitchell, K.E. & Dutton, J.A. (1981). Bifurcations from stationary to periodic solutions in a low-order mode of forced dissipative barotropic flow. *J. Atmos. Sci.* **38**, 690–716.

Miura, R.M. (1968). A remarkable explicit nonlinear transformation. *J. Math. Phys.* **9**, 1202–4.

Miura, R.M. (1976). The Korteweg-deVries equation: a survey of results. *SIAM Review* **18**, 412–59.

Miura, R.M. (Ed.) (1976). *Backlund Transformations. The Inverse Scattering Method, Solitons & Their Applications*: Lecture Notes in Mathematics **515**. Springer-Verlag.

Miura, R.M., Gardner, C.S., Kruskal, M.D. (1968). Existence of conservation laws & constants of motion. *J. Math. Phys.* **9**, 1204–9.

Mo, K.C. (1972). Theoretical prediction for the onset of widespread instability in conservative nonlinear oscillator systems. *Physica* **57**, 445–54.

Moffat, H.K. (1977). Six lectures on general fluid dynamics & two on hydromagnetic dynamo theory. In *Fluid Dynamics*, ed. R. Balian & J.L. Peube, pp. 151–233. Gordon & Breach.

Moffat, H.K. (1978). *Magnetic Field Generation in Electrically Conducting Fluids*. Cambridge Univ. Press.

Mokross, F. & Büttner, H. (1983). Thermal conductivity in the diatomic Toda lattice, *J. Phys.* **C16**, 4539–4546.

Montroll, E.W. (1978). On some mathematical models of social phenomena. In *Nonlinear Equations in Abstract Spaces*, ed. V. Lakshmikantham, pp. 161–216. Academic Press.

Moon, F.C. (1980). Experimental models for strange attractor vibrations in elastic systems. In *New Approach to Nonlinear Problems in Dynamics*, ed. P.J. Holmes. pp. 487–95. SIAM.

Moon, F.C. (1984). *Magneto–Solid Mechanics*. Wiley.

Moon, F.C. (1987). *Chaotic Vibrations*, Wiley.

Moore-Ede, M.C., Sulzman, F.M. & Fuller, C.A. (1982). *The Clocks that Time US*. Harvard Univ. Press.

Morozov, A.D. (1976). A complete qualitative investigation of duffing's equation. *Diff. Equations* **12**, 164–74.

Morse, P.M. & Feshbach, H. (1953). *Methods of Theoretical Physics*, McGraw-Hill.

Moser, J. (1962). On invariant curves of area-preserving mappings of an annulus. *Nachr. Akad. Wiss. Wiss. Gottingen Math. Phys. Kl.* **2**, 1–20.

Moser, J. (1967). Convergent series expansions for quasi-periodic motions. *Math. Annal.* **169**, 136–76.

Moser, J. (1968). Lectures on Hamiltonian systems. *Memoirs Amer. Math. Soc.* **81**, 1–60.

Moser, J. (1973). *Stable & Random Motions in Dynamical Systems*. Ann. Math. Studies, 77. Princeton Univ. Press.

Moser, J. (1975). Three integrable Hamiltonian systems connected with isospectral deformations. *Adv. in Math.* **16**, 197–220.

Moser, J. (1978). Nearly Integrable and Integrable Systems. In *Topics in Nonlinear Dynamics*, ed. S. Jorna, pp. 1–15, Amer. Inst. Phys.

Moser, J. (1981). *Integrable Hamiltonian Systems & Spectral Theory*. Academia Nazionale dei Lincei: Pisa.

Muncaster, R.G. & Zinnes, D.A. (1984). Phase portraits, singularities, and the dynamics of hostility in international relations. *SIAM J. Appl. Math.*

Musha, T. & Higuchi, H. (1977). *Proceedings of the Symposium on 1/f Fluctuations*. p. 189. Inst. Elect. Engineers: Tokoyo.

Nagumo, J. & Sato, S. (1972). On a response characteristic of a mathematical neuron model, *Kybernetik* **10**, 155–64.

Nakamura, A. (1979). A direct method of calculating periodic wave solutions to nonlinear evolution equations. I. Exact two–periodic wave solution. _J. Phys. Soc. Japan_ **47**, 170.

Nakazaea, H. (1970). On the lattice thermal conduction. _Prog. Theor. Phys. Suppl._ **45**, 231–62.

Nash, C. & Sen, S. (1983). _Topology and Geometry for Physicists._ Academic Press.

Nayfeh, A.H. (1986). Perturbation methods in nonlinear dynamics. In _Nonlinear Dynamics Aspects of Accelerators._ Lecture Notes in Physics **247**, ed. J.M. Jowett, M. Month, & S. Turner, pp. 238–314. Springer-Verlag.

Nayfeh, A.H. & Mook, D.T. (1979). Nonlinear oscillations. Wiley.

Nekhoroshev, N.N. (1971). Behavior of hamiltonian systems close to integrable, _Funct. Anal. Appl._ **5**, 338–9.

Nekhoroshev, N.N. (1977). An exponential estimate of the time of stability of near-integrable Hamiltonian systems, _Russ. Math. Survey_ **32**, 1–65.

Nemytskii, V.V. & Stepanov, V.V. (1960). _Qualitative Theory of Differential Equations._ Princeton Univ. Press.

von Neumann, J. (1966). _Theory of Self-Reproducing Automata_, ed. and completed by A.W. Burks. Univ. of Illinois Press: Urbana.

Newell, A.C. (1985). _Soliton in Mathematics and Physics._ SIAM.

Newell, A.C. & Redekopp, L.G. (1977). Breakdown of Zakharov–Shabat theory and soliton creation, _Phys. Rev. Lett._ **38**, 377–80.

Newhouse, S., Ruelle, D. & Takens, F. (1978). Occurrance of strange axiom A attractors near quasi–periodic flow on T^m, m > 3. _Commun. Math. Phys._ **64**, 35–40.

Newton, R.G. (1978). Three-dimensional solitons. _J. Math. Phys._ **19**, 1068–73.

Nicolis, C. & Nicolis, G. (1984). Is there a climatic attractor? _Nature_ **311**, 529–32.

Nicolis, G. (1971). Stability and dissipative structures in open systems for iron equilibrium. _Adv. Chem. Phys._ **19**, 209.

Nicolis, G. (Ed.) (1985). _Aspects of Chemical Evolution._ Adv. Chem. Phys. **55**, Wiley.

Nicolis, G. & Prigogine, I. (1978). _Self-Organization in Nonequilibrium Systems._ Wiley-Interscience.

Nicolis, J.S. (1986). _Dynamics of Hierarchical Systems._ Springer-Verlag.

Niesert, U., Harnasch, & Bresch, C. (1981). Origin of life between seylla and charybolis. _J. Mol. Evol._ **17**, 348–53.

Nitecki, Z. (1971). _Differential Dynamics._ M.I.T. Press.

Noid, D.W., Koszykowski, M.L. & Marcus, R.A. (1981). Quasiperiodicity and stochastic behavior in molecular, _Ann. Rev. Phys. Chem._ **31**, 267–309.

Normand, C., Pomeau, Y. & Velarde, M.G. (1977). Convective instability: a physicist's approach. _Rev. Mod. Phys._ **49**, 581–624.

Northcote, R.S. & Potts, R.B. (1964). Energy sharing and equilibrium for nonlinear systems. _J. Math. Phys._ **5**, 383.

Nosé, S. (1984). A molecular dynamics method for simulations in the canonical ensemble. _Molecular Phys._ **52**, 255–68.

Noyes, R.M. (1976). Oscillations in chemical systems. Corrected stoichiometry of the oregonator. _J. Chem. Phys._ **65**, 848–9.

Olds, C.D. (1963). _Continued Fractions._ Random House.

O'Malley, R.E. Jr. (1974). _Introduction to Singular Perturbations._ Academic Press.

O'Molley, R.E. Jr. (1976). Phase plane solutions to some singular perturbation problems. _J. Math. Anal. Appl._ **54**, 449–66.

O'Malley, R.E. Jr. (1976). _Asymtotic Methods and Singular Peturbations._ SIAM–AMS Proceedings 10. Amer. Math. Soc.

O'Malley, R.E. Jr. (1977). *Singular Perturbation Analysis for Ordinary Differential Equations.* Comm. Math. Institute Rijksuniversiteit: Utrecht.

Oono, Y. (1978). A Heuristic approach to Kolmogorov entropy as a disorder parameter, *Prog. Theor. Phys.* **60**, 1944–46.

Oono, Y. & Kohmoto, M. (1985). Discrete model of chemical turbulence. *Phys. Rev. Letters* **55**, 2927–31.

Oono, Y. & Osikawa, M. (1980). Chaos in nonlinear difference equations. I. *Prog. Theor. Phys.* **64**, 54–67.

Oono, Y. & Yeung, C. (1987). A cell dynamical system model of chemical turbulence. *J. Stat. Phys.* **48**, 593–644.

Orszag, S.A. (1977). Lectures on the statistical theory of turbulence. In *Fluid Dynamics*, ed. R. Balian & J.-L. Peube, pp. 237–368. Gordon and Breach.

Orszag, S.A. & Kells, L.C. (1980). Transition to turbulence in plane poiseuille and plane couette flow. *J. Fluid Mech.* **96**, 159–205.

Orszag, S.A. & Patera, A.T. (1981). Subcritical transition to turbulence in plane shear flows. In *Transition and Turbulence*, ed. R.E. Meyer. pp. 127–46. Academic Press.

Orszag, S. & Yakhot, V. (1986). Reynolds number scaling of cellular–automation hydrodynamics. *Phys. Rev. Letters* **56**, 1691–3.

Osborne, A. & Stuart, A.E.G. (1978). On the separability of the Sine–Gordon equation and similar quasilinear partial differential equations. *J. Math. Phys.* **19**, 1573–8.

Oseledec, V.I. (1968). A multiplicative ergodic theorem: Lyapunov characteristic numbers for dynamical systems. *Trans. Moscow Math. Soc.* **19**, 197.

Osikawa, M. & Oono, Y. (1981). Chaos in C°—endomorphism of interval. *Res. Inst. Math. Sci. Kyoto Univ.* **17**, 165–77.

Ostlund, S., Rand, D., Sethna, J. & Siggia, E. (1983). Universal properties of the transition from quasi–periodicity to chaos in dissipative systems. *Physica* **8D**, 303–42.

Othmer, N.G. & Scriven, L.E. (1971). Instability and dynamic pattern in cellular networks. *J. Theor. Biol.* **32**, 507–37.

Ovsjanikov (1958). Group and group invariant solutions of differential equations. *Dokl. Akad. Nauk. USSR* **3**, 439.

Pacault, A. & Vidal, C. (Eds.) (1979). *Synergetics*: From Equilibrium Springer-Verlag.

Packard, N.H. & Wolfram, S. (1985). Two-dimensional cellular automata, *J. Stat. Phys.* **38**, 901–46.

Packard, N.H., Crutchfield, J.P., Farmer, J.D. & Shaw, R.S. (1980). Geometry from a time series, *Phys. Rev. Letters* **45**, 712–6.

Paladin, G. & Vulpiani, A. (1987). Anomoulous scaling laws in multifractral objects, *Phys. Reports* **156**, 147–225.

Park, J.K., Steglitz, K. & Thurston, W.D. (1986). Soliton-like behavior in automata, *Physica* **19D**, 423–32.

Parker, E.N. (1975). The dynamo mechanism for the generation of large-scale magnetic fields. *Ann. N.Y. Acad. Sci.* **252**, 141–55.

Pars, L.A. (1983). *A Treatise on Analytical Dynamics.* Wiley.

Paulos, J.A. (1981). *Mathematics and Humor.* Univ. of Chicago Press.

E. Paulré, E. (Ed.) (1981). *System Dynamics and the Analysis of Change.* North-Holland. (Economic, medical, war, urban, etc., models.)

Pauwelussen, J.P. (1981). Nerve impulse propagation in a branching nerve system: a simple model. *Physica* **4D**, 67–88.

Pavelle, R., Rothstein, M. & Fitch, J. (1981). Computer algebra. *Sci. Amer.* **245** (6), 136.

Peitgen, H.-O. & Richter, P.H. (1986). *The Beauty of Fractals*, Springer-Verlag.

Peixoto, M.M. (Ed.) (1973). *Dynamical Systems*. Academic Press.

Percival, I. (1982). *Introduction to Dynamics*. Cambridge Univ. Press.

Perlmutter, A. & Scott, L.F. (Eds.) (1977). *The Significance of Nonlinearity in the Natural Sciences*. Plenum.

Perring, J.K. & Skyrme, T.H.R. (1962). A model unified field equation. *Nucl. Phys.* **31**, 550–5.

Peschel, J. & Mende, W. (1986). *The Predator–Prey Model*. Springer-Verlag.

Petrovsky. I.G. (1954). *Lectures on Partial Differential Equations*. Interscience Pub.

Petrosky, P. & Prigogine, I. (1982). The KAM invariant and Poincaré's theorem. In *Applications of Modern Dynamics to Celestial Mechanics and Astrodynamics*, ed. V. Szebehely, pp. 185–99, Reidel.

Pietronero, L. & Tosatti, E. (Eds.) (1986). *Fractals in Physics*, North-Holland.

Pimbley, G.H. Jr. (1969). *Eigenfunction Branches of Nonlinear Operators and Their Bifurcation*. Lecture Notes in Mathematics 104. Springer-Verlag.

Pines, D. (Ed.) (1985). *Emerging Syntheses in Science*. Santa Fe Institute.

Pippard, A.B. (1978). *The Physics of Vibration*. Cambridge Univ. Press.

Pippard, A.B. (1985). *Response and Stability*. Cambridge Univ. Press.

Pnevmatikos, S.N. (Ed.) (1985). *Singularities and Dynamical Systems*. Math. Studies **103**, North-Holland.

Pnevmatikos, S.N. (1985). Solitons in nonlinear atomic chains. In *Singularities and Dynamical Systems*, ed. S.N. Pnevmatikos, pp. 397–437. North-Holland.

Poincaré, H. (1892, 1893, 1894). *Les Methodes Nouvelles de la Mécanique Celeste*. Vols. 1–3. Gauthier–Villars: Paris. Reprinted by Dover, 1957. (Translation: *New Methods of Celestial Mechanics*. NASA, 1967.)

Poincaré, J.H. (1929). *The Foundations of Science*, Science Press.

Poston, T. & Stewart, I.N. (1976). *Taylor Expansions and Catastrophes*. Pitman.

Poston, T. & Stewart, I. (1978). *Catastrophe Theory and Its Applications*. Pitman.

Poston, T. & Woodcock, A.E.R. (1973). Zeeman's catastrophe machine. *Proc. Camb. Phil. Soc.* **74**, 217.

Poundstone, W. (1985). *The Recursive Universe*. Contemporary Books.

Peess. W.H., Flannery, B.F., Teukolsky, S.A. & Vetterling, W.T. (1986). *Numerical Recipes*. Cambridge Univ. Press.

Preston, C. (1983). *Iterates of Maps on an Interval*. Springer-Verlag.

Prigogine, I. (1962). *Non-Equilibrium Statistical Mechanics*, Interscience.

Prigogine, I. (1980). *From Being to Becoming*. Freeman.

Prigogine, I. & Lefever, R. (1968). Symmetry breaking instabilities in dissipative systems, II. *J. Chem. Phys.* **48**, 1695–700.

Prigogine, I. & Stengers, I. (1984). *Order Out of Chaos*. Bantam Books.

Rabinowitz, P.H. (Ed.) (1977). *Applications of Bifurcation Theory*. Academic Press.

Rajaraman, A. (1975). Some non-perturbative semi-classical methods in quantum field theory. *Phys. Rep.* **21C**, 227–313.

Ramajaran, R. (1982). *Solitons and Instantons: An Introduction to Solitons and Instantons in Quantum Field Theory*. North-Holland.

Ramani, A., Dorizzi, B., Grammaticos, B. & Bountis, T. (1984). Integrability and the painlevé property for low-dimensional systems. *J. Math. Phys.* **25**, 878–83.

Rannou, F. (1974). Numerical study of discrete plane area–preserving map. *Astron. Astrophys.* **31**, 289–301.

Rashevsky, N. (1938). *Mathematical Biophysics*. Univ. of Chicago Press. Reprint, Dover, 1960.

Räty, R., Isomäki, H.M. & von Boehm, J. (1984). Chaotic motion of a classical anharmonic oscillator. *Acta Polytechnica Scandinavica, Mech. Eng. Ser.* **85**, 1–30.

Rebbi, C. (1979). Solitons. *Sci. Amer.* **240**(2), 92–116.

Richtmyer, R.D. (1986). A study of the Lorenz attractor, science and computers, *Adv. Math. Suppl. Studies*, **10**, 207–19.

Risset, J.C. (1982). Stochastic processes in music and art. In *Stochastic Processes in Quantum Theory and Statistical Physics*. Lecture Notes in Physics **173**, ed. S. Albeverio, Ph. Combe, & M. Sirugue-Collin. Springer-Verlag.

Robbins, K.A. (1977). A new approach to subcritical instability and turbulent transitions in a simple Dynamo. *Math. Proc. Camb. Phil. Soc.* **82**, 309–25.

Robbins, K.A. (1979). Periodic solutions and bifurcation structure at high R in the Lorenz model. *J. Appl. Math.* **36**, 457–71.

Roberts, P.H. (1971). Dynamo theory. In *Lectures in Applied Mathematics*, ed. W.H. Reid, pp. 129–206. Amer. Math. Soc.

Roederer. J. (1979). *Introduction to the Physics and Psychophysics of Music*. Springer-Verlag.

Rogers, C. & Shadwick, W.F. (1982). *Backlund Transformations and Their Applications*. Academic Press.

Rohlin, V.A. (1964). Exact endomorphisms of a Lebesque space. *Trans. Amer. Math. Soc.* **39**, 1–36.

Rohlin, V.A. (1967). Lectures on the Entropy Theory of Measure-Preserving Transformations, Russian Math Survey **22**(5), 1–52.

Rolfsen, D. (1976). *Knots and Links*. Publish or Perish: Berkeley.

Rollins, R.W. & Hunt, E.R. (1982). Exactly solvable model of a physical system exhibiting universal chaotic behavior. *Phys. Rev. Letters* **49**, 1295–8.

Rosen, R. (1985). *Anticipatory Systems*, Pergamon.

Rosenbluth. M.N. (1972). Superadiabaticity in mirror machines. *Phys. Rev. Letters* **29**, 408–10.

Rössler, O.E. (1972). A principle for chemical multivibration. *J. Theor. Biol.* **36**, 413–7.

Rössler, O.E. (1977). Continuous chaos. In *Synergetics: A Workshop*, ed. H. Haken, pp. 184–97. Springer-Verlag.

Rössler, O.E. (1979a). Continuous chaos: four prototype equations. *Ann. N.Y. Acad. Sci.* **316**, 376–92.

Rössler, O.E. (1979b). Chaos. In *Structural Stability in Physics*, ed. W. Guttinger & H.E. Kemeier, pp. 290–309. Springer-Verlag.

Rössler, O.E. (1983). The chaotic hierarchy. *Z. Naturforsch.* **38a**, 788–801.

Roux, J.C., Simoyi, R.H. & Swinney, H.L. (1983). Observation of a strange attractor. *Physica* **8D**, 257–66.

Rubel, L.A. (1981). A universal differential equation. *Bull. Amer. Math. Soc.* New Sevies. **4**, 345–49.

Rubel, L.A., Digital simulation of analog computation and Church's thesis, *J. Symbolic Logic* (to appear).

Rubinstein, J. (1970). Sine-Gordon equation. *J. Math. Phys.* **11**, 258–66.

Rudakov, L.I. & Tsytovich, V.N. (1978). Strong langmuir turbulence. *Phys. Rep.* **40**, 1–73.

Ruelle, D. (1978). Dynamical systems with turbulent behavior. In *Mathematical Problems in Theoretical Physics*, ed. G. Del'Atonio, S. Doplicher & G. Jona-Lasinio, pp. 341–60. Lecture Notes in Physics 80. Springer-Verlag.

Ruelle, D. (1983). Small random perturbations and the definition of attractors. In *Geometric dynamics*. Lecture Notes in Mathematics, Eds. A Doid & B. Eckmann, pp. 663–76. Springer-Verlag.

Ruelle, D. & Takens, F. (1971). On the nature of turbulence. *Commun. Math. Phys.* **20**, 167–92, **23**, 343–4.

Russell, D.A., Hanson, J.D. & Ott, E. (1980). Dimension of strange attractors. *Phys. Rev. Letters* **45**, 1175–8.

Russell, J.S. (1844). Report on waves. In *Report of 14th Meeting, British Assoc. Adv. Sci.* pp. 311–90. Marray: London.

Ryutov, D.D. & Stupakov, G.V. (1978). Diffusion of resonance particles in ambipolar plasma traps. *Sov. Phys. Dokl.* **23**, 412–4.

Saaty, T. L., Alexander, J.M. (1981). *Thinking with Models*. Pergamon.

Saari, D.G. (1984). The ultimate of chaos resulting from weighted voting systems. *Adv. Appl. Math.* **5**, 286–308.

Sagdeev, R.Z. (Ed.) (1984). *Nonlinear and Turbulent Processes in Physics*, Vols. 1–3. Harwood Acad. Pub.

Sagdeev, R.Z., Usikov, D.A., & Zaslavsky, G.M. (1988). *Nonlinear Physics*. Harwood Academic Pub..

Saitô, N., Ooyama, N. & Aizawa, Y. (1970). Computer experiments on ergodicity problems in anharmonic lattice vibrations. *Prog. Theor. Phys. Suppl.* **45**, 209–30.

Saitô, N., Hirotomi, N. & Ichimura, A. (1975). The induction phenomenon and ergodicity in the anharmonic lattice vibration. *J. Phys. Soc. Japan* **39**, 1431–8.

Saltzman, B. (1962). Finite amplitude free convection as an initial value problem I. *J. Atmos. Sci.* **19**, 329–41.

Sanders, J.A. & Verhulst, F. (1984). *Averaging Methods in Nonlinear Dynamical Systems*. Appl. Math. Sciences. Springer-Verlag.

Schmidt, G. & Tondl, A. (1986). *Non-Linear Vibrations*, Cambridge Univ. Press.

Schmutz M. & Rueff, M. (1984). Bifurcations schemes of the Lorenz model. *Physica* **11D**, 167–78.

Schrödinger, E. (1914). Zur dynamik elastisch gekoppelter punktsysteme. *Ann. d. Phys.* **44**, 916–34.

Schult, R.L., Creamer, D.B., Henyey, F.S. & Wright, J.A. (1987). Symmetric and nonsymmetric coupled logistic maps, *Phys. Rev.* **A35**, 3115–18.

Schuster, H.G. (1984). *Deterministic Chaos*. Physik–Verlag: Weinheim.

Schuster, P. & Sigmund, K. (1983). Replicator dynamics. *J. Theor. Biol.* **100**, 533–8.

Schuur, P.C. (1986). *Asymptotic Analysis of Solition Problems*. Springer-Verlag.

Scott, A.C. (1970). *Active and Nonlinear Wave Propagation in Electronics*. Wiley.

Scott, A.C., Chu, F.Y.F. & McLaughlin, D.W. (1973). The soliton: a new concept in applied science. *Proc. IEEE* **61**, 1443–83.

Sedov, L.I. (1959). *Similarity and Dimensional Methods in Mechanics*. Academic Press.

Segur, H. (1986). Some open problems, *Physica* **18D**, 1–12.

Shannon, C.E. (1948). A mathematical theory of communication. *Bell. Tech. J.* **27**, 623–56.

Sharkovsky, A.N. (1964). Coexistence of the cycles of a continuous mapping of the line into itself. *Ukr. Mat. Zh.* **16**(1), 61–71.

Shaw, R. (1981). Strange attractors, chaotic behavior, and information flow. *Z. Naturforsch.* **36a**, 80–112.

Shaw, R. (1984). *The Dripping Faucet as a Model Chaotic System*. Aerial Press.

Shimada, I. & Nagashima, T. (1979). A numerical approach to ergodic problems of dissipative dynamical systems. *Prog. Theor. Phys.* **61**, 1605.

Siegel, C.L. & Moser, J.K. (1971). *Lectures on Celestial Mechanics*. Springer-Verlag.

Sigmund, K. (1983). The maximum principle for replicator equations. *J. Theor. Biol.* **100**, 63–72.

Simoyi, R.H., Wolf, A. & Swinney, H.L. (1982). One-dimensional dynamics in a multi-component chemical reaction. *Phys. Rev. Letters* **49**, 245–8.

Simó, C. (1979). On the Hénon-Pomeau attractor, *J. Stat. Phys.* **21**, 465–94.

Sinai, J.G. (1959). On the concept of entropy of a dynamical system. *Dokl. Akad. Nauk. SSSR* **124**, 768–76. (Russian.)

Sinai, Ya.G. (1977). *Introduction to Ergodic Theory.* Princeton Univ. Press.

Sinclair, R.M., Hosea, J.C. & Sheffield, G.V. (1970). A method for mapping a toroidal magnetic field by storage of phase stabilized electrons. *Rev. Sci. Inst.* **41**, 1552–9.

Sinhah, D.K. (Ed.) (1981). *Catastrophe Theory and Applications.* Wiley.

Singer, D. (1978). Stable orbits and bifurcation of maps of the interval. *SIAM J. Appl. Math.* **35**, 260–7.

Smale, S. (1963). Diffeomorphisms with many periodic points. In *Differential and Combinatorial Topology*, ed. S.S. Cairns, pp. 63–80. Princeton Univ. Press.

Smale, S. (1967). Differentiable dynamical systems. *Bull. Amer. Math. Soc.* **73**, 747–817.

Smale, S. (1974). A mathematical model of two cells via Turing's equation. *Lecture Notes in Applied Mathematics* **6**, 15–26. Amer. Math. Soc.

Smale, S. (1976). On the differential equations of species in competition. *J. Math. Biol.* **3**, 5–7.

Smale, S. (1977). Dynamical systems and turbulence. In *Turbulence Seminar.* Lecture Notes in Matehmatics **615**, 48–70. Springer-Verlag.

Smale, S. (1978). Review of catastrophe theory, by E.C. Zeeman. *Bull. Amer. Math. Soc.* **84**, 1360–8.

Smale, S. (1981). *Essays on Dynamical Systems, Economic Processes, and Related Topics.* Springer-Verlag.

Smale, S. (1981). The fundamental theorem of algebra and complexity theory. *Bull. Amer. Math. Soc.* New Series **4**, 1–36.

Smale, S. & Williams, R. (1976). The quantitative analysis of a difference equation of population growth. *J. Math. Biol.* **3**, 1–4.

Smart, D.R. (1974). *Fixed Point Theorems.* Cambridge Univ. Press.

Sparrow, C. (1982). *Bifurcations in the Lorenz Equation.* Lecture Notes in Applied Mathematics. Springer-Verlag.

Stanley, H.E. & Ostrowsky, N. (Eds.) (1986). *On Growth and Form.* Nijhoff.

Stapleton, H.J., Allen, J.P., Flynn, C.P., Stinson, D.G. & Kurtz, S.R. (1980). Fractal form of proteins. *Phys. Rev. Letters* **45**, 1456–9.

Stefan, O. (1977). A theorem of Sarkovskii on the existence of periodic orbits of continuous endomorphisms of the real line. *Commun. Math. Phys.* **54**, 237–48.

Stewart, I. (1982). Catastrophe theory in physics. *Rep. Prog. Phys.* **45**, 185–221.

Stine, J.R. & Noid, D.W. (1983). Method to determine the number of constants of the motion for multidimensional systems, *J. Phys. Chem.* **87**, 3039–42.

Stoker, J.J. (1950). *Nonlinear Vibrations.* Interscience Pub.

Stoker, J.J. (1957). *Water Waves.* Interscience Pub. (Appendix: On the Derivation of the Shallow Water theory, by K.O. Friedrichs.)

Størmer, C. (1955). *The Polar Aurora.* Clarendon Press: Oxford.

Stoutemyer, D.R. & Yun, D.Y.Y. (1980). Symbolic mathematical computation. In *Encyclopedia of Computer Science and Technology* **15**, Supplement, ed. J. Belzer, A.G. Holzman & A. Kent. Dekker.

Straffin, P.D. Jr. (1978). Periodic points of continuous maps. *Math. Mag.* **51**, 99–105.

Strangway, D.W. (1970). *History of the Earth's Magnetic Field.* McGraw-Hill.

Strauss, W.A. (1977). Existence of solitary waves in higher dimensions. *Commun. Math. Phys.* **55**, 149–62.

Stravrondis, O.N. (1972). *The Optics of Rays, Wavefronts and Caustics.* Academic Press.

Stuart, J.T. (1971). Nonlinear stability theory. *Ann. Rev. Fluid Mech.* **3**, 347–70.

Su, C.H. & Gardner, C.S. (1969). Derivation of the KdV equation and Burger's equation. *J. Math. Phys.* **10**, 536–9.

Sudarshan, E.C.G. & Mukunda, N. (1974). *Classical Dynamics: A Modern Perspective*. Wiley.

Sussman, H.J. (1977). A skeptic. *Science* **197**, 821.

Swindale, N.V. (1980). A model for the formation of ocular dominance stripes. *Proc. R. Soc. Lond.* **B208**, 243–64.

Swinney, H.L. & Gollub, J.P. (1978). The transition to turbulence. *Phys. Today* **31**(8), 41.

Swinney, H.L. & Gollub, J.P. (Eds.) (1981). *Hydrodynamic Instabilities and the Transition to Turbulence*. Topics in Applied Physics **45**, Springer-Verlag.

Swinney, H.W. & Gollub, J.P. (Eds.) (1985). *Hydrodynamic Instabilities and the Transition to Turbulence*. Topics in Applied Physics **45**. Springer-Verlag.

Sverdlove, R. (1977). Inverse problems for dynamical systems in the plane. In *Dynamical Systems*, eds A.R. Bednarek & L. Cesi, pp. 499–502, Academic Press.

Szebehely, V. (Ed.) (1982). *Applications of Modern Dynamics to Celestial Mechanics and Astrodynamics*. D. Reidel.

Szebehely, V. (1984). Review of concepts of stability. *Celestial Mech.* **34**, 49–64.

Szebehely, V. (1986). New nondeterministic celestial mechanics. In *Space Dynamics and Celestial Mechanics*, ed. K.B. Bhatnagar, pp. 3–14. D. Reidel.

Szlenk, W. (1984). *An Introduction to the Theory of Smooth Dynamical Systems*. Wiley.

Tabor, M. & Weiss, J. (1981). Analytic structure of the Lorenz system, *Phys. Rev.* **A24**, 2157–67.

Takeno, S. (Ed.) (1985). *Dynamical Problems in Soliton Systems*. Springer-Verlag.

Takens, F. (1973). *Introduction to Global Analysis*. Comm. Math. Inst., Rijksuniversiteit Utrecht.

Takens, F. (1981). Detecting strange attractors in turbulence. In *Dynamical Systems and Turbulence, Warwick 1980*. Lecture Notes in Mathematics **898**, ed. D.A. Rand & L.-S. Young. Springer-Verlag.

Takens, F. (1985). On the numerical determination of the dimensions of an attractor. In *Dynamical Systems and Bifurcations*, eds. B.L.J. Braaksma, H.W. Broer & F. Takens, pp. 99–106. Lect. Notes Math. **1125**, Springer-Verlag.

Testa, J., Perez, J. & Jeffries, C. (1982). Evidence of universal chaotic behavior of a driven nonlinear oscillator. *Phys. Rev. Letters* **48**, 714–7.

Thirring, W. (1978). *Classical Dynamical Systems*. Springer-Verlag.

Thom, R. (1969). Topological models in biology. *Topology* **8**, 313–35.

Thom, R. (1975). *Structural Stability and Morphogenesis*. W.A. Benjamin.

Thom, R. (1977). Structural stability, catastrophe theory and applied mathematics. *SIAM Review*, April.

Thompson, J.M.T. & Hunt, G.W. (1973). *A General Theory of Elastic Stability*. Wiley.

Thompson, J.M.T. & Hunt, G.W. (1975). Towards a unified bifurcation theory. *J. Appl. Math. Phys.* **26**.

Thompson, J.M.T. & Stewart, H.B. (1986). *Nonlinear Dynamics and Chaos*. Wiley.

Thornhill, S.G. & Haar, D. (1978). Langmuir turbulence and modulational instability. *Phys. Rep.* **43**, 43.

Toda, M. (1967). *J. Phys. Soc. Japan* **22**, 431–6; **23**, 501–6.

Toda, M. (1971). Waves in nonlinear lattice. *Prog. Theor. Phys. Suppl.* **45**, 174.

Toda, M. (1975). Studies of a non-linear lattice. *Phys. Rep.* **18C**, 1–124.

Toda, M. (1981). *Theory of Nonlinear Lattices*. Springer-Verlag.

Toffoli, T. (1977). *Cellular Automata Mechanics*. Technical Report 208. Computer and Communication Sciences Department, Univ. of Michigan.

Toffoli, T. (1982). Physics and computation. *Int. J. Theor. Phys.* **21**, 165–75.

Toffoli, T. (1984). Cellular automata as an alternative to (rather than an approximation of) differential equations modeling physics. *Physica* **10D**, 117–27.

Toffoli, T. & Margolus, N. (1987). *Cellular Automata: A New Environment for Modeling*, MIT Press.

Tomita, K. (1982). Chaotic response of nonlinear oscillators. *Phys. Rep.* **86**, 113–67.

Treve, Y.M. (1975). Theory of chaotic motion with application to controlled fusion research. In *Topics in Nonlinear Dynamics*, ed. S. Jorna, pp. 147–220. North-Holland.

Tricomi, F.G. (1957). *Integral Equations.* Interscience Pub.

Troy, W. (1977). A threshold phenomenon in the Field—Noyes model of the Belousov–Zhabotinskii reaction. *J. Math. Anal. Appl.* **58**, 233–48.

Tuck, J.L. & Menzel, M.T. (1972). The superperiod of the nonlinear weighted string (FPU) problem. *Adv. Math.* **9**, 339–407.

Tucker, A.W. & Bailey, H.S. Jr. (1950). Topology. *Sci. Amer.* **182**(1), 18–24.

Turing, A.M. (1952). The chemical basis of morphogenesis. *Phil. Trans. R. Soc. Lond.* **B237**, 37–72.

Turner, J.S., Roux, J.C., McCormick, W.D. & Swinney, H.L. (1981). Alternating periodic and chaotic regimes in chemical reaction-experiment and theory. *Phys. Letters* **85A**, 9–12.

Tyson, J.J. (1979). Oscillations, bistability, and echo waves. In Models of the Belousov–Zhobotinskii rgaction, ed. O. Gurrel & O.E. Rossler. *Ann. Y.Y. Acad. Sci.* **316**, 279–95.

Ueda, Y. (1979). Randomly transitional phenomena in systems governed by Duffing's equation. *J. Stat. Phys.* **20**, 181–96.

Uezu, T. (1983). Topology in dynamical systems, *Phys. Letters* **93A**, 161–6.

Uezu, T. & Aizawa, Y. (1982). Topological character of a periodic solution in three-dimensional ordinary differential equation systems. *Prog. Theor. Phys.* **68**, 1907–16.

Ulam, S.M. (1960). *A Collection of Mathematical Problems*, Wiley.

Ulam, S.M. (1965). Introduction to Studies of Nonlinear Problems by E. Fermi, J. Pasta & S.M. Ulam. In *Collected Papers of Enrico Fermi*, Vol. 2, Univ. Chicago Press.

Ulam, S.M. (1982). Transformations, Iterations and mixing flows. In *Dynamical Systems II*, eds. A.R. Bednarek and L. Cesari, pp. 419–26, Academic Press.

Ulam, S.M. (1983). *Adventures of a Mathematician*, Scribner.

Ulam, Stanislaw (1909–1984); *Los Alamos Science*, Number 15, Special Issue (1987).

Umberger, D.K. & Farmer, J.D. (1985). Fat fractals on the energy surface, *Phys. Rev. Letters* **55**, 661–64.

Vainberg, M.M. & Trenogin, V.A. (1974). *Theory of Branching of Solutions of Nonlinear Equations.* Noordhoff: Leyden.

van Bendegem, J.P. (1987). *Finite Empirical Mathematics; Outline of a Model.* Rijksuniversitent Gent.

Van der Blij, F. (1978). Some details of the History of the Korteweg-deVries equation, *Nieu Archief VoorWiskunde*(3) **26**, 54–64.

van der Pol, B. & van der Mark, J. (1927). Frequency demultiplication. *Nature* **120**, 363–4.

VanVelsen, J.F.C. (1978). On linear response theory and area preserving mappings. *Phys. Rep.* **41C**, 135–90.

Vendrik, M.C.M. (1979). A classification of phase diagrams by means of "elementary" catastrophe theory (ECT). I. *Physica* **99A**, 103–44.

Verhulst, F. (Ed.) (1979). *Asymptotic Analysis.* Lecture Notes in Mathematics **711**. Springer-Verlag.

Verhulst, F. (Ed.) (1983). *Asymptotic Analysis of Hamiltonian Systems.* Lecture Notes in Mathematics 985. Springer-Verlag.

Vichniac, G.Y. (1984a). Simulating physics with cellular automata. *Physica* **10D**, 96–116.

Vichniac, G.Y. (1984b). Instability in discrete algorithms and exact reversibility. *SIAM J. Alg. Disc. Meth.* **5**.

Vidal, C. & Pacault, A. (1982). Spatial chemical Structures: chemical waves. A. Review. In *Evolution of Order and Chaos*, ed. H. Haken, pp. 74–99. Springer-Verlag.

Vidal, C. & Pacault, A. (Eds.) (1984). *Nonequilibrium Dynamics in Chemical Systems.* Springer-Verlag.

Vidyasaga, M. (1978). *Nonlinear Systems Analysis.* Prentice-Hall.

Visscher, W.M. (1976). *Computer Studies of Transport Properties in Simple Models of Solids*, Meth. Comput. Phys. **15**, ed. G. Gilat, pp. 371–408. Academic Press.

Von Karmen, T. (1940). The engineer grapples with nonlinear problems (fifteenth Josiah Willard Gibbs lecture). *Bull. Amer. Math. Soc.* **46**, 615–83.

Wadati, M. (1972). The exact solution of the modified Korteweg-deVries equation. *J. Phys. Soc. Japan* **32**, 1681.

Wadati, M. (1975). Wave propagation in nonlinear lattice. *J. Phys. Soc. Japan* **38**, 673–80, 682–6.

Wahlquist, H.D. & Estabrook, F.B. (1973). Bäcklund transformations for solutions of the Kortewig-de Vries equation. *Phys. Rev. Letters* **31**, 1386.

Walker, G.H. & Ford, J. (1969). Amplitude instability and ergodic behavior for conservative nonlinear oscillator systems. *Phys. Rev.* **188**, 416–32.

Walker, J. (1978). Chemical systems that oscillate between one color and another. *Sci. Amer.* **239**, 152–60.

Waller, I. & Kapral, R. (1984). Spatial and temporal structure in systems of coupled nonlinear oscillators. *Phys. Rev.* **A30**, 2047–55.

Wasow, W. (1965). *Asymptotic Expansion for Ordinary Differential Equations.* Interscience.

Weiland, J. & Wilhemsson, H. (1977). *Coherent Non-Linear Interaction of Waves in Plasmas.* Pergamon.

Weinreich, G. (1979). The coupled motions of piano strings, *Scientific American* **240**(1), 118–27.

Weiss, L. (Ed.) (1972). *Ordinary Differential Equations.* Academic Press.

Wesfreid, J.E. & Zaleski, S. (Eds.) (1984). *Cellular Structures in Instabilities.* Lecture Notes in Physics **210**. Springer-Verlag.

Whiteman, K.J. (1977). Invariants and stability in classical mechanics. *Rep. Prog. Phys.* **40**, 1033–69.

Whitham, G.B. (1974). *Linear and Nonlinear Waves.* Wiley.

Whitney, H. (1936). Differentiable manifolds. *Ann. Math.* **37**, 645.

Whitney, H. (1944). The self-intersections of a smooth n-manifold in $2n$-space. *Ann. Math.* **45**, 220–46. The singularities of a smooth n-manifold in $(2n-1)$-space. *Ann. Math.* **45**, 247–93.

Whittaker, E.T. (1944). *Analytical Dynamics of Particles and Rigid Bodies.* Dover.

Wicken, J.S. (1987). *Evolution, Thermodynamics, and Information*, Oxford Univ. Press.

Wiener, N. & Rosenblueth, A. (1946). The mathematical formulation of the problem of conduction of impulses in a network of connected excitable elements, specifically in cardiac muscle. *Arch. Inst. Cardiol. Mexico* **16**, 205–65.

Wilhelmsson, H. (Ed.) (1979). Solitons in Physics, *Physica Scripta* **20**, 289–560.

Williams, R.F. (1979). The Structure of the Lorenz Attractor, *Pbl. Math. IHES* **50**, 73–99.

Winfree, A.T. (1980). *The Geometry of Biological Time.* Lecture Notes in Biomathematics 8. Springer-Verlag.

Winfree, A.T. (1987). *When Time Breaks Down.* Princeton Univ. Press.

Winternitz, P. (1983). Lie groups and solutions of nonlinear differential equations. In *Nonlinear Phenomena.* Lecture Notes in Physics **189**, ed. K.B. Wolf, pp. 265–325. Springer-Verlag.

Wintner, A. (1947). *The Analytic Foundations of Celestial Mechanics.* Princeton Univ. Press.

Wolf, A., Swift, J.B., Swinney, H.L. & Vastano, J.A. (1985). Determining Lyapunov exponents from a time series. Physica **16D**, 285–317.

Wolfram, S. (1982). Physics and computation. *Int. J. Theor. Phys.* **21**, 165–75.

Wolfram, S. (1983). Statistical mechanics of cellular automata. *Rev. Mod. Phys.* **55**, 601.

Wolfram, S. (1984). Universality and complexity in cellular automata, *Physica* **10D**, 1–35.

Wolfram, S. (1985). Twenty problems in the theory of cellular automata. *Physica Scripta* **T9**, 107–83.

Wolfram, S. (1986). *Theory and Applications of Cellular Automata*. World Scientific.

Wolfram, S. (1988). *Mathematica*, Wolfram Research Inc., Champaign, Illinois.

Wright, R. (1988). *Three Scientists and Their Gods*, Times Books.

Yahata, H. (1978). Temporal development of the Taylor vortices in a rotating fluid. *Prog. Theor. Phys. Suppl.* **64**, 165–85.

Yajima, N. (1985). Soliton resonance in plasmas. In *Dynamical Problems in Soliton Systems*, ed. S. Takeno, pp. 144–52. Springer-Verlag.

Yajima, N., Oikawa, M. & Satsuma, J. (1978). Interaction of ion-acoustic solitons in three-dimensional space. *J. Phys. Soc. Japan* **44**, 1711–4.

Yakubovich, V.A. & Starzhinskii, V.M. (1975). *Linear Differential Equations with Periodic Coefficients*, Vols. 1 and 2. Keter Publishing House: Jerusalem.

Yamada, T. & Fujisaka, H. (1983). Stability theory of synchronized motion in couple-oscillator systems II. *Prog. Theor. Phys.* **70**, 1240–8.

Yan-Qian, Y. *et al.* (1986). *Theory of Limit Cycles*. Amer. Math. Soc. Translation of Math. Monographs 66.

Yeung, C. (1988). Some problems on spatial patterns in nonequilibrium systems. *Thesis*, University of Illinois.

Yorke, J.A. & Yorke, E.D. (1979). Metastable chaos: the transition to sustained chaotic behavior in the Lorenz model. *J. Stat. Phys.* **21**, 263–77.

Yorke, J.A. & Yorke, E.D. (1981). Chaotic behavior and fluid dynamics. In *Hydrodynamic Instabilities and the Transition to Turbulence*, ed. H.L. Swinney & J.P. Gollub, pp. 77–94. Springgr-Verlag.

Zabusky, N.J. (1969). Nonlinear lattice dynamics and energy sharing. *J. Phys. Soc. Japan* **26**, Suppl.

Zabusky, N.J. (1981). Computational synergetics and mathematical innovation. *J. Comput. Phys.* **43**, 195–249.

Zabusky, N.J. & Kruskal, M.D. (1965). Interaction of "solitons" in a collisionless plasma and the recurrence of initial states. *Phys. Rev. Letters* **15**, 240–3.

Zahler, R.S. & Sussman, H.J. (1977). Claims and accomplishments of applied catastrophe theory. *Nature* **269**, 759–63. Review article.

Zakharov, V.E. (1972). Collapse of Langmuir waves. *Soc. Phys. JETP* **35**, 908–14.

Zakharov, V.E. (1974). On stochastization of one-dimensional chains of nonlinear oscillators. *Sov. Phys. JETP* **38**, 108–110.

Zakharov, V.E. & Manakov, S.V. (1977). Asymptotic behavior of non-linear wave systems integrated by the inverse scattering method. *Sov. Phys. JETP* **44**, 106–112.

Zakharov, V.E. & Shabat, A.B. (1972). Exact theory of two-dimensional self-focusing and one-dimensional self-modulation of waves in nonlinear media. *Sov. Phys. JETP* **34**, 62–9.

Zaslavsky, G.M. (1984). *Chaos in Dynamic Systems*. Harwood Academic Press.

Zeeman, E.C. (1977). *Catastrophe Theory*. Addison–Wesley. (Selected papers, 1972–1977.)

Zeeman, E.C. (1980). Population dynamics from game theory. In *Global Theory of Dynamical Systems*. Lecture Notes in Mathematics **814**, ed. Z. Nitecki & G. Robinson. Springer-Verlag.

Zehnder, E. (1973). Homoclinic points near elliptic fixed points. *Commun. Pure Appl. Math.* **26**, 131–92.

Zhabotinskii, A.M. (1967). *Oscillatory Processes in Biological and Chemical Systems*. Science Pub.: Moscow.

References by topics

TOPICS

Bäcklund transformations

Bifurcation, branching

Biological, ecological, and genetic systems

Catastrophe theory

Cellular automata

Chaotic dynamics (reviews; collections)

Chaotic dynamics (Theory)

Chaotic dynamics, experimental

Chemical oscillations, turbulence

Chemical systems (also see, Biological systems)

Circle map

Classical mechanics

Computer computations

Computer dynamics

Computer studies – nonchaotic dynamics

Computer studies of chaos, integrability, irreversibility, etc.

Coupled maps (patterns)

Difference equations (maps)

Dynamic systems

Dynamic entropies and information production (also see, chaotic dynamics)

Dynamo theory; Earth's magnetic field

Explosive dynamics

Fractals

Hydrodynamics

Integrable systems

Invariance properties and conservation laws

Inverse scattering method (also see, solitons)

Korteweg–deVries equation

Lattice dynamics

Lie theory of continuous groups

Lorenz equations

Lyapunov exponents

Mathematical definitions and terminology

Mathematics

Mechanical and electrical systems (also see, classical mechanics)

Meteorology

Modeling methodology

Nerve conduction

Neural nets

Nonlinear superposition (also see, Lie–Bäcklund transformations)

Nonsoliton asymptotic part

Optical phenomena

Painlevé conjecture

Patterns

Pattern dynamics (also see, Chemical turbulence, and coupled maps)

Related topics

Rössler models

Similarity analysis

Sine Gordon equation

Singular perturbation, asymptotic expansions

Social phenomena (see references on modeling methodology)

Solitons (also see, Korteweg–deVries, Bäcklund, Lattice dynamics and inverse scattering)

Solitons, direct method (Hirota)

Solitons, higher dimensional

Standard and area-preserving maps; microtron instability

Surveys

Synergetics

Turbulent-like dynamics (also see, chemical systems, and explosive dynamics)

Turbulence, onset

Bäcklund transformations

Anderson, R.L. & Ibraginov, N.H. (1979). *Lie–Bäcklund Transformations in Applications*, SIAM – contains historical references.

Lamb, G.L. Jr. (1974). Bäcklund transformations for certain nonlinear evolution equations, *J. Math. Phys.* **15**, 2157.

Martini, R. (Ed.) (1980). *Geometrical approaches to differential equations*, Lecture Notes in Mathematics **810**, Springer-Verlag.

Miura, R.M. (Ed.) (1976). *Bäcklund Transformations, Transformations, The Inverse Scattering Method, Solitons, and Their Applications*, Lecture Notes in Mathematics, **515**, Springer-Verlag.

Rogers, C. & Shadwick, W.F. (1980). *Bäcklund Transformations and Their Applications*, Academic Press.

Wahlquist, H.D. & Estabrook, F.B. (1973). Bäcklund transformations for solutions of the Korteweg–deVries equation, *Phys. Rev. Lett.* **31**, 1386.

Bifurcation, branching

Amann, H., Bazley, N. & Kirchgässner, K., (Eds.) (1981). *Applications of Nonlinear Analysis in the Physical Sciences*, Pitman.

Andronov, A.A., Leontovich, E.A., Gordon, I.I. & Maier, A.G. (1971). *Theory of Bifurcations of Dynamic Systems on a Plane*, US Dept. Commerce; Israel Program for Scientific Translations.

Arnold, V.I. (1972). Lectures on bifurcations in versal families, *Russian Math Survey*, **27**, 54–123.

Bardos, C. & Bessis, D. (Eds.) (1980). *Bifurcation Phenomena in Mathematical Physics and Related Topics*, D. Reidel Pub. Co.

Chow, S.N. & Hale, J.K. (1982). *Methods of Bifurcation Theory*, Springer-Verlag.

Crawford, J.D. & Omohundro, S. (1984). On the Global Structure of Period Doubling flows, *Physica* **13D**, 161–80.

Gurel, G. & Rössler, O.E. (Eds.) (1979). Bifurcation theory and applications in scientific disciplines, *Ann. NY Academy Science*, **316**.

Güttinger, S. (1986). Bifurcation geometry in physics, in *Frontiers of Nonequilibrium Statistical Physics*, eds. G.T. Moore and M.O. Scully, pp. 57–82, Plenum Press.

Holmes, P. & Williams, R.F. (1985). Knotted periodic orbits in suspensions of Smale's horseshoe: torus knots and bifurcation sequences, *Arch. Ration. Mech. Anal.* **90**, 115–94.

Iooss, G. & Joseph, D.D. (1980). *Elementary Stability and Bifurcation Theory*, Springer-Verlag.

Keller, J. B. & Antman, S. (Eds.) (1969). *Bifurcation Theory and Nonlinear Eigenvalue Problems*, W.A. Benjamin.

Marsden, J.E. & McCracken, M. (1976). *The Hopf Bifurcation and Its Applications*, Applied Math. Sci. **19**, Springer-Verlag.

Pimbley, G.H. Jr. (1969). *Eigenfunction Branches of Nonlinear Operators and Their Bifurcation*, Lecture Notes in Mathematics, **104**, Springer-Verlag.

Rabinowitz, P.H. (Ed.) (1977). *Applications of Bifurcation Theory*, Academic Press.

Takens, F. (1973). *Introduction to Global Analysis*, Comm. Math. Inst., Rijksuniversiteit Utrecht.

Thompson, J.M.T. & Hunt, G.W. (1975). Towards a unified bifurcation theory. *J. Appl. Math. Phys.* **26**.

Uezu, T. and Aizawa, Y. (1982). Topological character of a periodic solution in three-dimensional ordinary differential equation system, *Prog. Theor. Phys.* **68**, 1907–16.

Uezu, T. (1983). Topology in dynamical systems, *Phys. Letters* **93A**, 161–6.

Vainberg, M.M. & Trenogin, V.A. (1974). *Theory of Branching of Solutions of Nonlinear equations*, Noordhoff International Pub., Leyden.

Biological, ecological, and genetic systems

Albrecht, F., Gatzke, H., Haddad, A. & Wax, N. (1974). The dynamics of two interacting populations, *J. Math. Anal. and Appl.* **46**, 658–70.

Aubin, J.P., Saari, D. & Sigmund, K. (Eds.) (1985). *Dynamics of Macrosystems*, Notes in Economic and Mathematical Systems, **257**, Springer-Verlag.

Babloyantz, A. (1986). *Molecules, Dynamics and Life*, Wiley.

Blumenfeld, L.A. (1981). *Problems of Biological Physics*, Springer-Verlag.

Capasso, V. (1981). Periodic solutions for a system of nonlinear differential equations modeling the evolution of oro-faecal diseases. *In Nonlinear Differential Equations*, eds. P. deMottoni and L. Salvadori, Academic Press.

Cronin, J. (1981). *Mathematics of Cell Electrophysiology*, Lecture Notes in Applied Math., **63**, Marcel Dekker Inc., NY.

Dyson, F.J. (1982). A Model for the origin of life, *J. Mol. Evol.* **18**, 344–50.

Dyson, F.J. (1985). *Origins of Life*, Cambridge Univ. Press.

Ebeling, W. & Perschel, M. (Eds.) (1985). *Lotka–Volterra-Approach to Cooperation and Competition in Dynamic Systems*, Math. Res. **23**, Akademe-Verlag.

Eigen, M. (1971). Self-organization of matter and the evolution of biological macro-molecules, Naturwissenschaften **33a**, 465–522.

Eigen, M. & Schuster, P. (1979) *The Hypercycle: A Principle of Natural Self-organization*, Springer-Verlag.

Eigen, M. (1985). Macromolecular Evolution, In *Emerging Synthesis in Science* ed. D. Pines, pp. 25–69. The Santa Fe Institute.

Eigen, M., Gardiner, W.C. & Schuster, P. (1980). Hypercycles and compartments, *J. Theor. Biol.* **85**, 407–411.

Eigen, M., (1984). The Origin and Evolution of Life at the Molecular Level. In *Aspects of Chemical Evolution*, ed. G. Nicolis, Wiley, pp. 119–37.

Eigen, M. (1986). The Physics of Molecular Evolution, *Chemica Scripta* **26B**, 13–26.

Enns, R.H., Jones, B.L., Miura, R.M. & Rangnekar, S.S. (Eds.) (1981). *Nonlinear Phenomena in Physics and Biology*, Plenum Press.

Frauenthal, J.C. (1979). *Introduction to Population Modeling*, EDC/umap/55 Chapel St., Newton, Mass.

Freedman, H.I. (1980). *Deterministic Mathematical Models in Population Ecology*, Dekker.

Gilpin, M.E. (1973). Do hares eat lynx? *Amer. Nat.* **107**, 727–30.

Goel, N.S., Maitra, S.C. & Montroll, E.W. (1971). On the Volterra and other non-linear models of interacting populations, *Rev. Mod. Phys.* **43**, 231–76.

Goldanskii, V.I., Anikin, S.A., Avetisov, V.A. & Kuz.min, V.V. (1987). On the decisive role of chiral purity in the organization of the biosphere and its possible distruction, *Comments Mol. Cell. Biophys.* **4**, 79–88.

Grasman, J. & Veling, E.J.M. (1979). *Asymptotic Methods for the Volterra–Lotka Equation*, Lecture Notes in Mathematics, **711**, ed. F. Verhulst, Springer-Verlag.

Hofbauer, J. (1981). On the occurance of limit cycles in the Volterra–Lotka equation, *Nonlinear Analysis T.M.A.* **5**, 1003–7.

Hopf, F.A. & Hopf, F.W. (1986). Darwinian evolution in physics and biology: the dilemma of missing links. In *Frontiers of Nonequilibrium Statistical Physics*, eds. G.T. Moore & M.O. Scully, pp. 411–19, Plenum Press.

Hoppe, W., Lohmann, W., Markl, H. & Ziegler, H. (Eds.) (1983). *Biophysics*, Springer-Verlag.

Hoppensteadt, F.C. (Ed.) (1979). *Nonlinear Oscillations in Biology*, Lectures in Applied Mathematics **17**, AMS.

Hoppensteadt, F.C., *Mathematical Theories of Populations: Demographics, Genetics, and Epidemics*, SIAM.

Hoppensteadt, F.C. (Ed.) (1981). *Mathematical Aspects of Physiology*, Lectures in Applied Mathematics **19**, AMS.

Kolmogoroff, A. (1936). Sulla teoria di Volterra della lotta per l'esistenza, *Giornale dell'Istituto Italiano degli Attuari* **7**, 74–80.

Linkens, D.A., Biological systems. In *Modelling of Dynamical Systems*, Vol. 1, ed. N. Nicholson, pp. 193–223, P. Peregrinus.

Maynard Smith, J. (1986). *The Problems of Biology*, Oxford Univ. Press.

Maynard Smith, J. (1982). *Evolutionary Game Theory*, Cambridge Univ. Press.

Mayr, E. (1988). *Toward A New Philosophy of Biology*, Belknap Press of Harvard Univ. Press.

Meinhardt, H. (1982). *Models of Biological Pattern Formation*, (Academic Press.

Nicolis, G. (Ed.) (1985). *Aspects of Chemical Evolution*, Adv. Chem. Phys., **55**, Wiley.

Niesert, U., Harnasch, D. & Bresch, C. (1981). Origin of life between seylla and charybolis, *J. Mol. Evol.* **17**, 348–353.

Othmer, N.G. & Scriven, L.E. (1971). Instability and dynamic pattern in cellular networks, *J. Theor. Biol.* **32**, 507–37.

Peschel, M. & Mende, W. (1986). *The Predator–Prey Model*, Springer-Verlag.

Prigogine, I. & Lefever, R. (1968). Symmetry Breaking Instabilities in Dissipative Systems, II, *J. Chem. Phys.* **48**, 1695–700.

Rashevsky, N. (1938). *Mathematical Biophysics*, Univ. of Chicago Press, Reprint, Dover, 1960.

Schuster, P. & Sigmund, K. (1983). Replicaotr dynamics, *J. Theor. Biol.* **100**, 533–8.

Sigmund, K. (1983). The maximum principle for replicator equations, *J. Theor. Biol.* **100**, 63–72.

Smale, S. (1976). On the differential equations of species in competition, *J. Math. Biol.* **3**, 5–7.

Turing, A.M. (1952). The chemical bases of morphogenesis, *Phil. Trans. Roy. Soc.* (*London*) **B237**, 37–72.

Wicken, J.S. (1987). *Evolution, Thermodynamics, and Information*, Oxford Univ. Press.

Winfree, A.T. (1980). *The Geometry of Biological Time*, Biomathematics, **8**, Springer-Verlag.

Winfree, A.T. (1987). *When Time Breaks Down*, Princeton Univ. Press.

Zeeman, E.C. (1980). Population dynamics from game theory. In *Global Theory of Dynamical Systems*; Lecture Notes in Mathematics **814**, eds. Z. Nitecki & G. Robinson, Springer-Verlag.

Catastrophe theory

Arnold, V.I. (1984). *Catastrophe Theory*, Springer-Verlag.

Berry, M.V. Waves and Thom's Theorem, *Adv. Phys.* **25**, 1–26.

Berry, M.V. (1979). Catastrophe and stochasticity in semiclassical quantum mechanics; also, Catastrophe and fractal regimes in random waves. In *Structural Stability in Physics*, Springer-Verlag.

Berry, M. (1981). Singularities in Waves and Rays. In *Les Houches, Session XXXV, Physics of Defects*, eds. R. Balian, M. Kléman & J.-P. Poirier pp. 456–541, North Holland.

Chillingworth, D.R.J. (1976). Structural stability of mathematical models: the role of the catastrophe method. In *Mathematical Modelling*, eds. J.G. Andrews & R.R. McLone, pp. 231–58, Butterworths.

De Alfaro, V. & Rasetti, M. (1978). Structural stability theory and phases transition models, *Fortschritte der Physik* **26**, 143–73.

Fowler, D.H. (1972). The Riemann-Hugoniot Catastrophe and Van der Waal's Equation, *Towards a Theoretical Biology* **4**, 1–7, ed. C.H. Waddington.

Gilmore, R. (1981). *Catastrophe Theory for Scientists and Engineers,* John Wiley & Sons.

Guttinger, W. & Eikemeier, H. (Eds.) (1979). *Structural Stability in Physics,* Springer-Verlag.

Holmes, P.J. & Rand, D.A. (1976). The bifurcations of Duffing's equation: an application of catastrophe theory, *J. Sound and Vib.* **44**, 237–53.

Kolata, G.B. (1977). Catastrophe theory: the emperor has no clothes, *Science* **196**, 287.

Lu, Y.C. (1976). *Singularity Theory & an Introduction of Catastrophy Theory,* Springer-Verlag.

Paulos, J.A. (1981). *Mathematics and Humor,* University of Chicago Press.

Poston, T. & Stewart, I. (1978). *Catastrophe Theory and Its Applications,* Pitman.

Poston, T. & Stewart, I.N. (1976). *Taylor Expansions and Catastrophes,* Pitman.

Poston, T. & Woodcock, A.E.R. (1973). Zeeman's catastrophe machine, *Proc. Camb. Phil. Soc.* **74**, 217.

Russell, D.A., Hanson, J.D. & Ott, E. (1980). Dimension of strange attractors, *Phys. Rev. Letters* **45**, 1175–8.

Sinhah, D.K. (Ed.) (1981). *Catastrophe Theory and Applications,* John Wiley.

Smale, S. (1978). *Review of Catastrophe Theory,* by E.C. Zeeman, *Bull. Amer. Math. Soc.* **84**, 1360–8.

Stewart, I. (1982). Catastrophe theory in physics, *Rep. Prog. Phys.* **45**, 185–221.

Sussman, H. J. (1977). A skeptic, *Science* **197**, 821.

Thom, R. (1975). *Structural Stability and Morphogenesis,* Benjamin.

Thom, R. (1969). Topological models in biology, *Topology* **8**, 313–35.

Thom, R. (1977). *Structural Stability, Catastrophe Theory and Applied Mathematics,* SIAM Rev., April.

Thompson, J.M.T. & Hunt, G.W. (1973). A. *General theory of elastic stability,* Wiley.

Vendrik, M.C.M. (1979). A classification of phase diagrams by means of "elementary" catastrophe theory (ECT) I. *Physica* **99A**, 103–44.

Zahler, R.S. & Sussman, H.J. (1977). Claims and accomplishments of applied catastrophe theory (review article), *Nature* **269**, 759–63.

Zeeman, E.C. (1977). *Catastrophe Theory* (selected papers, 1972–1977), Addison–Wesley.

Cellular automata

Aizawa, Y. & Nishikawa, I. (1986). Toward the classification of the patterns generated by one-dimensional cell automata. In *Dynamical Systems and Nonlinear Oscillations,* ed. G. Ikegami, pp. 210–22, World Scientific.

Berlekamp, E.R., Conway, J.H. & Guy, R.K. (1982). *Winning Ways, for Your Mathematical Plays,* Vol. 2, Chap. 25, Academic Press.

Bennett, C.H. (1986). On the nature and origin of Complexity in discrete, homogeneous, locally-interacting systems, *Foundation Physics.* **16**, 585–92.

Burks, A.W. (Ed.) (1970). *Essays on Cellular Automata,* Univ. of Illinois Press, Urbana.

Complex Systems **4**, 545–839 (1987) (Lattice Gas Models).

Demongeot, J., Golés, E. & Tchuente, M. (Eds.) (1985). *Dynamical Systems and Cellular Automata,* Academic Press.

Dresden, M., A study of models in non-equilibrium statistical mechanics. In *Studies in Statistical Mechanics,* J. de Boer and G.E. Uhlenbeck, pp. 303–43, North Holland.

Farmer, D., Toffoli, T. & Wolfram, S. (Eds.) (1984). Cellular automata, *Physica D,* **10**, 1–246.

Feynman, R.P. (1967). *The Character of Physical Law,* MIT Press, Cambridge, pp. 57–8.

Feynman, R.P. (1982). Simulating physics with computers, *Int. J. Theor. Phys.* **21**, 467–8.

Fredkin, E. (1982). Digital Information Mechanics (talk given January, 1982, Moskito Island, BVI); Also see: *The Atlantic Monthly,* April, p. 29, 1988; Three Scientists and their Gods, by R. Wright, Times Books, 1988.

Fredkin, E. & Toffoli, T. (1982). Conservative Logic, *Int. J. Theor. Phys.* **21**, 219–53.

Frisch, U, Hasslacher, B. & Pomeau, Y. (1986). Lattice-gas automata for the Navier–Stokes equation, *Phys. Rev. Letters* **56**, 1505–8.

Gardner, M. (1970, 1971). Scientific American **223**(10) 120; **224**(2) 112; **224**(4) 117.

Gardner, M. (1983). *Wheels, Life, and Other Mathematical Amusements*, Freeman.

Jen, E. (1986). Global Properties of Cellular Automata, *J. Stat. Phys.* **43**, 219–42; Strings and pattern-recognizing properties of one-dimensional cellular automata, *ibid.* 243–65.

Kac, M. (1959). *Probability and Related Topics in Physical Sciences*, Interscience Pub.

Kolmogorov, A.N. (1984). Automata and Life. In *Cybernetics Today*, ed. I.M. Makarov, pp. 20–41, MIR Pub., Moscow.

Langton, C.G. (1984). Self-reproduction in cellular automata, *Physica* **10D**, 135–44.

Langton, C.G. (1986). Studying Artificial Life with Cellular Automata, *Physica* **22D**, 120–49.

Martin, O., Oldzko, A.M. & Wolfram, S. (1984). Algebraic properties of cellular automata, *Comm. Math. Phys.* **93**, 219–58.

Margolus, N. (1984). Physic-like models of computation, *Physica* **10D**, 81–95.

Margolus, N., Toffoli, T. & Vichniac, G. (1986). Cellular-automata supercomputers for fluid-dynamics modeling, *Phys. Rev. Letters* **56**, 1694–6.

Oono, Y. and Kohmoto, M. (1985). Discrete model of chemical turbulence, *Phys. Rev. Letters* **55**, 2947–51.

Oono, Y. & Yeung, C. (1987). A cell dynamical system model of chemical turbulence, *J. Stat. Phys.* **48**, 593–644.

Orszag, S. & Yakhot, V. (1986). Reynolds number scaling of cellular-automation hydrodynamics, *Phys. Rev. Letters* **56**, 1691–3.

Packard, N.H. & Wolfram, S. (1985). Two-dimensional cellular automata, *J. Stat. Phys.* **38**, 901–46.

Park, J.K., Steglitz, K. & Thurston, W.D. (1986). Soliton-like behaviour in automata, *Physica* **19D**, 423–32.

Poundstone, W. (1985). *The Recursive Universe*, Contemporary Books.

Toffoli, T. (1977). *Cellular Automata Mechanics*, Technical Report No. 208, Computer and Communication Sciences Department, Univ. of Michigan.

Toffoli, T. (1982). Physics and Computation, *Int. J. Theor. Phys.* **21**, 165–75.

Toffoli, T. (1984). Cellular automata as an alternative to (rather than an approximation of) differential equations modeling physics, *Physica* **10D**, 117–27.

Toffoli, T. & Margolus, N. *Cellular Automata: A New Environment for Modeling*, MIT Press.

Vichniac, G. Y. (1984). Simulating Physics with Cellular Automata, *Physics* **10D**, 96–116.

Vichniac, G.Y. (1984). Instability in discrete algorithms and exact reversibility, *SIAM J. Alg. Disc. Meth.* **5**.

Von Neumann, J. (1966). *Theory of Self-Reproducing Automata*, edited and completed by A.W. Burks, Univ. of Illinois Press, Urbana.

Wolfram, S. (1982). Physics and computation, *Int. J. Theor. Phys.* **21**, 165–75.

Wolfram, S. (1983). Statistical Mechanics of Cellular Automata, *Rev. Mod. Phys.* **55**, 601–44.

Wolfram S. (1985). Universality and complexity in cellular automata, *Physica* **10D**, 1–35.

Wolfram, S. (1985). Twenty problems in the theory of cellular automata, *Physica Scripta* **T9**, 107–83.

Wolfram, S. (1986). *Theory and Applications of Cellular Automata*, World Scientific Pub. Co.

Chaotic Dynamics (reviews; collections)

Abraham, N.B., Gollub, J.P. & Swinney, H.L. (1984). Testing nonlinear dynamics, *Physica* **11D**, 252–64.

Bai-lin, H. (1984). Chaos, World Scientific, Singapore.

Barnsley, M.F. & Demko, S.G. (Eds.) (1986). *Chaotic Dynamics and Fractals*, Academic Press.

Berge, P., Pomeau, Y., & Vidal, C. (1984). *Order Within Chaos; Towards a Deterministic Approach to Turbulence* Wiley.

Berry, M.V. (1978). Regular and Irregular Motion. In *Topics in Nonlinear Dynamics*, ed. S. Jorna, AIP **46**.

Bishop, A., Campbell, D. & Nicolaenko, B. (Eds.) (1981). *Nonlinear Problems: Present and Future*, North-Holland, Math. Studies **61**.

Campbell, D. & Rose, H. (Eds.) (1983). Order in chaos, *Physica* **7D**, 1–362.

Crutchfield, J.P., Farmer, J.D., Packard, N.H., & Shaw, R.S. (1986). Chaos, *Sci. Amer.* **255**(6), 46–57.

Cvitanović, P. (Ed.) (1984). *Universality in Chaos*, Adam Hilger Ltd., Bristol.

Devaney, R.L. (986). *An Introduction to Chaotic Dynamical Systems*, Benjamin.

Eckmann, J.-P. & Ruelle, D. (1985). Ergodic theory of chaos and strange attractors, *Rev. Mod. Phys.* **57**, 617–56.

Farmer, J.D., Ott, E. & Yorke, J.A. (1983). The dimension of chaotic attractors, *Physica* **7D**, 153–80.

Ford, J. (1983). How random is a coin toss?, *Physics Today*, April, 40–47.

Ford, J. (1989). What is chaos, that we should be mindful of it? In *The New Physics*, ed. P.C. Davis, Cambridge Univ. Press.

Garrido, L. (Ed.) (1983). *Dynamical Systems and Chaos*, Lect. Notes Physics **179**, Springer-Verlag.

Helleman, R.H.G. (1980). Self-generated chaotic behaviour in nonlinear mechanics. In *Fundamental Problems in Statistical Mechanics,* ed. E.G.D. Cohen, pp. 165–233, North-Holland.

Holden, A.V. (1986). *Chaos*, Princeton Univ. Press.

Holmes, R.J. & Moon, F.C. (1983). Strange attractors and chaos in nonlinear mechanics, *J. Appl. Mech.* **50**, 1021–32.

Iooss, G., Helleman, R.H.G. & Stora, R. (Eds.) (1983). *Chaotic Behaviour of Deterministic Systems*, North-Holland.

Moon, F.C. (1987). *Chaotic Vibrations*, Wiley.

Ruelle, D. (1980). Strange attractors, *Math. Intelligencer*, 126–37.

Schuster, H.G. (1984). *Deterministic Chaos*, Physik-Verlag, Weinheim.

Shaw, R. (1981). Strange attractors, chaotic behaviour, and information flow, *Z. Naturforsch* **36a**, 80–112.

Thompson, J.M.T. & Stewart, H.B. (1986). *Nonlinear Dynamics and Chaos*, Wiley.

Tomita, K. (1982). Chaotic response of nonlinear oscillators, *Phys. Reports* **86**, 113–67.

Zaslavsky, G.M. (1984). *Chaos in Dynamic Systems*, Harwood Academic Press.

Chaotic dynamics (theory)

Balazs, N.L. & Voros, A. (1986). Chaos on the pseudosphere, *Phys.Reports* **143**, No. 3 109–240.

Barnsley, M.F. (1986). Making chaotic dynamical systems to order. In *Chaotic Dynamics*, eds. M.F. Barnsley & S.G. Demko, pp. 55–68. Academic Press.

Barnsley, M.F. & Demko, S. (1985). Iterated function system and the global construction of fractals, *Proc. R. Soc. Lond.* **A399**, 243–75.

Barrow, J.D. (1982). Chaotic Behaviour in General Relativity, *Phys. Reports*, **85**, No. 1.

Benzinger, H.E., Burns, S.A. & Palmore, J.I. (1987). Chaotic complex dynamics and Newton's method, *Phys. Letters* **A119**, 441–6.

Cartwright, M.L. & Littlewood, J.E. (1945). On non-linear differential equations of the second order: I. The equation $\ddot{y} + k(1 - y^2)\,\dot{y} + y = b\lambda k\,\cos(\lambda t + a)$, k large, *J. London Math. Soc.* **20**, 180–9.

Dragt, A.J. & Finn, J.M. (1976). Insolubility of trapped particle motion in a magnetic dipole field, *J. Geophys. Res.* **81**, 2327–40.

Farmer, J.D., Ott, E. & Yorke, J.A. (1983). The dimension of chaotic attractors, *Physica* **7D**, 153–80.

Froehling, H., Crutchfield, J.P., Farmer, D., Packard, N.N. & Shaw, R. (1981). On determining the dimension of chaotic flows, *Physica* **3D**, 605–17.

Grassberger, P. & Procaccia, I. (1983). Measuring the strangeness of strange attractors, *Physica* **9D**, 189–202.

Hammel, S.M., Yorke, J.A. & Grebogi, C. (1987). Do numerical orbits of chaotic dynamical processes represent true orbits? *J. Complexity* **3**, 136–45.

Holmes, P.J. (1980). Averaging and chaotic motions in forced oscillations, *SIAM J. Appl. Math* **38**, 65–80 (correction in *New Approaches to Nonlinear Problems in Dynamics,* ed. P.J. Holmes, SIAM 1980.

Hirooka, H., Saito, N. & Ford, J. (1984). Chaos around hyperbolic fixed points, *J. Phys. Soc. Japan* **53**, 895–8.

Jensen, R.V. & Oberman, C.R. (1982). Statistical properties of dynamical systems, *Physica* **4D**, 183–96.

Levi, M. (1981). Qualitative analysis of the periodically forced relaxation oscillations, *Memoirs Amer. Math. Soc.* **32**, 244, 1–145.

Levinson, N. (1949). A second order differential equation with singular solutions, *Ann. Math.* **50**, 127–53.

Lichtenberg, A.J. & Lieberman, M.A. (1983). *Regular and Stochastic Motion,* Springer-Verlag.

Lorenz, E.N. (1984). The local structure of a chaotic attractors in four dimensions, *Physica* **13D**, 90–104.

McDonald, S.W., Grebogi, C., Ott, E. & Yorke, J.A. (1985). Fractal basin boundaries, *Physica D*, **17D**, 125–53.

Melnikov, V.K. (1963). On the stability of the center for time periodic perturbations, *Trans. Moscow Math. Soc.* **12**(1), 1–57.

Mistriotis, A.D. & Jackson, E.A. (1987). Transition to stochasticity for periodically perturbed area preserving system, *Physica Scripta* **35**, 97–104.

Motosov, A.D. (1976). A complete qualitative investigation of Duffing's equation, *Differential Eqs.* **12**, 164–74.

Newhouse, S.E. (1979). The abundance of wild hyperbolic sets and non-smooth stable sets for diffeomorphisms, *Publ. Math. I.H.E.S.* **50**, 101–51.

Oono, Y. & Osikawa, M. (1980). Chaos in nonlinear difference equations. I., *Prog. Theor. Phys.* **64**, 54–67.

Osikawa, M. & Oono, Y. (1981). Chaos in C°-endomorphism of interval, *Res. Inst. Math. Sci., Kyoto Univ.,* **17**, 165–77.

Osipov, A.V. (1976). Properties of solutions of Levinson's equations, *Diff. Eq.,* 1401–7.

Saari, D.G. (1984). The ultimate of chaos resulting from weighted voting systems, *Adv. Appl. Math.* **5**, 286–308.

Takens, F. (1985). On the numerical determination of the dimensions of an attractor. In *Dynamical Systems and Bifurcations*, eds. Lecture Notes in Mathematics **1125**, eds. B.L.J. Braaksma, H. Broer & F. Takens, pp. 99–106, Springer-Verlag.

Chaotic dynamics, experimental (also see turbulence)

Brandstäter, A., Swift, J., Swinney, H.L., Wolf, A., Farmer, J.D., Jen, E. & Crutchfield, P.J. (1983). Low-dimensional chaos in a hydrodynamic system, *Phys. Rev. Letters* **51**, 1442–5.

Creveling, H.F., dePaz, J.F., Baladi, J.Y. & Schoenhals, R.J. (1975). Stability characteristics of a single-phase free convection loop, *J. Fluid Mech.* **65**, 67.

Haken, H. (Ed.) (1981). *Chaos and Order in Nature*, Springer-Verlag.

Hunt, E.R. (1982). Comment on a driven nonlinear oscillator, *Phys. Rev. Letters* **49**, 1054 (1982); Testa, J., Perez, J., & Jeffries, C. *Respond, ibid.*, 1055.

Jeffries, C.D. (1985). Chaotic dynamics of instabilities in solids, *Physica Scripta* **T9**, 11–26.

Jensen, M.H., Kadanoff, L.P., Libchaber, A., Procaccia, I. & Stavans, J. (1985). Global universality at the onset of chaos: results of a forced Rayleigh–Bénard experiment, *Phys. Rev. Letters* **55**, 2798–801.

Kalia, R.K. & Vashista, P. (Eds.) (1982). *Melting, Localization, and Chaos*, North-Holland.

Lauterborn, W. & Crames, E. (1981). Subharmonic route to chaos observed in acoustics, *Phys. Rev. Letters* **47**, 1445–8.

Linsay, P.S. (1981). Period doubling and chaotic behaviour in a driven anharmonic oscillator, *Phys. Rev. Letters* **47**, 1349–52.

Moon, F.C. (1980). Experimental models for strange attractors vibrations in elastic systems. In *New Approach to Nonlinear Problems in Dynamics*, ed. P.J. Holmes, pp. 487–95, SIAM.

Nicolis, C. & Nicolis, G. (1984). Is there a climatic attractor?, *Nature* **311**, 529–32.

Rollins, R.W. & Hunt, E.R. (1982). Exactly solvable model of a physical system exhibiting universal chaotic behavior, *Phys. Rev. Letters* **49**, 1295–8.

Roux, J.C., Simoyi, R.H. & Swinney, H.L. (1983). Observation of a strange attractor, *Physica* **8D**, 257–66.

Shaw, R. (1984). *The Dripping Faucet as a Model Chaotic system*, Aerial Press.

Testa, J., Perez, J. & Jeffries, C. (1982). Evidence of universal chaotic behaviour of a driven nonlinear oscillator, *Phys. Rev. Letters* **48**, 714–7.

van der Pol, B. & van der Mark, J. (1927). Frequency demultiplication, *Nature* **120**, 363–4.

Chemical oscillations, turbulence

Belousov, B.P. (1959). *Sb. Ref. Radiats. Med.*, 1958, Medgiz, Moscow 145.

Epstein, I.R. (1984). New chemical oscillators, In *Non-Equilibrium Dynamics in Chemical Systems*, eds. C. Vidal & A. Pacault, pp. 24–34, Springer-Verlag.

Field, R.J. & Noyes, R.M. (1974). Oscillations in chemical systems. IV. Limit cycle behavior in a model of a real chemical reaction, *J. Chem. Phys.* **60**, 1877–84.

Glandsdorff, P. & Prigogine, I. (1971). *Thermodynamic Theory of Structure, Stability and Fluctuations*, Wiley.

Hastings, S.P. & Murray, J.D. (1975). The existence of oscillatory solutions in the field-noyes model of the Belousov-Zhabotinskii reactions, *J. Appl. Math.* **28**, 678–88.

Kuramoto, Y. (1983). *Chemical Oscillations, Waves and Turbulence*, Springer-Verlag.

Lefever, R. (1968). Stabilité des structures dissipatives, *Bull. Classe Sci., Acad. Roy. Belg.* **54**, 712–19.

Maselko, J. & Swinney, H. L. (1985). A complex transition sequence in the Belousov–Zhabotinskii reaction, *Physica Scripta* **T9**, 35–9.

Nicolis, G. (1971). Stability and dissipative structures in open systems far from equilibrium, *Adv. Chem. Phys.* **19**, 209–324.

Noyes, R.M. (1976). Oscillations in chemical systems. XIV. Corrected stoichiometry of the oregonator, *J. Chem. Phys.* **65**, 848–9.

Oono, Y. & Yeung, C. (1987). A. Cell dynamical system model of chemical turbulence, *J. Stat. Phys.* **48**, 593–644.

Rössler, O.E. (1972). A. Principle for chemical multivibration, *J. Theor. Biol.* **36**, 413–17.

Simayi, R.H., Wolf, A., & Swinney, H.L. (1982). One-dimensional dynamics in a multi-component chemical reaction, *Phys. Rev. Letters* **49**, 245–8.

Troy, S. (1977). A threshold phenomenon in the Field-Noyes model of the Belousov–Zhabotinskii reactions, *J. Math. Anal. Appl.* **58**, 233–48.

Turner, J.S., Roux, J.C., McCormick, W.D. & Swinney, H.L. (1981). Alternating periodic and chaotic regimes in chemical reaction-experiment and theory, *Phys. Letters* **85A**, 9–12.

Tyson, J.J. (1979). Oscillations, bistability, and echo waves in models of the Belousov–Zhabotinskii reaction, *Ann. NY Acad. Sci.* **316**, 279–95, eds. O. Gurel & O.E. Rössler.

Vidal, C. & Pacault, A. (Eds.) (1984). *Nonequilibrium dynamics in chemical systems*, Springer-Verlag.

Walker, J. (1978). Chemical systems that oscillate between one color and another, *Scientific American* **239**, 152–60.

Zhabotinskii, A.M. (1967). Oscillatory processes in biological and chemical systems, Science Publ., Moscow.

Chemical systems (also see, biological systems)

Ebert, K.H., Deuflhard, P. & Jäger, W. (1981). *Modelling of Chemical Reaction Systems*, Springer-Verlag.

Vidal, C. & Pacault, A. (1982). Spatial chemical structures, chemical waves. A reivew. In *Evolution of Order and Chaos*, eds. H. Haken, Springer-Verlag.

Circle map

Arnold, V.I. (1965). Small denominators. I. Mappings of the circumference onto itself, *Transl. Amer. Math. Soc. 2nd Series* **46**, 213–84.

Aronson, D.G., Chory, M.A., Hall, G.R. & McGehee, R.P. (1982). Bifurcations from an invariant circle for two-parameter families of maps of the plane, *Commun. Math. Phys.* **83**, 303–54.

Bélair, J. & Glass, L. (1985). Universality and self-similarity in the bifurcations of the circle maps, *Physica* **16D** (No. 2), 143–54.

Bohr, T., Bak, R. & Jensen, M.H. (1984). Transition to chaos by interaction of resonances in dissipative systems, *Phys. Rev.* **A30**, 1970–81.

Fein, A.P., Heutmaker, M.S. & Gollub, J.P. (1985). Scaling and transition from quasiperiodicity to chaos in ahydrodynamic system, *Physica Scripta* **T9**, 79–84.

Classical mechanics

Abraham, R. & Marsden, J.E. (1978). *Foundations of Mechanics*, second edition, Benjamin/Cummings – contains: A.N. Kolmogorov's 1954 address to the International Congress of Mathematicians: 'The General Theory of Dynamical Systems and Classical Mechanics'.

Arnold, V.I. & Avez, A. (1968). *Ergodic Problem of Classical Mechanics*, Benjamin.

Arnold, V.I. (1978). *Mathematical Methods of Classical Mechanics*, Springer-Verlag.

Arnold, V.I., Kozlov, V.V. & Neishtadt, A.I. (1988). Mathematical aspects of classical and celestial mechanics. In *Dynamical Systems*, III, ed. V.I. Arnold Springer-Verlag.

Bartlett, J.H. (1975). *Classical and Modern Mechanics*, Univ. of Alabama Press.

Benettin, G., Galgani, L. & Giorgilli, A. (1987). Realization of holonomic constraints and freezing of high frequency degrees of freedom in light of classical perturbation theory, *Commun. Math. Phys.* **113**, 87–103.

Birkhoff, G.D. (1917). Dynamical systems with two degrees of freedom, *Trans. Amer. Math. Soc.* **18**, 199–300.

Birkhoff, G.D. (1927). *Dynamical Systems*, Amer. Math. Soc.

Contopoulous, G. (1963). A classification of the integral of motion, *Ap. J.* **138**, 1297–305.

Fermi, E. (1923). Beweis dass ein mechanisches Normalsystem in Allgemeinen quasi-ergodisch ist, *Phys. Zeits.* **24**, 261–265.

Hagihara, Y. (1974). *Celestial Mechanics*, Vols I–V, Japan Soc. Promotion of Science, Tokyo Press.

Hoover, W.G. (1986). Molecular Dynamics, Lecture Notes in Physics **258**, Springer-Verlag.

Moser, J. (1968). Lectures on Hamiltonian systems, *Amer. Math. Soc. Memoirs*, No. 81, 1–60, Amer. Math. Soc.

Nosé, S. (1984). A molecular dynamics method for simulations in the canonical ensemble, *Molecular Phys.* **52**, 255–68.

Pars, L.A. (1968). *A Treatise on Analytical Dynamics*, J. Wiley.

Petrosky, P. & Prigogine, I. (1982). The KAM invariant and Poincaré's theorem. In *Applications of Modern Dynamics to Celestial Mechanics and Astrodynamics*, ed. V. Szebehely pp. 185–99, D. Reidel.

Poincaré, H. (1892, 1893, 1894). *Les Methodes Nouvelles de la Mecanique Celeste*, Vols. 1–3, Gauthier-Villars, Paris, Reprinted by Dover, 1957 (Translation: *New Methods of Celestial Mechanics*, NASA, 1967.

Siegel, C.L. & Moser, J.K. (1971). *Lectures on Celestial Mechanics*, Springer-Verlag.

Sudarshan, E.C.G. & Mukunda, N. (1974). *Classical Dynamics: A Modern Perspective*, Wiley.

Szebehely, V. (Ed.) (1982). *Applications of Modern Dynamics to Celestial Mechanics and Astrodynamics*, D. Reidel.

Szebehely, V. (1984). Review of concepts of stability, *Celestial Mech.* **34**, 49–64.

Szebehely, V. (1968). New Nondeterministic Celestial Mechanics. In *Space Dynamics and Celestial Mechanics*, eds. K.B. Bhatnagar, Reidel Pub. Co.

Thirring, W. (1978). *Classical Dynamical Systems*, Springer-Verlag.

Whittaker, E.T. (1944). *Analytical Dynamics of Particles and Rigid Bodies*, Dover.

Whiteman, K.J. (1977). Invariants and stability in classical mechanics, *Rep. Prog. Phys.* **40**, 1033–69.

Wintner, A. (1947). *The Analytic Foundations of Celestial Mechanics*, Princeton Univ. Press.

Computer computations

Pavelle, R., Rothstein, M. & Fitch, J. (1981). Computer algebra, *Sci. Amer.* **245**, No. 6, 136.

Press, W.H., Flannery, B.F., Teukolsky, S.A. & Vetterling, W.T. (1986). *Numerical Recipes*, Cambridge University Press.

Stoutemyer, D.R. & Yun, D.Y.Y. (1980). Symbolic mathematical computation In *Encyclopedia of Computer Science and Technology* **15**: Supplement, eds. J. Belzer, A.G. Holzman and A. Kent, Marcel Dekker.

Symbolic and Algebraic Computation; Eurosam '79, an International Symposium on Symbolic and Algebraic Manipulation. Lecture Notes in Computer Science **72**. Springer-Verlag, 1979.

Wang, P.S. (Ed.) (1981). *Proceedings of the 1981 ACM Symposium on Symbolic and Algebraic Computation.* Association for Computer Machinery.

Computer dynamics

Benettin, G.A., Casartelli, G.M. Galgani, L., Giorgilli, A. & Streleyn, J.M. (1978). On the reliability of numerical studies of stochasticity. *Nuovo Cimento* **44B**, 183–5.

Cutland, N. (1980). *Computability*, Cambridge Univ. Press.

Deutsch, D. (1985). Quantum theory, the Church–Turing principle and universal quantum computer, *Proc. R. Soc. Lond.* **A400**, 97–117.

Feynman, R.P. (1982). Simulating physics with computers, *Int. J. Theor. Phys.* **21**, 467–88.

Fredkin, E. & Toffoli, T. (1983). Conservative logic, *Int. J. Theor. Phys.* **21**, 219–53.

Lax, P.D. (1986). Mathematics and computing, *J. Stat. Phys.* **43**, 749–56.

Makarov, I.M. (Ed.) (1984). *Cybernetics Today*, MIR Publishers, Moscow.

McCauley, J.L. Jr. & Palmore, J.I. (1986). Computable Chaotic Orbits, *Phys. Letters A* **15**, 433

Toffoli, T. (1982). Physics and computation, *Int. J. Theor. Phys.* **21**, 165–75.

Computer studies – nonchaotic dynamics

Flaherty, J.R. & Hoppenstendt, F.C. (1978). Frequency entrainment or a forced Van der Pol oscillator, *Studies Applied. Math* **58**, 5–15.

Greenspan, D. (1980). *Arithmetic Applied Mathematics*, Pergamon Press.

Mistriotis, A.D. & Jackson, E.A. (1987). Transition to Stochasticity for a Periodically Perturbed Area Preserving System, *Physics Scripta* **35**, 97–104.

Størmer, C. (1955). *The Polar Aurora*, Oxford, Clarendon Press.

Zabusky, N.J. (1981). Computational Synergetics and Mathematical Innovation, *J. Comput. Phys.* **43**, 195–249.

Computer studies of chaos, integrability, irreversibility, etc.

Abraham, R. & Simó, C. (1985). Bifurcations and Chaos in Forced van der Pol Systems. In *Singularities and Dynamical Systems*, ed. S.N. Pnevmatikos, pp. 313–44, Elsevier Sci. Pub.

Channell, P.J. (1978). An illustrative example of the Chirikov criterion for stochastic instability. In *Topics in Nonlinear Dynamics, AIP Conf. No.* 46, ed. S. Jorna, pp. 248–59, AIP.

Chirikov, B.V. (1979). A universal instability of many-dimensional oscillator systems *Phys Reports* **52**, 265–376.

Chirikov, B.V., Keil, E. & Seisler, A.M. (1971). Stochasticity in many-dimensional Nonlinear Oscillator Systems, *J. Statistical Phys.* **3**, 307–21.

Chirikov, B.V., Izrailev, F.M. & Tryursky, V.A. (1973). Numerical experiments on the statistical behavior of dynamical systems with a few degrees of freedom, *Comp. Phys. Comm.* **5**, 11–16.

Chirikov, B.V., Ford, J. & Vivaldi, F. (1980). Some numerical studies of Arnold diffusion in a simple model. In *Nonlinear Dynamics and the Beam-Beam Interaction*, p. 323–40, AIP.

Ford, J. (1977). Relevance of Computer Experiments to the Pure Mathematics of Integrable and Ergodic Dynamical Systems. In *Dynamical Systems I–Warsaw*, pp. 75–92, Societe Math. de France.

Ford, J. (1973). The transition from analytic dynamics to statistical mechanics *Adv. Chem. Phys.* **24**, 155–85.

Ford, J. (1975). The statistical mechanics of classical analytic dynamics. In *Fundamental Problems in Statistical Mechanics*, ed. E. Cohen, pp. 215–55, North-Holland.

Helleman, R.H.G. (1977). On the Iterative Solution of a Stochastic mapping. In *Statistical Methods, Theory and Applications*, ed. U. Landman, Plenum.

Hénon, M. & Heiles, C. (1964). The applicability of the third integral of motion: some numerical experiments, *Astrophys. J.* **69**, 73–9.

Holmes, P. & Marsden, J.E. (1979). Qualitative techniques for bifurcation analysis of complex systems, *Ann. NY Acad. Sci.*, **316**, 608–22.

Kai, T. & Tomita, K. (1979). Stroboscopic phase protrait of a forced nonlinear oscillator, *Prog. Theor. Phys.* **61**, 54–73.

Lunsford, G.H. & Ford, J. (1972). On the stability of periodic orbits for nonlinear oscillator systems in regions exhibiting stochastic behavior, *J. Math. Phys.* **13**, 700–5.

Mo, K.C. (1972). Theoretical prediction for the onset of widespread instability in conservative nonlinear oscillator systems, *Physica* **57**, 445–54.

Morozov, A.D. (1976). A. Complete qualitative investigation of Duffing's equation, *Diff. Equations* **12**, 164–74.

Räty, R., Isomäki, H.M. & von Boehm, J. (1984). Chaotic motion of a classical anharmonic oscillator, *Acta Polytechnica Scandinavica, Mech. Eng. Ser.* **85**, 1–30.

Rössler, O.E. (1979). Chaos. In *Structural Stability in Physics*, eds. W. Guttinger & H.E. Kemeier, pp. 290–319, Springer-Verlag.

Simó, C. (1979). On the Hénon-Pomeau Attractor, *J. Stat. Phys.* **21**, 465–94.

Ueda, Y. (1979). Randomly Transitional Phenomena in Systems Governed by Duffing's Equation, *J. Stat. Phys.* **20**, 181–96.

Coupled maps (patterns)

Gu, Y., Tung, M., Yuan, J.-M., Feng, D.H. & Narducci, L.M. (1984). Crises and hysteresis in coupled logistic maps, *Phys. Rev. Letters* **52**, 701–4.

Hogg, T. & Huberman, B.A. (1984). Generic behavior of coupled oscillators, *Phys. Rev.* **A29**, 275–81.

Kaneko, K. (1983). Transition from torus to chaos accompanied by frequency lockings with symmetry breaking, *Prog. Theor. Phys.* **69**, 1427–42.

Kaneko, K. (1984). Period-doubling of kink-antikink patterns, *Prog. Theor. Phys.* **72**, 480–6.

Keeler, J.D. & Farmer, J.D. (1986). Robust space-time intermittency and $1/f$ noise, *Physica* **D23**, 413–35.

Waller, I. & Kapral, R. (1984). Spatial and temporal structure in systems of coupled nonlinear oscillators, *Phys. Rev.* **A30**, 2047–55.

Yamada, T. & Fujisaka, H. (1983). Stability theory of synchronized motion in couple-oscillator systems II, *Prog. Theor. Phys.* **70**, 1240–8.

Difference equations (maps)

Aubry, S. (1983). The twist map, the extended Frenkel–Kontrova model and the devil's staircase, *Physica* **7D**, 240–58.

Beau, W., Metzler, W. & Ueberla, A. (1987). *The Route to Chaos of Two Coupled Logistic Maps*, Univ. of Kassel.

Bowen, R. (1978). *On Axiom A Diffeomorphisms*, CBMS Regional Conference Series in Mathematics No. 35.

Chang, S.J. & McCown, J. (1985). Universality behaviors and fractal dimensions associated with M-furcations, *Phys. Rev.* **A31**, 3791–801.

Chernikov, A.A., Sagdeev, R.Z., Usikov, D.A., Zakharov, M.Yu. & Zaslavsky, G.M. (1987). Minimal chaos and stochastic webs, *Nature* **326** (9), 559–63.

Collet, P. & Eckmann, J-P. (1980) *Iterated Maps on the Interval as Dynamic Systems*, Birkhauser.

Coven, E.M., Kan, I. & Yorke, J.A. (1986). Pseudo-orbit shadowing in the family of tent maps.

Curry, J.H. (1979). On the Hénon transformation, *Commun. Math. Phys.* **68**, 129–40.

Feigenbaum, M.J. (1978). Quantitative universality for a class of nonlinear transformations, *J. Stat. Phys.* **19**, 25–52.

Feigenbaum, M.J. (1979). The universal metric properties of nonlinear transformations, *J. Stat. Phys.* **21**, 186–223.

Guckenheimer, J. (1977). On the bifurcation of maps of the internal, *Inventiones Math.* **39**, 165–78.

Guckenheimer, J., Oster, G. & Ipaktchi, A. (1977). The dynamics of density dependent population models, *J. Math. Biol.* **4**, 101–47.

Guckenheimer, J. (1979). The bifurcation of quadratic functions. In *Bifurcation Theory*, NY Acad. Sciences **316**, pp. 78–85.

Guckenheimer, J. (1979). Sensitive dependence to initial conditions for one dimensional maps, *Commun. Math. Phys.* **70**, 133–60.

Hénon, M. (1976). A two dimensional mapping with a strange attractor, *Comm. Math. Phys.* **50**, 69–77.

Hénon, M. (1969). Numerical study of quadratic area-preserving mappings, *Quart. Appl. Math.* **27**, 291–312.

Hénon, M. (1982). On the Numerical Computation of Poincaré Maps, *Physica* **5D**, 412–14.

Holmes, P.J. & Whitley, D.C. (1984). Bifurcations of one- and two-dimensional maps, *Phil. Trans. R. Soc. Lond.* **A311**, 43–102; Erratum, *ibid.*, 601.

Hsu, C.S. (1987). *Cell-To-Cell Mapping*, Springer-Verlag.

Ichikawa, Y.H., Kamimura, T. & Karney, C.F.F. (1983). Stochastic motion of particles in tandom mirror devices, *Physica* **6D**, 233–40.

Kaplan, J.L. & Yorke, J.A. (1979). Chaotic behavior of multidimensional difference equations. In *Functional Differential Equations*, eds. H.O. Peitgen & H.D.Waither, Lecture Notes in Mathematics **730**, pp. 204–27, Springer-Verlag.

Karney, C.F.F. (1983). Long-time correlations in the stochastic regime, *Physica* **8D**, 360–80.

Kuznetsov, S.P. (1986). Universality and scaling in the behavior of coupled Feigenbaum systems, *Radiophys. Quantum Electronics* **28**, 681–95.

Lasota, A. (1977). *Ergodic Problems in Biology, Dynamical Systems II-Warsaw*, pp. 239–50, Société Mathématique de Franc.

Li, T.Y. & Yorke, J.A. (1975). Period three implies chaos, *Amer, Math. Monthly* **82**, 983–92.

MacKay, R.S. (1986). Transition to chaos for area-preserving maps. In *Nonlinear Dynamics Aspects of Particle Accelerators*, eds. J.M. Jowett, M. Month & S. Turner, pp. 390–454, Springer.

Marotto, F.R. (1978). Snap-back repellers imply chaos in R^n, *J. Math. Anal. Appl.* **63**, 199–223.

May, R.M. (1976). Simple mathematical models with very complicated dynamics, *Nature* **261**, 459–67 (review article).

Metropolis, N., Stein, M. & Stein, P. (1973). On finite limit sets for transformations on the unit interval, *J. Combinatorial Theory* **15**, 25–44.

Osikawa, M. & Oono, Y., (1981). Chaos in C°-endomorphism of interval. *Inst. Math. Studies, Kyoto Univ.*, **17**, 165–177 (1981).

Preston, C. (1983). *Iterates of Maps on an Interval*, Springer-Verlag.

Rannou, F. (1974). Numerical study of discrete plane area-preserving map, *Astron. Astrophys.* **31**, 289–301.

Ryutov, D.D. & Stupakov, G.V. (1978). Diffusion of resonance particles in ambipolar plasma traps, *Sov. Phys. Dokl.* **23**, 412–14.

Schult, R.L., Creamer, D.B., Henyey, F.S. & Wright, J.A. (1987). Symmetric and nonsymmetric coupled logistic maps, *Phys. Rev.* **A35**, 3115–18.

Sharkovsky, A.N. (1964). Coexistence of the cycles of a continuous mapping of the line into itself, *UKr. Mat. Zh.* **16**(1), 61–71.

Singer, D. (1978). Stable orbits and bifurcation of maps of the interval, *SIAM J. Appl. Math.* **35**, 260–7.

Smale, S. & Williams, R. (1976). The quantitative analysis of a difference equation of population growth, *J. Math. Biol.* **3**, 1–4.

Stefan, O. (1977). A theorem of Sarkovskii on the existence of periodic orbits of continuous endomorphisms of the real line, *Commun. Math. Phys.* **54**, 237–48.

Straffin, P.D. Jr. (1978). Periodic points of continuous maps, *Math Mag.* **51**, 99–105.

Dynamic systems

Abraham, R.H. & Shaw, C.D. (1982). *Dynamics, The Geometry of behavior*; Part One: *Periodic Behavior*; (1983) Part Two: *Chaotic Behavior* (1984); Part Three: *Global Behavior*, Aerial Press.

Alekseev, V.M. & Yakobson, M.V. (1981). Symbolic dynamics and hyperbolic dynamic systems, *Phys. Rep.* **75**, 287–325.

Benettin, G., Galgani, L. & Giorgilla, A. (1985). A proof of Nekhoroshev's theorem for the stability times in nearly-integrable Hamiltonian systems, *Celest. Mech.* **37**, 1–25.

Bowen, R. (1975). *Methods of symbolic dynamics, Lecture Notes in Mathematics* **470**, Springer-Verlag.

Broomhead, D.S. & King, G.P. (1986). Extracting qualitative dynamics from experimental data, *Physica* **20D**, 217–36.

Diner, S., Fargue, D. & Lochak, G. (Eds.) (1986). *Dynamical Systems*, World Scientific.

Garrido, L. (Ed.) (1982). *Dynamic Systems and Chaos*, Lecture Notes in Physics. **179**, Springer-Verlag.

Golubev, V.V. (1953). *Lectures on Integration of the Equations of Motion of a Rigid Body about a Fixed Point*, State Publishing House Moscow.

Guckenheimer, J. & Holmes, P. (1983). *Nonlinear Oscillations, Dynamical Systems, and Bifurcations of Vector Fields*, Springer-Verlag.

Guckenheimer, J., Moser, J. & Newhouse, S.E. (1980). *Dynamical Systems*, Birkhäser.

Helleman, R.H.G. (Ed.) (1980). Nonlinear dynamics, *Ann. NY Acad. Sciences* **257**.

Hirsch, M.W. & Smale, S. (1974). *Differential Equations, Dynamical Systems and Linear Algebra*, Academic Press.

Hirsch, M.W. (1984). The dynamical systems approach to differential equations, *Bull. Amer. Math. Soc.* **11**, 1–64.

Irwin, M.C. (1980). *Smooth Dynamical Systems*, Academic Press.

Jorna, S. (Ed.) (1978). *Topics in Nonlinear Dynamics*, AIP Conference Proceedings, 46, AIP.

Kolmogorov, A.N. (1954). *The General Theory of Dynamical Systems and Classical Mechanics* (English translation in *Foundations of Mechanics* by R. Abraham & J.E. Marsden).

Lichtenberg, A.J. & Lieberman, M.A. (1983). *Regular and Stochastic Motion*, Springer-Verlag.

Markley, N.G., Martin, J.C. & Perrizo, W. (1978). *The Structure of Attractors in Dynamical Systems*, Springer-Verlag.

Moser, J. (1973). Stable and Random Motions in Dynamical Systems, *Ann. Math. Studies*, **77**, Princeton Univ. Press.

Nekhoroshev, N.N. (1971). Behavior of Hamiltonian systems close to integrable, *Funct. Anal. Appl.* **5**, 338–9.

Nekhoroshev, N.N. (1977). An exponential estimate of the time of stability of near-integrable Hamiltonian systems, *Russ. Math. Survey* **32**, 1–65.

Nicolis, J.S. (1986). Dynamics of hierarchical systems, Springer-Verlag.

Nitecki, Z. (1971). *Differential Dynamics*, MIT Press.

Packard, N.H., Crutchfield, J.P., Farmer, J.D. & Shaw, R.S. (1980). Geometry from a time series, *Phys. Rev. Letters* **45**, 712–16.

Peixoto, M.V. (Ed.) (1973). *Dynamical Systems*, Academic Press.

Percival, I. (1982). *Introduction to Dynamics*, Cambridge Univ. Press.

Pnevmatikos, S.N. (Ed.) (1985). *Singularities and Dynamical Systems*, Math. Studies, **103**, North-Holland.

Ruelle, D. (1983). Small random perturbations and the definition of attractors. In *Geometric Dynamics*, Lecture Notes in Mathematics, eds., A. Doid & B. Eckmann, pp. 663–76, Springer-Verlag.

Sinai, Ya. G. (1977). *Introduction to Ergodic Theory*, Princeton Univ. Press.

Smale, S. (1963). Diffeomorphisms with many periodic points. In *Differential and Combinatorial Topology*, ed. S.S. Cairns, Princeton Univ. Press, pp. 63–80.

Smale, S. (1967). Differentiable dynamical System, *Bull. Amer. Math. Soc.* **73**, 747–817.

Smale, S. (1981). *Essays on Dynamical Systems, Economic Processes, and Related Topics*, Springer-Verlag.

Szlenk, W. (1984). An Introduction to the Theory of Smooth Dynamical Systems, Wiley.

Sverdlove, R. (1977). Inverse problems for dynamical systems in the plane. In *Dynamical Systems*, eds. A.R. Bednarek & L. Cesi, pp. 499–502, Academic Press.

Three papers on dynamical systems: A.G. Kusnirenko, Problems in the general theory of dynamical systems on a manifold (1–42); A. B. Katok, Dynamical systems with hyperbolic structure (43–96); V.M. Alekseev, Quasirandom oscillations and qualitative questions in celestial mechanics (97–169); *AMS Translations* **116**, 1981.

Dynamic entropies and information production (also see, chaotic dynamics)

Adler, R.L., Konheim, A.G. & McAndrew, (1965). Topological entropy, *Trans. Amer. Math. Soc.* **114**, 309–19.

Block, L., Guckenheimer, J., Misiurewicz, M. & Young, L.S. (1980). Periodic points and topological entropy of one dimensional maps. In *Global Theory of Dynamical Systems*, eds. A. Dold & B. Eckmann, Lecture Notes in Mathematics, **819**, Springer-Verlag.

Dinaburg, E.I. (1971). On the relations among various entropy characteristics of dynamical systems, *Math. USSR Izvestija* **5**, 337–78.

Kolmogorov, A.N. (1958). A new invariant of transitive dynamical systems. *Dokl. Akad. Nauk SSSR* **119**, 861–5.

Ledrappier, F. & Young, L.-S. (1985). The metric entropy of diffeomorphisms Part I: Characterization of measures satisfying Pesin's entropy formula, *Ann. Math.* **122**, 509–39; Part II: Relations between entropy, exponents, and dimension, *ibid*, 540–74.

Misiurewicz, M. (1976). Topological conditional entropy, *Studia Mathematica* **55**, 175–200.

Misiurewicz, M. and Szlenk, W. (1980). Entropy of piecewise monotone mappings, *Studia Mathematica* **57**, 45–63.

Oono, Y. (1978). A. Heuristic approach to Kolmogorov entropy as a disorder parameter, *Prog. Theor. Phys.* **60**, 1944–6.

Rohlin, V.A. (1967). Lectures on the entropy theory of measure-preserving transformations, *Russian Math Survey* **22** (5), 1–52.

Shaw, R. (1981). Strange attractors, chaotic behavior, and information flow, *Z. Naturforsch.* **36a**, 80–112.

Sinai, Ja. G. (1959). On the concept of entropy of a dynamical system, *Dokl. Akad. Nauk. SSSR* **124**, 768–76 (Russian).

Dynamo theory: Earth's magnetic field

Bullard, E. (1978). The disk dynamo. In *Topics in Nonlinear Dynamics, AIP Conference Proceedings*, ed. S. Jorna, p. 373–89, AIP.

Busse, F.H. (1978). Magnetohydrodynamics of the Earth's dynamo, *Ann. Rev. Fluid Mech.* **10**, 435–62.

Carrigan, C.R. & Gubbins, D. (1979). The source of the Earth's magnetic field, *Sci. Amer.* **240** (2), 118.

Cook, A.E. & Roberts, P.H. (1970). The Rikitake two-disc dynamo system, *Proc. Camb. Phil. Soc.* **68**, 547–69.

Cox, A. (1982). Magnetostratigraphic time scale. In *Geologic Time Scale*, eds. W.B. Harland *et al.* Cambridge Univ. Press. p. 63.

Oziewonski, A.M. (Ed.) (1980). *Physics of the Earths Interior*, Proc. International School of Physics 'Enrico Ferni' North-Holland; articles by F.H. Busse (Motions in the Earth's core and the origin of geomagnesium pp. 493–507) and J.A. Jacobs (The evolution of the Earth's core and Geodynamics, pp. 508–30).

Gubbins, D. (1974). Theories of the geomagnetic and solar dynamos, *Rev. Geophys. Sp. Phys.* **12**, 137–54.

Hibberd, F.H. (1979). The origin of the Earth's magnetic field, *Proc. R. Soc. Lond.* **A369**, 31–45.

Hoffman, K.A. (1988). Ancient magnetic reversals: clues to the geodynamics, *Sci. Amer.* **258**(5), 76–83.

Inglis, D.R. (1981). Dynamo theory of the Earth's varying magnetic field, *Rev. Mod. Phys.* **53**, 481–96.

Jacobs, A.J. (1984). *Reversals of the Earth's Magnetic Field*, Heyden & Son, Philadelphia.

Macdonald, K.C. & Luyendyk, B.P. (1981). The Crest of the East Pacific Rise, *Sci. Amer.* **244**(5), 100–16.

Merrill, R.T. & McElhinny, M.W. (1983). *The Earth's Magnetic Field, Its History, Origin and Planetary Perspective*, International Geophysics Series, **32**, Academic Press.

Moffat, H.K. (1978). *Magnetic Field Generation in Electrically Conducting Fluids*, Cambridge Univ. Press.

Moffat, H.K. (1977). Six lectures on general fluid dynamics and two on hydromagnetic dynamo theory. In *Fluid Dynamics*, eds. R. Balian and J.L. Peube, p. 151–233, Gordon & Breach.

Parker, E.N. (1975). The dynamo mechanism for the generation of large-scale magnetic fields, *NY Acad. Sci.* **252**, 141–55.

Robbins, K.A. (1977). A new approach to subcritical instability and turbulent transitions in a simple dynamo, *Math. Proc. Camb. Phil. Soc.* **82**, 309–25.

Roberts, P.H. (1971). Dynamo theory. In *Lectures in Applied Mathematics*, ed. W.H. Reid, pp. 129–206 American Mathematical Society.

Strangway, D.W. (1970). *History of the Earth's magnetic field*, McGraw-Hill.

Explosive dynamics

Muncaster, R.G. & Zinnes, D.A. (1984). Phase portraits, singularities, and dynamics of hostility in international relations, *SIAM J. Applied Math*.

Rudakov, L.I. & Tsytovich, V.N. (1978). Strong Langmuir turbulence, *Phys. Reports* **40**, 1–73.

Thornhill, S.G. & ter Haar, D. (1978). Langmuir turbulence and modulational instability, *Phys. Reports* **43**, 43.

Weiland, J. & Wilhelmsson, H. (1977). *Coherent Non-linear Interaction of Waves in Plasmas*, Pergamon Press.

Zakharov, V.E. (1972). Collapse of Langmuir waves, *Soviet Physics JETP* **35**, 908–14.

Fractals

Berry, M.V. (1979). Diffractals, *J. Phys.* **A12**, 781.

Colvin, J.T. & Stapleton, H.J. (1985). Fractal and spectral dimensions of biopolymer chains: solvent studies of electron spin relaxation rates in myoglobin azide, *J. Chem. Phys.* **82**, 4699–706.

Falconer, K.J. (1985). *The Geometry of Fractal Sets*, Cambridge Univ. Press.

Frauenfelder, H., Petsko, G.A. & Tsenoglu, (1979). Temperature-dependent X-ray diffraction as a probe of protein structural dynamics, *Nature (London)* **280**, 558–63.

Grebogi, C., Ott, E. & Yorke, J.A. (1983). Fractal basin boundaries, long-lived chaotic and unstable-unstable pair bifurcations, *Phys. Rev. Letters* **50**, 935–8.

Halsey, T.C., Jensen, M.H., Kadanoff, L.P., Procassia, I. & Shraiman, B.I. (1986). Fractal measures and their singularities: the characterization of strange sets, *Phys. Rev.* **A33**, 1141–51.

Jackson, E.A. (1985). The Lorenz System: II. The homoclinic convolution of the stable manifolds, *Physica Scripta* **32**, 476–81.

Kaplan, J.L., Mallet-Paret, J. & Yorke, J.A. (1984). The Lyapunov dimension of a nowhere differentiable attracting torus, *Ergod. Th. and Dynam. Sys.* **4**, 261–81.

Mandelbrot, B.B. (1977). *Fractals* (*Form, Chance and Dimension*), W.H. Freeman.

Mandelbrot, B.B. (1983). *The Fractal Geometry of Nature*. W.H. Freeman.

Peitgen, H.-O. & Richter, P.H. (1986). *The Beauty of Fractals*, Springer-Verlag.

Pietronero, L. & Tosatti, E. (Eds.) (1986). *Fractals in Physics*, North-Holland.

Stapleton, H.J., Allen, J.P., Flynn, C.P. Stinson, D.G. & Kurtz, S.R. (1980). Fractal form of proteins, *Phys. Rev. Letters* **45**, 1456–9.

Umberger, D.K. & Farmer, J.D. (1985). Fat fractals on the energy surface, *Phys. Rev. Letters* **55**, 661–4.

Hydrodynamics

Balian, R. & Peube, J.L. (Eds.) (1977). *Fluid Dynamics*, Gordon and Breach.

Bergé, P. & Dubois, M. (1984). Rayleigh–Bénard convection, *Contemp. Phys.* **25**, 535–82.

Birkhoff, G. (1955). *Hydrodynamics*, Dover, Inc.

Burgers, J.M. (1948). A mathematical model illustrating the theory of turbulence. In *Advances in Applied Mechanics*, **1**, eds. R. von Mises and T. von Kármán, pp. 171–99, Academic Press.

Chandrasekhar, S. (1961). *Hydrodynamic and Hydromagnetic Stability*, Oxford Univ. Press.

Drazin, P.G. & Reid, W.H. (1981) *Hydrodynamic Stability*, Cambridge Univ. Press.

Jeffrey, A. & Taniuti, T. (1964). *Nonlinear Wave Propagation*, Academic Press.

Joseph, D. (1976). *Nonlinear Stability of Fluid Motions*, Springer-Verlag.

Lamb, H. (1945). Hydrodynamics Dover, Inc.

Landau, L.D. & Lifshitz, E.M. (1959). *Fluid Mechancs*, Addison-Wesley Publishing Co.

Saltzman, B. (1962). Finite amplitude free convection as an initial value problem I, *J. Atoms. Sci.* **19**, 329–41.

Stoker, J.J. (1957). Water Waves, Interscience Publishers. (Appendix: On the Derivation of the Shallow Water theory, by K. O. Friedrichs.)

Swinney, H.W. & Gollub, J.P. (Eds.) (1985). *Hydrodynamic Instabilities and the Transition to Turbulence*, Topics Appl. Phys. **45**, Springer-Verlag.

Whitham, G.B. (1974). *Linear and Nonlinear Waves*, Wiley.

Integrable systems

Calogero, F. (1983). Integrable dynamical systems and related mathematical results. In *Nonlinear Phenomena* Lecture Notes Physics **189**, ed. K.B. Wolf, Springer-Verlag, pp. 49–95.

Casati, G., Chirikov, B.V. & Ford, J. (1980). Marginal local instability of quasi-periodic motion, *Phys. Lett.* **77A**, 91–4.

Moser, J. (1975). Three integrable Hamiltonian systems connected with isospectral deformations, *Adv. in Math.* **16**, 197–220.

Moser, J. (1978). Nearly Integrable and integrable systems. In *Topics in Nonlinear Dynamics*, ed. S. Jorna., Amer. Inst. Phys.

Moser, J. (1981). *Integrable Hamiltonian Systems and Spectral Theory*, Academia Nazionale dei Lincei, Pisa.

Invariance properties and conservation laws

Denman, H.H. (1965). Invariance and conservation laws in classical mechanics, *J. Math. Phys.* **6**, 1611–16.

Hill, E.L. (1951). Hamilton's principle and the conservation theorems of mathematical physics, *Rev. Mod. Phys.* **23**, 253–60.

Houtappel, R.M.F., Van Dam, H. & Wigner, E.P. (1965). The conceptual basis and use of the geometric invariance principles, *Rev. Mod. Phys.* **37**, 595–632.

Inverse scattering method (also see, solitons)

Ablowitz, M.J., Kaup, D.K., Newell, A.C. & Segur, H. (1974). The inverse scattering transform-fourier analysis for nonlinear problems, *Studies Appl. Math.* **53**, 249–315.

Ablowitz, M.J., Kaup, D.J. & Newell, A.C. (1974). Coherent pulse propagation, A dispersive irreversible phenomena, *J. Math. Phys.* **15**, 1852–8.

Ablowitz, M.J. & Segur, H. (1981). *Solitons and the Inverse Scattering Transform* SIAM.

Ablowitz, M.J. & Fokas, A.S. (1983). Comments on the inverse scattering transform and related nonlinear evolution equations. In *Nonlinear phenomena*, Lecture Notes Physics **189**, ed. K.B. Wolf, Springer-Verlag, pp. 4–23.

Bednar, J.B. *et al.* (Eds.) (1983). *Conference on Inverse Scattering: Theory and Applications*, SIAM.

Conference on the theory and application of solitons. In *Rocky Mountain J. Math.* **8**, 1–428 (1978) eds. H. Flaschka and D.W. McLaughlin.

De Santo, J.A., Saenz, A.W. & Zacharg, W.W. (Eds.) (1980). *Mathematical Methods and Applications of Scattering Theory*, Lecture Notes in Physics **130**, Springer-Verlag.

Eckhaus, W. & Van Harten, A. (1981). *The Inverse Scattering Transformation and the Theory of Solitons,*, North-Holland.

Flaschka, H. (1976). On the Toda Lattice, *Prog. Theor. Phys.* **51**, 703–16.

Flaschka, H., Discrete & periodic illustrations of some aspects of the inverse method, *ibid*, pp. 441–66.

Flaschka, H. & Newell, A.C. (1975). Integrable systems of nonlinear evolution equations. In *Dynamic Systems, Theory and Applications*, ed. J. Moser, Springer-Verlag, pp. 355–440.

Gardner, C.S., Greene, J.M., Krushal, Krushal, M.D. & Miura, R.M. (1967, 1974). Method for solving the Korteweg-de Veries equation, *Phys. Rev. Letters* **19**, 1095–7; Korteweg-deVries equation and generalization. VI. methods for exact solution, *Comm. Pure Appl. Math*, **27**, 97–133.

Kaup, D.J. (1976). The Three-Wave Interaction–A Nondispersive Phenomena, Studies Appl. Math. **55**, 9–44.

Kay, I. & Moses, H.E. (1982). *Inverse Scattering Papers 1955–1963*, Vol. XII, *Lie Groups, History Frontiers and Applications*, Mathematical Science.

Manakov, S.V. (1975). Complete integrability and stochastization of discrete dynamical systems, *Sov. Phys. JETP*, **40**, 269–74.

McLaughlin, D.W. (1975). Four examples of the Inverse method as a canonical transformation, *J. Math. Phys.* **16**, 96–9.

Miura, R.M. (1976). The Korteweg–deVries equation: a survey of results, *SIAM Review* **18**, 412–59.

Perlmutter, A. & Scott, L.F. (Eds.) (1977). *The significance of Nonlinearity in the Natural Sciences*, Plenum Press.

Ramajaran, R. (1982). *Solitons and Instantons: An Introduction to Solitons and Instantons in Quantum Field Theory*, North-Holland.

Scott, A.C., Chu, F.Y.F. & McLaughlin, D.W. (1973). The soliton: a new concept in applied science, *Proc. IEEE* **61**, 1443–83.

Zakharov, V.E. & Shabat, A.B. (1972). Exact theory of two-dimensional self-focusing and one-dimensional self-modulation of waves in nonlinear media, *Sov. Phys. JETP* **34**, 62–9.

Zakharov, V.E. (1974). On stochastization of one-dimensional chains of nonlinear oscillators, *Sov. Phys. JETP* **38**, 108–10.

Korteweg–deVries equation

Gardner, C.S., Greene, J.M., Kruskal, M.D. & Miura, R.M. (1967). A method for solving the Korteweg–deVries equation, *Phys. Rev. Letters* **19**, 1095–7.

Korteweg, D.J. & deVries, G. (1895). On the change of form of long wves advancing in a rectangular canal, and a new type of long stationary waves, *Phil. Mag.* **39**, 422–43.

Kruskal, M. (1982). The Korteweg–deVries equation. In *Laser-Plasma Interaction*, eds. R. Balian & J.C. Adam, pp. 788–808, North-Holland.

Kruskal, M.D., Miura, R.M., Gardner, C.S. & Zabusky, N.J. (1970). V. Uniqueness and nonexistence of polynomial conservation laws, *J. Math. Phys.* **11**, 952–60.

Lax, P.D. (1968). Integrals of Nonlinear Equations of Evolution and Solitary Waves, *Comm. Pure Appl. Math.* **21**, 467–90.

Miura, R.M. (1968). I.A. remarkable explicit nonlinear transformation, *J. Math. Phys.* **9**, 1202–4.

Miura, R.M., Gardner, C.S. & Kruskal, M.D. (1968). II. Existence of conservation laws and constants of motion, *J. Math. Phys.* **9**, 1204–9.

Su., C.H. & Gardner, C.S. (1969). III. Derivation of the KdV equation and Burgers equation, *J. Math. Phys.* **10**, 536–9.

Van der Blij, F. (1978). Some details of the history of the Korteweg-deVries equation, *Nieu Archief VoorWiskunde* (3) **26**, 54–64.

Wadati, M. (1972). The exact solution of the modified Korteweg–deVries equation, *J. Phys. Soc. Japan* **32**, 1681.

Wadati, M. & Toda, M. (1972). The exact N-soliton solution of the Korteweg-deVries equation, *J. Phys. Soc. Japan* **32**, 1403–11.

Zabusky, N.J. & Kruskal, M.D. (1965). Interaction of 'solitons' in a collisionless plasma and the recurrence of initial states, *Phys. Rev. Letters* **15**, 240–3.

Lattice dynamics

Bocchieri, P., Scotti, A., Bearzi, B. & Loinger, A. (1970). Anharmonic chain with Lennard–Jones interaction, *Phys. Rev.* **A2**, 2013–19.

Budinsky, N. & Bountis, T. (1983). Stability of Nonlinear Modes and Chaotic Properties of 1D Fermi-Pasta-Ulam Lattices, *Physica* **8D**, 445–54.

Casati, G. & Ford, J. (1975). Stochastic transition in the unequal-mass Toda lattice, *Phys. Rev.* **A12**, 1702–9.

Fermi, E., Pasta, J.R. & Ulam S.M. (1965). Studies of nonlinear problems, I. Los Alamos Report LA-1940, May 1955. In *Collected Works of E. Fermi*, Vol. 2, pp. 978–88, University of Chicago Press.

Fesser, K., McLaughlin, D.W., Bishop, A.R. & Holian, B.L. (1985). Chaos and nonlinear modes in perturbed Toda chain, *Phys. Rev.* **A31**, 2728–31 (1985).

Flaschka, H. (1974). The Toda lattice I. Existence of integrals, *Phys. Rev.* **B9**, 1924–5.

Ford, J. (1961). Equipartition of energy for nonlinear systems, *J. Math. Phys.* **2**, 387–93.

Ford, J., Stoddard, S.D. & Turner, J.S. (1973). On the integrability of the Toda lattice, *Prog. Theor. Phys.* **50**, 1547–60.

Froeschlé, C. (1978). Connectance of dynamical systems with increasing number of degrees of freedom, *Phys. Rev.* **A18**, 277.

Hénon, H. (1974). Integrals of the Toda lattice, *Phys. Rev.* **B9**, 1921–3.

Jackson, E.A. (1963). Nonlinear coupled oscillators. I. Perturbation theory: ergodic problem, *J. Math. Phys.* **4**, 551–8; Nonlinear coupled oscillators. II. Comparison of theory with computer solutions. *J. Math. Phys.* **4**, 686–700.

Jackson, E.A., Pasta, J.R. & Waters, J.F. (1968). Thermal conductivity of one-dimensional lattices, *J. Comput. Phys.* **2**, 202–7.

Jackson, E.A. (1978). Nonlinearity and irreversibility in lattice dynamics, *Rocky Mount. J. Math.* **8**, 127–46.

Kruskal, M.D. & Zabusky, N.J. (1964). Stroboscopic perturbation procedure for treating a class of nonlinear wave equations, *J. Math. Phys.* **5**, 231–44.

Livi, R., Pettini, M., Ruffo, S., Sparpaglione, M. & Vulpiani, A. (1983). Relaxation to different stationary states in the Fermi–Pasta–Ulam model, *Phys. Rev.* **A28**, 3544–52.

Livi, R., Pettini, M., Ruffo, S., Sparpaglione, M. & Vulpiani A. (1985). Equipartition threshold in nonlinear large Hamiltonian systems: The Fermi–Pasta–Ulam model, *Phys. Rev.* **A31**, 1039–45.

MacDonald, R.A. & Tsai, D.H. (1978). Molecular dynamics calculations of energy transport in crystalline solids, *Phys. Reports* **46**, 1–48.

Manakov, S.V. (1974). *Sov. Phys. JETP* **40**, 269.

Maradudin, A.A., Montroll, E.W., Weiss, G.H. & Ipatova, I.P. (1971). *Theory of Lattice Dynamics in the Harmonic Approximation*, second eds. Academic Press.

Mokross, F. & Büttner H. (1983). Thermal conductivity in the diatomic Toda lattice, *J. Phys.* **C16**, 4539–46.

Nakazaea, H. (1970). On the Lattice Thermal Conduction, *Suppl. Prog. Theor. Phys.* **45**, 231–62.

Northcote, R.S. & Potts, R.B. (1964). Energy sharing and equilibrium for nonlinear systems, *J. Math. Phys.* **5**, 383.

Pnevmatikos, S.N. (1985). Solitons in nonlinear atomic chains. In *Singularities and Dynamical Systems,* ed. S.N. Pnevmatikos, 397–437 North-Holland.

Saitô, N., Hirotomi, N. & Ichimura, A. (1975). The induction phenomenon and ergodicity in the anharmonic lattice vibration, *J. Phys. Soc. Japan* **39**, 1431–8.

Saitô, N., Ooyama, N. & Aizawa, Y. (1970). Computer experiments on ergodicity problems in anharmonic lattice vibrations, *Suppl. Prog. Ther. Phys.* **45**, 209–30.

Schrödinger, E. (1914). Zur Dynamik elastisch gekoppelter Punktsysteme, *Ann. d. Phys.* **44**, 916–34.

Toda, M. (1975). Studies of a non-linear lattice, *Phys. Reports* **18C**, 1–124.

Toda, M. (1967). Vibration of a chain with nonlinear interaction, *J. Phys. Soc. Japan* **22**, 431–36; Wave propagation in anharmonic lattices, **23**, 501–6.

Toda, M. (1971). Waves in nonlinear lattice, *Prog. Theor. Phys. Suppl. No. 45*, 174.

Toda, M. (1981). *Theory of Nonlinear Lattices*, Springer-Verlag.

Tuck, J.L. & Menzel, M.T. (1972). The superperiod of the nonlinear weighted string (FPU) problem, *Adv. Math.* **9**, 339–407.

Visscher, W.M. (1976). *Computer studies of Transport properties in Simple models of solids, Meth. Comput. Phys.* **15**, ed. G. Gilat, pp. 371–408, Academic Press.

Wadati, M. (1975). Wave propagation in nonlinear lattice. I, II., *J. Phys. Soc. Japan* **38**, 673–80, 682–6.

Walker, G.H. & Ford, J. (1969). Amplitude instability and ergodic behavior for conservative nonlinear oscillator systems, *Phys. Rev.* **188**, 416–32.

Zabusky, N.J. (1969). Nonlinear lattice dynamics and energy sharing, *J. Phys. Soc. Japan* **26**, Suppl.

Lie theory of continuous groups

Campbell, J.E. (1903). *Introductory on Lie's Theory of finite Continuous Transformations Groups*, Oxford, Clarendon Press (reprint: Chelsea, 1966).

Chevalley, C. (1946). *Theory of Lie Groups*, Princeton Univ. Press.

Coggeshall, S.V. & Axford, R.A. (1986). Lie group invariance properties of radiation hydrodynamic equations and their associated similarity solutions, *Phys. Fluids* **28**, 2398–20.

Cohen, A. (1911). *Introduction to the Lie Theory of One-Parameter Groups*, D.C. Heath & Co.

Eisenhart, L.P. (1933). *Continuous Groups of Transformations*, Princeton Univ. Press.

Hill J.M. (1982). *Solution of Differential Equations by Means of One-Parameter Groups*, Pitman.

Ince, E.L. (1956). *Ordinary Differential Equations*, Dover, Chap. IV.

Lie, S. (1893). *Vorlesungen ueber Continuierliche Gruppen mit Geometrischen und Anderen Anwendugen*, B.C. Teubner, Leipzig.

Winternitz, P. (1983). Lie groups and solutions of nonlinear differential equations. In Nonlinear Phenomena Lecture Notes in Physics, **189**, ed. K.B. Wolf pp. 265 25 Springer-Verlag.

Lorenz equations

Alfsen, K.H. Frøyland, J. (1984). *Systematics of the Lorenz Model at $\sigma = 10$*, Univ. of Oslo, Institute of Physics, Norway.

Birman, J.S. & Williams, R.F. (1983). Knotted periodic orbits in dynamical systems-I: Lorenz's equations, *Topology* **22**, 47–82.

Curry, J.H. (1978). A generalized Lorenz system, *Comm. Math. Phys.* **60**, 193–204.

Fowler, A.C. & McGuinness, M.J. (1982). A description of the Lorenz attractor at high Prandtl number, *Physica* **5D**, 149–82.

Fowler, A.C. & McGuiness, M.J. (1982). Hysteresis in the Lorenz equations, *Phys. Letters* **92A**, 103–6.

Frøyland, J. & Alfsen, K.H. (1984). Lyapunov–exponent spectra for the Lorenz model, *Phys. Rev.* **A29**, 2928–31.

Jackson, E.A. (1985). The Lorenz system: I. The global structure of its stable manifolds, *Physica Scripta* **32**, 469–75; The Lorenz system: II. The homoclinic convolution of the stable manifolds, *Physica Scripta* **32**, 476–81.

Kaplan, J.L. Yorke, J.A. (1979). The onset of chaos in a fluid flow model of Lorenz, *Ann. NY Acad. Sci.* **316**, 400–7.

Kaplan, J.L. & Yorke, J.A. (1979). Preturbulence: a regime observed in a fluid flow model of Lorenz, *Commun. Mathn., Phys.* **67**, 93–108.

Lanford, O.E. (1977). In *Turbulence Seminar*; Lecture Notes in Mathematics **615**, eds. P. Bernard & T. Rativ p. 114, Springer-Verlag.

Lorenz, E.N. (1979). On the prevalence of aperiodicity in simple systems. In *Global Analysis*, Lecture Notes in Mathematics **755**, eds. M. Grmela & J.E. Marsden, pp. 53–75, Springer-Verlag.

Lorenz, E.N. (1963). Deterministic nonperiodic flow, *J. Atoms. Sci.* **20**, 130–41; The mechanics of vacillation, *ibid.*, 448–64.

Mammerville, P. & Pomeau, Y. (1979). Intermittency and the Lorenz model, *Phys. Letters* **75A**, 1–2.

Manneville, P. & Pomeau, Y. (1980). Different ways to turbulence in dissipative dynamic sytems, *Physica* **1D**, 219–26.

McGuiness, M.J. (1985). A computation of the limit capacity of the Lorenz attractor, *Physica* **16D**, 265–75.

Richtmyer, R.D. (1986). A study of the Lorenz attractor, science and computers, *Adv. Math. Suppl. Studies*, **10**, 207–19.

Robbins, K.A. (1979). Periodic solutions and bifurcation structure at high R in the Lorenz model, *J. Appl. Math* **36**, 457–71.

Schmutz, M. & Rueff, M. (1984). Bifurcations schemes of the Lorenz model, *Physica* **11D**, 167–78.

Sparrow, C. (1980). *Bifurcations in the Lorenz Equation*, Lecture Notes in Applied Mathematics Springer-Verlag.

Uezu, T. & Aizawa, Y. (1982). Topological character of a periodic solution in three-dimensional ordinary differential equation system, *Prog. Theor. Phys.* **68**, 1907–16.

Williams, R.F. (1979). The structure of the Lorenz attractor, *Publ. Math. I.H.E.S.* **50**, 73–99.

Yorke, J.A. & Yorke, E.D. (1981). Chaotic behavior and fluid dynamics. In *Hydrodynamic Instabilities and the Transition to Turbulence*, eds. H.L. Swinney & J.P. Gollub, Springer-Verlag, pp. 77–94.

Yorke, J.A. & Yorke, E.D. (1979). Metastable chaos: the transition to sustained chaotic behavior in the Lorenz model, *J. Stat. Phys.* **21**, 263–77.

Lyapunov exponents

Bennetin, G., Galgani, L., Giorgilli, A. & Strelcyn, J.-M. (1980). Lyapunov characteristic exponents for smooth dynamical systems and for Hamiltonian systems: a method for computing all of them, *Meccanica* **15**, 9.

Benettin, G., Galgani, L. & Strelcyn, J.M. (1976). Kolmogorov entropy and numerical experiments, *Phys. Rev.* **A14**, 2338–45.

Feit, S.D. (1978). Characteristic exponents and strange attractors, *Commun. Math. Phys* **61**, 249–60.

Oseledec, V.I. (1968). A multiplicative ergodic theorem. Lyapunov characteristic numbers for dynamical systems, *Trans. Moscow Math. Soc.* **19**, 197.

Shimada, I. & Nagashima, T. (1979). A numerical approach to ergodic problem of dissipative dynamical systems, *Prog. Theor. Phys* **61**, 1605.

Wolf, A., Swift, J.B., Swinney, H.L. & Vastano, J.A. (1985). Determining Lyapunov exponents from a time series, *Physica* **16D**, 285–317.

Mathematical definitions and terminology

Choquet-Bruhat, Y., Dewitt-Morette, C. & Dillard-Bleick, M. (1977). *Analysis, Manifolds and Physics*, North Holland.

Iyanaga, S. & Kawada, Y. (Eds.) (1980). *Encyclopedic Dictionary of Mathematics*, Math. Soc. Japan; MIT Press.

Mathematics

Ames, W.F. (1965, 1972). *Nonlinear Partial Differential Equations in Engineering*, Vols I and II Academic Press.

Appel, K & Haken, W. (1977). Every Planar Map is Four Colorable, I. Discharging; II. With J. Koch. Reducibility. *Iu. J. Math* **21**, 429–90; 491–567.

Appel, K. & Haken, W. (1986). The four color proof suffices, *Math. Intelligencer* **8**, 10–20.

Arnold, V.I. (1961). The stability of the equilibrium position of a Hamiltonian system of ordinary differential equations in the general elliptic case, *Soviet Math. Dokl.* **2**, 247–9.

Arnold, V.I. (1963). Small denominators and problems of stability of motion in classical and celestial mechanics, *Russian Math. Surveys* **18**, No. 6, 85–189.

Arnold, V.I. (1964). *Dokl. Akad. Nauk. SSSR* **156**, 9.

Arnold, V.I. (1978). *Ordinary Differential Equations*, M.I.T. Press.

Arnold, V.I. (1983). *Geometrical Methods in the Theory of Ordinary Differential Equations*, Springer-Verlag.

Barber, M.N. & Ninham, B.W. (1970). *Random and Restricted Walks*, Gordon and Breach.

Billingsley, P. (1965). *Ergodic Theory and Information*, Wiley.

Birkhoff, G.D. (1913). Proof of Poincaré's geometric theorem, *Trans. A.M.S.* **14**, 14–22.

Birkhoff, G.D. (1922). Surface Transformations and their Dynamical Application, *Acta. Math.* **43**, 1–119 (1922).

Bogoliubov, N & Mitropolsky, Y.A. (1961). *Asymptotic Methods in the Theory of Nonlinear Oscillations*, Gordon and Breach Science Publishers.

Bogoliubov, N.N., Mitropoliskii, Ju.A. & Samoilenko, A.M. (1976). *Methods of Accelerated Convergence in Nonlinear Mechanics*, Springer-Verlag.

Borrelli, R.L. & Coleman, C.S. (1987). *Differential Equations, A Modeling Approach*, Prentice-Hall.

Bröcker, Th. & Jänich, K. (1982). *Introduction to Differential Topology*, Cambridge Univ. Press.

Brown, M. & Neumann, W.D. (1977). Proof of the Poincaré-Birkhoff Fixed Point Theorem, *Mich. Math. J.* **24**, 21–31.

Buck, R.C. (1981). The Solutions to a Smooth PDE can be Dense in C(1), *J. Diff. Eq.* **41**, 239–244.

Cesari, L. (1963). *Asymptotic Behavior and Stability Problems in Ordinary Differential Equations*, Academic Press.

Cesari, L. (1976). *Functional Analysis, Nonlinear Differential Equations, and the Alternative Method*. Lecture Notes in Pure and Applied Mathematics **19**, pp. 1–197 Dekker.

Chinn, W.G. & Steenrod, N.E. (1966). *First Concepts of Topology*, MAA.

Coddington, E.A. & Levinson, N. (1955). *Theory of Ordinary Differential equations*, McGraw-Hill Book Co.

Cunningham, W.J. (1964). *Introduction to Nonlinear Analysis*, McGraw-Hill Book Co.

De Vogelaere, R. (1958). On the structure of symmetric period solutions of conservative systems, with applications. In *Concentrations to the Theory of Nonlinear Oscillations*, **4**, ed. S. Lefshetz, pp. 53–84.

Duffin, R.J. (1981). Rubel's Universal Differential Equation, *Proc. Natl. Acad. Sci. USA* **78**, 4661–2.

Eminhizer, C.H., Helleman, R.H.G. & Montroll, E.W. (1976). On a convergent nonlinear perturbation theory without small denominators or secular terms, *J. Math. Phys.* **17**, 121–40.

Francis, G.K. (1987). *A Topological Picturebook*, Springer-Verlag.

Furstenberg, H. (1983). Poincaré recurrence and number theory. In *The Mathematical Heritage of Henri Poincaré*, Part 2, pp. 193–216, American Math. Soc.

Gantmacher, R.R. (1959). *The Theory of Matricies*, Chelsa Pub. Co.

Goursat, E. (1964). *A Course in Mathematical Analysis*, Dover.

Guillemin, V. & Pollack, A. (1974). *Differential Topology*, Prentice-Hall.

Hale, J. (1977). *Theory of Functional Differential Equations*, Springer-Verlag.

Hale, J.K. (1980). *Ordinary Differential Equations*, R.E. Krieger Pub. Co.

Hardy, G.E. (1916). Weierstrass's non-differentiable functions, *Tran. Amer. Math. Soc.* **17**, 301_5.

Hartman, P. (1964). Ordinary Differential Equations, Wiley Reprinted by Hartman, 1973. ²

Hirsch, M.W. & Smale, S. (1974). *Differential Equations, Dynamical Systems and Linear Algebra*, Academic Press.

Hurewicz, W. & Wallman, H. (1948). *Dimension Theory*, Princeton Univ. Press.

Ince, E.L. (1956). Ordinary Differential Equations, Dover.

Lancaster, P. (1985). *Theory of Matricies*, Academic Press.

LaSalle, J.P. & Lefschetz S. (Eds.) (1963). *Nonlinear Differential Equations and Nonlinear Mechanics*, Academic Press.

Lefschetz, S. (1959). Differential Equations; Geometric Theory, Interscience.

Levinson, N. & Smith, O.K. (1942). A general equation for relaxation oscillations, *Duke Math. J.* **9**, 287–403.

MacKay, R.S. & Percival, I.C. (1985). Converse KAM: theory and practice, *Commun. Math. Phys.* **98**, 469–512.

Marsden, J.E. & McCracken, M. (1976). *The Hopf Bifurcation and its Applications*, Applied Math. Sci. **1**, Springer-Verlag.

Mather, J.N. (1984). Non-existence of Invariant Circles, *Ergod. Th & Dynam. Sys.* **4**, 301–9.

McLachlan, N.W. (1964). *Theory and Application of Mathieu Functions*, Dover.

Moser, J. (1962). On Invariant Curves of Area-preserving mappings of an Annulus, *Nachr. Akad. Wiss Göttingen Math. Phys. Kl.* **2**, 1–20.

Moser, J. (1967). Convergent Series Expansions for Quasi-periodic Motions, *Math. Annalen* **169**, 136–76.

Morse, P.M. & Feshbach, H. (1953). *Methods of Theoretical Physics*, McGraw-Hill.

Nash, C. & Sen S. (1983). *Tolopogy and Geometry for Physicists*, Academic Press.

Nayfeh, A.H. (1986). Perturbation methods in nonlinear dynamics. In *Nonlinear Dynamics Aspects of Accelerators*, eds. J.M. Jowett, Month, M. & Turner, S. Lecture Notes in Physics **247**, pp. 238–314, Springer-Verlag.

Nemytskii, V.V. & Stepanov, V.V. (1960). *Qualitative Theory of Differential Equations*, Princeton Univ. Press.

Nitecki, Z. (1971). *Differentiable Dynamics*, MIT Press.

Olds, C.D. (1963). *Continued Fractions*, Random House.

Rolfsen, D. (1976). *Knots and Links*, Publish or Perish, Inc., Berkeley, CA.

Rohlin, V.A. (1964). Exact endomorphisms of a Lebesque space, *Amer. Math. Soc. Trans.* **39**, 1–36.

Rubel, L.A. (1981). A universal differential equation, *Bull. (NS) Amer. Math. Soc.* **4**, 345–9.

Sanders, J.A. & Verhulst, F. (1985). *Averaging Methods in Nonlinear Dynamical Systems*, Springer-Verlag.

Shannon, C.E. (1948). A mathematical theory of communication, *Bell Tech. J.* **27**, 623–56.

Smale, S. (1981). The fundamental theorem of algebra and complexity Theory, *Bull. Amer. Math. Soc. (N.S.)* **4**, 1–6.

Smart, D.R. (1974). *Fixed Point Theorems*, Cambridge Univ. Press.

Takens, F. (1981). Detecting strange attractors in turbulence. In *Dynamical Systems and Turburlence, Warwick 1980*. Lecture Notes in Mathematics **898**, eds D.A. Rand & L.-S. Yang, Springer-Verlag.

Tucker, A.W. & Bailey, H.S. Jr. (1950). Topology, *Sci. Amer.* **182**(1), 18–24.

Ulam, S.M. (1960). *A Collection of Mathematical Problems* Wiley.

Ulam, S.M. (1982). Transformations, iterations and mixing flows. In *Dynamical Systems* II eds. A.R. Bednarek & L. Cesari, Academic Press, pp. 419–26.

Weiss, L. (Ed.) (1972). *Ordinary Differential Equations*, Academic Press.

Whitney, H. (1936). Differentiable manifolds, *Ann. Math.* **37**, 645–80.

Whitney, H. (1944). The self-intersections of a smooth n-manifold in $2n$-space, *Ann. Math.* (2) **45**, 220–46.

Whitney, H. (1944). The singularities of a smooth n-manifold in $(2n-1)$-space, *Ann. Math.* (2) **45**, 247–93.

Yakubovich, V.A. & Starzhinskii, V.M. (1985). *Linear Differential Equations with Periodic Coefficients*, Vols. 1 and 2, Keter Publishing House, Jerusalem.

Yan-Qian, Y. & Cai Sui-lin, *et al.* (1986). *Theory of Limit Cycles*, Amer. Math. Soc.; Translation of Math. Monographs **66**.

Zehnder, E. (1973). Homoclinic points near elliptic fixed points, *Common. Pure Appl. Math.* **26**, 131–82.

Mechanical and electrical system (also see, classical mechanics)

Andronov, A.A. Khaikin, S.E. & Vitt, A.A. (1966). *Theory of Oscillators*, Pergamon Press.
  Arnold, V.I. & Avez, A. (1968). *Ergodic Problems of Classical Mechanics*, W.A. Benjamin, Inc.
Hayashi, C. (1985). *Nonlinear Oscillations in Physical Systems*, Princeton Univ. Press.
Kryloff, N. Bogliuboff, N. (1947). *Introduction to Non-Linear Mechanics*, Annals of Math. Studies, Princeton Univ. Press.
Minorsky, N. (1947). *Introduction to Nonlinear Mechanics*, J.W. Edwards, Ann Arbor, Mich.
Minorsky, N. (1962). *Non-Linear Oscillations*, Van Nostrand Co.
Moon, F.C. (1984). *Magneto-Solid Mechanics*, Wiley.
Neyfeh, A.H. & Mook, D.T. (1979). Nonlinear Oscillations, J. Wiley & Sons.
Pippard, A.B. (1978). *The Physics of Vibration*, Cambridge Univ. Press.
Schmidt, G. & Tondl, A. (1986). *Non-Linear Vibrations*, Cambridge Univ. Press.
Scott, A.C. (1970). *Active and Nonlinear Wave Propagation in Electronics*, Willey & Sons, Inc.
Stoker, J.J. (1950). *Nonlinear Vibrations*, Interscience Publishers.
Vidyasaga, M. (1978). *Nonlinear Systems Analysis*, Prentice-Hall.

Meteorology

Charney, J.G. & DeVore, J.G. (1979). Multiple flow equilibrium in the atmosphere and blocking, *J. Atmos. Sci.* **36**, 1205–16.
Lorenz, E.N. (1980). Attractor sets and quasi-geostrophic equilibrium, *J. Atmos. Sci.* **37**, 1685–1699.
Mitchell, K.E. & Dutton J.A. (1981). Bifurcations from stationary to periodic solutions in a low-order mode of forced dissipative barotropic flow, *J. Atmos. Sci.* **38**, 690–716.

Modeling methodology

Andrews, J.G. & McLone, R.R. (Eds.) (1976). *Mathematical Modeling*, Butterworths.
Aris, R. (1978). *Mathematical Modelling Techniques*, Pitman.
Aubert, K.E. (1984). *Spurious mathematical modelling, Math. Intelligencer* **6**, 54–60.
Dym, C.L. & Ivey, E.S. (1980). *Principles of Mathematical Modeling*, Academic Press.
Farmer, D., Lapedes, A., Packard, N. & Wendroff, B. (Eds.) (1986). Evolution, games and learning; models for adaptation in machines and nature, *Physica* **22D**, Nos. 1–3.
Grabert, H. (1982). *Projection Operator Techniques in Nonequilibrium Statistical Mechanics*, Springer Tracts in Modern Physics, **95**, Springer-Verlag.
Paulré E. (Ed.) (1981). *System Dynamics and the Analysis of Change*, North-Holland.
Saaty, T.L. & Alexander, J.M. (1981). *Thinking with Models*, Pergamon.

Nerve conduction

Adey, W.R. (1975). Evidence for cooperative mechanisms in the susceptibility of cerebral tissue to environmental and intrinsic electric fields. In *Functional Linkage in Biomolecular Systems*, eds. F.O. Schmit, D.M. Schneider, and D.M. Crothers, Raven Press, NY.
Basar, E., Durusan, R., Gorder, A. & Ungan, P. (1969). Combined Dynamics of E.E.G. and evoked potential, *Biol. Cybernetics* **34**, 21.
Coley, J.W. & Dodge, F.A. (1966). Digital computer solutions for excitation and propagation of the nerve impulse, *Biphys. J.* **6**, 583–99.
Fitzhugh, R. Impulses and physiological states in theoretical models of nerve membrane, *Biophys. J.* **1**, 445–64.
Fitzhugh, R. (1969). Mathematical models of excitation and propagation in nerves. In *Biological Engineering*, ed. H.P. Schwan, McGraw-Hill.

Harmon, L.D. (1961). Studies with artificial neurons, I: Properties and functions of an artificial neuron, *Kybernetik* **1**, 89–101.

Hodgkin, A.L. & Huxley, A.F. (1952). A quantitative description of membrane current and its application to conduction and excitation in nerves, *J. Physiol.* **117**, 500–544.

Krinskii, V.I., Pertsov, A.M. & Reshetilov, A.N. (1972). Investigation of one mechanism of origin of the etopic focus of excitation in modified Hodgkin–Huxley equations, *Biofizika* **17**, 271–7.

Nagumo, J. & Sato, S. (1972). On a response characteristic of a mathematical neuron model, *Kybernetik* **10**, 155–64.

Pauwelussen, J.P. (1981). Nerve impulse propagation in a branching nerve system: a simple model, *Physica* **4D**, 67–88.

Wiener, N. & Rosenblueth, A. (1946). The mathematical formulation of the problem of conduction of impulses in a network of connected excitable elements, Specifically in Cardiac Muscle, *Arch. Inst. Cardiol. Mexico* **16**, 205–65.

Neural nets

Amari, S. & Arbib, M.A. (eds.) (1982) *Competition and Cooperation in Neutral Nets*, Lecture Notes in Biomathematics **45**, Springer-Verlag.

Clark, J.W., Winston, J.V. & Rafelski, J. (1984). Self-organization of neural networks. *Phys. Letters* **10A**, 207–11.

Clark, J.W., Rafelski, J. & Winston, J.V. (1985). Brain without mind: computer simulation of neural networks with modifiable neuronal interactions, *Phys. Reports* **123**(4), 215–73.

Cooper, L.N. (1973). A Possible Organization of Animal Memory and Learning, in *Proceedings of the Nobel Symposium on Collective Properties of Physical Systems*, **24**, eds. B. Lundquist and L. Lundquist, 252–64.

Ermentrout, E.B. & Cowan, J.D. (1979). A mathematical theory of visual hallucination patterns. *Biol. Cybernetics* **34**, 137–50.

Ermentrout, E.B. & Cowan, J.D. (1979). Temporal oscillations in neuronal nets, *J. MAth. Biology* **7**, 265–80.

Ermentrout, E.B. & Cowan, J.D. (1980). Large scale spatially organized activity in neural nets. *SIAM J. Appl. Math* **38**, 1–21.

Hopfield, J.J. (1982). Neural networks and physical systems with emergent collective computational abilities, *Proc. Natl. Acad. Sci. USA* **79**, 2554–8.

Hopfield, J.J. (1984). Neurons with graded response have collective computational properties like those of two-state neurons, *Proc. Natl. Acad. Sci. USA* **81**, 3088–92.

Little, W.A. (1974). The existence of persistent states in the brain, *Math. Biosci.* **19**, 101–20.

Nonlinear superposition (also see Lie–Bäcklund transformations)

Ames, W.F. (1978). Nonlinear superposition for operator equations. In *Nonlinear Equations in Abstract spaces,* ed. V. Lakshmikantham, pp. 43–66, Academic Press, pp. 43–66.

Anderson, R.L., Harnad, J. & Winternitz, P. (1982). Systems of ordinary differential equations with nonlinear superposition principles, *Physica* **4D**, 164–82.

Inselberg, A. (1972). Superpositions for nonlinear operators. I. Strong superpositions and linearizability, *J. Math. Anal. Appl.* **40**, 494–508.

Nonsoliton asymptotic part

Ablowitz, M.J. & Segur, H. (1977). Asymptotic solutions of the Korteweg–deVries equation, *Studies Appl. Math.* **57**, 13–44.

Manakov, S.V. (1974). Nonlinear Fraunhofer diffraction, *Soviet Phys. JETP* **38**, 693–6.

Schuur, P.C. (1986). *Asymptotic Analysis of Soliton Problems*, Springer-Verlag.

Zakharov, V.E., Monakov, S.V. (1977). Asymptotic behavior of non-linear wave systems integrated by the inverse scattering method, *Soviet Phys. JETP* **44**, 106–12.

Optical phenomena

Abraham, E. & Smith, S.D. (1982). Optical bistability and related devices, *Rep. Prog. Phys.* **45**, 815–85.

Eglund, J.C., Snapp, R.R. & Schieve, W.C. (1984). Fluctuations, instabilities and chaos in laser-driven nonlinear ring cavity. In *Progress in Optics*, **21**, ed. E. Wolf, pp. 374–441, North-Holland.

Lugiato, L.A. (1984). Theory of optical bistability. In *Progress in Optics*, **21**, ed. E. Wolf, pp. 71–211, North-Holland.

Optical bistability, dynamical nonlinearity, and photonic logic, *Phil. Trans. Roy. Soc. Lond.* **313** 1984).

Painlevé conjecture

Ablowitz, M.J. & Segur, H. (1977). Exact linearization of a Painlevé trancendent, *Phys. Rev. Letters* **38**, 1103–6.

Ablowitz, M.J., Ramani, A. & Segur, H. (1978). Nonlinear evolution equations and ordinary differential equations, *Lett. Nuovo. Cimento* **23**, 333–8.

Ablowitz, M.J., Ramani, A. & Segur, H. (1980). A connection between nonlinear evolution equations and ordinary differential equations of P-type. I. *J. Math. Phys.* **21**, 715–21.

Bountis, T., Segur, H. & Vivaldi, F. (1982). Integrable Hamiltonian systems and the Painlevé property, *Phys. Rev.* **A25**, 1257–64.

Bountis, T.C. (1984). A. Singularity analysis of integrability of chaos in dynamical systems. In *Singularities and Dynamical Systems*, ed. S. Pnevmatikos, North-Holland.

Bountis, T. & Segur, H. (1982). *Logarithmic singularities and chaotic behavior in Hamiltonian systems*, AIP Conference Proceedings, No. 88 eds. M. Tabor and Y.M. Treve, pp. 279–92, AIP.

Chang, Y.F., Tabor, M. & Weiss, J. (1982). Analytic structure of the Hénon–Heiles Hamiltonian in integrable and nonintegrable regimes. *J. Math. Phys.* **23**, 531–8.

Dorizzi, B. Grammatieos, B. & Ramani A. (1983). A New Class of Integrable Systems, *J. Math. Phys.* **24**, 2282–2288.

Dorizzi, B., Grammaticos, B. & Ramani, A. (1984). Explicit integrability for Hamiltonian systems and the Painlevé conjecture, *J. Math. Phys.* **25**, 481–5.

Ramani, A., Dorizzi, B., Grammaticos, B. & Bountis, T. (1984). Integrabillity and the Painlevé property for low-dimensional systems, *J. Math. Phys.* **25**, 878–83.

Tabor, M. & Weiss, J. (1981). Analytic structure of the Lorenz system, *Phys. Rev.* **A24**, 2157–67.

Patterns

Bard, J.B.L. (1981). A model for generating aspects of zebra and other mammalian coat patterns, *J. Theor. Biol.* **93**, 363–85.

Boon, J.P. & Noullez, A. (1986). Development, growth, and form in living systems. In *On Growth and Form*, eds. H.F. Stanley & N. Ostrowsky, Martinus Nijhoff Pub.

Gierer, A. & Meinhardt, H. (1972). A theory of biological pattern formation, *Kybernetik* **12**, 30–9.

Glansdorff, P. & Prigogine, I. (1971). *Thermodynamic Theory of Structure Stability and Fluctuations*, Wiley-Interscience.

Hasegawa, A. (1985). Self-organization processes in continuous media, *Adv. in Phys.* **34**, 1–42.

Martinez, H.M. (1972). Morphogenesis and chemical dissipative structures, *J. Theor. Biol.* **36**, 479–501.

Meinhardt, H. (1982). *Models of Biological Pattern Formation*, academic Press.

Nicolis, G. & Prigogine, I. (1977). *Self-Organization in Nonequilibrium Systems*, Wiley-Interscience.

Stanley, H.E. & Ostrowsky, N. (Eds.) (1986). *On Growth and Form*, Martinus Nijhoff Pub.

Swindale, N.V. (1980). A model for the formation of ocular dominance stripes, *Proc. Roy. Soc. London* **B208**, 243–64.

Turing, A.M. (1952). The Chemical Basis of Morphogenesis, *Phil. Trans. Roy. Soc. London* **B327**, 37–72.

Pattern dynamics (also see, chemical turbulence, and coupled maps)

Bishop, G. (1985). Pattern selection and low-dimensional chaos in systems of coupled nonlinear oscillation. In *Dynamical Problems in Soliton Systems*, ed. S. Takeno, pp. 250–7, Springer-Verlag.

Ciliberto, S. & Gollub, J.P. (1984). Pattern competition leads to chaos, *Phys. Rev. Letters*, 922–5.

Gollub, J.P. & Meyer, C.W. (1983). Symmetry-breaking instabilities on a fluid surface, *Physica* **6D**, 337–346.

Kaneko, K. (1983). Transition from torus to chaos accompanied by frequency lockings with symmetry breaking, *Prog. Theor. Phys.* **69**, 1427–42.

Kapral, R. (1985). Pattern formation in two-dimensional arrays of coupled, discrete-time oscillators, *Phys. Rev.* **A31**, 3868–79.

Smale, S. (1974). A Mathematical Model of two cells via Turing's equation, *AMS Lecture Notes in Appl. Math.* **6**, 15–26.

Wesfreid, J.E. & Zaleski, S. (Eds.) (1984). *Cellular Structures in Instabilities*, Lecture Notes in Physics **210**, Springer-Verlag.

Winfree, A.T. (1987). When Time Breaks Down, Princeton Univ. Press.

Zhabotinsky, A.M. & Zaikin, A.N. (1973). Autowave processes in distributed chemical systems, *J. Theor. Biol.* **40**, 45–61.

Related topics

Anderson, P. W., arrow, K.J. & Pines, D. (Eds.) (1988). *The Economy as an Evolving Complex System; Santa Fe Institute Studies in the Sciences of Complexity*, Addison-Wesley.

Berry, M.V. (1977). Semi-classical mechanics in phase space: a study of Wigner's function, *Phil. Trans. Roy. Soc. London*, **287**, 237–71.

Bluman, G.W. & Cole, J.D. (1974). *Simulating Methods for Differential Equations*, Springer-Verlag.

Bohm, D. (1984). *Causality and Chance in Modern Physics*, Univ. of Pennsylvania Press.

Bohm, D. & Peat, F.D. (1987). *Science, Order, and Creativity*, Bantam.

Brillouin, L. (1963). *Science and Information Theory*, Academic Press.

Brillouin, L. (1964). *Scientific Uncertainty, and Information*, Academic Press.

Bruce, J.W., Giblin, P.J. & Gibson, C.G. (1984). Caustics through the looking glass, *Math. Intelligence* **6**, 47–58.

Chapman, S. & Cowling, T.R. (1970). *The Mathematical Theory of Nonuniform Gases*, Cambridge Univ. Press.

Davidson, M. (1983). *Uncommon Sense*, J.P. Tarcher Inc., Los Angeles.

Davis, R. (1988). *The Cosmic Blueprint*, Simon and Schuster.

deAlfaro, V. & Rastti, M. (1978). Structural stability theory and phase transition models, *Fortschrette der Physik* **26**, 143–73.

deGroot, S.R. (1961). *Thermodynamics of Irreversible Processes*, North-Holland.

Gardner, M. (1981). The Laffer curve and other laughs in current economics, *Sci. Amer.* **245**(6), 18.

Gardner, M. (1986). *Knotted Doughnuts and Other Mathematical Entertainments*, W.H. Freeman.

Hawking, S.W. (1988). *A Brief History of Time*, Bantam.

Heims, S.J. (1984). *John Von Neumann and Norbert Wiener*, MIT Press.

Jackson, E.A. (1984). Radiation reaction dynamics in an electromagnetic wave and constant magnetic field, *J. Math. Phys.*, 1584–91.

Kac, M. (1959). *Probability and Related Topics in Physical Sciences*, Interscience.

Khinchin, A.I. (1949). *Mathematical Foundations of Statistical Mechanics*, Dover.

Laslow, E. (1987). *Evolution, The Grand Synthesis*, Shambhala Pub., Boston.

Leggett, A.J. (1987). *The Problems of Physics*, Oxford Univ. Press.

Leisegang, S. (1953). Zum Astigmatismus von Electronenlinsen, *Optik* **10**, 5–14.

Makarov, I.M. (Ed.) (1984). *Cybernetics Today*, Mir Publishers.

Moore-Ede, M.C., Sulzman, F.M. & Fuller, C.A. (1982). *The Clocks that Time Us*, Harvard Univ. Press.

Mistriotis, M.D. (1986). From Determinism to Stochasticity, Thesis, Dept. Phys., Univ. Illinois, Urbana.

Nicolis, G. & Prigogine, I. (1978). *Self-Organization in Nonequilibrium Systems*, Wiley-Interscience.

Pines, D. (Ed.) (1985). *Emerging Syntheses in Science*, Santa Fe Institute.

Pippard, A.B. (1985). *Response and Stability*, Cambridge Univ. Press.

Poincaré, J.H. (1929). *The Foundations of Science*, Science Press.

Prigogine, I. (1962). *Non-Equilibrium Statistical Mechanics*, Interscience.

Prigogine, I. & Stengers, I. (1984). *Order Out of Chaos*, Bantam Books.

Prigogine, I. (1980). *From Being to Becoming*, Freeman.

Roederer, J. (1979). *Introduction to the Physics and Psychphysics of Music*, Springer-Verlag.

Risset, J.C. (1982). Stochastic processes in music and art. In *Stochastic Processes in Quantum Theory and Statistical Physics*, Lecture Notes in Physics, **173**, eds. S. Albeverio, Ph. Combe, & M. Sirugue-Collin, Springer-Verlag.

Rosen, R. (1985). *Anticipatory Systems*, Pergamon.

Sinclair, R.M., Hosea, J.C. & Sheffield, G.V. (1970). A method for mapping a toroidal magnetic field by storage of phase stabilized electrons, *Rev. Sci. Instruments*, **41**, 1552–9.

Stravrondis, O.N. (1972). *The Optics of Rays, Wavefronts and Caustics*, Academic Press.

Ulam, Stanislaw (1909–1984); *Los Alamos Science*, Number 15, Special Issue (1987).

Ulam, S.M. (1965). Introduction to studies of nonlinear problems by E. Fermi, J. Pasta and S.M. Ulam. In *Collected Papers of Enrico Fermi*, Vol. 2, Univ. Chicago Press.

Ulam, S.M. (1983). *Adventures of a Mathematician*, Scribner.

VanVelsen, J.F.C. (1978). On linear response theory and area preserving mappings, *Physics Reports* **41C**, 135–90.

Rössler models

Crutchfield, J., Farmer, D., Packard, N., Shaw, R., Jones, G. & Donnelly, J. (1980). Power spectral analysis of a dynamical system, *Phys. Letters* **76A**, 1–4.

Farmer, D., Crutchfield, J., Froehling, H., Packard, N. & Shaw, R. (1980). Power spectra and mixing properties of strange attractors, *Ann. NY Acad. Sci.* **357**, 453–72.

Fraser, S. & Karpal, R. (1982). Analysis of flow hysteresis by a one-dimensional map, *Phys. Rev.* **A25**, 3223–33.

Kolata, G. (1984). Order out of chaos in computers, *Science* **223**, 917–9.

Rössler, O.E. (1983). The chaotic hierarchy, *Z. Naturforsch.* **38a**, 788–801.

Rössler, O.E. (1979). Continuous chaos – Four Prototype Equations, *Ann. NY Acad. Sci.* **316**, 376–92.

Rössler, O.E. (1977). Continuous chaos. In *Synergetics* – A Workshop, ed. H. Haken, Springer-Verlag, pp. 184–97.

Similarity analysis

Birkhoff, G. (1955). *Hydrodynamics*, Dover Pub., Inc.

Bluman, G.W. & Cole, J.D. (1974). *Similarity Methods for Differential Equations*, Springer-Verlag.

Dickson, L.E. (1924). Differential equations from the Group Standpoint, *Annal of Math.* **25**, 287.

Hansen, A.G. (1964) *Similarity Analyses of Boundary Value Problems in Engineering*, Prentice-Hall, Inc.

Ovsjanikov, L.V. (1958). Group and group invariant solutions of differential equations, *Dokladi Akad. Nauk. USSR* **118**, 439–42.

Sedov, L.I. (1959). *Similarity and Dimensional Methods in Mechanics*, Academic Press.

Sine Gordon equation

Ablowitz, M.J., Kaup, D.J., Newell, A.C. & Segur, H. (1973). MEthod for solving the Sine–Gordon equations, *Phys. Lett.* **30**, 1262–4.

Barone, A., Esposito, F. Magee, C.J. & Scott, A.C. (1971). Theory and applications of the Sine–Gordon equation, *Nuovo Cimento, Series 2*, **1**, 227–67 (Contains extensive references to all applications).

Lamb, G.L. Jr. (1971). Analytical descriptions to ultrashort optical pulse propagation in resonant media, *Rev. Mod. Phys.* **43**, 99–124.

Osborne, A. & Stuart, A.E.G. (1978). On the separability of the Sine–Gordon equation and similar quasilinear partial differential equations, *J. Math. Phys.* **19**, 1573–8.

Perring, J.K. & Skyrme, T.H.R. (1962). A model unified field equation, *Nucl. Phys.* **31**, 550–5.

Rajaraman, A. (1975). Some non-perturbative semi-classical methods in quantum field theory, *Physics Reports* **21C**, 227–313.

Rubinstein, J. (1970). Sine–Gordon equation, *J. Math. Phys.* **11**, 258–66.

Singular perturbation, asymptotic expansions

Asymptotic Methods and Singular Perturbations, SIAM-AMS Proceedings, **10**, AMS, 1976.

Cole, J.D. (1968). *Perturbation Methods in Applied Mathematics*, Blaisdell.

Dorodnitsyn, A.A. (1947). Asymptotic solution of the van der Pol equation, *Inst. Mech. Acad. Sci., USSR*, **11**, (Russian).

Eckhaus, W. (1973). Matched asymptotic expansions and singular perturbations, North-Holland.

Eckhaus, W. (1977). Formal approximations and singular perturbations, *SIAM Review* **19**.

Erdélyi, A. (1956). *Asymptotic Expansions*, Dover.

Kruskal, M. (1962). Asymptotic theory of Hamiltonian systems, *J. Math. Phys.* **3**, 806–28.

LaSalle, J. (1949). Relaxation oscillations, *Quart. J. App. Math.* **7**, 1.

Lebowitz, N.R. & Schaar, R. (1975). Exchange of stabilities in autonomous systems, *Studies Appl. Math.* **54**, 229–60.

Meyer, R.E. & Parter, S.V. (Eds.) (1980). *Singular Perturbations and Asymptotics*, Academic Press.

O'Malley, R.E. Jr. (1974). *Introduction to Singular Perturbations*, Academic Press.

O'Malley, R.E. Jr. (1977). *Singular Perturbation Analysis for Ordinary Differential Equations*, Comm. Math. Institute Rijksuniversiteit, Utrecht. Netherlands.

O'Malley, R.E. Jr. (1976). Phase plane solutions to some singular perturbation problems, *J. Math.*

Sanders, J.A. & Verhulst, F. (1984). *Averaging Methods in Nonlinear Dynamical Systems* Appl. Math. Sciences, Springer-Verlag.

Verhulst, F. (1983). *Asymptotic Analysis of Hamiltonian Systems*, Lecture Notes in Mathematics **985**, ed. F. Verhulst, Springer-Verlag.

Verhulst, F. (Ed.) (1979). *Asymptotic Analysis*, Lecture Notes in Mathematics, **711**, Springer-Verlag.

Wasow, W. *Asymptotic Expansion for Ordinary Differential Equations*, Interscience.

Social phenomena (see references on modeling methodology)

Aubin, J.P., Saari, D. & Sigmund, K. (Eds.) (1985). *Dynamics of Macrosystems*, Lecture Notes in Economic and Mathematical Systems **257**, Springer-Verlag.

Gazis, D.C. (1967). Mathematical Theory of Automobile Traffic, *Science* **157**, 273–81.

Gazis, D.C. (Ed.) (1974). *Traffic science*, Wiley.

Meadows, D.L. & Meadows, D.H. (1974). *The Limits to Growth*, Universe Books, NY.

Meadows, D.L. & Meadows, D.H. (Eds.) (1973). *Toward Global Equilibrium*, Wright–Allen Press, Cambridge, Mass.

Meadows, D.L. *et al.* (1974). *Dynamics of Growth in a Finite World*, Wright–Allen Press, Cambridge, Mass.

Montroll, E.W. (1978). On some mathematical models of social phenomena. In *Nonlinear Equations in Abstract Spaces*, ed. V. Lakshmikantham, pp. 161–216, Academic Press.

Musha, T. & Higuchi, H. (1977). In *Proceedings of the Symposium on l/f Fluctuations*, p. 189, Inst. Elect. Engineers, Tokyo, Japan.

Muncaster, R.G. & Zinnes, D.A. (1984). Phase portraits, singularities, and the dynamics of hostility in international relations, *SIAM J. Applied Math.*

Saari, D.G. (1984). The ultimate of chaos resulting from weighted voting systems, *Adv. App. Math.* **5**, 286–308.

Solitons (also see, Korteweg–deVries, Bäcklund, lattice dynamics and inverse scattering)

Ablowitz, M.J. & Segur, H. (1981). *Solitons and the Inverse Scattering Transform*, SIAM.

Bishop, A.R. & Schneider, T. (Eds.) (1978). *Solitons and Condensed Matter Physics*, Springer-Verlag.

Bulllough, R.K. & Dodd, R.K. (1979). Solitons in physics: basic concepts; solitons in mathematics. In *Structural Stability in Physics*, eds. W. Guttinger and H. Eikemier, pp. 219– 53. Springer-Verlag.

Bullough, R.K. & Caudrey, P.J. (Eds.) (1980). *Solitons*, Topics in Current Physics, **17**, Springer-Verlag.

Campbell, D.K., Newell, A.C., Schrieffer, R.J. & Segur, H. (Eds.) (1986). Solitons and Coherent Structures, *Physica* **D18**, 1–480.

Davydov, A.S. (1985). *Solitons in Molecular Systems*, D. Reidel.

Dodd, R.K., Eilbeck, J.C., Gibbon, J.D. & Morris, H.C. (1982). *Solitons and Nonlinear Wave Equations*, Academic Press.

Eilenberger, G. (1981). *Solitons*, Springer-Verlag.

Flascka, H. & McLaughlin, P.W. (Eds.) (1978). Conference on the theory and application of solitons, *Rocky Mountain J. Math.* **8**, 1–428.

Ichikawa, Y.H. & Ino, K.H. (1985). Lax-pair operators for squared-sum and squared-difference eigenfunctions, *J. Math. Phys.* **26**, 1976–8.

Lamb, G.L. Jr. (1980). *Elements of Soliton Theory*, Wiley-Interscience.

Lonngren, K. & Scott, A. (Eds.) (1978). *Solitons in Action*, Academic Press.

Makhankov, V.G. (1978). Dynamics of classical solitons (in non-integrable systems), *Physics Reports* **35**, 1

Manakov, S.V. & Zakhorov, V.E. (Eds.) (1981). Soliton Theory, *Physica* **3D**, Nos 1 and 2.

Newell, A.C. (1985) *Soliton in Mathematics and Physics*, SIAM.

Newell, A.C. & Redekopp, L.G. (1977). Breakdown of Zakharov–Shabat theory and soliton creation, *Phys. Rev. Lett.* **38**, 377–80.

Rajaraman, R. (1982). *Solutions and Instantons: An Introduction to Solitons and Instantons in Quantum Field Theory*, North-Holland.

Russell, J.S. (1844). *Report on Waves, Report of 14th Meeting, British Assoc. Adv. Sci.*, pp. 311–90, John Murray, London.

Sagdeev, R.Z. (Ed.) (1984). *Nonlinear and Turbulent Processes in Physics*, Vol. 2, Harwood Acad. Pub.

Scott, A.C., Chu, F.Y.F. & McLaughlin, D.W. (1973). The soliton: a new concept in applied science, *Proc. IEEE* **61**, 1443–83.

Segur, H. (1986). Some open problems, *Physica* **18D**, 1–12.

Takeno, S. (Ed.) (1985). *Dynamical Problems in Soliton Systems* Springer-Verlag.

Wilhelmsson, H. (Ed.) (1979). Solitons in physics, *Physica Scripta* **20** (3/4).

Zakharov, V.E. & Shabat, A.B. (1972). Exact theory of two-dimensional self-focusing and one-dimensional self-modulation of waves in nonlinear media, *Soviet Phys. JETP* **34**, 62–69.

Solitons, direct method (Hirota)

Ablowitz, M.J. & Satsuma, J. (1978). Solitons and rational solutions of nonlinear evolution equations, *J. Math. Phys.* **19**, 2180–6.

Hirota, R. (1985). Fundamental properties of the binary operators in soliton theory and their generalization. In *Dynamic Problems in Soliton Systems*, ed. S. Takeno, pp. 42–9, Springer-Verlag.

Hirota, R. (1971). Exact solution of the Korteweg–deVries equation for multiple collisions of solitons, *Phys. Rev. Lett.* **27**, 1192–4.

Hirota, R. (1976). Direct methods of finding exact solutions of nonlinear evolution equations. In *Bäcklund Transformations*, ed. R.M. Miura Lecture Notes in Mathematics **515**, Springer-Verlag.

Hirota, (1985). Direct methods in soliton theory. In *Solitons: Topics of Modern Physics*, eds. R.K. Bullough & P.J. Candrey. Springer-Verlag.

Nakamura, A. (1979). A direct method of calculating periodic wave solutions to nonlinear evolution equations. I. Exact two-periodic wave solution, *J. Phys. Soc. Japan*, **47**, 170.

Solitons, higher dimensional

Ablowitz, MJ. (1981). Remarks on nonlinear evolution equations and inverse scattering transform. In *Nonlinear Phenomena in Physics and Biology*, eds. R.H. Enns *et al.*

Ablowitz, M.J. & Nachman, A.I. (1986). Multidimensional nonlinear evolution equations and inverse scattering, *Physica* **18D**, 223–41.

Anderson, D.L.T. (1971). Stability of time-dependent particle-like solutions in nonlinear field theories II, *J. Math. Phys.* **12**, 945–52.

Davydov, A.S. (1985). *Solitons in Molecular Systems*, D. Reidel.

Fokas, A.S., & Ablowitz, M.J. (1983). The inverse scattering transform for multi-dimensional (2 + 1) problems. In *Nonlinear Phenomena*, Lecture Notes in Physics **189**, ed. K.B. Wolf, pp. 139–54. Springer-Verlag.

Gibbon, J.D., Freeman, N.C. & Johnson, R.S. (1978). Correspondence between the classical gd^4.

Double and single Sine–Gordan equations for three-dimensional solitons, *Physics Letters* **65A**, 3800–4.

Kadomtsev, B.B. & Petviashvili, V.I. (1970). *Sov. Phys. Doklady*, **15**, 539–41.

Kako, F. & Yajima, N. (1980). Interaction of ion–acoustic solitons in two-dimensional space, *J. Phys. Soc. Japan.* **49**, 2063–71.

Newton, R.G. (1978). Three-dimensional solitons, *J. Math. Phys.* **19**, 1068–73.

Rebbi, C. (1979). Solitons, *Sci. Amer.* **240**(2), 92–116.

Strauss, W.A. (1977). Existence of solitary waves in higher dimensions, *Commun. Math. Phys.* **55**, 149–62.

Yajima, N., Oikawa, M. & Satsuma, J. (1978). Interaction of ion–acoustic solitons in three-dimensional space, *J. Phys. Soc. Japan* **44**, 1711–14.

Yajima, N. (1985). Soliton resonance in plasmas. In *Dynamical Problems in Soliton Systems*, ed. S. Takeno, pp. 144–52, Springer-Verlag.

Standard and area-preserving maps: microtron instability

Chirikov, B.V. (1979). A universal instability of many-dimensional oscillator systems *Physics Reports* **52**, 265–376.

Collet, P., Eckmann, J.-P. & Koch, H. (1981). On universality for area-preserving maps of the plane, *Physica* **3D**, 457–67.

Greene, J.M. (1986). How a swing behaves, *Physica* **18D**, 427–7.

Greene, J.M. (1979). A method of determining a stochastic transition, *J. Math. Phys.* **20**, 1183–201.

Greene, J.M., MacKay, R.S., Vivaldi, F. & Feigenbaum, M.J. (1981). Universal behavior in families of area-preserving maps, *Physica* **3D**, 468–86.

Ichikawa, Y.H., Kamimura, T. & Hatori, T. (1987). Stochastic diffusion in the standard map. *Physica* **29D**. 247–55.

Jowett, J.M., Month, M. & Turner, S. (Eds.) (1985). *Nonlinear Dynamics Aspects of Particle Accelerators*, (Lecture Notes in Physics, **247**, Springer-Verlag.

MacKay, R.S. (1986). Transitions to chaos in area-preserving maps. *Nonlinear Dynamics Aspects of Particle Accelerators*, eds. J.M. Jowett, M. Month, and S. Turner, pp. 390–454, Springer-Verlag.

Melekhin, V.N. (1976). Phase dynamics of particles in a microtron and the problem of stochastic instability of nonlinear systems, *Sov. Phys. JETP* **41**, 803–8.

Sinclair, R.M., Hoser, J.C. & Sheffield, G.V. (1970). A method for mapping a torodial magnetic field by storage of phase stabilized electrons, *Rev. Sci. Instruments* **141**, 1552–9.

Surveys, general

Abraham, R. & Shaw, C. Dynamics: *The Geometry of Behavior;* Part one: Periodic Behavior (1982); Part two: Chaotic Behavior (1983); Part three: *Global Behavior* (1983); Part four: *Bifurcation Behavior* (1984), Aerial Press, Santa Cruz, CA.

Davis, H.T. (1962). *Introduction to Nonlinear Differential and Integral Equations*, Dover.

Diner, S., Fargue, D. & Lochak, G. (Eds.) (1986). *Dynamical Systems: A Renewal of Mechanism*, World Scientific.

Hirsch, M.W. (1984). The dynamical systems approach to diferential equations, *Bull. Amer. Math. Soc.* **11**, 1–64.

Von Karmen, T. (1940). The engineer grapples with nonlinear problems (Fifteenth Josiah Willard Gibbs Lecture) *Bull. Amer. Math. Soc.* **46**, 615–83.

Synergetics

Haken, H. (1977). *Synergetics, An Introduction*, Springer-Verlag; Vol. 2: *A Workshop* (*ibid*).

Haken, H. *Advanced Synergetics*, Springer-Verlag.

Kozak, J.J. (1979). *Nonlinear Problems in the Theory of Phase Transitions*, Adv. Chem. Phys. **40**, pp. 229–368, Interscience Pub.

Pacault, A. & Vidal, C. (Eds.) (1979). *Synergetics: Far from Equilibrium*, Springer-Verlag.

Zabusky, N.J. (1981). Computational synergetics and mathematical innovation, *J. Comput. Phys.* **43**, 195–249 (1981).

Turbulent-like dynamics (also see, chemical systems, and explosive dynamics)

Barenblatt, G.I., Iooss, G. & Joseph, D.D. (Eds.) (1983). *Nonlinear Dynamics and Turbulence*, Pitman.

Berry, M.V. (1978). Regular & Irregular motion. In *Topics in Nonlinear Dynamics*, ed. S. Jorna, pp. 16–120, AIP. Conf. Proc. No. 46, AIP.

Frisch, U. (1986). Fully developed turbulence: where do we stand? In *Dynamical Systems, A Renewal of Mechanism*, eds. S. Diner, D. Fargue & G. Lochak, pp. 13–28, World Scientific.

Hopf, E. (1948). A mathematical example displaying features of turbulence, *Comm. Pure Appl. Math.* **1**, 303–22.

Lorenz, E.N. (1964). The problem of deducing the climate from the governing equations, *Tellus* **16**, 1–11.

Martin, P.C. (1975). The onset of turbulence: a review of recent developments in theory and experiment, in *Proceedings of the International Conference on Statistical Physics, Budapest 1975*, eds. L. Pal & P. Szepfalusy, 69–96, North-Holland.

McLaughlin, J. (1976). Successive bifurcations leading to stochastic behavior, *J. Stat. Phys.* **15**, 307–26.

McLaughlin, J.B. & Martin P.C. (1975). Transition to turbulence in a statistically stressed fluid system, *Phys. Rev.* **A12**, 186–203.

Normand, C. & Pomeau, Y. (1977), convective instability: a physicist's approach, *Rev. Mod. Phys.* **49**, 581–624.

Orszag, S.A. (1977). Lectures on the statistical theory of turbulence. In *Fluid Dynamics*, eds. R. Balian J.-L. Peube, Gordon and Breach, pp. 237–368.

Ruelle, D. & Takens, F. (1971). On the nature of turbulence, *Comm. Math. Phys.* **20**, 167–92, **23**, 343–4.

Ruelle, D. (1978). Dynamical systems with turbulent behavior. In *Mathematical Problems in Theoretical Physics*, eds. G. Del'Atonio, S. Doplicher & G. Jona-Lasinio. Lecture Notes in Physics, **80**, Springer-Verlag.

Smale, S. (1977). Dynamical systems and turbulence. In *Turbulence Seminar*, Lecture Notes in Mathematics, **615**, Springer-Verlag, 48–70.

Stuart, J.T. (1971). Nonlinear stability theory, *Ann. Rev. Fluid Mech.* **3**, 347–70.

Swinney, H.W. & Gollub, J.P. (Eds.) (1985). *Hydrodynamic Instabilities and the Transition to Turbulence*, Springer-Verlag.

Treve, Y.M. (1975). Theory of chaotic motion with application to controlled fusion research. In *Topics in Nonlinear Dynamics*, ed. S. Jorna, pp. 147–220, North-Holland.

Turbulence, onset

A. Experimental

Ahlers, G. & Behringer, R. (1978). Evolution of turbulence from the Rayleigh–Benard instability, *Phys. Rev. Lett.* **40**, 712–16.

Ahlers, G., Behringer, R.P. (1978). The Rayleigh–Benard instability and the evolution of turbulence, *Prog. Theor. Phys. Suppl.* **64**, 186–201.

Behringerm, R.P. & Ahlers, G. (1977). Heat transport and critical slowing down near the Rayleigh–Benard instability in cylindrical containers, *Phys. Letters* **62A**, 329–31.

Benjamin, T.B. (1978). Bifurcation phenomena in steady flows of a viscous fluid I. theory, II. Experiment, *Proc. Roy. Soc.* **A359**, 1–26, 27–43.

Berge, P. (1979). *Experiments on Hydrodynamic Instabilities and the Transition to Turbulence*, Lecture Notes in Physics, **104**, pp. 289–308, Springer-Verlag.

Busse, F.H. (1980). Convections in a rotating layer: a simple case of turbulence, *Science* **208**, 173–5.

Coles, D. (1965). Transition in circular couette flow, *J. Fluid Mech.* **21**, 385–425.

Fenstermacher, P.R., Swinney, H.L. & Gollub, J.P. (1979). Dynamic instabilities and the transition to chaotic Taylor Vortex flow, *J. Fluid Mech.* **94**, 103–28.

Fenstermacher, P.R., Swinney, H.L., Benson, S.V. & Gollub, J.P. (1979). Bifurcations to periodic, quasiperiodic and chaotic regimes in rotating and convecting fluids, *Ann. NY Acad. Sci.* **316**, 652–6.

Gollub, J.P. & Freilich, M.H. (1974). Optical heterodyne study of the Taylor instability, in a Rotating fluid, *Phys. Rev. Lett.* **33**, 1465–8.

Gollub, J.P. & Swinney, H.L. (1975). Onset of turbulence in a rotating fluid, *Phys. Rev. Lett.* **35**, 927–30.

Gollub, J.P. & Benson, S.V. (1978). Chaotic response to periodic perturbation of convecting fluid, *Phys. Rev. Lett.* **41**, 948.

Gollub, J.P. & Meyer, C.W. (1983). Symmetry-breaking instabilities on a fluid surface, *Physica* **6D**, 337–46.

Koschmieder, E. L. (1981). Experimental aspects of hydrodynamic instabilities. In *Order and Fluctuations in Equilibrium and Nonequilibrium Statistical Mechanics*, eds. G. Nicolis, G. Dewel & J.W. Turner, pp. 159–88, Wiley.

Meyer, R.E. (Ed.) (1981). *Transition and Turbulence*, Academic Press.

Swinney, H.L. & Gollub, J.P. (Eds.) (1985). *Hydrodynamic Instabilities and the Transition to Turbulence*, Topics in Applied Physics, **45**, Springer-Verlag.

Swinney, H.L. & Gollub, J.P. (1978). The transition to turbulence, *Phys. Today* **31**(8) 41.

B. Theoretical (also see, Lorenz equations)

Drazin, P.G. & Reid, W.H. (1981). *Hydrodynamic Stability*, Cambridge Univ. Press.

Eckmann, J.-P. (1981). Roads to turbulence in dissipative dynamical systems, *Rev. Mod. Phys.* **53**, 643.

Feigenbaum, M.J. (1979). The onset spectrum of turbulence, *Phys. Letters* **74A**, 375–8.

Newhouse, S., Ruelle, D. & Takens, F. (1978). Occurrence of strange axiom *A* attractors near quasi-periodic flows on T^m, $m \geqslant 3$. *Commun. Math. Phys.* **64**, 35–40.

Orszag, S.A. & Kells, L.C. (1980). Transition to turbulence in plane poiseccille and plane couette flow, *J. Fluid Mech.* **96**, 159–205.

Orszag, S.A. & Patera, A.T. (1981). Subcritical transition to turbulence in plane shear flows. In *Transition and Turbulence*, ed. R.E. Meyer, pp. 127–46, Academic Press.

Packard, N.H., Crutchfield, J.P., Farmer, J.D., & Shaw, R.S. (1980). Geometry from a time series, *Phys. Rev. Lett.* **45**, 712–16.

Ruelle, D. (1978). Dynamical Systems with Turbulent Behavior. In *Mathematical Problems in Theoretical Physics*, eds. G. Dell'Antonio, S. Doplicher & G. Jona-Lasino, Lecture Notes in Physics, **80**, pp. 341–60, Springer-Verlag.

Sagdeev, R.Z. (Ed.) (1984). *Nonlinear and Turbulent Processes in Physics*, Vols. 1, 2, and 3, Harwood Acad. Pub.

Takens, F. (1981). *Detecting Strange Attractors in Turbulence in Dynamical Systems and Turbulence, Warwick 1980*, Lecture Notes in Mathematics **898**, eds. D.A. Rand & L.-S. Young, Springer-Verlag.

Yahata, H. Temporal Development of the Taylor vortices in a Rotating Fluid, *Prog. Theor. Phys. Suppl.* **64**, 165–85.

References added at 1991 reprinting

Abraham, N.B. (Ed.) (1989). *Quantitative Measures of Complex Dynamical Systems*. Plenum Press.

Amit, D.J. (1989). *Modeling Brain Function*. Cambridge Univ. Press.

Barnsley, M. (1988). *Fractals Everywhere*. Academic Press.

Barrow, J.D. (1988). *The World within the World*. Clarendon Press, Oxford.

Barrow, J.D. & Tipler, F.J. (1988). *The Anthropic Cosmological Principle*. Oxford Univ. Press.

Bedford, T. & Swift, J. (Eds.) (1988). *New Directions in Dynamical Systems*. Cambridge Univ. Press.

Bloxham, J. & Gubbins, D. (1989). The evolution of the Earth's magnetic field. *Scientific American* **261**, #6, 68–75.

Cairns-Smith, A.G. (1982). *Genetic Takeover, and the Mineral Origins of Life*. Cambridge Univ. Press.

Chang, S.J. & Wright, J. (1981). Transitions and distribution functions for chaotic systems. *Phys. Rev.* **A22**, 1419–33.

Clark, J.W., Rafelski, J. & Winston, W. (1985). Brain without mind: computer simulation of neural networks with modifiable neuronal interactions. *Phys. Reports* **123**, 215–73.

Croom, F.H. (1989). *Principles of Topology*, Sanders.

Deutsch, D. (1985). Quantum theory, the Church–Turing principle and universal quantum computer. *Proc. R. Soc. Lond.* **A400**, 97–117.

Duncan, R. & Weston-Smith, M. (Eds.) (1978). *The Encyclopaedia of Ignorance*, Pergamon Press.

Edelman, G.M. (1987). *Neural Darwinism*. Basic Books.

Escande, D.F. (1985). Stochasticity in classical hamiltonian systems: universal aspects. *Phys. Reports* **121** (3 & 4), 165–261.

Fox, R.F. (1988). *Energy and the Evolution of Life*. W.H. Freeman.

Fox, S. (1988). *The Emergence of Life*. Basic Books.

Freeman, W.J. (1991). The physiology of perception. *Scientific American* **264** (2), 78–85.

Frisch, U. & Orszag, S.A. (1990). Turbulence: challenges for theory and experiment. *Physics Today* **43**, #1, 24–32.

Froyland, J. (1983). Lyapunov exponents for multidimensional orbits. *Phys. Letters* **97A**, 8–10.

Gelperin, A. & Tank, D.W. (1990). Odour-modulated collective network oscillations of olfactory interneurons in a terrestrial mollusc. *Nature* **345**, 437–40.

Goel, N.S. & Thompson, R.L. (1988). *Computer Simulations of Self-organization in Biological Systems*. Macmillan.

Goldberger, A.L., Rigney, D.R. & West, B.J. (1990). Chaos and fractals in human physiology. *Scientific American* **262**, #2, 40–9.

Harrison, L.G. (1987). What is the status of reaction-diffusion theory thirty-four years after Turing? *J. Theor. Biol.* **125**, 369–84.

Huberman, B.A. & Crutchfield, J.R. (1979). Chaotic states of anharmonic systems in periodic fields. *Phys. Rev. Letters* **43**, 1743–7.

Jackson, E.A. & Mistriotis, A.D. (1989). Thermal conductivity of one- and two-dimensional lattices. *J. Phys.: Condens. Matter* **1**, 1223–38.

Jensen, M.H., Kadanoff, L.P., Libchaber, A., Procaccia, I. & Stavans, J. (1985). Global universality at the onset of chaos: results of a forced Rayleigh–Benard experiment. *Phys. Rev. Letters* **55**, 2798–801.

Landford, O.E. (1987). Circle mappings. pp. 1–17. In *Recent Developments in Mathematical Physics*, ed. H. Mitter & L. Pitner. Springer-Verlag.

Langton, C.G. (Ed.) (1989). *Artificial Life* (Vol. VI, Santa Fe Institute), Addison-Wesley.

Marchall, C. (1990). *The Three-body Problem.* Elsevier.

Moon, F.C. (1987). *Chaotic Vibrations*, Wiley-Interscience.

Newhouse, S. (1979). The abundance of wild hyperbolic sets and non-smooth stable sets for diffeomorphisms. *Publ. IHES* **50**, 101–51.

Paladin, G. & Vulpiani, A. (1987). Anomalous scaling laws in multifractal objects. *Phys. Reports* **156**, 147–225.

Palmore, J.I. & McCauley, J.L. (1987). Shadowing by computable chaotic orbits. *Phys. Letters* **A122**, 399–402.

Peliti, L. & Vulpiani, A. (Eds.) (1988). *Measures of Complexity*, Springer-Verlag.

Popper, K. (1968). *The Logic of Scientific Discovery*, Harper and Row.

Sagdeev, R.Z., Usikov, D.A. & Zaslavsky, G.M. (1988). *Nonlinear Physics.* Harwood Academic.

Schaffer, W.M. (1988). Perceiving order in the chaos of nature. In *Evolution of Life Histories of Mammals*, ed. M.S. Boyce, pp. 313–50.

Schroeder, M. (1990). *Fractals, Chaos, Power Laws.* W.H. Freeman.

Skarda, C.A. & Freeman, W.J. (1987). How brains make chaos in order to make sense of the world, *Behavioral and Brain Sciences* **10**, 161–95.

Svirezhev. Yu. M. & Passekov, V.P. (1990). *Fundamentals of Mathematical Evolutionary Genetics.* Kluwer.

Thompson, J.M.T. & Stewart, H.B. (1986). *Nonlinear Dynamics and Chaos.* Wiley.

van Bendegem, J.P. (1987). *Finite Empirical Mathematics; Outline of a Model.* Rijsuniversitent Gent.

Zeeman, E.C. (1988). *Stability of Dynamical Systems, Nonlinearity* **1**, 115–55.

Index